Matrix Computations
and Semiseparable Matrices

Matrix Computations and Semiseparable Matrices

Volume I: **Linear Systems**

Raf Vandebril
Department of Computer Science
Catholic University of Louvain

Marc Van Barel
Department of Computer Science
Catholic University of Louvain

Nicola Mastronardi
M. Picone Institute
for Applied Mathematics, Bari

The Johns Hopkins University Press
Baltimore

The Johns Hopkins University Press
2715 North Charles Street
Baltimore, Maryland 21218-4363
www.press.jhu.edu

Library of Congress Cataloging-in-Publication Data

Vandebril, Raf, 1978–
   Matrix computations and semiseparable matrices / Raf Vandebril,
Marc Van Barel, Nicola Mastronardi.
      v. cm.
   Includes bibliographical references and indexes.
   Contents: v. 1. Linear systems
   ISBN-13: 978-0-8018-8714-7 (hardcover : alk. paper)
   ISBN-10: 0-8018-8714-3 (hardcover : alk. paper)
   1. Semiseparable matrices. 2. Matrices—Data processing.
3. Numerical analysis. I. Barel, Marc van, 1960– II. Mastronardi, Nicola,
1962– III. Title.
   QA188.V36 2007
   512.9'434—dc22   2007030657

A catalog record for this book is available from the British Library.

Contents

Preface   xiii

Notation   xvii

I   Introduction to semiseparable and related matrices   1

1   Semiseparable and related matrices: definitions and properties   5
     1.1   Symmetric semiseparable and related matrices   7
     1.2   Relations between the different symmetric definitions   12
            1.2.1   Known relations   12
            1.2.2   Common misunderstandings about generator repre-
                     sentable semiseparable matrices   13
            1.2.3   A theoretical problem with numerical consequences   14
            1.2.4   Generator representable semiseparable and semisep-
                     arable matrices   16
            1.2.5   Semiseparable plus diagonal and quasiseparable ma-
                     trices   21
            1.2.6   Summary of the relations   25
            1.2.7   More on the pointwise convergence   26
     1.3   Unsymmetric semiseparable and related matrices   28
     1.4   Relations between the different "unsymmetric" definitions . . .   31
            1.4.1   Extension of known symmetric relations   31
            1.4.2   Quasiseparable matrices having a symmetric rank
                     structure   34
            1.4.3   Summary of the relations   36
     1.5   Relations under inversion   37
            1.5.1   The nullity theorem   37
            1.5.2   The inverse of semiseparable and tridiagonal matri-
                     ces   40
            1.5.3   The inverse of a quasiseparable matrix   43
            1.5.4   The inverse of a semiseparable plus diagonal matrix   44
            1.5.5   Summary of the relations   47
     1.6   Conclusions   50

2   The representation of semiseparable and related matrices   53
   2.1   Representations 55
        2.1.1   The definition of a representation 56
        2.1.2   A representation is "just" a representation 57
        2.1.3   Covered representations 58
   2.2   The symmetric generator representation 59
        2.2.1   The representation for symmetric semiseparables . . 59
        2.2.2   Application to other classes of matrices 62
   2.3   The symmetric diagonal-subdiagonal representation 65
        2.3.1   The representation for symmetric semiseparables . . 66
        2.3.2   The representation for symmetric quasiseparables . 69
   2.4   The symmetric Givens-vector representation 71
        2.4.1   The representation for symmetric semiseparables . . 72
        2.4.2   Examples 74
        2.4.3   Retrieving the Givens-vector representation 74
        2.4.4   Swapping the representation 77
        2.4.5   Application to other classes of matrices 78
   2.5   The symmetric quasiseparable representation 79
        2.5.1   The representation for symmetric semiseparables . . 80
        2.5.2   Application to other classes of matrices 82
   2.6   Some examples 82
   2.7   The unsymmetric generator representation 84
        2.7.1   The representation for semiseparables 84
        2.7.2   Application to other classes of matrices 87
   2.8   The unsymmetric Givens-vector representation 90
        2.8.1   The representation for semiseparables 90
        2.8.2   Application to other classes of matrices 91
   2.9   The unsymmetric quasiseparable representation 92
        2.9.1   The representation for semiseparables 92
        2.9.2   Application to other classes of matrices 92
   2.10   The decoupled representation for semiseparable matrices 93
        2.10.1   The decoupled generator representation 94
        2.10.2   The decoupled Givens-vector representation 94
   2.11   Summary of the representations 95
   2.12   Are there more representations? 97
   2.13   Some algorithms related to representations 98
        2.13.1   A fast matrix vector multiplication 98
        2.13.2   Changing representations 101
        2.13.3   Computing the determinant of a semiseparable ma-
                 trix in $\mathcal{O}(n)$ flops 104
   2.14   Conclusions 106

3   Historical applications and other topics   109
   3.1   Oscillation matrices 110
        3.1.1   Introduction 111
        3.1.2   Definition and examples 113

3.1.3   The inverse of a one-pair matrix 116
3.1.4   The example of the one-pair matrix 120
3.1.5   Some other interesting applications 121
3.1.6   The connection with eigenvalues and eigenvectors . 121
3.2   Semiseparable matrices as covariance matrices 123
3.2.1   Covariance calculations 123
3.2.2   The multinomial distribution 125
3.2.3   Some other matrices 128
3.3   Discretization of integral equations 130
3.4   Orthogonal rational functions 133
3.5   Some comments 134
3.5.1   The name "semiseparable" matrix 135
3.5.2   Eigenvalue problems 135
3.5.3   References to applications 140
3.6   Conclusions 141

II   Linear systems with semiseparable and related matrices   143

4   Gaussian elimination   149
4.1   About Gaussian elimination and the LU-factorization 151
4.2   Backward substitution 152
4.3   Inversion of triangular semiseparable matrices 153
4.3.1   The inverse of a bidiagonal 154
4.3.2   The inverse of lower semiseparable matrices 156
4.3.3   Examples 157
4.4   Theoretical considerations of the LU-decomposition 159
4.5   The LU-decomposition for semiseparable matrices 163
4.5.1   Strongly nonsingular matrices 163
4.5.2   General semiseparable matrices 165
4.6   The LU-decomposition for quasiseparable matrices 171
4.6.1   A first naive factorization scheme 171
4.6.2   The strongly nonsingular case, without pivoting . . 174
4.6.3   The strongly nonsingular case 176
4.7   Some comments 178
4.7.1   Numerical stability 178
4.7.2   A representation? 179
4.8   Conclusions 179

5   The QR-factorization   181
5.1   About the QR-decomposition 182
5.2   Theoretical considerations of the QR-decomposition 183
5.3   A QR-factorization of semiseparable matrices 186
5.3.1   The factorization using Givens transformations . . . 186
5.3.2   The generators of the factors 189
5.3.3   The Givens-vector representation 191

    5.3.4   Solving systems of equations 192
5.4   A QR-factorization of quasiseparable matrices 192
5.5   Implementing the QR-factorization 194
5.6   Other decompositions 199
    5.6.1   The URV-decomposition 199
    5.6.2   Some other orthogonal decompositions 202
5.7   Conclusions 204

6   A Levinson-like and Schur-like solver   205
6.1   About the Levinson algorithm 206
    6.1.1   The Yule-Walker problem and the Durbin algorithm 207
    6.1.2   The Levinson algorithm 208
    6.1.3   An upper triangular factorization of the inverse . . . 209
6.2   Generator representable semiseparable plus diagonal matrices . 210
    6.2.1   The class of matrices 210
    6.2.2   A Yule-Walker-like problem 211
    6.2.3   A Levinson-type algorithm 215
    6.2.4   An upper triangular factorization of the inverse . . . 218
    6.2.5   Some general remarks 219
6.3   A Levinson framework 219
    6.3.1   The matrix block decomposition 220
    6.3.2   Simple $\{p_1, p_2\}$-Levinson conform matrices 222
    6.3.3   An upper triangular factorization 231
    6.3.4   The look-ahead procedure 232
6.4   Examples 234
    6.4.1   Givens-vector-represented semiseparable matrices . . 234
    6.4.2   Quasiseparable matrices 235
    6.4.3   Tridiagonal matrices 236
    6.4.4   Arrowhead matrices 237
    6.4.5   Unsymmetric structures 239
    6.4.6   Upper triangular matrices 239
    6.4.7   Dense matrices 240
    6.4.8   Summations of Levinson-conform matrices 240
    6.4.9   Matrices with errors in structures 241
    6.4.10   Companion matrices 242
    6.4.11   Comrade matrix 243
    6.4.12   Fellow matrices 244
6.5   The Schur algorithm 245
    6.5.1   Basic concepts 245
    6.5.2   The Schur reduction 247
    6.5.3   The Schur complement of quasiseparable matrices . 248
    6.5.4   A Schur-like algorithm for quasiseparable matrices . 250
    6.5.5   A more general framework for the Schur reduction . 253
6.6   Conclusions 256

7   Inverting semiseparable and related matrices   257

7.1	Known factorizations	258
	7.1.1	Inversion via the QR-factorization	258
	7.1.2	Inversion via the LU-factorization	259
	7.1.3	Inversion via the Levinson algorithm	259
7.2	Direct inversion methods	261
	7.2.1	The inverse of a symmetric tridiagonal matrix . . .	261
	7.2.2	The inverse of a symmetric semiseparable matrix . .	264
	7.2.3	The inverse of a tridiagonal matrix	267
	7.2.4	The inverse of a specific semiseparable matrix . . .	271
7.3	General formulas for inversion	274
7.4	Scaling of symmetric positive definite semiseparable matrices . .	277
7.5	Decay rates for the inverses of tridiagonal matrices	278
	7.5.1	M-matrices	278
	7.5.2	Decay rates for the inverse of diagonally dominant tridiagonal M-matrices	280
	7.5.3	Decay rates for the inverse of diagonally dominant tridiagonal matrices	281
7.6	Conclusions	285

III Structured rank matrices   287

8   Definitions of higher order semiseparable matrices   293

8.1	Structured rank matrices	294
8.2	Definition of higher order semiseparable and related matrices . .	298
	8.2.1	Some inner structured rank relations	298
	8.2.2	Semiseparable matrices	300
	8.2.3	Quasiseparable matrices	301
	8.2.4	Generator representable semiseparable matrices . . .	302
	8.2.5	Extended semiseparable matrices	304
	8.2.6	Hessenberg-like matrices	305
	8.2.7	Sparse matrices	306
	8.2.8	What is in a name	307
8.3	Inverses of structured rank matrices	310
	8.3.1	Inverse of structured rank matrices	310
	8.3.2	Some particular inverses	316
8.4	Generator representable semiseparable matrices	318
	8.4.1	Generator representation	318
	8.4.2	When is a matrix generator representable?	320
	8.4.3	Decomposition of structured rank matrices	328
	8.4.4	Decomposition of semiseparable and related matrices	338
8.5	Representations	342
	8.5.1	Givens-vector representation	343
	8.5.2	Quasiseparable representation	343
	8.5.3	Split representations	344
8.6	Conclusions	345

9 **A QR-factorization for structured rank matrices** **347**

9.1   A sequence of Givens transformations from bottom to top . . . 350

    9.1.1   Annihilating Givens transformations on lower rank 1 structures 351

    9.1.2   Arbitrary Givens transformations on lower rank 1 structures 354

    9.1.3   Givens transformations on lower rank structures . . 356

    9.1.4   Can the value of p in $\Sigma_l^{(p)}$ be larger than 1? 360

    9.1.5   Givens transformations on upper rank 1 structures . 363

    9.1.6   Givens transformations on upper rank structures . . 367

    9.1.7   Givens transformations on rank structures 369

    9.1.8   Other directions of sequences of Givens transformations 372

    9.1.9   Examples 375

    9.1.10   Summary 378

9.2   Making the structured rank matrix upper triangular 379

    9.2.1   Annihilating completely the rank structure 380

    9.2.2   Expanding the zero rank structure 380

    9.2.3   Combination of ascending and descending sequences 381

    9.2.4   Examples 383

    9.2.5   Solving systems of equations 386

    9.2.6   Other decompositions 386

9.3   Different patterns of annihilation 387

    9.3.1   The leaf form for removing the rank structure . . . 389

    9.3.2   The pyramid form for removing the rank structure . 389

    9.3.3   The leaf form for creating zeros 391

    9.3.4   The diamond form for creating zeros 391

    9.3.5   Theorems connected to Givens transformations . . . 392

    9.3.6   The $\wedge$-pattern 395

    9.3.7   The $\times$-pattern 395

    9.3.8   The $\vee$-pattern 396

    9.3.9   Givens transformations in the $\vee$-pattern 397

9.4   Rank-expanding sequences of Givens transformations 400

    9.4.1   The Givens transformation 400

    9.4.2   Rank-expanding Givens transformations on upper rank 1 structures 402

    9.4.3   Existence and the effect on upper rank structures . 405

    9.4.4   Global theorem for sequences from bottom to top . 415

9.5 QR-factorization for the Givens-vector representation 416

    9.5.1   The Givens-vector representation 416

    9.5.2   Applying transformations on the Givens-vector representation 418

    9.5.3   Computing the rank-expanding Givens transformations 420

9.6   Extra material 425

    9.6.1   A rank-expanding QR-factorization 425

		9.6.2	Parallel factorization	426
		9.6.3	QZ-factorization of an unstructured matrix	429
	9.7		Multiplication between structured rank matrices	430
		9.7.1	Products of structured rank matrices	431
		9.7.2	Examples	432
	9.8		Conclusions	433

10 A Gauss solver for higher order structured rank systems 435

	10.1		A sequence of Gauss transformation matrices without pivoting .	437
		10.1.1	Annihilating Gauss transforms on lower rank 1 structures	437
		10.1.2	Arbitrary Gauss transforms on lower rank structures	440
		10.1.3	Gauss transforms on upper rank structures	442
		10.1.4	A sequence of Gauss transforms from bottom to top	443
		10.1.5	Ascending and descending Gauss transforms	444
		10.1.6	Zero-creating Gauss transforms	445
		10.1.7	Rank-expanding Gauss transforms	446
		10.1.8	A sequence of Gauss transforms from top to bottom	448
	10.2		Effect of pivoting on rank structures	448
		10.2.1	The effect of pivoting on the rank structure	449
		10.2.2	Combining Gauss transforms with pivoting	451
		10.2.3	Sequences of transformations involving pivoting . . .	452
		10.2.4	Numerical stability	453
	10.3		More on sequences of Gauss transforms	454
		10.3.1	Upper triangular Gauss transforms	454
		10.3.2	Transformations on the right of the matrix	455
	10.4		Solving systems with Gauss transforms	456
		10.4.1	Making the structured rank matrix upper triangular	456
		10.4.2	Gauss solver and the LU-factorization	458
		10.4.3	Other decompositions	459
	10.5		Different patterns of annihilation	461
		10.5.1	The graphical representation	461
		10.5.2	Some standard patterns	463
		10.5.3	Some theorems connected to Gauss transforms . . .	464
	10.6		Conclusions	470

11 A Levinson-like solver for structured rank matrices 473

	11.1	Higher order generator representable semiseparable matrices . .	474
	11.2	General quasiseparable matrices	475
	11.3	Band matrices	476
	11.4	Unsymmetric structures	478
	11.5	Summations of Levinson-conform matrices	478
	11.6	Conclusions	479

12 Block quasiseparable matrices 481

| | 12.1 | Definition . . . . | 481 |

12.2   Factorization of the block lower/upper triangular part 482
12.3   Connection to structured rank matrices 483
12.4   Special cases 484
12.5   Multiplication of a block quasiseparable matrix by a vector . . . 484
12.6   Solver for block quasiseparable systems 485
12.7   Block quasiseparable matrices and descriptor systems 487
12.8   Conclusions 490

13 $\mathcal{H}$, $\mathcal{H}^2$ and hierarchically semiseparable matrices   **491**
13.1 $\mathcal{H}$-matrices or hierarchical matrices 491
13.2 $\mathcal{H}^2$-matrices 495
13.3   Hierarchically semiseparable matrices 495
13.4   Other classes of structured rank matrices 496
13.5   Conclusions 497

14   Inversion of structured rank matrices   **499**
14.1   Banded Toeplitz matrices 500
14.2   Inversion of (generalized) Hessenberg matrices 511
14.3   Inversion of higher order semiseparable and band matrices . . . 515
       14.3.1   Strict band and generator representable semisepara-
                   ble matrices 515
       14.3.2   Band and semiseparable matrices 518
14.4   Block matrices 521
14.5   Quasiseparable matrices 525
14.6   Generalized inverses 527
14.7   Conclusions 529

15   Concluding remarks & software   **531**
15.1   Software 531
15.2   Conclusions 532

Bibliography   **533**

Author/Editor Index   **557**

Subject Index   **565**

Preface

In this book we study the class of semiseparable matrices, which is nowadays a "hot topic". Matrices arising in a wide variety of applications, e.g., quasiseparable, hierarchically semiseparable, $\mathcal{H}$-matrices, structured rank matrices, etc., are all extensions of the most simple case: the semiseparable matrices. A good understanding of the most simple form is essential for developing algorithms and methods for the more general types of matrices. Therefore, we give general and precise definitions of structured rank matrices, including the class of semiseparable matrices. We emphasize suitable representations of this class and we study algorithms for these matrices in comparison with tridiagonal matrices, which are the inverses of semiseparable ones. The book mainly aims at two goals. First we want to get the reader familiar with the class of structured rank matrices, including, e.g., quasiseparable and semiseparable matrices, and second we want to present the most common system solvers adapted for the class of structured rank matrices. In the book we restricted ourselves to working with real matrices (unless stated otherwise). Most of the results have however direct generalizations towards the complex case.

The book contains rather basic numerical linear algebra material, in the sense that no real applications are considered. This does not mean, however, that it is not valuable for engineers. Semiseparable matrices are currently used in engineering applications, see, e.g., signal processing, video processing, state space models. In lots of these applications there is the need for computing eigenvalues and/or for solving systems with semiseparable matrices in a fast and accurate way.

The book provides a solid background for all people interested in semiseparable matrices and their generalizations. This background ranges from theorems predicting the structure of the inverses and LU and QR-decompositions of very general semiseparable structures; it provides analysis of the different representations; it presents an overview of references of old and recent material in the field, and also higher order and hybrid variants of structured rank matrices are discussed. Moreover, most of the presented algorithms are available as implementations in MATLAB, thereby giving the reader tools to test the different examples given in the book. In this point of view the book can be seen as a reference book providing hints, clues and essential tools for a lot of problems involving semiseparable matrices.

The reader can use the book in two different ways. Due to the extensive index list in the back and the selective reading paragraphs, he can use the book as a reference book, searching thereby only the definitions or theorems he is interested in. The book however is written with the aim to provide a self-contained manuscript for solving structured rank systems of equations.

The book is divided in three main parts. To get the reader acquainted with matrices such as semiseparable, quasiseparable, generator representable semiseparable, we dedicated a complete part to the definition and representation of these matrices for the easy case, the case of semiseparability rank 1. The differences and similarities as well as structures under inversion are all studied in detail.

The second part focuses to system solving, again restricted to these most simple classes of matrices. The LU-decomposition, the QR-decomposition, and a Levinson-like solver for these matrices are discussed.

The class of structured rank matrices is of course not limited to these classes of semiseparability rank 1. In the third part of the book we therefore provide a general framework for defining structured rank matrices. Based on this unified framework we present the generalizations of the QR-decomposition, the LU-factorization and the Levinson solver for these matrices. In this part of the book also attention is paid to so-called hybrid variants. Pointers are presented to block quasiseparable, $\mathcal{H}$-matrices, and other types of special structured rank matrices.

Bibliography and author/editor index

After each section in the book, references, including a small summary of each cited paper, related to that section are presented.

The overall bibliography list contains bold face numbers in braces at the end of each source. These numbers point to the pages in which the corresponding manuscript is cited.

Besides a subject index an author/editor index is included. All cited authors and editors are included with references to their citations and occurences in the bibliography. Authors' names in upper case refer to manuscripts in which the author is the first author, when the name is written in lower case this refers to the manuscripts of which the author is a coauthor.

Additional resource

Together with the book a webpage has been developed. This page can be considered as an additional resource for those interested in or working in this area.

The site will contain an update of references. This means additional references, which were not known to us at the moment of writing but closely related to some of the proposed subjects. Updates of the references will be for some of the manuscripts that are submitted for publication.

Also errata and additional material related to the book will be posted here. With additional material we mean interesting additional examples, exciting new developments in the area, and so forth.

Several of the algorithms proposed or explained in this book can be downloaded from this resource.

All this information can be found at:
`http://www.cs.kuleuven.be/~mase/books/`

Acknowledgments

This book could never have been written without the help of many people. In the first place we thank Gene Golub for helping us with publishing the book and for being a wonderful host during Raf's visits. Hugo Woerdeman provided us with interesting and helpful information related to the section on completion problems. We thank the Ph.D. students who started together with Raf and were involved in our initial research related to structured rank matrices: Gianni Codevico and Ellen Van Camp. Currently we are involved in an exciting research environment, the MASE-team (MAtrices having StructurE), consisting of Steven, Yvette and Katrijn. We also thank Trevor Lipscombe from The Johns Hopkins University Press for his positive reaction to the idea of the book and for his patience.

Raf is very grateful to Els, for creating a loving and warm environment at home. It is always a pleasure coming home after another day of writing at the office.

Marc expresses his gratitude to his wife Maria and his sons Andreas and Lucas for their continuing love, support and patience, especially when he was revising parts of the book late into the evening.

Nicola thanks his wife Teresa for her patience and love.

We would like to emphasize that this book is also due to several fruitful discussions with many friends and colleagues at meetings, conferences and other occasions. We especially want to thank the following persons for their interactions and contributions to our work: G. Ammar, R. Bevilacqua, D. Bini, A. Bultheel, S. Chandrasekaran, P. Dewilde, Y. Eidelman, D. Fasino, L. Gemignani, I. Gohberg, B. Gragg, M. Gu, W. Hackbusch, A.M. Kaashoek, T. Kailath, F.T. Luk, G. Meurant, V. Olshevsky, V. Pan, B. Plestenjak, L. Reichel, M. Schuermans, G. Strang, E. Tyrtyshnikov, P. Van Dooren, S. Van Huffel and P. Van gucht.

We also acknowledge all the organizations who supported us during the writing of the book.

- The Fund for Scientific Research–Flanders (FWO–Vlaanderen) for giving a Post-doctoral fellowship (Postdoctoraal Onderzoeker) to the first author and for supporting the projects:

  - G.0078.01 (SMA: Structured Matrices and their Applications);

  - G.0176.02 (ANCILA: Asymptotic aNalysis of the Convergence behavior of Iterative methods in numerical Linear Algebra);

  - G.0184.02 (CORFU: Constructive study of Orthogonal Functions);

  - G.0455.0 (RHPH: Riemann-Hilbert problems, random matrices and Padé-Hermite approximation).

- The Research Council K.U.Leuven for the projects:

  - OT/00/16 (SLAP: Structured Linear Algebra Package);

  - OT/05/40 (Large rank structured matrix computations).

- The Belgian Programme on Interuniversity Poles of Attraction, initiated by the Belgian State, Prime Minister's Office for Science, Technology and Culture, project IUAPV-22 (Dynamical Systems and Control: Computation, Identification & Modelling)

- The research of the third author was partially supported by

  - MIUR, grant number 2004015437;
  - The short-term mobility program, Consiglio Nazionale delle Ricerche;
  - VII Programma Esecutivo di Collaborazione Scientifica Italia–Comunità Francese del Belgio, 2005–2006.

Notation

Throughout the book we use the following notation, unless stated otherwise.

$A = (a_{i,j})_{i,j}$ A general (or a quasiseparable) matrix A with elements $a_{i,j}$.

$A(i:j,k:l)$ Submatrix of A, consisting of rows i up to and including j, and columns k up to and including l.

$A(\alpha;\beta)$ ....   Submatrix of A, consisting of indices out of the set α and β.

$\alpha \times \beta$ ...... $\alpha \times \beta$ denotes the product set $\{(i,j)|i \in \alpha, j \in \beta\}$, with α and β sets.

B ....   A band matrix.

D ....   A diagonal matrix.

$\mathcal{D}$ ....   The class of diagonal matrices.

$\det(A)$ ....   The determinant of a matrix A.

$\operatorname{diag}(\mathbf{d})$ ....   Denotes a diagonal matrix, with as diagonal elements, the elements from the vector $\mathbf{d}$.

G ....   A Givens transformation.

H ....   A (generalized) Hessenberg matrix.

I_k ....   The identity matrix of size $k \times k$.

inv ....   E.g. $\mathcal{A}_{inv}$: the subclass of invertible matrices from the class $\mathcal{A}$.

L ....   A lower triangular matrix (with regard to the LU-factorization).

M ....   An elementary Gaussian transformation (with regard to the LU-factorization).

P ....   A permutation matrix (with regard to the LU-factorization).

$\operatorname{r}(\Sigma;A)$ ....   The structured rank of the matrix A, with structure Σ.

$r_{\mathcal{A}}$ ....   A representation map for the class $\mathcal{A}$.

$\operatorname{rank}(A)$ ..   The rank of a matrix A.

Σ ....   Defines a structure.

$s_{\mathcal{A}}$ ....   A map for retrieving the representation of a class $\mathcal{A}$, in some sense the inverse of $r_{\mathcal{A}}$.

S ....   A semiseparable matrix.

$S(\mathbf{u}, \mathbf{v})$............. A symmetric generator representable semiseparable matrix with generators $\mathbf{u}$ and $\mathbf{v}$.

$S(\mathbf{u}, \mathbf{v}, \mathbf{p}, \mathbf{q})$........ An unsymmetric generator representable semiseparable matrix with generators $\mathbf{u}, \mathbf{v}, \mathbf{p}$ and $\mathbf{q}$.

$\mathcal{S}$................. The class of semiseparable matrices.

$\mathcal{S}^{(d)}$.............. The class of semiseparable plus diagonal matrices.

$\mathcal{S}^{(g)}$.............. The class of generator representable semiseparable matrices.

$\mathcal{S}^{(g,d)}$............. The class of generator representable semiseparable plus diagonal matrices.

$\mathcal{S}^{(s)}$.............. The class of diagonal-subdiagonal representable semiseparable matrices.

sym............... E.g. $\mathcal{A}_{sym}$: the subclass of symmetric matrices from the class $\mathcal{A}$.

$t_{i:j}$............... An abbreviation for $t_{i:j} = t_i t_{i-1} \ldots t_j$ if $i \geq j$.
An abbreviation for $t_{i:j} = t_i t_{i+1} \ldots t_j$ if $i \leq j$.

T................. A tridiagonal or a Toeplitz matrix.

$\mathcal{T}$................. The class of tridiagonal matrices.

$\mathcal{T}^{(i)}$.............. The class of irreducible tridiagonal matrices.

$\mathrm{tril}(A, p)$.......... Denotes the lower triangular part of the matrix A, below and including subdiagonal p.

$\mathrm{triu}(A, p)$.......... Denotes the upper triangular part of the matrix A, above and including superdiagonal p.

Q................. A unitary(orthogonal) matrix (with regard to the QR-factorization).

$\mathcal{Q}$................. The class of quasiseparable matrices.

$\mathcal{Q}^{(g)}$.............. The class of generator representable quasiseparable matrices.

$\mathcal{Q}^{(s)}$.............. The class of quasiseparable matrices with a symmetric rank structure.

R................. An upper triangular matrix (with regard to the QR-factorization).

U................. An upper triangular matrix (with regard to the LU-factorization).

$\mathbf{u}^T = [u_1, u_2, \ldots, u_n]$ A column vector $\mathbf{u}$, with entries u_i.
All vectors are column vectors and denoted in bold.

$\mathbf{u}(i : j)$............. A subvector of $\mathbf{u}$, consisting of the elements i up to and including j.

Z................. A Hessenberg-like matrix.

Part I

Introduction to semiseparable and related matrices

This first part of the book is merely an extended introduction to the class of semiseparable and related matrices. In the past decade researchers rediscovered in some sense the class of structured rank matrices. Because semiseparable matrices were never before uniformly introduced, everyone started to use these matrices in slightly different forms. Classes of matrices, such as quasiseparable, semiseparable plus diagonal, generator representable semiseparable, block semiseparable and hierarchically semiseparable, all find their origin in the past few years. Unfortunately this diversion and the expanding number of publications related to structured rank matrices often leads to confusion, misunderstandings and misinterpretations, because there is no uniformity in nomenclature. For example in this text we use the term generator representable semiseparable; often, however, these matrices are simply considered as being semiseparable.

In this context we dedicated a complete part to the definition and the representation of semiseparable and related matrices. It is not common, but in this book, we clearly distinguish between the definition of a matrix and its representation. Often there is not really a difference between a class of matrices satisfying a definition and a class of matrices satisfying certain representation properties. First, let us illustrate this with an example. Let us define a class of matrices having all elements zero except the elements on the diagonal and name this class the class of diagonal matrices. Second, we define a class of matrices depending on only one vector; this is what we name the representation vector. To construct a matrix from this vector we simply put all the elements of this vector on the diagonal of a matrix and put all other elements zero. It is clear that both classes of matrices are identical, but the viewpoint is different. In the first case we defined the class of matrices by putting some constraints on the structure of the matrix. In the second case we explicitly constructed a mapping to retrieve the matrix given an initial vector. The second case does not necessarily reveal information on the structure of the matrix, whereas the first case, does not necessarily reveal information on how to store or represent the matrix efficiently. As there is an intrinsic difference between the definition and the representation, we wrote two chapters, one related to the definition and one related to the representation.

In the first chapter we closely investigate the definitions of several structured rank matrices by posing constraints on the structure of the matrix. We define the classes of semiseparable, semiseparable plus diagonal, quasiseparable and generator semiseparable matrices. Based on these structural constraints we investigate in detail the relations between all these different classes. A clear definition will avoid confusion in the remainder of the book. Finally we also investigate the structure of these structured rank matrices under inversion.

The second chapter focuses attention on the representation of the matrices proposed in the first chapter. Different types of representations are proposed such as the generator form, the diagonal-subdiagonal representation, the Givens-vector representation and the quasiseparable representation. Detailed investigations are made showing the possibilities and limitations of each of these representations, both in the symmetric and in the unsymmetric case.

The third and last chapter of this part of the book shows some historical applications of semiseparable and related matrices. Also an overview of references to topics not covered in this volume is given, such as eigenvalue and inverse eigenvalue problems. Eigenvalue and inverse eigenvalue problems related to structured rank matrices will be discussed in Volume II of this book.

✎ *After each introduction to a chapter or part there will be an paragraph marked with a pencil. The intention of this paragraph is to highlight the main goals and issues of the chapter considered. Keywords will be presented as well as a list of the most significant theorems, properties and examples in the chapter. For the introduction of this part the keywords and most important issues in the chapters are presented only briefly.*

This paragraph is intended for those readers who do not want to read the book from the beginning until the very end, but who intend to directly focus on those topics that interest them the most. part of their interest. Let us provide the essential information of this part.

The first part of the book is an extended "introduction" to structured rank matrices of the simple kind. With the simple kind, the rank 1 case is meant, the higher order cases are considered starting from Part III. The keywords in this part of the book are definitions and representations of structured rank matrices (having semiseparability rank 1). The matrices are defined in a formal mathematical setting and their possible representations for actual implementations are discussed.

The highlights in the first chapter are the definitions of the different classes of structured rank matrices such as semiseparable, Hessenberg-like, semiseparable plus diagonal and quasiseparable, both symmetric and nonsymmetric. The differences and similarities between the classes are discussed. Finally the nullity theorem is discussed, making statements on the rank structure of the inverse of structured rank matrices.

The main issues in Chapter 2 are the possible representations for structured rank matrices. The Givens-vector, generator and quasiseparable representation can be considered as the most important and commonly used ones.

The final chapter of this part contains historical examples and references. Reading this part is not essential for a full understanding of the remainder of the book.

Chapter 1

Semiseparable and related matrices: definitions and properties

In this first chapter of the book we will pay special attention to the definition of semiseparable matrices and closely related classes such as 'generator representable' semiseparable, quasiseparable and semiseparable plus diagonal matrices. For the first half of the book (Part I and Part II), we only consider the easy classes, i.e., the classes of semiseparability rank 1. Higher order structured rank matrices will be dealt with in chapters in Part III.

In the first section, only symmetric matrices are considered. Definitions are presented, covering the classes of semiseparable, semiseparable plus diagonal, generator representable semiseparable and quasiseparable matrices. Semiseparable and quasiseparable matrices are defined in a structured rank way, i.e., certain subblocks of these matrices need to obey certain rank conditions. The class of generator representable semiseparable matrices is defined, as the name already insinuates, by explicitly defining the structure of the elements of this matrix in terms of the generators.

Section 1.2 investigates in more detail the relations between all the above defined classes in the symmetric case. Firstly, some misunderstandings concerning generator representable semiseparable matrices are posed. Secondly, a numerical problem, related to generator representable matrices is shown, which is brought into relation with the fact that the class of generator representable semiseparable matrices is not closed under pointwise convergence. Further on, the relations between all these classes of matrices is investigated with regard to the pointwise limit. Theorems are presented, stating that the closure of the class of generator representable semiseparable matrices is the class of semiseparable matrices and also that the closure of the class of semiseparable plus diagonal matrices is the class of quasiseparable.

The third section covers the nonsymmetric definitions, presented analogously as in the first section. Moreover, the relations between the different symmetric classes as investigated in Section 1.2 are now expanded towards the nonsymmetric case.

The last section of this chapter is concerned with the structure of all the presented matrices under inversion. We will prove the nullity theorem in two different ways. The nullity theorem states that the nullity of a subblock of a certain invertible matrix A has the same nullity as the complementary subblock in the matrix A^{-1}. First we prove the theorem as Fiedler and Markham did it in [114], by multiplying the partitioned matrix A by its correspondingly partitioned inverse, thereby revealing properties of the subblocks, connected to the partitioning. Afterwards we enhance the ideas from Barrett and Feinsilver [15] to provide the reader with another proof of the theorem, based on the determinants of matrices. Because the formulation of the theorem is quite abstract, two small corollaries will be formulated, which will give more immediate insight into the structured rank of matrices and their inverses. Based on these theorems easy proofs will be provided, stating that the inverse of semiseparable matrices gives tridiagonal matrices, the inverse of a generator representable matrix is an irreducible tridiagonal and that the class of quasiseparables is closed under inversion. The inverse of a semiseparable plus diagonal matrix is treated differently than the other ones. In general the inverse of a semiseparable plus diagonal matrix, with a nonzero diagonal matrix, is again a semiseparable plus diagonal matrix, having as diagonal the inverse of the original diagonal.

✎ *This chapter defines the different classes of matrices in a uniform mathematical way, based on rank restrictions posed on subblocks. Definition 1.1 and the following graphical representation present two commonly used objects in this book: a definition of a structured rank matrix based on the rank of subblocks and the graphical matrix representation with elements* $\times$ *and* $\boxtimes$*, which will be frequently reconsidered when explaining algorithms. This definition introduces the class of semiseparable matrices. Example 1.2 is important for showing the differences between definitions later on. Definition 1.3, Definition 1.4 and Definition 1.5 discuss, respectively, the rank structure of a generator representable, semiseparable plus diagonal and quasiseparable matrix. In particular, the example of the structure of the generator representable semiseparable matrix based on the two generators* $\mathbf{u}$ *and* $\mathbf{v}$ *is also important. Example 1.6 shows that the class of quasiseparable matrices covers the other classes.*

In Section 1.2 relations between the symmetric definitions are discussed. Theorem 1.12 and Proposition 1.13 contain the most important results, stating the relation between a generator representable semiseparable matrix and a semiseparable matrix. Example 1.14 illustrates these results.

Statements concerning the closure of these classes of matrices are also made. In case one is only interested in system solving with a matrix for which the representation is given beforehand, these results are not essential. They are useful for people who want to approximate a given matrix by, e.g., a semiseparable, or a quasiseparable. These theorems on the closure provide limitations for each of the presented classes. For results on this closure Theorem 1.15 and Theorem 1.17 are the important ones.

Section 1.3 defines the unsymmetric analogues of the structured rank matrices. For a generator representable matrix the example following Definition 1.23 gives the main idea. Remark that that the rank structure in the upper and lower triangular share the same diagonal.

Section 1.4 discusses the closure of the unsymmetric classes. The same remark holds as above. This is a must only for those interested in approximations or iterative procedures. Definition 1.29 is important for defining quasiseparable matrices, having a particular structure with its impact in Theorem 1.30.

People wanting to invert structured rank matrices can find direct rank relations between subblocks of a matrix A and its inverse A^{-1} in Section 1.5. Corollary 1.36 provides a ready-to-use formula, which is the basis for all proofs in this section. The remaining sections consider different types of structured rank matrices with examples. A summary of all these results is given in Section 1.5.5.

1.1 Symmetric semiseparable and related matrices

The main aim of this first chapter of the book is to present some insight into the class of semiseparable matrices. To start we will focus on the class of semiseparable matrices of semiseparability rank 1. This is of course the most easy subclass in the field of semiseparables, just as the tridiagonals are in the class of band matrices. Submatrices of semiseparable matrices satisfy certain rank conditions, therefore they are often also named structured rank matrices.[1] In this section we will define semiseparable matrices by stating which parts of the matrices have to satisfy the particular rank conditions.

First we will define the most simple semiseparable and related matrices, namely the symmetric ones of semiseparability rank 1. We will gradually build up the different types of structured rank matrices, such as generator representable, quasiseparable and semiseparable plus diagonal. In a later chapter in the book, we will also consider structured rank matrices of higher order semiseparability ranks. Initially we start by introducing the most simple cases, gradually, as our knowledge increases, we will introduce more and more difficult types of structures.

In the remainder of the book, while speaking about a semiseparable matrix, the most easy case of rank 1 is always meant. If we consider higher order cases, it will be mentioned explicitly.

Definition 1.1. *A matrix S is called a symmetric semiseparable matrix if all submatrices taken out of the lower and upper triangular part[2] of the matrix are of rank 1 and the matrix is symmetric.*

Or, this means that a symmetric matrix $S \in \mathbb{R}^{n \times n}$ is symmetric semiseparable, if the following relation is satisfied:

$$\text{rank}\,(S(i:n, 1:i)) \leq 1 \text{ with } i = 1, \dots, n.$$

---

[1]Structured rank or rank structured, both terms are frequently used.

[2]The lower(upper) triangular part of the matrix denotes the part below(above) and including the diagonal. The part below(above) the diagonal is addressed as the strictly lower(upper) triangular part.

With $S(i : j, k : l)^3$, we denote that part of the matrix, ranging from row i up to and including row j, and column k up to and including column l.

We will illustrate this with a figure and some examples. A matrix is symmetric semiseparable, if all subblocks marked by the $\boxtimes$ taken out of the lower triangular part in this 5×5 example have rank at most 1. We remark that this concerns only the rank structure for the lower triangular part of the matrix. By symmetry, this rank structure also holds for the upper triangular part. The marked subblocks in the following figures of the semiseparable matrix should have rank at most 1:

$$
\begin{bmatrix} \boxtimes & \times & \times & \times & \times \\ \boxtimes & \times & \times & \times & \times \\ \boxtimes & \times & \times & \times & \times \\ \boxtimes & \times & \times & \times & \times \\ \boxtimes & \times & \times & \times & \times \end{bmatrix},
\begin{bmatrix} \times & \times & \times & \times & \times \\ \boxtimes & \boxtimes & \times & \times & \times \\ \boxtimes & \boxtimes & \times & \times & \times \\ \boxtimes & \boxtimes & \times & \times & \times \\ \boxtimes & \boxtimes & \times & \times & \times \end{bmatrix},
\begin{bmatrix} \times & \times & \times & \times & \times \\ \times & \times & \times & \times & \times \\ \boxtimes & \boxtimes & \boxtimes & \times & \times \\ \boxtimes & \boxtimes & \boxtimes & \times & \times \\ \boxtimes & \boxtimes & \boxtimes & \times & \times \end{bmatrix},
$$

$$
\begin{bmatrix} \times & \times & \times & \times & \times \\ \times & \times & \times & \times & \times \\ \times & \times & \times & \times & \times \\ \boxtimes & \boxtimes & \boxtimes & \boxtimes & \times \\ \boxtimes & \boxtimes & \boxtimes & \boxtimes & \times \end{bmatrix} \text{ and }
\begin{bmatrix} \times & \times & \times & \times & \times \\ \times & \times & \times & \times & \times \\ \times & \times & \times & \times & \times \\ \times & \times & \times & \times & \times \\ \boxtimes & \boxtimes & \boxtimes & \boxtimes & \boxtimes \end{bmatrix}.
$$

Example 1.2 The following three matrices are symmetric semiseparable matrices according to the above definition above:

$$
\begin{bmatrix} 1 & 2 & 3 \\ 2 & 2 & 3 \\ 3 & 3 & 3 \end{bmatrix},
\begin{bmatrix} 1 & & \\ & 2 & \\ & & 3 \end{bmatrix} \text{ and }
\begin{bmatrix} 1 & 2 & \\ 2 & 2 & \\ & & 3 \end{bmatrix}.
$$

The nonshown entries are assumed to be zero.   ∎

The following, frequently used definition for semiseparable matrices, is the so-called generator definition[4] (see, e.g., [54, 105, 210, 284]). Hence we name the matrices satisfying this definition, 'generator representable semiseparable' matrices.

Definition 1.3. *A matrix S is called a symmetric generator representable semiseparable matrix if the lower triangular part of the matrix is coming from a rank 1 matrix and the matrix S is symmetric.*

This means that, a matrix S is symmetric generator representable semiseparable if (for $i = 1, \ldots, n$):

$$
\mathrm{tril}(S) = \mathrm{tril}(\mathbf{uv}^T),
$$
$$
\mathrm{triu}(S) = \mathrm{triu}(\mathbf{vu}^T),
$$

where $\mathbf{u}$ and $\mathbf{v}$ are two column vectors of length n. With $\mathrm{triu}(\cdot)$ and $\mathrm{tril}(\cdot)^5$, we denote, respectively, the upper triangular and the lower triangular part of the matrix.

---

[3]This is MATLAB style notation. MATLAB is a registered trademark of The MathWorks, Inc.

[4]In our book, we name these matrices in this way. Often these matrices are simply referred to as semiseparable matrices. (See also the notes and references at the end of this section.)

[5]The commands triu($\cdot$) and tril($\cdot$) are defined similarly to the MATLAB commands.

A symmetric generator representable semiseparable matrix has the following structure, with $\mathbf{u}^T = [u_1, u_2, \ldots, u_n]$ and $\mathbf{v}^T = [v_1, v_2, \ldots, v_n]$:

$$S = \begin{bmatrix} u_1v_1 & u_2v_1 & u_3v_1 & \cdots & u_nv_1 \\ u_2v_1 & u_2v_2 & u_3v_2 & \cdots & u_nv_2 \\ u_3v_1 & u_3v_2 & u_3v_3 & & \vdots \\ \vdots & \vdots & & \ddots & \\ u_nv_1 & u_nv_2 & u_nv_3 & \cdots & u_nv_n \end{bmatrix}. \tag{1.1}$$

Reconsidering Example 1.2, one can easily see that the first matrix is symmetric generator representable semiseparable, with generators $\mathbf{u}^T = [1, 2, 3]$ and $\mathbf{v}^T = [1, 1, 1]$, but the second and third matrices are not. More information on the relation between these definitions will be given in Chapter 2, where the representation of all these matrices will be investigated in more detail. In the remainder of the text, we will denote a generator representable matrix S, with generators $\mathbf{u}, \mathbf{v}$ as $S(\mathbf{u}, \mathbf{v})$.

Quite often the class of semiseparable matrices or 'generator representable' semiseparable matrices, is not large enough to deal with specific problems; the class of semiseparable plus diagonal matrices is sometimes required.

Definition 1.4. *A matrix S is called a symmetric semiseparable plus diagonal matrix if it can be written as the sum of a diagonal and a symmetric semiseparable matrix.*

This type of matrix arises, for example, in the discretization of particular integral equations (see, e.g., [196, 197]) and Section 3.3 in Chapter 3.

Finally, we will define a last important class of structured rank matrices, closely related to the class of semiseparable ones and to the class of semiseparable plus diagonal matrices, namely the class of quasiseparable matrices[6].

Definition 1.5. *A matrix S is called a symmetric quasiseparable matrix if all the subblocks, taken out of the strictly lower triangular part of the matrix (respectively, the strictly upper triangular part) are of rank 1, and the matrix S is symmetric.*

This means that a symmetric matrix $S \in \mathbb{R}^{n \times n}$ is symmetric quasiseparable, if the following relation is satisfied (for $i = 1, \ldots, n - 1$):

$$\text{rank}\,(S(i+1 : n, 1 : i)) \leq 1 \text{ for } i = 1, \ldots, n - 1.$$

Speaking about quasiseparable matrices, one might say that the lower triangular part is of quasiseparable form, meaning that the diagonal is not includable in the low rank structure. Or one might say that the strict lower triangular part is of

---

[6]We note that in general quasiseparable matrices are not defined in this way, by specifying the rank structure of subblocks, but by specifying the structure of each element separately by parameters [93].

semiseparable form. Moreover if we say that a specific part of a matrix is of semi-separable form, it means that this specific part satisfies the semiseparable structural constraints.

It is clear that the class of quasiseparable matrices is the most general one. It includes the class of semiseparable matrices, the class of generator representable semiseparable matrices and the class of semiseparable plus diagonal matrices. The only difference with the class of semiseparable matrices is the inclusion of the diagonal. The following section will go into more detail concerning the difference between quasiseparable and semiseparable plus diagonal matrices. Let us present some examples of quasiseparable matrices.

Example 1.6 The following matrices are all symmetric quasiseparable matrices, and none of them are either semiseparable or generator representable semiseparable:

$$
\begin{bmatrix} 0 & 1 & 2 \\ 1 & 0 & 2 \\ 2 & 2 & 0 \end{bmatrix},
\begin{bmatrix} 1 & 1 & 0 \\ 1 & 2 & 2 \\ 0 & 2 & 3 \end{bmatrix} \text{ and }
\begin{bmatrix} 1 & 1 & 0 & 1 \\ 1 & 2 & 0 & 2 \\ 0 & 0 & 3 & 1 \\ 1 & 2 & 1 & 4 \end{bmatrix}.
$$

Moreover, the second matrix is of quasiseparable form, but is not of semiseparable plus diagonal form.   ∎

In this section, we defined different classes of semiseparable matrices. In the next section, we will investigate in more detail the relations between them. Constraints will be derived under which some of the classes will coincide.

Notes and references

In this section we will provide some references concerning the earliest occurrences of semi-separable and related matrices. Most of the references will be repeated afterwards, when they were discussed in more detail. To our knowledge, the first appearances of semiseparable matrices, named 'one-pair' or 'single-pair' matrices at that time, are the following ones. For the sake of 'historical' completeness all translations of these references are included.

☞ M. Fiedler and T. L. Markham. Generalized totally nonnegative matrices. *Linear Algebra and its Applications*, 345:9–28, 2002.

☞ F. R. Gantmacher and M. G. Kreĭn. Sur les matrices oscillatoires et complètement non négatives. *Compositio Mathematica*, 4:445–476, 1937. (In French).

☞ F. R. Gantmacher and M. G. Kreĭn. *Oscillyacionye matricy i yadra i malye kolebaniya mehaničeskih sistem. [Oscillation matrices and kernels and small oscillations of mechanical systems.].* Moscow-Leningrad, 1941. (In Russian).

☞ F. R. Gantmacher and M. G. Kreĭn. *Oscillyacionye matricy i yadra i malye kolebaniya mehaničeskih sistem. [Oscillation matrices and kernels and small oscillations of mechanical systems.].* Gosudarstv. Isdat. Tehn.-Teor. Lit., Moscow-Leningrad, second edition, 1950. (In Russian).

☞ F. R. Gantmacher and M. G. Kreĭn. *Oszillationsmatrizen, Oszil-*
lationskerne und kleine Schwingungen mechanischer Systeme. Wissenschaftliche Bearbeitung der deutschen Ausgabe: Alfred Stöhr. Mathematische Lehrbücher und Monographien, I. Abteilung, Bd. V. Akademie-
Verlag, Berlin, 1960. (In German).

☞ F. R. Gantmacher and M. G. Kreĭn. Oscillation matrices and kernels and
small vibrations of mechanical systems. Technical Report AEC-tr-448, Off.
Tech. Doc., Dept. Commerce, Washington, DC, 1961.

☞ F. R. Gantmacher and M. G. Kreĭn. *Oscillation matrices and kernels*
and small vibrations of mechanical systems. AMS Chelsea Publishing,
Providence, Rhode Island, revised edition, 2002.

In their 1937 paper, Gantmacher and Kreĭn prove that the inverse of a symmetric Jacobi
matrix (this corresponds to a symmetric irreducible[7] tridiagonal matrix) is a one-pair
matrix (this corresponds to a symmetric generator representable semiseparable matrix) via
explicit calculations. In the original book (1941), the theory of the previously published
paper is included and it is often referred to as the first book in which the inverse of an
irreducible tridiagonal matrix is explicitly calculated. The book of 1941, was published in
Russia and revised in 1950. Several translations were made of it, first into the German
tongue (1960) and afterwards (1961) also into an English technical report. The complete
book as it appeared in 2002 is based on the three references (1950, 1960, 1961). More
information on these one-pair matrices and their proposed inversion method can be found
in Chapter 3.

From the 1950s, several people tried to invert tridiagonal, semiseparable and related
matrices (the inverse of a tridiagonal matrix is, as we will later prove, of semiseparable
form). These references are among the earliest trying to invert these matrices in various
fields of mathematics. These publications consider integral equations, statistics and boundary value problems. More detailed information can be found in upcoming Section 1.5, on
inversion.

☞ E. Asplund. Inverses of matrices a_{ij} which satisfy $a_{ij} = 0$ for $j > i + p$.
Mathematica Scandinavica, 7:57–60, 1959.

☞ S. O. Asplund. Finite boundary value problems solved by Green's matrix.
Mathematica Scandinavica, 7:49–56, 1959.

☞ W. J. Berger and E. Saibel. On the inversion of continuant matrices.
Journal of the Franklin Institute, 256:249–253, 1953.

☞ S. N. Roy and A. E. Sarhan. On inverting a class of patterned matrices.
Biometrika, 43:227–231, 1956.

Quasiseparable matrices were recently introduced and investigated in the following
two manuscripts.

☞ Y. Eidelman and I. C. Gohberg. On a new class of structured matrices.
Integral Equations and Operator Theory, 34:293–324, 1999.

☞ E. E. Tyrtyshnikov. Mosaic ranks for weakly semiseparable matrices. In
M. Griebel, S. Margenov, and P. Y. Yalamov, editors, *Large-Scale Scientific*
Computations of Engineering and Environmental Problems II, volume 73
of *Notes on numerical fluid mechanics,* pages 36–41. Vieweg, 2000.

---

[7]A tridiagonal matrix is called irreducible, if its subdiagonal and superdiagonal elements are
different from zero.

In the first manuscript, Eidelman and Gohberg investigate a generalization of the class of semiseparable matrices, namely the class of quasiseparable matrices, in its most general form (this will be investigated in Chapter 12 of this book). They show that the class of quasiseparable matrices is closed under inversion, and they present a linear complexity inversion method. Tyrtyshnikov names the class of quasiseparable matrices weakly semiseparable matrices.

Historically, semiseparable matrices have been defined in different, quite often inconsistent ways. One can divide these definitions into two mainstreams: the 'Linear Algebra' form and the 'Operator Theory' definition. From an operator theoretical viewpoint, semiseparable matrices can be considered as coming from the discretization of semiseparable kernels (see also Chapter 3). The resulting matrix is a generator representable semiseparable matrix. Hence, quite often, in the literature, one speaks about a semiseparable matrix, but one means a 'generator representable' matrix, as defined in this section. From a linear algebra viewpoint, semiseparable matrices are considered as the inverses of tridiagonal matrices. In this book we use the 'Linear Algebra' formulation. We will however always clearly state which type of semiseparable matrix we are working with.

1.2 Relations between the different symmetric definitions

The main goal of this section is to provide more information concerning the relations of the previously defined classes of symmetric semiseparable, generator representable semiseparable, semiseparable plus diagonal, tridiagonal and quasiseparable matrices. Our interest extends out to the inclusions among these sets and the closure of these classes under a certain norm.

1.2.1 Known relations

This section summarizes the known relations between the different types of matrices. Let us identify the different classes of matrices considered here as follows:

$$\mathcal{T}_{sym} = \{A \in \mathbb{R}^{n \times n} \mid A \text{ is a symmetric tridiagonal matrix}\},$$

$$\mathcal{T}_{sym}^{(i)} = \{A \in \mathbb{R}^{n \times n} \mid A \text{ is an irreducible symmetric tridiagonal matrix}\},$$

$$\mathcal{S}_{sym} = \{A \in \mathbb{R}^{n \times n} \mid A \text{ is a symmetric semiseparable matrix}\},$$

$$\mathcal{D} = \{A \in \mathbb{R}^{n \times n} \mid A \text{ is a diagonal matrix}\},$$

$$\mathcal{S}_{sym}^{(d)} = \{A \in \mathbb{R}^{n \times n} \mid A \text{ is a symmetric semiseparable plus diagonal matrix}\},$$

$$= \{A \in \mathbb{R}^{n \times n} \mid A = S + D, \text{ with } S \in \mathcal{S}_{sym}, D \in \mathcal{D}\},$$

$$\mathcal{S}_{sym}^{(g)} = \{A \in \mathbb{R}^{n \times n} \mid A \text{ is a symmetric generator}$$
$$\text{representable semiseparable matrix}\},$$

$$\mathcal{Q}_{sym} = \{A \in \mathbb{R}^{n \times n} \mid A \text{ is a symmetric quasiseparable matrix}\}.$$

With $_{sym}$, we denote the restriction of each of these classes to the symmetric matrices. Later on we will also use the notation $_{inv}$ for the restriction to the class of invertible matrices. We clearly have the following inclusions among these sets of matrices (in the generic case, i.e., n large enough[8]):

$$\mathcal{Q}_{sym} \supsetneq \mathcal{S}_{sym}^{(d)} \supsetneq \mathcal{S}_{sym} \supsetneq \mathcal{S}_{sym}^{(g)},$$

$$\mathcal{Q}_{sym} \supsetneq \mathcal{T}_{sym} \supsetneq \mathcal{T}_{sym}^{(i)}.$$

We also have

$$\mathcal{S}_{sym} \supsetneq \mathcal{D}, \quad \mathcal{T}_{sym} \supsetneq \mathcal{D},$$

$$\mathcal{S}_{sym}^{(g)} \not\supseteq \mathcal{D}, \quad \mathcal{T}_{sym}^{(i)} \not\supseteq \mathcal{D}.$$

With $\mathcal{A} \supsetneq \mathcal{B}$ we denote that $\mathcal{B}$ is a subset of $\mathcal{A}$ but distinct. To illustrate that the inclusions are strict, we present some examples.

Example 1.7 The matrix S_1 belongs to $\mathcal{S}_{sym}$, but not to $\mathcal{S}_{sym}^{(g)}$

$$S_1 = \begin{bmatrix} 1 & & \\ & 2 & \\ & & 3 \end{bmatrix}.$$

The matrix S_2 belongs to $\mathcal{S}_{sym}^{(d)}$, but not to $\mathcal{S}_{sym}$

$$S_2 = \begin{bmatrix} 1 & 2 & 3 \\ 2 & 2 & 3 \\ 3 & 3 & 3 \end{bmatrix} - \begin{bmatrix} 1 & & \\ & 2 & \\ & & 3 \end{bmatrix} = \begin{bmatrix} 0 & 2 & 3 \\ 2 & 0 & 3 \\ 3 & 3 & 0 \end{bmatrix}.$$

The matrix S_3 belongs to $\mathcal{Q}_{sym}$, but not to $\mathcal{S}_{sym}^{(d)}$

$$S_3 = \begin{bmatrix} 1 & 2 & 0 \\ 2 & 2 & 3 \\ 0 & 3 & 3 \end{bmatrix}.$$

And finally, the matrix S_2, from above, belongs to $\mathcal{Q}_{sym}$, but it is not tridiagonal and hence does not belong to $\mathcal{T}_{sym}$. The strict inclusion between the class of irreducible tridiagonal matrices, and the class of tridiagonal matrices is straightforward. ∎

Before investigating in more detail the relations of the closure of the above defined sets, we will treat some important issues related to the generator representable semiseparable matrices, such as common misunderstandings and numerical stability of retrieving the representation.

1.2.2 Common misunderstandings about generator representable semiseparable matrices

The following examples are included to illustrate some common misunderstandings about generator representable semiseparable matrices. It will be shown that the inverse of a symmetric tridiagonal is not always a symmetric generator representable semiseparable matrix or a symmetric generator representable semiseparable matrix plus a diagonal, as often stated in publications.

---

[8]For example $n = 1$ generates trivial equalities between all classes.

Example 1.8 Several papers state that the inverse of a tridiagonal matrix is a generator representable semiseparable matrix. However, consider the following matrix:

$$T_1 = \begin{bmatrix} 0 & 1 & 0 \\ 1 & 0 & 0 \\ 0 & 0 & 1 \end{bmatrix}.$$

This is clearly a nonsingular symmetric tridiagonal matrix. According to the statement above, its inverse should be a symmetric semiseparable matrix representable with two generators $\mathbf{u}$ and $\mathbf{v}$. Matrix T_1 is its own inverse, and one cannot represent this matrix with two generators $\mathbf{u}$ and $\mathbf{v}$. When we expand this class to the class of generator representable semiseparable plus diagonal matrices, we can represent the matrix T_1 in this way, but this is not the case for all the inverses of symmetric tridiagonal matrices.

For example, consider the following matrix T_2, which is a block combination of the matrix from above:

$$T_2 = \begin{bmatrix} 0 & 1 & 0 & 0 & 0 & 0 \\ 1 & 0 & 0 & 0 & 0 & 0 \\ 0 & 0 & 1 & 0 & 0 & 0 \\ 0 & 0 & 0 & 0 & 1 & 0 \\ 0 & 0 & 0 & 1 & 0 & 0 \\ 0 & 0 & 0 & 0 & 0 & 1 \end{bmatrix}.$$

The reader can verify that the inverse of this nonsingular symmetric tridiagonal matrix cannot be represented by two generators $\mathbf{u}$ and $\mathbf{v}$, or by two generators and a diagonal. ∎

The next section considers a numerical problem of the generator representable semiseparable matrices. The nature of this problem is the nonclosedness of this class.

1.2.3   A theoretical problem with numerical consequences

The following example deals with the numerical instability of the representation with generators. Mainly, this problem arises, when the representation needs to be retrieved given a certain semiseparable matrix. After the example we will show that this numerical problem is inherent to the definition of generator representable semiseparable matrices.

Example 1.9 Suppose a symmetric 5×5 matrix A is given with the following eigenvalues: $[1, 2, 3, 100, 10^4]$. Reducing the matrix A to a similar symmetric semiseparable matrix[9] (the procedure for doing that is explained in the second volume of this book and in [282]) generates the following matrix S (using 16 decimal digits of precision in MATLAB):

---

[9]The reduction algorithm possesses a subspace iteration property, thereby creating small elements in the bottom row.

$$\begin{bmatrix} 1.6254 & -5.0812 \cdot 10^{-1} & -8.0163 \cdot 10^{-2} & -6.3532 \cdot 10^{-5} & 5.2384 \cdot 10^{-11} \\ -5.0812 \cdot 10^{-1} & 1.4259 & 2.2496 \cdot 10^{-1} & 1.7829 \cdot 10^{-4} & -1.4597 \cdot 10^{-10} \\ -8.0163 \cdot 10^{-2} & 2.2496 \cdot 10^{-1} & 2.9485 & 2.3368 \cdot 10^{-3} & -1.9152 \cdot 10^{-9} \\ -6.3532 \cdot 10^{-5} & 1.7829 \cdot 10^{-4} & 2.3368 \cdot 10^{-3} & 9.9999 \cdot 10^{1} & -8.1957 \cdot 10^{-5} \\ 5.2384 \cdot 10^{-11} & -1.4597 \cdot 10^{-10} & -1.9152 \cdot 10^{-9} & -8.1957 \cdot 10^{-5} & 1.0000 \cdot 10^{4} \end{bmatrix}.$$

Although this matrix is semiseparable it can clearly be seen that the last entry of the diagonal already approximates the largest eigenvalue. Representing this matrix with the traditional generators $\mathbf{u}$ and $\mathbf{v}$ gives us the following possible vectors for $\mathbf{u}$ and $\mathbf{v}$, respectively:

$$\begin{bmatrix} 3.1029 \cdot 10^{10} & -9.6998 \cdot 10^{9} & -1.5302 \cdot 10^{9} & -1.2128 \cdot 10^{6} & 1.0000 \end{bmatrix}^{T}$$

and

$$\begin{bmatrix} 5.2384 \cdot 10^{-11} & -1.4597 \cdot 10^{-10} & -1.9152 \cdot 10^{-9} & -8.1957 \cdot 10^{-5} & 1.0000 \cdot 10^{4} \end{bmatrix}^{T}.$$

Because the second element of $\mathbf{v}$ is of the order 10^{-10} and is constructed by summations of elements, whose magnitude is of order 1, we can expect that this element has a precision of only 6 significant decimal digits left. Trying to reconstruct the matrix S with the given generators $\mathbf{u}$ and $\mathbf{v}$ will therefore create large relative errors in the matrix (e.g., up to $\approx 10^{-2}$). This means that this representation looses approximately 14 decimal digits. The digit loss in this example is unacceptable. ∎

This problem can be explained rather easily, and it is inherent to the representation connected with the definition. Because the diagonal matrices do not belong to the class of matrices represented by the generators $\mathbf{u}$ and $\mathbf{v}$, the representation of the matrix, based on $\mathbf{u}$ and $\mathbf{v}$ above can never be very good. Very large and very small numbers can be seen in the vectors $\mathbf{u}$ and $\mathbf{v}$ to try to compensate the fact that the matrix is almost block diagonal.[10]

To investigate more in detail this numerical problem, we will first look at the class of tridiagonal matrices. A first important property of symmetric tridiagonal matrices is that they are defined by the diagonal and subdiagonal elements. In fact storing $2n - 1$ elements is enough to reconstruct the tridiagonal matrix. The following property is important in several applications. Suppose we have a symmetric tridiagonal matrix T and apply the QR algorithm, to compute the eigenvalues of this matrix. Doing so we get a sequence of tridiagonal matrices

$$T^{(0)} \rightarrow T^{(1)} \rightarrow \cdots \rightarrow T^{(n)} \rightarrow \cdots$$

converging towards a (block) diagonal matrix. This diagonal matrix also belongs to the class of tridiagonal matrices. To formulate this more precisely, we need the definition of pointwise convergence:

---

[10]One can compute $\mathbf{u}$ and $\mathbf{v}$ differently than presented above, thereby reducing the error; nevertheless, the accuracy problem remains.

Definition 1.10. *The pointwise limit of a collection of matrices $A_\epsilon \in \mathbb{R}^{n \times n}$ (if it exists) for $\epsilon \to \epsilon_0$, with $\epsilon, \epsilon_0 \in \mathbb{R}$ and with the matrices A_ϵ as*

$$A_\epsilon = \begin{bmatrix} (a_{1,1})_\epsilon & \cdots & (a_{1,n})_\epsilon \\ \vdots & \ddots & \vdots \\ (a_{n,1})_\epsilon & \cdots & (a_{n,n})_\epsilon \end{bmatrix}$$

is defined as:

$$\lim_{\epsilon \to \epsilon_0} A_\epsilon = \begin{bmatrix} \lim_{\epsilon \to \epsilon_0} (a_{1,1})_\epsilon & \cdots & \lim_{\epsilon \to \epsilon_0} (a_{1,n})_\epsilon \\ \vdots & \ddots & \vdots \\ \lim_{\epsilon \to \epsilon_0} (a_{n,1})_\epsilon & \cdots & \lim_{\epsilon \to \epsilon_0} (a_{n,n})_\epsilon \end{bmatrix}.$$

Let us summarize the properties of tridiagonal matrices mentioned above in a corollary.

Corollary 1.11. *The class of symmetric tridiagonal matrices T has the following properties:*

- *They can be represented by order $\mathcal{O}(n)$[11] information if the size of the matrix is $n \times n$.*

- *The class of tridiagonal matrices is closed for pointwise convergence.*

If we look again at the class of generator representable semiseparable matrices, we can clearly see that they can be represented by $\mathcal{O}(n)$ information. This is the main reason why authors choose to use this definition. Nevertheless, as we already mentioned, the set of diagonal matrices is not included in the class of generator representable semiseparable matrices, and we can easily construct a sequence of generator representable semiseparable matrices (e.g., via the QR-algorithm) converging to a diagonal. This means that the class of generator representable semiseparable matrices is not closed under pointwise convergence. One might correctly wonder now about what is the connection between semiseparable and generator representable semiseparable matrices with respect to the pointwise limit.

1.2.4   Generator representable semiseparable and semiseparable matrices

We will prove in this section that the class of semiseparable matrices is closed under the pointwise convergence, and even more that the closure of the generator representable semiseparable matrices is indeed the class of semiseparable matrices. First a theorem and a proposition are needed.

---

[11]With $f(n)$ is of order $\mathcal{O}(n)$, we mean that two constants k and c exist such that for every $n \geq k$, $f(n) \leq cg(n)$.

As already mentioned before in this book, the submatrix of the matrix A consisting of the rows $i, i+1, \ldots, j-1, j$ and the columns $k, k+1, \ldots, l-1, l$ is denoted using the MATLAB-style notation $A(i:j, k:l)$, the same notation style is used for the elements $i, \ldots, j$ in a vector $\mathbf{u}$: $\mathbf{u}(i:j)$ and the (i,j)th element in a matrix A: $A(i,j)$.

Theorem 1.12. *Suppose S is a symmetric semiseparable matrix of size n. Then, S is not a generator representable semiseparable matrix if and only if there exist indices i, j with $1 \leq j \leq i \leq n$ such that $S(i,j) = 0$, $S(i, 1:i) \neq 0$ and $S(j:n, j) \neq 0$.*

Proof. Suppose we have a symmetric semiseparable matrix S such that the element $S(i,j) = 0$, $S(i, 1:i) \neq 0$ and $S(j:n, j) \neq 0$ with $1 \leq j \leq i \leq n$. If S would be representable with two generators $\mathbf{u}^T = [u_1, u_2, \ldots, u_n]$ and $\mathbf{v}^T = [v_1, v_2, \ldots, v_n]$, this means that $S(i,j) = u_i v_j$. Hence, $u_i = 0$ or $v_j = 0$. If $u_i = 0$ this implies that

$$S(i, 1:i) = 0$$

because

$$\begin{aligned} S(i, 1:i) &= u_i[v_1, \ldots, v_i] \\ &= 0[v_1, \ldots, v_i] \\ &= 0. \end{aligned}$$

If u_i is different from zero this means that v_j equals zero, implying

$$\begin{aligned} S(j:n, j) &= v_j[u_j, \ldots, u_n]^T \\ &= 0[u_j, \ldots, u_n]^T \\ &= 0. \end{aligned}$$

This leads to a contradiction. The other direction of the proof can be given in a similar way. □

The next proposition shows how the class of symmetric generator representable semiseparable matrices can be embedded in the class of symmetric semiseparable matrices.

Proposition 1.13. *Suppose a symmetric semiseparable matrix S is given that cannot be represented by two generators. Then this matrix can be written as a block diagonal matrix, for which all the blocks are symmetric semiseparable matrices representable with two generators.*

Proof. It can be seen that a matrix S cannot be represented by two generators (e.g., $\mathbf{u}$ and $\mathbf{v}$), if and only if (see Theorem 1.12)

$$\exists k : 1 \leq k \leq n, \exists l : 1 \leq l \leq k \text{ such that } S(k,l) = 0$$
$$\exists i : l \leq i < k \text{ such that } S(i,l) \neq 0$$
$$\exists j : l < j \leq k \text{ such that } S(k,j) \neq 0.$$

$$
\begin{array}{ccc}
l & j & k \\
\downarrow & \downarrow & \downarrow
\end{array}
$$

$$
\begin{array}{l}
\\
\\
l \rightarrow \\
\\
i \rightarrow \\
\\
k \rightarrow \\
\\
\end{array}
\left[
\begin{array}{ccccccc}
\ddots & & \vdots & & \vdots & & \\
& \ddots & \vdots & & \vdots & & \\
\cdots & \cdots & \ddots & & \vdots & & \\
& & \times & \ddots & \vdots & & \\
\cdots & \cdots & 0 & \times & \ddots & & \\
& & \vdots & & & &
\end{array}
\right] .
$$

Suppose now that the element $S(\hat{i}, l) \neq 0$, with $l \leq \hat{i} < k$ and all $S(i, l) = 0$ for $\hat{i} < i < k$. The rank 1 assumption on the blocks implies that $S(i, j) = 0$, for all $\hat{i} < i \leq n$ and $1 \leq j < \hat{i} + 1$. This means that our matrix can be divided into two diagonal blocks. This procedure can be repeated until all the diagonal blocks are representable by two generators.   □

Example 1.14 Consider the following matrix

$$
S = \begin{bmatrix}
1 & 1 & 0 & 0 \\
1 & 1 & 0 & 0 \\
0 & 0 & 2 & 2 \\
0 & 0 & 2 & 2
\end{bmatrix} .
$$

This matrix can be divided into two generator representable blocks:

$$
S(\mathbf{u}_1, \mathbf{v}_1) = \begin{bmatrix} 1 & 1 \\ 1 & 1 \end{bmatrix} \text{ and } S(\mathbf{u}_2, \mathbf{v}_2) = \begin{bmatrix} 2 & 2 \\ 2 & 2 \end{bmatrix},
$$

where $\mathbf{u}_1^T = \mathbf{v}_1^T = \mathbf{u}_2^T = [1, 1]$ and $\mathbf{v}_2^T = [2, 2]$. When considering, for example, a diagonal matrix, the previous property states that we should divide the matrix completely into 1×1 blocks, which are naturally representable with two generators.   ∎

This theorem will come in handy when formulating proofs for general semiseparable matrices; using it, one can divide the matrix in different blocks, all having an explicit representation as generator representable semiseparable matrices.

    We are now ready to formulate the most important theorem of this section. Clearly seen in the following proof is the situation when problems arise with the definition of semiseparable matrices in terms of the generators. Denote the pointwise closure of a class of matrices $\mathcal{A}$ as $\overline{\mathcal{A}}$. To avoid an overloaded notation we denote with $\overline{\mathcal{S}}_{sym}^{(g)}$ the set $\mathcal{S}_{sym}^{(g)}$; this assumption in notation is used for the closure of all defined classes.

Theorem 1.15. *The pointwise closure of the class of symmetric semiseparable matrices representable by two generators $\mathcal{S}_{sym}^{(g)}$ is the class of symmetric semiseparable matrices $\mathcal{S}_{sym}$. This means that*

$$\overline{\mathcal{S}}_{sym}^{(g)} = \mathcal{S}_{sym},$$

and the class $\overline{\mathcal{S}}_{sym} = \mathcal{S}_{sym}$ is closed under pointwise convergence.

Proof.

$\Rightarrow$ Suppose a family of semiseparable matrices representable with two generators is given:

$$S(\mathbf{u}(\epsilon), \mathbf{v}(\epsilon)) \in \mathbb{R}^{n \times n} \text{ for } \epsilon \in \mathbb{R} \text{ and } \epsilon \to \epsilon_0,$$

such that the pointwise limit exists:

$$\lim_{\epsilon \to \epsilon_0} S(\mathbf{u}(\epsilon), \mathbf{v}(\epsilon)) = S \in \mathbb{R}^{n \times n}.$$

It will be shown that this matrix belongs to the class of semiseparable matrices. It is known that $\lim_{\epsilon \to \epsilon_0} (u_i(\epsilon) v_j(\epsilon)) \in \mathbb{R}$. (Note that this does not imply that $\lim_{\epsilon \to \epsilon_0} u_i(\epsilon), \lim_{\epsilon \to \epsilon_0} v_j(\epsilon) \in \mathbb{R}$, which can lead to numerically unsound problems when representing these semiseparable matrices with two generators $\mathbf{u}, \mathbf{v}$.) It remains to prove that, $\forall i \in \{2, \ldots, n\}$:

$$\text{rank} \left(\lim_{\epsilon \to \epsilon_0} (S(\mathbf{u}(\epsilon), \mathbf{v}(\epsilon))(i : n, 1 : i)) \right) = \text{rank}(S(i : n, 1 : i)) \leq 1.$$

We have ($\forall i \in \{2, \ldots, n\}$):

$$\text{rank}(S(i : n, 1 : i)) = \max\{\text{rank}(M) \,|\, M \text{ is a nonempty}$$
$$\text{square submatrix of } S(i : n, 1 : i)\}.$$

Let us take a certain nonempty 2×2 square submatrix M of $S(i : n, 1 : i)$. Denote the corresponding square submatrix of $S(\mathbf{u}(\epsilon), \mathbf{v}(\epsilon))$ with $M(\epsilon)$. We know that $\forall \epsilon : \text{rank}(M(\epsilon)) \leq 1$, i.e., $\det(M(\epsilon)) = 0$. The determinant is continuous and therefore we have:

$$\det(M) = \det \left(\lim_{\epsilon \to \epsilon_0} M(\epsilon) \right)$$
$$= \lim_{\epsilon \to \epsilon_0} \det(M(\epsilon))$$
$$= 0.$$

This means that $\text{rank}(M) \leq 1$, as this has to be valid for all 2×2 submatrices. This leads directly to

$$\text{rank}(S(i : n, 1 : i)) \leq 1,$$

which proves one direction of the theorem.

$\Leftarrow$ Suppose a semiseparable matrix S is given such that it cannot be represented by two generators. Then there exists a family $S(\mathbf{u}(\epsilon), \mathbf{v}(\epsilon))$ with $\epsilon \to \epsilon_0$ such that

$$\lim_{\epsilon \to \epsilon_0} S(\mathbf{u}(\epsilon), \mathbf{v}(\epsilon)) = S.$$

According to Proposition 1.13 the matrix can be written as a block diagonal matrix, consisting of generator representable diagonal blocks. Assume, in our case, that we have 2 diagonal blocks (more diagonal blocks can be dealt with in an analogous way), i.e., S has the following structure:

$$S = \left[\begin{array}{cc} S(\mathbf{u}, \mathbf{v}) & 0 \\ 0 & S(\mathbf{s}, \mathbf{t}) \end{array} \right].$$

In a straightforward way we can define the generators $\mathbf{u}(\epsilon), \mathbf{v}(\epsilon)$:

$$\mathbf{u}(\epsilon)^T = \left[\frac{u_1}{\epsilon}, \ldots, \frac{u_k}{\epsilon}, s_1, \ldots, s_l \right]$$
$$\mathbf{v}(\epsilon)^T = \left[\epsilon v_1, \ldots, \epsilon v_k, t_1, \ldots, t_n \right].$$

Clearly, the following limit converges (it becomes even more clear when looking at the explicitly constructed matrix in Equation (1.1)) :

$$\lim_{\epsilon \to 0} S(\mathbf{u}(\epsilon), \mathbf{v}(\epsilon)) = S.$$

This proves the theorem.

We know now that $\mathcal{S}_{sym}$ is closed under the pointwise convergence, as $\overline{\mathcal{S}}_{sym} = \overline{\overline{\mathcal{S}}}_{sym}^{(g)} = \overline{\mathcal{S}}_{sym}^{(g)} = \mathcal{S}_{sym}$. $\qquad\qquad\square$

The proof shows that the limit

$$\lim_{\epsilon \to 0} S(\mathbf{u}(\epsilon), \mathbf{v}(\epsilon)) = S$$

exists, but the limits of the generating vectors

$$\mathbf{u}(\epsilon)^T = \left[\frac{u_1}{\epsilon}, \ldots, \frac{u_k}{\epsilon}, s_1, \ldots, s_l \right],$$
$$\mathbf{v}(\epsilon)^T = \left[\epsilon v_1, \ldots, \epsilon v_k, t_1, \ldots, t_n \right]$$

do not necessarily exist. In fact for $\epsilon \to 0$ some elements of $\mathbf{u}(\epsilon)$ can become extremely large, while some elements of $\mathbf{v}(\epsilon)$ can become extremely small. This is the behavior observed in Example 1.9.

Let us present a small example for the theorem.

Example 1.16 Reconsider the matrix from Example 1.14. Define the generators $\mathbf{u}(\epsilon)$ and $\mathbf{v}(\epsilon)$ for example as follows:

$$\mathbf{u}(\epsilon)^T = \left[\frac{1}{\epsilon}, \frac{1}{\epsilon}, 1, 1 \right] \text{ and } \mathbf{v}(\epsilon)^T = [\epsilon, \epsilon, 2, 2].$$

Taking the limit of the generator representable semiseparable matrix $S(\mathbf{u}(\epsilon), \mathbf{v}(\epsilon))$ gives us:

$$\lim_{\epsilon \to 0} S(\mathbf{u}(\epsilon), \mathbf{v}(\epsilon)) = \lim_{\epsilon \to 0} \begin{bmatrix} 1 & 1 & \epsilon & \epsilon \\ 1 & 1 & \epsilon & \epsilon \\ \epsilon & \epsilon & 2 & 2 \\ \epsilon & \epsilon & 2 & 2 \end{bmatrix} = \begin{bmatrix} 1 & 1 & 0 & 0 \\ 1 & 1 & 0 & 0 \\ 0 & 0 & 2 & 2 \\ 0 & 0 & 2 & 2 \end{bmatrix},$$

which is exactly the matrix we wanted to obtain. ∎

We proved in this last theorem that the class of semiseparable matrices is in fact an extension of the class of generator representable semiseparable matrices. Moreover we proved that the class of semiseparable matrices is closed under pointwise convergence. Looking back at Corollary 1.11, one might wonder now if it is also possible to represent the class of semiseparable matrices using $\mathcal{O}(n)$ parameters. This is the subject of the next chapter. First we take a closer look at the class of semiseparable plus diagonal and quasiseparable matrices followed by the unsymmetric definitions of semiseparable and related matrices.

1.2.5 Semiseparable plus diagonal and quasiseparable matrices

In the previous sections we investigated in detail the relations between semiseparable and generator representable semiseparable matrices. The class of quasiseparable matrices is however clearly different from the class of semiseparable matrices. It is clear, for example, that a tridiagonal matrix is also quasiseparable and all semiseparable plus diagonal matrices are quasiseparable, whereas a tridiagonal matrix is never semiseparable. In this section we will investigate the relations between all these definitions for the symmetric case.

This section will also be concerned with a class of matrices of the following form $\mathcal{S}_{sym} + D$, where D is a diagonal matrix:

$$\mathcal{S}_{sym} + D = \{A \in \mathbb{R}^{n \times n} \mid A = S + D, \text{ with } S \in \mathcal{S}_{sym}\}.$$

This is a subclass of the class of semiseparable plus diagonal matrices, namely these matrices having a fixed diagonal D.

This class is an important class for the development of, for example, QR-algorithms for semiseparable plus diagonal matrices, as a QR-step on a semiseparable plus diagonal matrix maintains its structure (see [32, 75, 78, 104]), and moreover maintains also the diagonal. This means that the QR-algorithm generates matrices in the class $\mathcal{S}_{sym} + D$, with a specific diagonal D.

Adapting the scheme as presented in Subsection 1.2.1 gives us ($\forall D \in \mathcal{D}$):

$$\mathcal{Q}_{sym} \supsetneq \mathcal{S}_{sym}^{(d)} \supseteq \mathcal{S}_{sym} + D \supsetneq \mathcal{D}$$

$$\mathcal{Q}_{sym} \supsetneq \mathcal{T}_{sym} \supsetneq \mathcal{T}_{sym}^{(i)}.$$

The main theorem, which we will prove in this section, states:

$$\overline{\mathcal{S}}_{sym}^{(d)} = \mathcal{Q}_{sym}.$$

This has some important consequences. For example, if every matrix of $\mathcal{Q}_{sym}$ can be represented, as the limit of a sequence of matrices in $\mathcal{S}_{sym}^{(d)}$, this implies that a tridiagonal matrix can be represented as the limit of a sequence of semiseparable plus diagonal matrices. Moreover, this means that this class of semiseparable plus diagonal matrices is not closed under pointwise convergence. Considering the class of semiseparable plus diagonal matrices with a fixed diagonal, however, the class will be closed under the pointwise limit. Let us prove these two statements.

Theorem 1.17. *Consider the class of matrices $\mathcal{S}_{sym} + D$, where $D \in \mathcal{D}$. Then we have that $\overline{\mathcal{S}_{sym} + D} = \mathcal{S}_{sym} + D$.*

Proof. The class of matrices $\mathcal{S}_{sym} + D$ is of the following form:

$$\mathcal{S}_{sym} + D = \{A \in \mathbb{R}^{n \times n} \,|\, A = S + D \text{ with } S \in \mathcal{S}_{sym}\}.$$

It remains to prove that, if we consider a sequence of converging matrices $S_\epsilon + D$ in $\mathcal{S}_{sym} + D$, the limit

$$\lim_{\epsilon \to \epsilon_0} (S_\epsilon + D)$$

will also belong to the class $\mathcal{S}_{sym} + D$.
Because the diagonal is fixed, and the limit exists, we can shift through the limit in the formulation. This implies that the limit of the matrices S_ϵ also exists.

$$\lim_{\epsilon \to \epsilon_0} (S_\epsilon + D) = \lim_{\epsilon \to \epsilon_0} S_\epsilon + D$$
$$= S + D.$$

This limit belongs to $\mathcal{S}_{sym} + D$ as the class $\mathcal{S}_{sym}$ is closed under pointwise convergence. $\square$

The class of semiseparable plus diagonal matrices (with varying diagonal) is however not closed under pointwise convergence, as the following example illustrates.

Example 1.18 Consider the following tridiagonal matrix, which is the limit of a sequence of semiseparable plus diagonal matrices. Note that, to obtain this tridiagonal matrix, the diagonal also needs to change.

$$\lim_{\epsilon \to \infty} \left(\begin{bmatrix} a & b & \frac{1}{\epsilon} \\ b & \epsilon b d & d \\ \frac{1}{\epsilon} & d & e \end{bmatrix} + \begin{bmatrix} 0 & 0 & 0 \\ 0 & c - \epsilon b d & 0 \\ 0 & 0 & 0 \end{bmatrix} \right) = \begin{bmatrix} a & b & \\ b & c & d \\ & d & e \end{bmatrix}$$

The matrices on the left are of semiseparable plus diagonal form, whereas the limit is a tridiagonal matrix, which does not belong to the class of semiseparable plus diagonal matrices. ∎

Before proving the main theorem of this subsection, namely that $\overline{\mathcal{S}}_{sym}^{(d)} = \mathcal{Q}_{sym}$, we show which quasiseparable matrices are not of semiseparable plus diagonal form.

Proposition 1.19. *A symmetric quasiseparable matrix $A \in \mathcal{Q}_{sym}$ is not representable as a semiseparable plus diagonal matrix if and only if there exists an $i \in \{2, \ldots, n-1\}$, such that $A(i+1, i-1) = 0$ and the elements $A(i,i)$, $A(i, i-1)$ and $A(i+1, i)$ are all different from zero.*

Proof. (The proof is closely related to the proof of Proposition 1.13; we do not repeat the same proof, but present the main ideas.) We know by definition that all the subblocks taken out of the quasiseparable matrix, strictly below the diagonal, and strictly above the diagonal, have rank at most 1. This means that strictly below the diagonal everything is already of the correct form. The only problem can occur if one is not able to write the matrix as the sum of a semiseparable plus a diagonal. This means that if one is not able to subtract a value out of the diagonal of the quasiseparable matrix, such that the remaining diagonal element can be included in the semiseparable structure, below the diagonal. (Due to symmetry, we only consider the lower triangular part.) Let us therefore consider the following blocks $A(i : i+1, i-1 : i)$ of the matrix A (for $i = 2, \ldots, n-1$) which we write as:

$$A(i : i+1, i-1 : i) = \begin{bmatrix} a & b \\ c & d \end{bmatrix}.$$

We want to write this matrix now in the following form:

$$\begin{bmatrix} a & d_i \\ c & d \end{bmatrix} + \begin{bmatrix} 0 & b - d_i \\ 0 & 0 \end{bmatrix},$$

such that the left matrix is of rank 1. (If all elements a, c and d are zero, one can choose the element $d_i = 0$.) This means that $cd_i = ad$. A problem might only occur, if both a and d are different from zero, and $c = 0$, then it is not possible to find d_i. These assumptions correspond exactly to the demands posed in the theorem. $\qquad\square$

If one has a problem as in Proposition 1.19, the elements at this particular position are in tridiagonal form. Suppose we have, for example, the following matrix, which cannot be written as a semiseparable plus diagonal matrix:

$$A = \begin{bmatrix} \times & & & & & \\ \boxtimes & \times & & & & \\ \boxtimes & \boxtimes & \times & & & \\ \boxtimes & 0 & \boxtimes & \times & & \\ \boxtimes & \boxtimes & \boxtimes & \boxtimes & \boxtimes & \\ \boxtimes & \boxtimes & \boxtimes & \boxtimes & \boxtimes & \times \end{bmatrix},$$

where the elements $\times$ denote arbitrary matrix elements, and the elements $\boxtimes$ denote that part of the matrix for which all the subblocks are of rank 1 (or this part is in semiseparable form). As the element just above and the element next right to the zero are nonzero, this imposes some extra constraints on the structure. Because every subblock taken out of the part denoted with ⊠ has to be of rank at most 1, we get:

$$
A = \begin{bmatrix}
\times & & & & & \\
\boxtimes & \times & & & & \\
\boxtimes & \boxtimes & \times & & & \\
0 & 0 & \boxtimes & \times & & \\
0 & 0 & \boxtimes & \boxtimes & \boxtimes & \\
0 & 0 & \boxtimes & \boxtimes & \boxtimes & \times
\end{bmatrix},
$$

which means that all elements in the lower left block have to be zero. Hence, that specific part of the matrix can be considered in tridiagonal form.

Theorem 1.20. *The pointwise closure of the class $\mathcal{S}^{(d)}_{sym}$ of symmetric semiseparable plus diagonal matrices is the class $\mathcal{Q}_{sym}$ of symmetric quasiseparable matrices. This means that*

$$
\overline{\mathcal{S}}^{(d)}_{sym} = \mathcal{Q}_{sym},
$$

and the class $\overline{\mathcal{Q}}_{sym} = \mathcal{Q}_{sym}$ is closed under pointwise convergence.

Proof.

⇒ This direction of the theorem can be proved completely similarly as the arrow ⇒ in the proof of Theorem 1.15. The only difference is that the subblocks considered here do not include the diagonal. Here one should prove that the rank of all submatrices taken out of the strictly lower triangular part, or the strictly upper triangular part, remains bounded by 1.

⇐ Suppose a symmetric quasiseparable matrix A is given such that it is not a semiseparable plus diagonal matrix. Then it needs to be proved that there exists a family $S_\epsilon + D_\epsilon$ with $\epsilon \to \epsilon_0$ such that

$$
\lim_{\epsilon \to \epsilon_0} (S_\epsilon + D_\epsilon) = A.
$$

According to Proposition 1.19 we know the specific structure of a matrix that is not representable as the sum of a semiseparable matrix plus a diagonal. Let us consider such a matrix of dimension 5×5, with only one element causing problems. The general case can be dealt with in a similar fashion.

This means that the matrix A can be written in the following form:

$$
A = \begin{bmatrix}
\times & & & & \\
\boxtimes & \times & & & \\
\boxtimes & \boxtimes & \times & & \\
0 & 0 & \boxtimes & \times & \\
0 & 0 & \boxtimes & \boxtimes & \times
\end{bmatrix} = \begin{bmatrix}
\times & & & & \\
\boxtimes & \times & & & \\
a & b & \times & & \\
0 & 0 & d & \times & \\
0 & 0 & c & \boxtimes & \times
\end{bmatrix},
$$

with $a, b, d, c \in \mathbb{R}$ and b and d necessarily different from zero. Making the zero block slightly different from zero solves the problem. Let us therefore define the quasiseparable matrices A_ϵ, which only differ from the matrix A, for the lower left 2×2 block as

$$
A_\epsilon = \begin{bmatrix}
\times & & & & & \\
\boxtimes & \times & & & & \\
a & b & \times & & & \\
\frac{ac}{\epsilon} & \frac{bc}{\epsilon} & c & \times & & \\
\frac{ad}{\epsilon} & \frac{bd}{\epsilon} & d & \boxtimes & \times &
\end{bmatrix}.
$$

The complete lower triangular part is still of the quasiseparable form. It is clear that

$$
\lim_{\epsilon \to \infty} A_\epsilon = A,
$$

because every matrix A_ϵ is now representable as a semiseparable plus diagonal matrix

$$
A_\epsilon = S_\epsilon + D_\epsilon,
$$

we get

$$
\lim_{\epsilon \to \infty} A_\epsilon = \lim_{\epsilon \to \infty} (S_\epsilon + D_\epsilon) = A,
$$

which proves the theorem.

$\square$

Having this theorem proved, all the relations concerning inclusions and closures are known. To conclude, we summarize these results.

1.2.6 Summary of the relations

Let us reconsider, in brief, all the relations investigated in this section. Please note that the classes of matrices considered here contain only the symmetric matrices. The relations between nonsymmetric matrices will be investigated in Section 1.4. Among the different classes of matrices, defined as in the previous section, we get the following relations.

$$
\mathcal{Q}_{sym} \supsetneq \mathcal{S}_{sym}^{(d)} \supsetneq \mathcal{S}_{sym} \supsetneq \mathcal{S}_{sym}^{(g)},
$$
$$
\mathcal{Q}_{sym} \supsetneq \mathcal{S}_{sym}^{(d)} \supsetneq \mathcal{S}_{sym} + D \supsetneq \mathcal{D},
$$
$$
\mathcal{Q}_{sym} \supsetneq \mathcal{T}_{sym} \supsetneq \mathcal{T}_{sym}^{(i)}.
$$

We also have

$$
\mathcal{S}_{sym} \supsetneq \mathcal{D}, \quad \mathcal{T}_{sym} \supsetneq \mathcal{D},
$$
$$
\mathcal{S}_{sym}^{(g)} \not\supseteq \mathcal{D}, \quad \mathcal{T}_{sym}^{(i)} \not\supseteq \mathcal{D}.
$$

Moreover, we know that the classes $\mathcal{Q}_{sym}, \mathcal{S}_{sym} + D, \mathcal{S}_{sym}, \mathcal{T}_{sym}$ and $\mathcal{D}$ are closed under the pointwise convergence (for every $D \in \mathcal{D}$) and the classes $\mathcal{S}_{sym}^{(g)}, \mathcal{T}_{sym}^{(i)}$

and $\mathcal{S}^{(d)}$ are not. We also have the following relations, with regard to the pointwise closure of these classes:

$$\overline{\mathcal{S}}^{(d)}_{sym} = \overline{\mathcal{Q}}_{sym} = \mathcal{Q}_{sym},$$
$$\overline{\mathcal{S}}^{(g)}_{sym} = \overline{\mathcal{S}}_{sym} = \mathcal{S}_{sym},$$
$$\overline{\mathcal{T}}^{(i)}_{sym} = \overline{\mathcal{T}}_{sym} = \mathcal{T}_{sym}.$$

In fact, we did not prove the result for the class of irreducible tridiagonal matrices, but this is straightforward.

1.2.7   More on the pointwise convergence

In this section attention is paid to the closure of classes, with regard to the pointwise convergence. We focus on the pointwise convergence, but in reality all norms on $\mathbb{R}^{n \times m}$ are equivalent. We will briefly mention here some results related to different vector and matrix norms (see, e.g., [152]) and their relations to the convergence of vectors and/or matrices.

A norm is a function $\| \cdot \| : \mathcal{R} \to \mathbb{R}$, where $\mathcal{R} = \mathbb{R}^n$ in the vector case, and $\mathcal{R} = \mathbb{R}^{n \times m}$ in the matrix case. The function $\| \cdot \|$ satisfies the following conditions:

$$\begin{cases} \|x\| \geq 0 & x \in \mathcal{R}, \quad (\|x\| = 0 \Leftrightarrow x = 0), \\ \|x + y\| \leq \|x\| + \|y\| & x, y \in \mathcal{R}, \\ \|cx\| = |c|\|x\| & x \in \mathcal{R}, c \in \mathbb{R}. \end{cases}$$

Some popular vector norms over $\mathbb{R}^n$ are the following ones:

$$\|\mathbf{x}\|_1 = \left(|x_1| + |x_2| + \cdots + |x_n|\right),$$
$$\|\mathbf{x}\|_2 = \left(|x_1|^2 + |x_2|^2 + \cdots + |x_n|^2\right)^{1/2},$$
$$\|\mathbf{x}\|_\infty = \max_{1 \leq i \leq n} |x_i|.$$

These norms are equivalent as the following inequalities illustrate:

$$\|\mathbf{x}\|_\infty \leq \|\mathbf{x}\|_1 \leq \sqrt{n}\|\mathbf{x}\|_2 \leq n\|\mathbf{x}\|_\infty.$$

Moreover, all norms over $\mathbb{R}^n$ are equivalent. A set of vectors $\mathbf{x}(\epsilon)$ is said to converge to a vector $\mathbf{x}$, if

$$\lim_{\epsilon \to \epsilon_0} \|\mathbf{x}(\epsilon) - \mathbf{x}\| = 0.$$

It is clear that, due to the equivalence of norms, the choice of norm does not play any role in this definition. One can also clearly see that the pointwise limit is related to the infinity norm $\| \cdot \|_\infty$ over $\mathbb{R}^n$.

As $\mathbb{R}^{n \times m}$ is isomorphic to $\mathbb{R}^{nm}$ we know that all norms over $\mathbb{R}^{n \times m}$ are also equivalent to each other. Therefore, the attention put on the pointwise convergence in this section was not necessary. Any kind of norm could have been considered without having changed the results of this section. The pointwise limit used for the convergence of matrices is related to the norm (take $A \in \mathbb{R}^{n \times m}$)

$$\|A\|_{max} = \max_{1 \leq i \leq n, 1 \leq j \leq m} |a_{ij}|.$$

To conclude we present some common matrix norms:

$$\|A\|_1 = \max_{1 \leq j \leq m} \sum_{i=1}^{n} |a_{ij}|,$$

$$\|A\|_\infty = \max_{1 \leq i \leq n} \sum_{j=1}^{m} |a_{ij}|,$$

$$\|A\|_F = \sqrt{\sum_{i}^{n} \sum_{j}^{m} |a_{ij}|^2}.$$

These norms are also equivalent

$$\|A\|_{max} \leq \|A\|_2 \leq \sqrt{mn} \, \|A\|_{max},$$

$$\frac{1}{\sqrt{m}} \, \|A\|_F \leq \|A\|_2 \leq \|A\|_F,$$

$$\frac{1}{\sqrt{m}} \, \|A\|_\infty \leq \|A\|_2 \leq \sqrt{n} \, \|A\|_\infty,$$

$$\frac{1}{\sqrt{n}} \|A\|_1 \leq \|A\|_2 \leq \sqrt{m} \, \|A\|_1.$$

Notes and references

The relation between semiseparable matrices and generator representable matrices was studied in a similar fashion as presented here in the following manuscript.

☞ R. Vandebril, M. Van Barel, and N. Mastronardi. A note on the representation and definition of semiseparable matrices. *Numerical Linear Algebra with Applications*, 12(8):839–858, October 2005.

The authors presented an overview of the differences between generator representable semiseparable matrices and semiseparable matrices. Also a new type of representation, the Givens-vector representation for semiseparable matrices, was presented.

More on the relations between norms and convergence can be found in several basic linear algebra books such as the following.

☞ J. W. Demmel. *Applied Numerical Linear Algebra*. SIAM, 1997.

☞ G. H. Golub and C. F. Van Loan. *Matrix Computations*. The Johns Hopkins University Press, Baltimore, Maryland, third edition, 1996.

☞ L. N. Trefethen and D. Bau. *Numerical Linear Algebra*. SIAM, 1997.

1.3   Unsymmetric semiseparable and related matrices

In Section 1.1, we introduced some important classes of symmetric semiseparable and related matrices. Their inner relations were investigated in Section 1.2. In this section, we will reconsider the definitions from Section 1.1 and adapt them towards the unsymmetric case. Even though the generalizations are quite often trivial, we do define them for completeness.

The generalization towards an unsymmetric semiseparable matrix just makes a distinction between the lower and upper triangular part of this matrix.

Definition 1.21. *A matrix S is called a semiseparable matrix if all submatrices taken out of the lower and upper triangular part of the matrix are of rank 1.*

This means that for a semiseparable matrix S (for $i = 1, \ldots, n$):

$$\operatorname{rank}(S(i:n, 1:i)) \leq 1,$$
$$\operatorname{rank}(S(1:i, i:n)) \leq 1.$$

Example 1.22 The following matrices are semiseparable.

$$\begin{bmatrix} 1 & 3 & 9 \\ 2 & 2 & 6 \\ 3 & 3 & 3 \end{bmatrix}, \quad \begin{bmatrix} 1 & 1 & 1 \\ & 2 & 2 \\ & & 3 \end{bmatrix} \quad \text{and} \quad \begin{bmatrix} 1 & 4 & \\ 2 & 2 & \\ & & 3 \end{bmatrix}.$$

∎

The definition of a not necessarily symmetric, generator representable semiseparable matrix, naturally splits up the lower and upper triangular part of the matrix. The diagonal however should be includable in both the lower and upper triangular part of the matrix.

Definition 1.23. *A matrix S is called a generator representable semiseparable matrix if the lower and upper triangular parts of the matrix are coming from a rank 1 matrix.*

In formulas, this means that a matrix S is generator representable semiseparable if

$$\operatorname{tril}(S) = \operatorname{tril}(\mathbf{uv}^T),$$
$$\operatorname{triu}(S) = \operatorname{triu}(\mathbf{pq}^T),$$

where $\mathbf{u}, \mathbf{v}, \mathbf{p}$ and $\mathbf{q}$ are four column vectors of length n. A generator representable semiseparable matrix has the following structure:

$$S(\mathbf{u}, \mathbf{v}, \mathbf{p}, \mathbf{q}) = \begin{bmatrix} u_1 v_1 & q_2 p_1 & q_3 p_1 & \cdots & q_n p_1 \\ u_2 v_1 & u_2 v_2 & q_3 p_2 & \cdots & q_n p_2 \\ u_3 v_1 & u_3 v_2 & u_3 v_3 & \ddots & \vdots \\ \vdots & \vdots & \vdots & \ddots & q_n p_{n-1} \\ u_n v_1 & u_n v_2 & u_n v_3 & \cdots & u_n v_n \end{bmatrix},$$

with $q_i p_i = u_i v_i$ for all $i = 1, \ldots, n$. This last constraint means that the diagonal is includable in both the upper and lower triangular rank structure.

Reconsidering Example 1.22, one can clearly see that the first matrix is generator representable, whereas the last one is not. The second matrix has the upper triangular part of generator representable form, but the lower not.

In a natural way we can define semiseparable plus diagonal matrices. The definition is not included. Finally, the class of quasiseparables is defined as follows.

Definition 1.24. *A matrix S is called a quasiseparable matrix if all the subblocks, taken out of the strictly lower triangular part of the matrix (respectively, the strictly upper triangular part) are of rank 1.*

Thus, a matrix A is quasiseparable, if the following relations are satisfied (for $i = 1, \ldots, n-1$):

$$\text{rank}\,(S(i+1:n, 1:i)) \leq 1,$$
$$\text{rank}\,(S(1:i, i+1:n)) \leq 1.$$

Example 1.25 We know that, for example, general tridiagonal and semiseparable matrices are also quasiseparable, but we can also have matrices of the following form:
$$\begin{bmatrix} 1 & 1 & 1 \\ 1 & 1 & 1 \\ 0 & 1 & 1 \end{bmatrix} \text{ and } \begin{bmatrix} 1 & 3 & 3 \\ 4 & 2 & 3 \\ & 5 & 3 \end{bmatrix}.$$

These are matrices for which the lower triangular part is either semiseparable or tridiagonal, and the same holds for the upper triangular part.

For example, a unitary Hessenberg matrix is quasiseparable but is not necessarily semiseparable nor tridiagonal. In Section 1.5 of this chapter, Theorem 1.44, we will prove that a unitary Hessenberg matrix is of quasiseparable form. ∎

In the next section, we will investigate in more detail the inner relations between all these classes of matrices.

Notes and references

Unsymmetric semiseparable and related matrices are quite often mentioned in one breath with their representation. Let us therefore only briefly consider some manuscripts concerning unitary Hessenberg matrices. More information on the other type of matrices can be found in Chapter 2, where the different types of matrices are reconsidered, closely related to their representation.

It is not so difficult to prove that unitary Hessenberg matrices are of quasiseparable form. Computing the QR-factorization of the unitary Hessenberg matrix involves $n-1$ Givens transformations from top to bottom factoring the matrix in a product of Givens transformations and, due to maintaining of the unitary structure, a diagonal. This means that the upper triangular part of the unitary matrix will also be of low rank form.

A lot of attention has been paid by several authors to developing QR-algorithms for unitary Hessenberg matrices. As there are lots of publications related to unitary Hessenberg matrices we only list few of them.

Gragg, in particular, has intensively studied unitary Hessenberg matrices. The first QR-algorithm for unitary Hessenberg matrices, of complexity $\mathcal{O}(n^2)$, based on the Schur parameterization was presented in the following manuscripts.

☞ W. B. Gragg. The QR algorithm for unitary Hessenberg matrices. *Journal of Computational and Applied Mathematics*, 16:1–8, 1986.

The stability of this method was analyzed by Stewart.

☞ G. W. Stewart. Stability properties of several variants of the unitary Hessenberg QR-algorithm in structured matrices in mathematics. In *Computer Science and Engineering, II (Boulder, CO, 1999)*, volume 281 of *Contemp. Math.*, pages 57–72. Amer. Math. Soc., Providence, RI, 2001.

In the following manuscripts the authors investigate the convergence of the QR-algorithm applied on a unitary Hessenberg matrix. New variants of the QR-algorithm are provided. A divide-and-conquer method is also presented.

☞ W. B. Gragg and L. Reichel. A divide and conquer algorithm for the unitary eigenproblem. In M. T. Heath, editor, *Hypercube multiprocessors 1987*, pages 639–647, Philadelphia, 1987. SIAM.

☞ T. L. Wang and W. B. Gragg. Convergence of the shifted QR algorithm, for unitary Hessenberg matrices. *Mathematics of Computation*, 71(240):1473–1496, 2002.

☞ T. L. Wang and W. B. Gragg. Convergence of the unitary QR algorithm with unitary Wilkinson shift. *Mathematics of Computation*, 72(241):375–385, 2003.

As already mentioned, unitary Hessenberg matrices have the upper triangular part in the matrix of semiseparable form. In the following paper by Gemignani the quasiseparable structure of the unitary Hessenberg matrix is exploited to develop a QR-algorithm. Moreover the presented algorithm is also valid for unitary Hessenberg plus rank 1 matrices. The presented method transforms the matrix into a hermitian semiseparable plus diagonal matrix via the Möbius transform, then a QR-method is applied for computing its eigenvalues. This manuscript also contains many references related to the QR-algorithm for unitary Hessenberg matrices.

☞ L. Gemignani. A unitary Hessenberg QR-based algorithm via semiseparable matrices. *Journal of Computational and Applied Mathematics*, 184:505–517, 2005.

Unitary Hessenberg matrices and semiseparable matrices also arise in inverse eigenvalue problems. More information can be found in the following manuscripts. These last two references will be revisited in more detail in Volume 2 of this book. There, attention will be paid to inverse eigenvalue problems.

☞ M. Van Barel, D. Fasino, L. Gemignani, and N. Mastronardi. Orthogonal rational functions and diagonal plus semiseparable matrices. In F. T. Luk, editor, *Advanced Signal Processing Algorithms, Architectures, and Implementations XII*, volume 4791 of *Proceedings of SPIE*, pages 167–170, 2002.

☞ M. Van Barel, D. Fasino, L. Gemignani, and N. Mastronardi. Orthogonal rational functions and structured matrices. *SIAM Journal on Matrix Analysis and its Applications*, 26(3):810–829, 2005.

1.4   Relations between the different "unsymmetric" definitions

In this section we will investigate in detail the relations between the different classes of not necessarily symmetric matrices, such as semiseparable and others. First, we investigate straightforward extensions of the results of the symmetric matrices. Second, we will see that we need to define a new class of matrices, namely the quasiseparable matrices having a symmetric rank structure. The inverse of a semiseparable plus diagonal matrix will namely be in the class of quasiseparable matrices having a symmetric rank structure.

1.4.1   Extension of known symmetric relations

As in the symmetric case, we define the different classes of matrices as follows:

$$\mathcal{T} = \{A \in \mathbb{R}^{n \times n} \mid A \text{ is a tridiagonal matrix}\},$$
$$\mathcal{T}^{(i)} = \{A \in \mathbb{R}^{n \times n} \mid A \text{ is an irreducible tridiagonal matrix}\},$$
$$\mathcal{S} = \{A \in \mathbb{R}^{n \times n} \mid A \text{ is a semiseparable matrix}\},$$
$$\mathcal{D} = \{A \in \mathbb{R}^{n \times n} \mid A \text{ is a diagonal matrix}\},$$
$$\mathcal{S}^{(d)} = \{A \in \mathbb{R}^{n \times n} \mid A \text{ is a semiseparable plus diagonal matrix}\},$$
$$= \{A \in \mathbb{R}^{n \times n} \mid A = S + D, \text{ with } S \in \mathcal{S}, D \in \mathcal{D}\},$$
$$\mathcal{S}^{(g)} = \{A \in \mathbb{R}^{n \times n} \mid A \text{ is a generator}$$
$$\text{representable semiseparable matrix}\},$$
$$\mathcal{Q} = \{A \in \mathbb{R}^{n \times n} \mid A \text{ is a quasiseparable matrix}\}.$$

We already have the following relations:

$$\mathcal{Q} \supsetneq \mathcal{S}^{(d)} \supsetneq \mathcal{S} \supsetneq \mathcal{S}^{(g)},$$
$$\mathcal{Q} \supsetneq \mathcal{T} \supsetneq \mathcal{T}^{(i)}.$$

We also have

$$\mathcal{S} \supsetneq \mathcal{D}, \quad \mathcal{T} \supsetneq \mathcal{D},$$
$$\mathcal{S}^{(g)} \not\supseteq \mathcal{D}, \quad \mathcal{T}^{(i)} \not\supseteq \mathcal{D}.$$

Clearly this is a straightforward extension of the results for symmetric matrices. Several theorems posed in Section 1.2 have their analogue in the nonsymmetric case. Straightforward proofs are not included.

The treatise related to theoretical and numerical problems related to generator representable semiseparable matrices extends in a natural manner to the unsymmetric case. Similarly, one can translate the theorems stating which semiseparable or semiseparable plus diagonal matrices are not generator representable semiseparable, respectively, quasiseparable anymore; the only difference is that one has to deal with the upper and lower triangular part separately now. Translating the theorems above towards the nonsymmetric case, it is an easy exercise to prove the following relations:

$$\overline{\mathcal{S}}^{(g)} = \mathcal{S},$$
$$\overline{\mathcal{T}}^{(i)} = \mathcal{T},$$
$$\overline{\mathcal{Q}} = \mathcal{Q},$$
$$\overline{\mathcal{S} + D} = \mathcal{S} + D, \quad \forall D \in \mathcal{D}.$$

Hence the classes $\mathcal{S}, \mathcal{T}, \mathcal{S} + D$ and $\mathcal{Q}$ are closed, under any norm.

One question remains unanswered: what about the class $\mathcal{S}^{(d)}$; is the closure of this class the class of quasiseparable matrices? The derivation is not straightforward anymore, as the upper and lower triangular parts of the matrix $\mathcal{Q}$ are not related through the diagonal or via symmetry anymore. This means that if we can rewrite the matrix $\mathcal{Q}$ as $\hat{\mathcal{Q}} + D$ in such a manner that the diagonal of $\hat{\mathcal{Q}}$ is includable in the lower triangular part, this does not necessarily imply the inclusion of the diagonal in the upper triangular part of $\hat{\mathcal{Q}}$, which was important for the case of symmetric matrices.

Let us investigate in more detail the closure of the class of semiseparable plus diagonal matrices.

Theorem 1.26. *The closure of the class of semiseparable plus diagonal matrices $\mathcal{S}^{(d)}$ incorporates the class of tridiagonal matrices $\mathcal{T}$.*

Proof. We have to prove that $\mathcal{T} \subseteq \overline{\mathcal{S}}^{(d)}$. We will prove this statement for a 3×3 matrix. The general $n \times n$ case is similar. Suppose we have the following tridiagonal matrix T:

$$T = \begin{bmatrix} a & b & \\ c & d & e \\ & f & g \end{bmatrix}.$$

This matrix can be written as the limit of a sequence of semiseparable plus diagonal matrices, with a varying diagonal:

$$\lim_{\epsilon \to \infty} \left(\begin{bmatrix} a & b & \frac{be}{\epsilon} \\ c & \epsilon & e \\ \frac{cf}{\epsilon} & f & g \end{bmatrix} + \begin{bmatrix} 0 & & \\ & d - \epsilon & \\ & & 0 \end{bmatrix} \right) = \begin{bmatrix} a & b & \\ c & d & e \\ & f & g \end{bmatrix}.$$

This proves the statement. □

In contrast to the symmetric case, the closure of the class of semiseparable plus diagonal matrices is not equal anymore to the class of quasiseparable matrices. This is because the quasiseparable matrices do not have any connection with their diagonal. This means that the lower and upper triangular structures of the quasiseparable matrices do not need to be related in any way. Let us illustrate this with an example.

Example 1.27 Suppose the following 3×3 quasiseparable matrix A is given

$$A = \begin{bmatrix} a & b & c \\ d & e & f \\ 0 & g & h \end{bmatrix},$$

with all elements a, b, c, d, e, f, g and h different from zero. Suppose the matrix A can be written as the limit of a sequence of semiseparable plus diagonal matrices $S_\epsilon + D_\epsilon$, for $\epsilon \to \epsilon_0$. This will lead to a contradiction. As the limit equals A, the lower left element of S_ϵ, defined here as α_ϵ, has to go to zero, for ϵ going to ϵ_0. To obtain that the lower left 2×2 block is of rank 1, we have to change the diagonal. This means that

$$S_\epsilon + D_\epsilon = \begin{bmatrix} a & \times & \times \\ d_\epsilon & \frac{d_\epsilon g_\epsilon}{\alpha_\epsilon} & \times \\ \alpha_\epsilon & g_\epsilon & h \end{bmatrix} + \begin{bmatrix} 0 & & \\ & e - \frac{d_\epsilon g_\epsilon}{\alpha_\epsilon} & \\ & & 0 \end{bmatrix},$$

where the elements marked by $\times$ are not yet defined, $\alpha_\epsilon \to 0$, $d_\epsilon \to d$ and $g_\epsilon \to g$ for $\epsilon \to \epsilon_0$.

The upper right 2×2 block of the matrix S_ϵ also needs to be of rank 1 and its limit needs to be the upper right 2×2 block of the matrix A. This means that our matrix $S_\epsilon + D_\epsilon$, will be of the following form:

$$S_\epsilon + D_\epsilon = \begin{bmatrix} a & b_\epsilon & c_\epsilon \\ d_\epsilon & \frac{d_\epsilon g_\epsilon}{\alpha_\epsilon} & f_\epsilon \\ \alpha_\epsilon & g_\epsilon & h \end{bmatrix} + \begin{bmatrix} 0 & & \\ & e - \frac{d_\epsilon g_\epsilon}{\alpha_\epsilon} & \\ & & 0 \end{bmatrix},$$

with $b_\epsilon \to b$, $c_\epsilon \to c$ and $f_\epsilon \to f$ for $\epsilon \to \epsilon_0$. Moreover, we also know that the upper right 2×2 block of the matrix S_ϵ needs to be of rank 1, which gives us the following relation:

$$b_\epsilon f_\epsilon = \frac{d_\epsilon g_\epsilon}{\alpha_\epsilon} c_\epsilon.$$

Rewriting this equation towards c_ϵ gives us

$$\frac{b_\epsilon f_\epsilon \alpha_\epsilon}{d_\epsilon g_\epsilon} = c_\epsilon.$$

Taking the limit on both sides of the equation leads to $c = 0$. This contradicts our primary assumptions, namely $c \neq 0$. Hence this matrix A cannot be written as the limit of a sequence of semiseparable plus diagonal matrices. We remark that the resulting limiting matrix has some kind of symmetry in the rank structure. ∎

This means that we already have the following relations:

$$\mathcal{T} \subsetneq \overline{\mathcal{S}}^{(d)},$$
$$\mathcal{S}^{(d)} \subsetneq \overline{\mathcal{S}}^{(d)},$$
$$\overline{\mathcal{S}}^{(d)} \subsetneq \mathcal{Q}.$$

The relation $\mathcal{T} \cup \mathcal{S}^{(d)} = \overline{\mathcal{S}}^{(d)}$ is not correct, however, as the following example illustrates.

Example 1.28 Consider the following 4×4 semiseparable plus diagonal matrix $S_\epsilon + D_\epsilon$:

$$
S_\epsilon + D_\epsilon = \begin{bmatrix} 1 & 3 & 9 & \frac{3}{\epsilon} \\ 2 & 2 & 6 & \frac{2}{\epsilon} \\ 3 & 3 & 3\epsilon & 1 \\ \frac{1}{\epsilon} & \frac{1}{\epsilon} & 1 & 4 \end{bmatrix} + \begin{bmatrix} 1 & & & \\ & 1 & & \\ & & 1-3\epsilon & \\ & & & 1 \end{bmatrix}.
$$

The limit gives us the following matrix:

$$
\lim_{\epsilon \to \infty} (S_\epsilon + D_\epsilon) = \begin{bmatrix} 2 & 3 & 9 & 0 \\ 2 & 3 & 6 & 0 \\ 3 & 3 & 1 & 1 \\ 0 & 0 & 1 & 5 \end{bmatrix},
$$

which is neither a tridiagonal, nor a semiseparable plus diagonal matrix. The matrix however is quasiseparable. Moreover, we remark once more that the resulting matrix has some kind of symmetry in the rank structure. ∎

This means that the closure of $\mathcal{S}^{(d)}$ will be a specific subclass of the quasiseparables. This class has a specific type of symmetry related to the rank structure of the matrix.

1.4.2 Quasiseparable matrices having a symmetric rank structure

We will call this class the class of quasiseparables, with a symmetric rank structure and denote it as $\mathcal{Q}^{(s)}$. Let us define this class more precisely.

Definition 1.29. *A quasiseparable matrix A is said to have a symmetric rank structure if for every $i = 2, \ldots, n-1$ one (or both) of the following conditions hold:*

C1. *There exists a $d_i \in \mathbb{R}$, such that the matrix $A_i = A - d_i I$ satisfies the following relations*

$$
\operatorname{rank}(A_i(i : n, 1 : i)) \leq 1,
$$
$$
\operatorname{rank}(A_i(1 : i, i : n)) \leq 1.
$$

C2. *The relations $A(i+1 : n, 1 : i-1) = 0$ and $A(1 : i-1, i+1 : n) = 0$ are satisfied.*

This means that the rank structure of the upper triangular part corresponds to the rank structure of the lower triangular part. This means that if subtracting a diagonal element makes a certain block in the lower triangular part (including the diagonal) of rank 1, the upper triangular part should also have a corresponding block (including the diagonal) of rank 1. If none such diagonal element exists, this means that there should be a specific zero block in the lower triangular and upper triangular parts. Note that for the symmetric case we automatically have that $\mathcal{Q}_{sym}^{(s)} = \mathcal{Q}_{sym}$.

Let us illustrate the class of $\mathcal{Q}^{(s)}$ with some examples. Semiseparable, tridiagonal and semiseparable plus diagonal matrices all belong to this class. We stress once more that the matrix itself does not need to be symmetric, but only the involved rank structure to some extent. For example, upper triangular semiseparable and bidiagonal matrices also belong to this class. The reader can try to construct sequences of semiseparable plus diagonal matrices converging to, e.g., a bidiagonal matrix to verify this statement.

Let us formulate and prove the main theorem of this section.

Theorem 1.30. *The closure of the class of semiseparable plus diagonal matrices is the class of quasiseparable matrices having a symmetric rank structure. This means:*

$$\overline{\mathcal{S}}^{(d)} = \mathcal{Q}^{(s)}$$

Proof. An easy exercise reveals that for every $A \in \mathcal{Q}^{(s)}$ there exists a sequence of semiseparable plus diagonal matrices converging to this matrix A.

Let us prove the other direction. Assume we have a sequence of converging semiseparable plus diagonal matrices, $S_\epsilon + D_\epsilon$, converging to A (for $\epsilon \to \epsilon_0$). Let us prove that A belongs to the class of symmetric quasiseparables. We know that for every $i = 1, \ldots, n$:

$$A_{ii} = \lim_{\epsilon \to \epsilon_0} ((S_\epsilon)_{ii} + (D_\epsilon)_{ii}).$$

We make a distinction now. Assume $i \in \{2, \ldots, n-1\}$.

- Assume that $\lim_{\epsilon \to \epsilon_0}((D_\epsilon)_{ii}) = d_i$, i.e., that the limit of that specific diagonal element of the matrix D_ϵ exists. This means that both terms need to converge, and hence also the limit $\lim_{\epsilon \to \epsilon_0}((S_\epsilon)_{ii}) = S_{ii}$ needs to exist. This leads in a natural way to the following two relations, where we define $A_i = A - d_i I$:

$$\lim_{\epsilon \to \epsilon_0} (S_\epsilon(i:n, 1:i)) = A_i(i:n, 1:i),$$
$$\lim_{\epsilon \to \epsilon_0} (S_\epsilon(1:i, i:n)) = A_i(1:i, i:n).$$

  Combining this with techniques used in the proof of Theorem 1.15, we get that

$$\text{rank}\,(A_i(i:n, 1:i)) \leq 1,$$
$$\text{rank}\,(A_i(1:i, i:n)) \leq 1.$$

This proves that if the limit $\lim_{\epsilon \to \epsilon_0}((D_\epsilon)_{ii})$ exists, Condition *(C1)* in Definition 1.29 is always satisfied. If we are able to prove that, in case the limit does not exist, Condition *(C2)* is always satisfied, the proof is finished.

- Assume that $\lim_{\epsilon \to \epsilon_0}((D_\epsilon)_{ii}) = \pm\infty$, i.e., that the limit does not exist. Moreover, as the limit of the sum exists, we know that also $\lim_{\epsilon \to \epsilon_0}((S_\epsilon)_{ii}) = \mp\infty$. As for every ϵ the matrix S_ϵ needs to be semiseparable, this poses some constraints on the matrix. The block $S_\epsilon(i : n, 1 : i)$ therefore needs to be of rank 1; this means that the determinant of every 2×2 subblock taken out of this matrix has to be zero. For every k and l, where $l = i+1, \ldots, n$ and $k = 1, \ldots, i-1$, this means that:

$$\det \begin{bmatrix} (S_\epsilon)_{ik} & (S_\epsilon)_{ii} \\ (S_\epsilon)_{lk} & (S_\epsilon)_{li} \end{bmatrix} = (S_\epsilon)_{ik}(S_\epsilon)_{li} - (S_\epsilon)_{ii}(S_\epsilon)_{lk}.$$

As we know that the element $(S_\epsilon)_{ii} \to \mp\infty$, this implies that for all k, l the elements $(S_\epsilon)_{lk} \to 0$. A similar construction can be made for the block in the upper triangular part. Hence we obtain that $A(i + 1 : n, 1 : i - 1) = 0$ and also $A(1 : i - 1, i + 1 : n) = 0$, which means that Condition *(C2)* is satisfied.

The two possible options for the convergence of a sequence of semiseparable plus diagonal matrices states that the resulting limiting matrix A exactly obeys the conditions of Definition 1.29. If a diagonal element of D_ϵ converges, Condition *(C1)* of the definition is satisfied. If a diagonal element tends to go to $\pm\infty$, Condition *(C2)* is always satisfied. □

Concerning the proof, we have the following remark. If a diagonal element tends to go to infinity, we know that Condition *(C2)* is always satisfied, this does not mean, however, that Condition *(C1)* is not satisfied, and vice versa. One condition satisfied, does not exclude the other condition. It is possible that Condition *(C1)* and Condition *(C2)* are satisfied at the same time, as the following, trivial example shows.

Example 1.31 The limit of the following sequence of converging semiseparable plus diagonal matrices gives a matrix that satisfies Conditions *(C1)* and *(C2)* from the definition, in diagonal element 2:

$$\lim_{\epsilon \to \infty} \begin{bmatrix} 1 & & \\ & \epsilon & \\ & & 1 \end{bmatrix} + \begin{bmatrix} 1 & & \\ & -\epsilon & \\ & & 1 \end{bmatrix} = \begin{bmatrix} 1 & & \\ & 0 & \\ & & 1 \end{bmatrix}.$$

This example concludes this section. It is clear now that, for example, a generic unitary Hessenberg matrix cannot be written as the limit of a sequence of semiseparable plus diagonal matrices. In [133], it is shown that one can transform a unitary Hessenberg matrix via a Möbius transformation into a semiseparable plus diagonal matrix.

1.4.3   Summary of the relations

To conclude, the relations between the different classes of unsymmetric semiseparable and related matrices are briefly repeated.

$$\mathcal{Q} \supsetneq \mathcal{Q}^{(s)} \supsetneq \mathcal{S}^{(d)} \supsetneq \mathcal{S} \supsetneq \mathcal{S}^{(g)},$$
$$\mathcal{Q} \supsetneq \mathcal{Q}^{(s)} \supsetneq \mathcal{S}^{(d)} \supsetneq \mathcal{S} + D \supsetneq \mathcal{D},$$
$$\mathcal{Q} \supsetneq \mathcal{Q}^{(s)} \supsetneq \mathcal{T} \supsetneq \mathcal{T}^{(i)}.$$

We also have

$$\mathcal{S} \supsetneq \mathcal{D}, \quad \mathcal{T} \supsetneq \mathcal{D},$$
$$\mathcal{S}^{(g)} \not\supsetneq \mathcal{D}, \quad \mathcal{T}^{(i)} \not\supsetneq \mathcal{D}.$$

The classes $\mathcal{Q}, \mathcal{Q}^{(s)}, \mathcal{S} + D, \mathcal{S}, \mathcal{T}$ and $\mathcal{D}$ are closed, for every $D \in \mathcal{D}$. The classes $\mathcal{S}^{(g)}, \mathcal{T}^{(i)}$ and $\mathcal{S}^{(d)}$ are not closed. The following relations hold

$$\overline{\mathcal{S}}^{(d)} = \overline{\mathcal{Q}}^{(s)} = \mathcal{Q}^{(s)},$$
$$\overline{\mathcal{S}}^{(g)} = \overline{\mathcal{S}} = \mathcal{S},$$
$$\overline{\mathcal{T}}^{(i)} = \overline{\mathcal{T}} = \mathcal{T}.$$

To conclude we note that

$$\mathcal{Q}^{(s)}_{sym} = \mathcal{Q}_{sym}.$$

In fact, we did not prove the result for the class of irreducible tridiagonal matrices, but this is straightforward.

1.5 Relations under inversion

It has already been mentioned in the text that the inverse of a semiseparable matrix is a tridiagonal one. In this section, we will prove this statement. Moreover, we will also investigate the inverses of irreducible tridiagonals, of quasiseparable matrices and of semiseparable plus diagonal matrices. In this section, we will not present algorithms for computing the inverse but theoretical predictions of the rank structure of the inverse. Algorithms for computing the inverse can be found in Part II of the book.

1.5.1 The nullity theorem

In this section we will prove the nullity theorem in two different ways. Although this theorem is not so widely spread, it can easily be used to derive several interesting results about structured rank matrices and their inverses. It was formulated for the first time by Gustafson [167] for matrices over principal ideal domains. In [114], Fiedler and Markham translated this abstract formulation to matrices over a field. Barrett and Feinsilver formulated theorems close to the nullity theorem in [13, 15]. Based on their observations we will provide an alternative proof of this theorem. The theorem will be followed by some small corollaries. In the following subsections we will apply these corollaries to classes of matrices closely related to semiseparable matrices, such as tridiagonal, generator representable, semiseparable plus diagonal and quasiseparable matrices.

Definition 1.32. *Suppose a matrix $A \in \mathbb{R}^{m \times n}$ is given. The nullity $\mathrm{n}(A)$ is defined as the dimension of the right null space of A. This means*

$$\mathrm{n}(A) = n - \mathrm{rank}\,(A)\,.$$

Theorem 1.33 (The nullity theorem). *Suppose we have the following invertible matrix $A \in \mathbb{R}^{n \times n}$ partitioned as*

$$A = \left[\begin{array}{cc} A_{11} & A_{12} \\ A_{21} & A_{22} \end{array} \right]$$

with A_{11} of size $p \times q$. The inverse B of A is partitioned as

$$B = \left[\begin{array}{cc} B_{11} & B_{12} \\ B_{21} & B_{22} \end{array} \right]$$

with B_{11} of size $q \times p$. Then the nullities $\mathrm{n}(A_{11})$ and $\mathrm{n}(B_{22})$ are equal.

Proof (From [114]). Suppose $\mathrm{n}(A_{11}) \leq \mathrm{n}(B_{22})$. If this is not true, we can prove the theorem for the matrices

$$\left[\begin{array}{cc} A_{22} & A_{21} \\ A_{12} & A_{11} \end{array} \right] \text{ and } \left[\begin{array}{cc} B_{22} & B_{21} \\ B_{12} & B_{11} \end{array} \right],$$

which are also each other's inverse. Suppose $\mathrm{n}(B_{22}) > 0$ otherwise $\mathrm{n}(A_{11}) = 0$ and the theorem is proved.

When $\mathrm{n}(B_{22}) = c > 0$, then there exists a matrix F with c linearly independent columns, such that $B_{22}F = 0$. Hence, multiplying the following equation to the right by F

$$A_{11}B_{12} + A_{12}B_{22} = 0,$$

we get

$$A_{11}B_{12}F = 0. \tag{1.2}$$

Applying the same operation to the relation:

$$A_{21}B_{12} + A_{22}B_{22} = I$$

it follows that $A_{21}B_{12}F = F$, and therefore $\mathrm{rank}\,(B_{12}F) \geq c$. Using this last statement together with Equation (1.2), we derive

$$\mathrm{n}(A_{11}) \geq \mathrm{rank}\,(B_{12}F) \geq c = \mathrm{n}(B_{22}).$$

Together with our assumption $\mathrm{n}(A_{11}) \leq \mathrm{n}(B_{22})$, this proves the theorem. $\square$

This provides us with the first proof of the theorem. Before providing an alternative proof, we briefly illustrate that the theorem can also be adapted to nonleading submatrices.

Suppose we have a matrix A with its inverse B partitioned as in the theorem. Then we have also that the nullities of the blocks A_{12} and B_{12} are equal to each other. This can be seen rather easily by applying the nullity theorem to the matrices PB and AP, which are each other's inverses, and where P is a block matrix of the following form:

$$P = \begin{bmatrix} 0 & I \\ I & 0 \end{bmatrix}.$$

With the leading zero block of dimension $q \times q$. The matrices PB and AP are each other's inverses and have the blocks A_{12} and B_{12} in the correct position for applying the nullity theorem.

The alternative proof is based on some lemmas and makes use of determinantal formulas. Let A be an $m \times n$ matrix. Denote with M the set of numbers $\{1, 2, \ldots, m\}$ and with N the set of numbers $\{1, 2, \ldots, n\}$. Let α and β be nonempty subset of M and N, respectively. Then, we denote with the matrix $A(\alpha; \beta)$ the submatrix of A with row indices in α and column indices in β. Let us denote with $|\alpha|$ the cardinality[12] of the corresponding set α.

Lemma 1.34 ([131, p. 13]). *Suppose A is an $n \times n$ invertible matrix and α and β two nonempty sets of indices in $N = \{1, 2, \ldots, n\}$, such that $|\alpha| = |\beta| < n$. Then, the determinant of any square submatrix of the inverse matrix $B = A^{-1}$ satisfies the following equation*

$$|\det B(\alpha; \beta)| = \frac{1}{|\det(A)|} |\det A(N\backslash\beta; N\backslash\alpha)|.$$

With $N\backslash\beta$ the difference between the sets N and β is meant (N minus β).

The theorem can be seen as an extension of the standard determinantal formula for calculating the inverse of a matrix, in which each element is determined by a minor in the original matrix. This lemma already implies the nullity theorem for square subblocks and nullities equal to 1, since this case is equivalent with the vanishing of a determinant. The following lemma shows that we can extend this argument also to the general case, i.e., every rank condition can be expressed in terms of the vanishing of certain determinants.

Lemma 1.35. *Suppose $A \in \mathbb{R}^{n \times n}$ is a nonsingular matrix and $n \geq |\alpha| \geq |\beta|$. The following three statements are equivalent:*

1. $n\,(A(\alpha; \beta)) \geq d$.

2. $\det A(\alpha'; \beta') = 0$ for all $\alpha' \subseteq \alpha$ and $\beta' \subseteq \beta$ and $|\alpha'| = |\beta'| = |\beta| - d + 1$.

3. $\det A(\alpha'; \beta') = 0$ for all $\alpha \subseteq \alpha'$ and $\beta \subseteq \beta'$ and $|\alpha'| = |\beta'| = |\alpha| + d - 1$.

---

[12]The cardinality of a set means the number of elements in that set.

Proof. The arrows (1) $\Leftrightarrow$ (2) and (1) $\Rightarrow$ (3) are straightforward. The arrow (3) $\Rightarrow$ (1) makes use of the nonsingularity of the matrix A. Suppose the nullity of $A(\alpha; \beta)$ to be less than d. This would mean that there exist $|\beta| - d + 1$ linearly independent columns in the block $A(\alpha; \beta)$. Therefore $A(\alpha; N)$ has rank less then $|\alpha|$, implying the singularity of the matrix A. □

An alternative proof of the nullity theorem can be derived easily by combining the previous two lemmas. In [260], Strang proves a related result and he comments on different ways to prove the nullity theorem.

The following corollary is a straightforward consequence of the nullity theorem.

Corollary 1.36 (Corollary 3 in [114]). *Suppose $A \in \mathbb{R}^{n \times n}$ is a nonsingular matrix, and α, β are nonempty subsets of N with $|\alpha| < n$ and $|\beta| < n$. Then*

$$\text{rank}\left(A^{-1}(\alpha; \beta)\right) = \text{rank}\left(A(N \backslash \beta; N \backslash \alpha)\right) + |\alpha| + |\beta| - n.$$

Proof. By permuting the rows and columns of the matrix A, we can always move the submatrix $A(N \backslash \beta; N \backslash \alpha)$ into the upper left part A_{11}. Correspondingly, the submatrix $B(\alpha; \beta)$ of the matrix $B = A^{-1}$ moves into the lower right part B_{22}. We have

$$\text{n}\left(A_{11}\right) = n - |\alpha| - \text{rank}\left(A_{11}\right)$$
$$\text{n}\left(B_{22}\right) = |\beta| - \text{rank}\left(B_{22}\right)$$

and because $\text{n}\left(A_{11}\right) = \text{n}\left(B_{22}\right)$, this proves the corollary. □

When choosing $\alpha = N \backslash \beta$, we get

Corollary 1.37. *For a nonsingular matrix $A \in \mathbb{R}^{n \times n}$ and $\alpha \subseteq N$, we have:*

$$\text{rank}\left(A^{-1}(\alpha; N \backslash \alpha)\right) = \text{rank}\left(A(\alpha; N \backslash \alpha)\right).$$

This corollary states in fact that the rank of all blocks of the matrix just below and just above the diagonal will be maintained under inversion. In the next section we will use the previously obtained results about the ranks of complementary blocks of a matrix and its inverse to prove the rank properties of the inverse for some classes of structured rank matrices.

1.5.2   The inverse of semiseparable and tridiagonal matrices

We will prove, using the nullity theorem, that the inverse of an invertible semiseparable matrix is always a tridiagonal matrix. Also the relation between irreducible tridiagonal matrices and generator representable matrices under inversion will be examined.

Theorem 1.38. *The inverse of an invertible tridiagonal matrix is a semiseparable matrix and vice versa.*

Proof. We will only prove that the inverse of a tridiagonal matrix is a semiseparable one (the other way around is similar). In the following figures we see a tridiagonal matrix on the left and its inverse on the right. We have that the rank of the left block (denoted with zeros) plus 1 equals the rank of the right block (denoted with ⊠), according to Corollary 1.36. The indices, denoting the different subblocks are shown above the matrices.

Let us denote with N the set of indices $\{1, 2, \ldots, n\}$. We have the following blocks related to each other: the lower left 3×1 zero block of T and the lower left 4×2 block of T^{-1}:

$$\alpha = \{3, 4, 5\} \text{ and} \qquad\qquad N \backslash \beta = \{2, 3, 4, 5\} \text{ and}$$
$$\beta = \{1\} \qquad\qquad\qquad\qquad N \backslash \alpha = \{1, 2\}$$

$$T = \begin{bmatrix} \times & \times & & & \\ \times & \times & \times & & \\ 0 & \times & \times & \times & \\ 0 & & \times & \times & \times \\ 0 & & & \times & \times \end{bmatrix} \quad \leftrightarrow \quad T^{-1} = \begin{bmatrix} \times & \times & \times & \times & \times \\ \boxtimes & \boxtimes & \times & \times & \times \\ \boxtimes & \boxtimes & \times & \times & \times \\ \boxtimes & \boxtimes & \times & \times & \times \\ \boxtimes & \boxtimes & \times & \times & \times \end{bmatrix} ;$$

the lower left 2×2 block of T and the lower left 3×3 block of T^{-1}:

$$\alpha = \{4, 5\} \text{ and} \qquad\qquad N \backslash \beta = \{3, 4, 5\} \text{ and}$$
$$\beta = \{1, 2\} \qquad\qquad\qquad N \backslash \alpha = \{1, 2, 3\}$$

$$T = \begin{bmatrix} \times & \times & & & \\ \times & \times & \times & & \\ & \times & \times & \times & \\ 0 & 0 & \times & \times & \times \\ 0 & 0 & & \times & \times \end{bmatrix} \quad \leftrightarrow \quad T^{-1} = \begin{bmatrix} \times & \times & \times & \times & \times \\ \times & \times & \times & \times & \times \\ \boxtimes & \boxtimes & \boxtimes & \times & \times \\ \boxtimes & \boxtimes & \boxtimes & \times & \times \\ \boxtimes & \boxtimes & \boxtimes & \times & \times \end{bmatrix} ;$$

and the lower left zero 1×3 block of T related to the lower left 2×4 block of T^{-1}:

$$\alpha = \{5\} \text{ and} \qquad\qquad N \backslash \beta = \{4, 5\} \text{ and}$$
$$\beta = \{1, 2, 3\} \qquad\qquad\qquad N \backslash \alpha = \{1, 2, 3, 4\}$$

$$T = \begin{bmatrix} \times & \times & & & \\ \times & \times & \times & & \\ & \times & \times & \times & \\ & & \times & \times & \times \\ 0 & 0 & 0 & \times & \times \end{bmatrix} \quad \leftrightarrow \quad T^{-1} = \begin{bmatrix} \times & \times & \times & \times & \times \\ \times & \times & \times & \times & \times \\ \times & \times & \times & \times & \times \\ \boxtimes & \boxtimes & \boxtimes & \boxtimes & \times \\ \boxtimes & \boxtimes & \boxtimes & \boxtimes & \times \end{bmatrix} .$$

This clearly shows that all the marked submatrices on the right have rank 1. This means that the matrix on the right is semiseparable. This is only done for the lower triangular part, a similar construction can be made for the upper triangular part. □

Using the nullity theorem, one can also easily prove that the inverse of an irreducible tridiagonal matrix is a generator representable semiseparable matrix.

Corollary 1.39. *The inverse of an invertible irreducible tridiagonal matrix is a generator representable semiseparable matrix and vice versa.*

Proof. We only prove one direction of the corollary, the other direction is similar. Suppose our tridiagonal matrix T is not irreducible, then we will show by an example that its inverse also cannot be generator representable. Let us use the same notation as used throughout the previous proof and consider our matrix T of dimension 5×5. The zero block on the left corresponds to the block marked with $\boxtimes$ on the right.

$$\alpha = \{3, 4, 5\} \text{ and} \qquad N \backslash \beta = \{3, 4, 5\} \text{ and}$$
$$\beta = \{1, 2\} \qquad\qquad N \backslash \alpha = \{1, 2\}$$

$$\begin{bmatrix} \times & \times & & & \\ \times & \times & \times & & \\ 0 & 0 & \times & \times & \\ 0 & 0 & \times & \times & \times \\ 0 & 0 & & \times & \times \end{bmatrix} \quad \leftrightarrow \quad \begin{bmatrix} \times & \times & \times & \times & \times \\ \times & \times & \times & \times & \times \\ \boxtimes & \boxtimes & \times & \times & \times \\ \boxtimes & \boxtimes & \times & \times & \times \\ \boxtimes & \boxtimes & \times & \times & \times \end{bmatrix},$$

Moreover, due to Corollary 1.37, we know that the rank of these two blocks should be the same. This means that the block marked with $\boxtimes$ on the right is a zero block. Combining this with the knowledge gained in Theorem 1.12 and Proposition 1.13, we get that the matrix on the right is not generator representable. Similarly, the nullity theorem can be used in the other direction. $\square$

Example 1.40 Suppose a tridiagonal matrix T_1 is given, with its inverse S_1:

$$T_1 = \frac{1}{5} \begin{bmatrix} 2 & 1 & 0 & 0 \\ 1 & 2 & 1 & 0 \\ 0 & 1 & 2 & 1 \\ 0 & 0 & 1 & 2 \end{bmatrix} \text{ and } S_1 = T_1^{-1} = \begin{bmatrix} 4 & -3 & 2 & -1 \\ -3 & 6 & -4 & 2 \\ 2 & -4 & 6 & -3 \\ -1 & 2 & -3 & 4 \end{bmatrix}.$$

Clearly, the tridiagonal matrix is irreducible, and hence, the semiseparable matrix S_1 is of generator representable form. The next matrix has the upper triangular part still of generator representable form, but not the lower triangular part.

$$T_2 = \begin{bmatrix} 2 & 1 & 0 & 0 \\ 1 & 2 & 1 & 0 \\ 0 & 0 & 2 & 1 \\ 0 & 0 & 1 & 2 \end{bmatrix} \text{ and } S_2 = T_2^{-1} = \frac{1}{9} \begin{bmatrix} 6 & -3 & 2 & -1 \\ -3 & 6 & -4 & 2 \\ 0 & 0 & 6 & -3 \\ 0 & 0 & -3 & 6 \end{bmatrix}.$$

Finally, the last matrix is not generator representable in either the upper or lower triangular part.

$$T_3 = \frac{1}{2} \begin{bmatrix} 2 & 0 & 0 & 0 \\ 1 & 2 & 1 & 0 \\ 0 & 0 & 2 & 1 \\ 0 & 0 & 1 & 2 \end{bmatrix} \text{ and } S_3 = T_3^{-1} = \frac{1}{3} \begin{bmatrix} 3 & 0 & 0 & 0 \\ \frac{-3}{2} & 3 & -2 & 1 \\ 0 & 0 & 4 & -2 \\ 0 & 0 & -2 & 4 \end{bmatrix}.$$

■

In the last example we illustrate that the inverse of a lower bidiagonal matrix is a lower triangular semiseparable matrix. The proof is straightforward, using the nullity theorem.

Example 1.41 We have the following matrix S_4, which is lower triangular and semiseparable, and its inverse:

$$S_4 = \begin{bmatrix} 1 & 0 & 0 & 0 & 0 \\ 1 & 1 & 0 & 0 & 0 \\ 1 & 1 & 1 & 0 & 0 \\ 1 & 1 & 1 & 1 & 0 \\ 1 & 1 & 1 & 1 & 1 \end{bmatrix} \text{ and } B_4^{-1} = T_4 = \begin{bmatrix} 1 & 0 & 0 & 0 & 0 \\ -1 & 1 & 0 & 0 & 0 \\ 0 & -1 & 1 & 0 & 0 \\ 0 & 0 & -1 & 1 & 0 \\ 0 & 0 & 0 & -1 & 1 \end{bmatrix}.$$

∎

The structure of the inverse of a semiseparable and generator representable semiseparable matrix is known now. Let us investigate now the class of semiseparable plus diagonal and quasiseparable matrices. We will start with the class of quasiseparable matrices, as this is easier than the semiseparable plus diagonal case.

1.5.3   The inverse of a quasiseparable matrix

In this section we will prove that the inverse of a quasiseparable matrix is again a quasiseparable matrix.

Theorem 1.42.   *The inverse of an invertible quasiseparable matrix is again a quasiseparable matrix.*

Proof.   Using Corollary 1.37, we know that the rank of blocks, strictly below the diagonal is maintained. This clearly states that the inverse of an invertible quasiseparable matrix is again a quasiseparable one, as this is the only constraint posed on quasiseparable matrices.   □

Example 1.43 The following matrix A_1 is not semiseparable nor tridiagonal. The matrix is quasiseparable, however. Clearly also the inverse will be quasiseparable:

$$A_1 = \begin{bmatrix} 1 & 2 & 2 & 2 & 2 \\ 1 & 1 & 2 & 2 & 2 \\ 1 & 1 & 1 & 2 & 2 \\ 1 & 1 & 1 & 1 & 2 \\ 1 & 1 & 1 & 1 & 1 \end{bmatrix} \quad A_1^{-1} = \begin{bmatrix} -1 & 0 & 0 & 0 & 2 \\ 1 & -1 & 0 & 0 & 0 \\ 0 & 1 & -1 & 0 & 0 \\ 0 & 0 & 1 & -1 & 0 \\ 0 & 0 & 0 & 1 & -1 \end{bmatrix}.$$

One can clearly see that, for the matrix A_1, the lower triangular part is of semiseparable form, hence the inverse A_1^{-1} will have the lower triangular part of tridiagonal form. This means that in the lower triangular part only one subdiagonal is different from zero. Consider the matrix A_2, for which we cannot include the diagonal in the structure of either the lower and/or upper triangular part:

$$
A_2 = \begin{bmatrix} 1 & 2 & 2 & 2 & 2 \\ 2 & 1 & 2 & 2 & 2 \\ 2 & 2 & 1 & 2 & 2 \\ 2 & 2 & 2 & 1 & 2 \\ 2 & 2 & 2 & 2 & 1 \end{bmatrix} \quad A_2^{-1} = \frac{1}{9} \begin{bmatrix} -7 & 2 & 2 & 2 & 2 \\ 2 & -7 & 2 & 2 & 2 \\ 2 & 2 & -7 & 2 & 2 \\ 2 & 2 & 2 & -7 & 2 \\ 2 & 2 & 2 & 2 & -7 \end{bmatrix}.
$$

Clearly the inverse is of quasiseparable form. The matrix A_2 can be considered as a semiseparable plus diagonal matrix, which we will investigate in the next section. The matrix A_3, however, is quasiseparable and not of semiseparable plus diagonal form:

$$
A_2 = \begin{bmatrix} 0 & 1 & 1 & 1 & 1 \\ 2 & 0 & 1 & 1 & 1 \\ 2 & 2 & 0 & 1 & 1 \\ 2 & 2 & 2 & 0 & 1 \\ 2 & 2 & 2 & 2 & 0 \end{bmatrix} \quad A_2^{-1} = \frac{1}{30} \begin{bmatrix} -14 & 8 & 4 & 2 & 1 \\ 2 & -14 & 8 & 4 & 2 \\ 4 & 2 & -14 & 8 & 4 \\ 8 & 4 & 2 & -14 & 8 \\ 16 & 8 & 4 & 2 & -14 \end{bmatrix}.
$$

The inverse is again a quasiseparable matrix.   ∎

It was stated before that a unitary Hessenberg matrix is quasiseparable. Using the knowledge we have gained by the nullity theorem the proof is straightforward.

Theorem 1.44.   *A unitary Hessenberg matrix H is quasiseparable.*

Proof. We know that $HH^T = I$. As the matrix H is Hessenberg, this means that the lower triangular part is quasiseparable. Moreover, we know, due to the nullity theorem, that also the lower triangular part of H^T therefore needs to be quasiseparable. This completes the proof.   □

1.5.4   The inverse of a semiseparable plus diagonal matrix

In this section we will prove that the inverse of a semiseparable plus diagonal matrix is again a semiseparable plus diagonal matrix. Moreover, the diagonals of both matrices are closely related to each other.

To prove this statement, we will first formulate a lemma, which is an extension of the nullity theorem. The presented lemma is a weak formulation of more general results which were presented in [76].

Lemma 1.45.   *Suppose we have an invertible matrix A, for which $A \in \mathbb{R}^{n \times n}$. If there exists a $d_i \in \mathbb{R}$ different from zero, such that the matrix $A_i = A - d_i I$ satisfies the following relation:*

$$
\mathrm{rank}\,(A_i(i:n, 1:i)) = r,
$$

then we have that the matrix $\hat{A}_i = A^{-1} - d_i^{-1}I$ satisfies

$$\text{rank}\left(\hat{A}_i(i:n,1:i)\right) = r.$$

Proof. If we partition the matrix A as

$$A = \begin{bmatrix} A_{11} & A_{12} & A_{13} \\ A_{21} & A_{22} & A_{23} \\ A_{31} & A_{23} & A_{33} \end{bmatrix},$$

with $A_{11} = A(1:i-1,1:i-1)$, $A_{22} = A_{i,i}$ and $A_{33} = A(i+1:n,i+1:n)$.
Let B:

$$B = \begin{bmatrix} A_{11} & A_{12} & A_{13} \\ & d_i & A_{23} \\ & & A_{33} \end{bmatrix},$$

then we have that $A - B$ is a matrix of rank r. Hence,

$$A - B = UV^T,$$

with $U, V \in \mathbb{R}^{n \times r}$.
Using the following relation (see, e.g., [152, Section 2.1.3])

$$A^{-1} - B^{-1} = -B^{-1}(A - B)A^{-1},$$

leads to the observation, that $A^{-1} - B^{-1}$ also needs to be of rank r. Moreover, due to the structure of B^{-1}, the theorem holds.[13]   □

Using the above theorem, it is easy to prove the following for the inverse of diagonal plus semiseparable matrices, for which the diagonal is invertible.

Theorem 1.46. *Suppose $S + D$ to be a semiseparable plus diagonal matrix, with the diagonal D invertible. Then its inverse will also be a semiseparable plus diagonal matrix $\hat{S} + \hat{D}$ with $\hat{D} = D^{-1}$.*

Proof. The proof is straightforward, using Lemma 1.45.   □

Let us illustrate this, with some examples of inverses of semiseparable plus diagonal matrices.

Example 1.47   • Reconsidering matrix A_2 from Example 1.43, one can see that it is a semiseparable plus diagonal matrix $A_2 = S_2 + D_2$, with diagonal equal to $D_2 = -I$, with I the identity matrix. The inverse matrix A_2^{-1} can be written as $A_2^{-1} = \hat{S}_2 + \hat{D}_2$, where $\hat{D}_2$ equals $-I$.

---

[13]Note that for reasons of simplicity we assumed the matrix B to be invertible. A continuity argument can be used in case nonsingularity is not satisfied. Adding an infinitesimally small perturbation to A_{11} and A_{33} will not disturb the rank structure.

- Consider for example the inverse of the matrix

$$
S_2 + D_2 = \begin{bmatrix} 1 & 0 & 0 & 0 & 0 \\ 1 & 1 & 0 & 0 & 0 \\ 1 & 1 & 1 & 0 & 0 \\ 1 & 1 & 1 & 1 & 0 \\ 1 & 1 & 1 & 1 & 1 \end{bmatrix} + \begin{bmatrix} 1 & & & & \\ & 2 & & & \\ & & 4 & & \\ & & & 2 & \\ & & & & 1 \end{bmatrix},
$$

which is a semiseparable plus diagonal matrix, and its inverse is of the form

$$
(S_2+D_2)^{-1} = \frac{-1}{(4.5.9)} \begin{bmatrix} 90 & 0 & 0 & 0 & 0 \\ 30 & 30 & 0 & 0 & 0 \\ 12 & 12 & 9 & 0 & 0 \\ 16 & 16 & 12 & 30 & 0 \\ 16 & 16 & 12 & 30 & 90 \end{bmatrix} + \begin{bmatrix} 1 & & & & \\ & 1/2 & & & \\ & & 1/4 & & \\ & & & 1/2 & \\ & & & & 1 \end{bmatrix}
$$

which is clearly semiseparable plus diagonal.

- Consider the following 3×3 matrix

$$
S_3 + D_3 = \begin{bmatrix} 3 & 1 & 1 \\ 1 & 2 & 1 \\ 1 & 1 & 3 \end{bmatrix}.
$$

Its inverse is of the following form:

$$
(S_3 + D_3)^{-1} = \frac{-1}{12} \begin{bmatrix} 1 & 2 & 1 \\ 2 & 4 & 2 \\ 1 & 2 & 1 \end{bmatrix} + \begin{bmatrix} 1/2 & & \\ & 1 & \\ & & 1/2 \end{bmatrix},
$$

which is a semiseparable plus diagonal matrix with as diagonal the inverse of the matrix D_3.

∎

In all of the preceding examples, we assumed the diagonal to be invertible. Let us illustrate with an example of what happens if a diagonal element is zero.

Example 1.48 Let us reconsider the matrix S_2 from Example 1.47, but change the diagonal into $D_2 = \text{diag}([1, 2, 0, 2, 1])$. The inverse will have the following form:

$$
(S_2 + D_2)^{-1} = \frac{-1}{6} \begin{bmatrix} 3 & 0 & 0 & 0 & 0 \\ 1 & 1 & 0 & 0 & 0 \\ 2 & 2 & -6 & 0 & 0 \\ 0 & 0 & 2 & 1 & 0 \\ 0 & 0 & 2 & 1 & 3 \end{bmatrix} + \begin{bmatrix} 1 & & & & \\ & 1/2 & & & \\ & & 0 & & \\ & & & 1/2 & \\ & & & & 1 \end{bmatrix}.
$$

Clearly the nonzero elements of the diagonal are inverted, while the zero elements have their impact on the appearance of zero blocks, just as in the semiseparable case. ∎

The next theorem provides the structure of the inverse, in case one of the diagonal elements of the semiseparable plus diagonal matrix is zero.

Theorem 1.49.

Suppose $S + D$ to be a semiseparable plus diagonal matrix with the diagonal D. Then its inverse will be a matrix of the form $(S + D)^{-1} = A + \hat{D}$. With $\hat{D}$ a diagonal matrix, with diagonal elements $\hat{d}_i = 1/d_i$ if $d_i \neq 0$ and $\hat{d}_i = 0$ if $d_i = 0$. The matrix A has the following structure (for the lower triangular part):

$$\text{rank} \left(A(i : n, 1 : i) \right) = 1 \; \text{ if } d_i \neq 0, \; \text{ for } i = 1, \dots, n,$$
$$\text{rank} \left(A(i + 1 : n, 1 : i - 1) \right) = 0 \; \text{ if } d_i = 0, \; \text{ for } i = 2, \dots, n - 1,$$

the upper triangular part is similar.

Proof. The proof is a combination of Lemma 1.45 and Corollary 1.36. □

Reconsidering now Example 1.48, one can predict the structure of the inverse.

1.5.5 Summary of the relations

We will briefly present a summary of the discussed relations under inversion. If a class of matrices $\mathcal{A}$ is given, we denote the subset of $\mathcal{A}$ consisting of the invertible matrices as $\mathcal{A}_{inv}$.

In this section we proved most of the results below. The proof that the class $\mathcal{Q}^{(s)}$ is closed under inversion, is left to the reader.

$$\left(\mathcal{T}_{inv}^{(i)} \right)^{-1} = \mathcal{S}_{inv}^{(g)},$$
$$\left(\mathcal{T}_{inv} \right)^{-1} = \mathcal{S}_{inv},$$
$$\left(\mathcal{Q}_{inv} \right)^{-1} = \mathcal{Q}_{inv},$$
$$\left(\mathcal{Q}_{inv}^{(s)} \right)^{-1} = \mathcal{Q}_{inv}^{(s)},$$
$$((\mathcal{S} + D)_{inv})^{-1} = \left((\mathcal{S} + D^{-1})_{inv} \right), \text{ for } D \text{ invertible},$$
$$(\mathcal{S}^{(d)})_{inv}^{-1} \subsetneq \mathcal{Q}_{inv}^{(s)}.$$

Notes and references

As already mentioned, the first formulation of the nullity theorem was due to Gustafson.

☞ W. H. Gustafson. A note on matrix inversion. *Linear Algebra and its Applications*, 57:71–73, 1984.

He formulated the theorem for matrices over principal ideal domains. The nullity theorem, as presented here, was formulated by Fiedler and Markham, who translated the abstract formulation of Gustafson. The theorem, as illustrated in the text, is valuable for predicting structures of the inverse of structured rank matrices.

☞ M. Fiedler and T. L. Markham. Completing a matrix when certain entries of its inverse are specified. *Linear Algebra and its Applications*, 74:225–237, 1986.

And also Barrett and Feinsilver formulated theorems close to the nullity theorem.

☞ W. W. Barrett. A theorem on inverses of tridiagonal matrices. *Linear Algebra and its Applications*, 27:211–217, 1979.

☞ W. W. Barrett and P. J. Feinsilver. Inverses of banded matrices. *Linear Algebra and its Applications*, 41:111–130, 1981.

Barrett (1979) formulates another type of theorem connected to the inverses of tridiagonal matrices. In most of the preceding papers one assumed the sub- and superdiagonal elements of the corresponding tridiagonal matrix to be different from zero. In this paper only one condition is left. It is assumed that the diagonal elements of the symmetric semiseparable matrix are different from zero. Moreover, the proof is also suitable for nonsymmetric matrices. The theorems presented in this paper are very close to the final version result, stating that the inverse of a tridiagonal matrix is a semiseparable matrix, satisfying the rank definition. The paper of 1981 contains important results concerning the inverses of band matrices. Barrett and Feinsilver provide a general framework as presented in this section. General theorems and proofs considering the vanishing of minors when looking at the matrices and their inverses are given, thereby characterizing the complete class of band and semiseparable matrices, without excluding cases in which zeros appear. The results are a straightforward consequence of the paper of 1979. The authors also investigate Toeplitz matrices having a banded inverse (for an example of the inverse of a Toeplitz matrix discussed in their manuscript, see Section 14.1 in Chapter 14). Also the manuscript [14], contains information on the inverse of a tridiagonal matrices. They provide a probabilistic proof that the inverse of a tridiagonal matrix satisfies certain low rank constraints. (See Section 3.2 in Chapter 3.)

Fiedler published more papers in which he exploited the power of the nullity theorem to predict the structure of factorizations, inverses, etc., of specific structured rank matrices.

☞ M. Fiedler. Structure ranks of matrices. *Linear Algebra and its Applications*, 179:119–127, 1993.

☞ M. Fiedler and T. L. Markham. Rank-preserving diagonal completions of a matrix. *Linear Algebra and its Applications*, 85:49–56, 1987.

In the manuscript of 1987, Fiedler proves that the off-diagonal rank of a matrix is maintained under inversion. The off-diagonal rank can be seen as the maximum rank of all submatrices taken out of the matrix without the diagonal. More on general rank structures can be found in Part III of this book. In the manuscript of 1993, more general theorems connected to the structured ranks of matrices and their inverses are presented (see also Section 8.1 in Chapter 8).

The authors Vandebril, Van Barel and Mastronardi presented in the following manuscript an alternative proof, included in this section, for proving the nullity theorem. Moreover they also translated the nullity theorem towards the LU- and QR-decomposition of rank structured matrices. The nullity theorems for these decompositions make it possible to predict the rank structure of the L, U and Q factors.

☞ R. Vandebril and M. Van Barel. A note on the nulllity theorem. *Journal of Computational and Applied Mathematics*, 189:179–190, 2006.

Brualdi and Massey generalize some of the results of the manuscript of Fiedler (1987) for structures in which the diagonal is not included in their manuscript.

☞ R. A. Brualdi and J. J. Q. Massey. More on structure-ranks of matrices. *Linear Algebra and its Applications*, 183:193–199, 1993.

Rózsa, Romani and Bevilacqua provide results on the inverses of matrices satisfying certain rank conditions. Moreover they also provide an alternative proof of the nullity theorem, via the Schur complement and limits. (See also Section 8.3 in Chapter 8.)

☞ P. Rózsa, F. Romani, and R. Bevilacqua. On generalized band matrices and their inverses. In D. J. Brown, M. T. Chu, D. C. Ellison, and R. J. Plemmons, editors, *Proceedings of the Cornelius Lanczos International Centenary Conference*, volume 73 of *Proceedings in Applied Mathematics*, pages 109–121, Philadelphia PA, 1994. SIAM Press.

Strang and Nguyen investigate in this paper the nullity theorem as provided by Fiedler and Markham.

☞ G. Strang and T. Nguyen. The interplay of ranks of submatrices. *SIAM Review*, 46(4):637–646, 2004.

Theoretical remarks, an alternative proof and comments on the original papers are provided.

Reconsidering the results of Section 1.5.2 with the ones of Section 1.2 and Section 1.4, one sees that there is a correspondence between the class of generator representable semiseparable and irreducible tridiagonal matrices. Moreover, the class of irreducible tridiagonal matrices is not closed, but its closure is the class of tridiagonal matrices just as the closure of the class of generator representable matrices is the class of semiseparable matrices.

In a nontheoretical way, i.e., by explicitly computing the inverse, it is proved that the inverse of a quasiseparable matrix is again a quasiseparable matrix in the following articles.

☞ Y. Eidelman and I. C. Gohberg. On a new class of structured matrices. *Integral Equations and Operator Theory*, 34:293–324, 1999.

☞ Y. Eidelman and I. C. Gohberg. Fast inversion algorithms for a class of block structured matrices. *Contemporary Mathematics*, 281:17–38, 2001.

In both the papers linear inversion methods are presented; see also Chapter 12 for more information on quasiseparable, block quasiseparable and inversion methods.

The results used for the proof of maintenance of the semiseparable plus diagonal structure under inversion is a weak formulation of more general results proved in:

☞ S. Delvaux and M. Van Barel. Structures preserved by matrix inversion. *SIAM Journal on Matrix Analysis and its Applications*, 28(1):213–228, 2006.

This paper investigates in more detail several structures of matrices, which are maintained under inversion, including also displacement structures.

☞ G. Meurant. A review of the inverse of symmetric tridiagonal and block tridiagonal matrices. *SIAM Journal on Matrix Analysis and its Applications*, 13:707–728, 1992.

In this paper by Meurant an extensive list of references connected to the inverse of semi-separable and band matrices is included. Some results concerning the inverse of symmetric tridiagonal and block tridiagonal matrices are reviewed, based on the Cholesky decomposition. Also results concerning the decay of the elements of the inverse are obtained. The results of this manuscript are discussed in more detail in Section 7.5 in Chapter 7.

A generalization of the nullity theorem towards the generalized inverse can be found in [10]; see Section 14.6 in Chapter 14.

In fact a more general theorem than the nullity theorem as presented here exists. This theorem is closely related to completion problems. This theorem, its relation to completion problems and a derivation of how one can see the nullity theorem as a corollary of it is presented in the notes and references of Section 8.4, Chapter 8.

1.6   Conclusions

In this chapter we defined different classes of structured rank matrices. Either symmetric and nonsymmetric, semiseparable, generator representable semiseparable, quasiseparable and semiseparable plus diagonal matrices were defined in terms of blocks of the matrices, satisfying certain rank conditions. We took off with the class of symmetric matrices, as this is the easiest class, and we investigated the relations between these different classes. It was shown that the most general class of matrices was the class of the quasiseparables, containing both semiseparable and tridiagonal matrices. Interesting was also the fact that the closure of the class of symmetric semiseparable matrices resulted in the class of quasiseparables, and the closure of the generator representable semiseparable matrices resulted in the class of semiseparable matrices. The different classes can be extended towards the nonsymmetric case. Interesting to mention is the coupling between the upper and lower triangular part for generator representable and semiseparable matrices. Both the upper and lower triangular part include the diagonal in their structure. The relations between the nonsymmetric classes are almost the same as for the symmetric case. The most general class was again the class of quasiseparables, covering the class of semiseparables and tridiagonals. The closure of the class of generator representable semiseparable matrices was again the class of semiseparable matrices. But for the nonsymmetric case the closure of the class of semiseparable plus diagonal matrices was not the class of quasiseparable matrices, but a special subclass, namely the class of quasiseparable matrices having a symmetric rank structure. The final section of this chapter investigated the relations of these matrices with regard to to inversion. Firstly we proved the nullity theorem in two different ways. The nullity theorem relates the rank of certain subblocks in a matrix A to the rank of certain subblocks in its inverse A^{-1}, making it a valuable tool for working with rank structured matrices. Using the nullity theorem, it is an easy exercise to prove that the inverse of a semiseparable matrix is a tridiagonal matrix, the inverse of a generator representable semiseparable matrix is an irreducible tridiagonal matrix, and the inverse of a quasiseparable matrix is a quasiseparable matrix. The class of semiseparable plus diagonal matrices behaves differently. If the diagonal is invertible, the inverse of a semiseparable plus diagonal matrix is again a semiseparable plus diagonal matrix, with as diagonal the inverse of the original diagonal. Zero elements in the diagonal, are linked in the inverse with the vanishing of certain subblocks.

To conclude we might say that we defined several different types of semiseparable and related matrices and we investigated these classes with regard to each other, the closure and inversion.

The definitions of the matrices as presented in this first section, proved to be valuable tools for deriving structural properties of these matrices. However, the fact that a certain part of a matrix satisfies a certain rank condition does not provide us with any information concerning an efficient representation of this low rank block. Therefore, we will investigate in the second chapter of this book several possible ways for representing low rank parts in matrices.

Chapter 2

The representation of semiseparable and related matrices

In the previous chapter it was shown that, when one wants to solve the eigenvalue problem by means of the QR-algorithm, the definition of semiseparable matrices with generators has some disadvantages. Therefore we proposed the more elaborate definition, in terms of the structured rank. This class of semiseparable matrices is closed under any suitable chosen norm, just as the class of quasiseparable and tridiagonal matrices is closed. However, these classes of matrices can only be used efficiently if we have also an efficient representation as indicated by Corollary 1.11. The representations for the different classes of matrices as defined in the previous chapter are the subject of this chapter.

Different types of representations will be investigated, e.g., the generator representation, the representation with a diagonal and a subdiagonal, the representation with a sequence of Givens transformations and a vector and finally the quasiseparable representation. In fact all the above-mentioned representations are just specific parameterizations for a part of a matrix having low rank. Hence we can use all the representations above for representing all the different classes of matrices previously defined. Moreover we will briefly discuss the question: "Are there more representations then the ones discussed in this chapter?"

Before defining the different types of representations, we indicate properly what in this book is meant by a 'representation'. We base this definition on the representation of tridiagonal matrices, using only the diagonal, subdiagonal and the superdiagonal. The viewpoint of a representation is 'just' a representation is also clarified. In brief one could summarize this as stating that the choice of a representation for a specific matrix depends on external parameters and is not necessarily always linked to the intrinsic matrix structure. Under external parameters, one might classify the problem, the application one is dealing with, knowledge about the problems origin, ...

In Sections 2.2 to 2.5, four types of representations for the symmetric case are considered. One considers the generator representation, the diagonal subdiagonal representation, the Givens-vector representation and the quasiseparable representation. Each of these representations is investigated with regard to the definition of a representation as proposed before, and we try to use this representation for efficiently representing the class of semiseparable matrices. Moreover, for every type of representation we investigate in which way the representation can be used for representing the other classes of structured rank matrices from the first chapter. We conclude the part of the symmetric representations with Section 2.6, containing some examples.

The nonsymmetric case is more difficult than the symmetric one and is therefore covered in the Sections 2.7 to 2.9. The diagonal subdiagonal representation is not covered anymore, as it will not be used for representing nonsymmetric semiseparable or related matrices. The last type of representation considered is a special type of representation for the nonsymmetric case, the so-called decoupled representation. This representation explicitly does not take into account the extra structure posed on the diagonal by, e.g., semiseparable and generator representable semiseparable matrices. In fact this representation uses order $\mathcal{O}(n)$ more parameters, but it covers also a slightly more general class than the class of semiseparable and/or generator representable semiseparable matrices. This representation is also quite often used for actual implementations of algorithms related to structured rank matrices.

In the last section of this chapter, several algorithms are presented for the different types of representations. Firstly, it is shown that the different representations for a specific matrix admit an $\mathcal{O}(n)^{14}$ multiplication between this matrix and a vector. Secondly, some manners are presented to change from one representation to another.

✎ *This chapter focuses on some different representations for structured rank parts in matrices. Subsection 2.1.3 briefly explains in words the different types of representation. This gives a good idea of the different types. The representations are first applied to semiseparable matrices to clearly distinguish between the types of representations. At the end of each section covering a representation, the representation is adapted for quasiseparable, semiseparable plus diagonal and other matrices.*

In Section 2.2 the generator representation is explained. Equation (2.1) presents the structure of the matrix in case this representation is used. We remark that the generator representation, as already shown in the previous chapter, cannot represent all semiseparable matrices, only the generator representable semiseparable ones, hence the name.

The diagonal subdiagonal representation discussed in Section 2.3 is merely included for completeness and is not used throughout the remainder of the book.

Section 2.4 discusses the Givens-vector representation which is used throughout the book. In particular, the first subsection (Subsection 2.4.1) contains the

---

[14]In this book we consider flops (floating point operations). A flop is any of the following operations: $+,-,*,/$. Some manuscripts consider flams (floating point addition and multiplication), this is on operation of the form $ax + y$ (axpy). Therefore, when speaking about operations in this book, we mean flops.

most essential results. How do I construct the semiseparable part in a matrix using the Givens-vector representation. Subsection 2.4.4 discusses the two variants of the Givens-vector representation. These variants can be extremely useful in implementations to reduce the complexity. The idea on how to change between the two variants is also discussed.

Section 2.5 discusses the quasiseparable representation, which is also widespread. It is important to know that this representation is able to represent semiseparable parts in matrices, which is not at all obvious looking at the form of the representation. The representation is presented in Definition 2.14 and it is proved in Theorem 2.15 that this representation is indeed capable of representing semiseparable matrix parts.

In order to distinguish clearly between the possibilities of the different representations, one can consider some of the examples discussed in Section 2.6.

The unsymmetric representations discussed in Sections 2.7 to 2.12 are in some sense straightforward generalizations of the symmetric case. For some types, such as the generator representation, different forms exist. But essentially the forms depicted in the following equations illustrate the different possibilities: Equation (2.12) (a variation on the generator representation for invertible matrices is shown in Equation (2.13)) and Equation (2.19). The decoupled representation as discussed in Section 2.10 neglects the diagonal connection in semiseparable matrices, e.g., Equation (2.22) shows the decoupled Givens-vector representation.

The chapter concludes with some algorithms related to the different types of representations. These algorithms are not essential for the remainder of the book, but might come in handy when implementing specific methods. Subsection 2.13.1 discusses a fast matrix-vector multiplication for different types of representations. Subsection 2.13.2 discusses how to interchange between representations; these algorithms are more interesting from a theoretical viewpoint than from a practical one. The last presented algorithm derives an $O(n)$ method for computing the determinant of a Givens-vector represented semiseparable matrix.

2.1 Representations

In this section, firstly the definition of a representation of a set of matrices is given. For some simple classes, such as tridiagonal matrices and/or band matrices it is straightforward how one should represent them. Just neglect the zero elements, order the remaining elements, such that their position in the matrix is known, and one has a sparse representation of the matrix. For a matrix not necessarily having a sparse structure, such as the structured rank matrices, which are often dense, this is not trivial. Even though these matrices can be represented with a small number of parameters, the choice of these parameters is not straightforward and not always unique. Therefore, we have to define some conditions that these parameters should satisfy. In the next sections we put forward some representations and their relations with classes of matrices, which will be used throughout the book.

2.1.1   The definition of a representation

Before investigating different possible representations, it is necessary to define what is exactly meant by a representation of a class of matrices. Although this is intuitively clear, we want a precise definition. Based on this definition we will take a close look at several representations proposed in this book, for representing the matrices previously defined in Chapter 1.

We define a representation based on a map as follows:

Definition 2.1. *An element $v \in \mathcal{V}$ is said to represent an element $u \in \mathcal{U}$ if there is a map r*

$$r : \mathcal{V} \subseteq \mathcal{X} \to \mathcal{U} \subseteq \mathcal{W},$$

where $\mathcal{X}$ and $\mathcal{W}$ are two vector spaces containing the sets $\mathcal{V}$ and $\mathcal{U}$, respectively. The following conditions are satisfied:

- *where $\mathcal{X}$ and $\mathcal{W}$ can be chosen, such that $\dim(\mathcal{X}) \leq \dim(\mathcal{W})$;*

- *$r(\mathcal{V}) = \mathcal{U}$, i.e., the map is surjective;*

- *$\exists$ a map $s : \mathcal{U} \to \mathcal{V}$ such that $r|_{s(\mathcal{U})}$ is bijective and $r(s(u)) = u$, $\forall u \in \mathcal{U}$, such that $r(v) = u$.*

According to the definition it is possible that $\dim(\mathcal{X}) = \dim(\mathcal{W})$. It is however obvious that we want the dimension of $\mathcal{X}$ to be as small as possible, as we will use the representation v instead of the element u for implementing algorithms connected to the set $\mathcal{U}$. In fact the map $s : \mathcal{U} \to \mathcal{V}$ always exists, but it is included in the definition, to show the importance of a map, which will in fact, return the representation for a given matrix. In this chapter different types of representations will be given, and they will be used throughout the remainder of the text. The map r is called a representation map of the set $\mathcal{U}$. The element $v \in s(\mathcal{U}) \subseteq \mathcal{V}$ for which $r(v) = u$ with $u \in \mathcal{U}$ is called a representation of u.

To check if this definition suits our needs, we investigate the following map, when studying tridiagonal matrices. As defined in Chapter 1 of the book, $\mathcal{T}$ denotes the class of tridiagonal matrices.

We have the following map:

$$r_{\mathcal{T}} : \mathbb{R}^{n-1} \times \mathbb{R}^n \times \mathbb{R}^{n-1} \to \mathcal{T}$$

$$(\mathbf{d}^{(l)}, \mathbf{d}, \mathbf{d}^{(u)}) \mapsto T = \begin{bmatrix} d_1 & d_1^{(u)} & 0 & & & \\ d_1^{(l)} & d_2 & d_2^{(u)} & \ddots & & \\ 0 & d_2^{(l)} & \ddots & \ddots & 0 & \\ & & \ddots & \ddots & \ddots & d_{n-1}^{(u)} \\ & & & 0 & d_{n-1}^{(l)} & d_n \end{bmatrix}.$$

It can clearly be seen that all the properties from Definition 2.1 are satisfied. The map $s_{\mathcal{T}}$ can be defined very easily as $s_{\mathcal{T}} = r_{\mathcal{T}}^{-1}$. This states the fact that we can use the diagonal and subdiagonal of a symmetric tridiagonal matrix to represent it.

2.1.2 A representation is "just" a representation

The title of this section, is a summary of the considerations made in this section. Let us clarify what we mean. Consider the following example.

Example 2.2 The symmetric generator representable matrix $S \in \mathcal{S}_{sym}^{(g)}$, can be represented using different vectors. Consider the vectors $\mathbf{u}_1 = [0, 2, 3]^T$, $\mathbf{v}_1 = [2, 4, 2]^T$, $\mathbf{u}_2 = [0, 4, 6]^T$ and $\mathbf{v}_2 = [1, 2, 1]^T$. Then we have the following equivalent representations.

$$S = S(\mathbf{u}_1, \mathbf{v}_1) = S(\mathbf{u}_2, \mathbf{v}_2) = \begin{bmatrix} 0 & 4 & 6 \\ 4 & 8 & 12 \\ 6 & 12 & 6 \end{bmatrix}.$$

■

These two sets of generators represent exactly the same matrix. Moreover, if we just want a representation of this matrix, there is no reason to prefer one set of vectors above the other one, even though they are different. But is it really necessary to make a choice? Perhaps the choice is already made depending on the type of application. If a certain application presents to us a system to solve represented with generators, like the matrix above, it would perhaps not necessarily be a good choice to transfer the generators to another type of representation.

The main aim of a representation is to represent the low rank part of these semiseparable and related matrices in an *efficient way*. To represent this low rank part, there are *different possible ways*, as will be shown in this chapter. Knowing that we have different sets of matrices and different possible representations, there is *no deterministic rule* to link a specific type of representation to a specific set of matrices. Hence a set of matrices such as semiseparable and/or quasiseparable is not intrinsically connected to a representation. The choice of representation to be used for a certain class of matrices depends, for example, on the following criteria:

- What is the stability of the representation, with regard to the problem one would like to solve (e.g., solve a system, compute the eigenvalues, rank revealing and so on.)?;

- What is the number of parameters used by the representation, and how many parameters do intrinsically determine the set of matrices (e.g., a quasiseparable matrix is generically determined by more parameters than a 'generator representable' semiseparable matrix)?;

- What type of application is considered, or more precisely, does the application already put forward the use of a specific representation (e.g., state space models [84] are related to the quasiseparable representation, whereas the discretization of a semiseparable kernel is related to a generator representable representation [7])?;

- Do algorithms for this problem and this representation exist?;

- Is the cost of changing the representation to another representation beneficial?

One might conclude by stating that a representation is simply a manner to deal with a specific problem for a specific matrix. Hence, *"A representation is 'just' a representation and not a justification."*

In the remainder of this chapter we will consider 4 types of representations. The generator representation, the diagonal-subdiagonal representation, the Givens-vector representation and the quasiseparable representation. Linking the 4 classes of matrices as defined in the previous chapter with these 4 representations, we get $4^2 = 16$ possible combinations. For the lack of space, we will not cover them all! Therefore, we restrict the theoretical investigations in the following sections to the applicability of these representations with regard to the class of semiseparable matrices. At the end of every section, we illustrate the use of these representations for the other classes of semiseparable related matrices. Finally we quickly discuss the existence of other types of representations.

Throughout the book, we will always clearly state with what class of matrices and which type of representation we are working with.

2.1.3   Covered representations

Let us briefly summarize the four main types of representations covered in this chapter. To avoid making this discussion unnecessarily complicated we implicitly assume in this subsection that we are talking about symmetric semiseparable matrices.

Generator representation

This representation assumes in fact that the low rank part in the rank structured matrix is coming from a rank 1 matrix. Hence we can use the two vectors defining this rank 1 matrix for representing the low rank part in this matrix. Unfortunately, as will be shown, this is not sufficient for representing the complete class of semiseparable matrices.

Diagonal-subdiagonal representation

As the title reveals, this representation only uses the elements from the diagonal and the subdiagonal for representing the complete lower triangular part of semiseparable form. To reconstruct an element in the matrix a certain zig-zag pattern needs to be followed. Unfortunately, to be able to reconstruct the complete matrix some elements need to be different from zero. Hence, this representation also is incapable of representing the complete class of semiseparable matrices.

Givens-vector representation

This representation, as will be shown, will be able to cover all semiseparable matrices. The Givens rotations store essentially the dependencies between succeeding rows, while the vector poses an extra weight upon each column. Moreover, it will be discussed in Chapter 5 that this representation is closely related to the QR-factorization of the involved matrix.

Quasiseparable representation

The quasiseparable representation can be seen as a more general formulation of the Givens-vector representation. It relaxes the condition that the dependencies between the rows should be expressed as Givens rotations. Hence the representation is more general, offers a great flexibility, but it is also not uniquely determined anymore. In fact one can also consider the generator and Givens-vector representations as a special form of the quasiseparable representation.

As the easiest type of representation involves the symmetric matrices, we will, as in the previous chapter, first cover the symmetric case.

Notes and references

In the previous section, we clearly distinguished between the definition of a set of matrices and its representation. In practice, however, researchers seldom make this distinction. Quite often they solve a particular problem involving a specific class of structured rank matrices, and afterwards extend the class of matrices in their definition towards the widest applicable range for this solver. This leads in many cases to a representation of matrices which is taken as the basis for the definition. For example, the class of semiseparable plus diagonal matrices is often redefined to fit a certain context. Even though the definition of semiseparable plus diagonal clearly indicates that the diagonal in the semiseparable part should be includable in the structure, quite often this is neglected. In this way, one has in fact a quasiseparable matrix. Also the class of semiseparable plus diagonal matrices is often defined as a matrix for which the strictly lower triangular and the strictly upper triangular part are coming from a rank 1 matrix. This is even a stronger restriction, as now we get a specific subclass of the class of quasiseparable matrices. Readers should carefully check consistency of the definition with the chosen representation when reading manuscripts, as one is often working with hybrid variants without adjusting the name of the class of matrices, or a clear definition of what type of matrices are considered in the manuscript.

2.2 The symmetric generator representation

Because the class of symmetric semiseparable matrices is easier to deal with, we start our investigations by restricting ourselves to the symmetric case. First, we define the representation based on the generators. Second, we apply this representation to the class of symmetric semiseparable plus diagonal matrices, and symmetric quasiseparable matrices.

2.2.1 The representation for symmetric semiseparables

Let us look now for a representation for the class of symmetric semiseparable matrices $\mathcal{S}_{sym}$.

First we investigate if the map corresponding to the definition of generator representable semiseparable matrices (Definition 1.3) satisfies the properties of a representation map for $\mathcal{S}_{sym}$. This map $r_{\mathcal{S}_{sym}^{(g)}}$ is defined in the following natural way:

$$r_{\mathcal{S}_{sym}^{(g)}} : \mathbb{R}^n \times \mathbb{R}^n \to \mathcal{S}_{sym} \tag{2.1}$$

$$(\mathbf{u}, \mathbf{v}) \mapsto \mathrm{tril}(\mathbf{u}\mathbf{v}^T) + \mathrm{triu}(\mathbf{v}\mathbf{u}^T, 1).$$

The first condition on the map: $\dim(\mathbb{R}^n \times \mathbb{R}^n) \leq \dim(\mathbb{R}^{n\times n})$ is satisfied (for $n \geq 2$). Moreover the dimension of $\mathbb{R}^n \times \mathbb{R}^n$ is much smaller than the one of $\mathbb{R}^{n\times n}$ if $n \gg 1$. The surjectivity condition, however, leads to problems, e.g., the matrix

$$S_1 = \begin{bmatrix} 0 & 1 & 0 \\ 1 & 0 & 0 \\ 0 & 0 & 1 \end{bmatrix}$$

belongs to the class of symmetric semiseparable matrices $\mathcal{S}_{sym}$ but $S_1 \notin r_{\mathcal{S}_{sym}^{(g)}} (\mathbb{R}^n \times \mathbb{R}^n)$. Therefore this representation can never be used to represent the complete class of semiseparable matrices, as we already know from the previous chapter. Therefore we adapt the target set to the class of symmetric generator representable semiseparable matrices, hence also the name. Redefine $r_{\mathcal{S}_{sym}^{(g)}}$ as

$$r_{\mathcal{S}_{sym}^{(g)}} : \mathbb{R}^n \times \mathbb{R}^n \to \mathcal{S}_{sym}^{(g)}$$

$$(\mathbf{u}, \mathbf{v}) \mapsto \mathrm{tril}(\mathbf{u}\mathbf{v}^T) + \mathrm{triu}(\mathbf{v}\mathbf{u}^T, 1).$$

This restriction makes $r_{\mathcal{S}_{sym}^{(g)}}$ a representation map for the class $\mathcal{S}_{sym}^{(g)}$ because the surjectivity condition is clearly satisfied. We will now search for a map $s_{\mathcal{S}_{sym}^{(g)}}$. The inverse of $r_{\mathcal{S}_{sym}^{(g)}}$ can not be chosen as $s_{\mathcal{S}_{sym}^{(g)}}$ because it is no longer a map as shown in the following example:

Example 2.3 Taking the following two vectors: $\mathbf{u}_1 = [1, 2, 3]^T$ and $\mathbf{v}_1 = [2, 2, 2]^T$, the matrix $r_{\mathcal{S}_{sym}^{(g)}} (\mathbf{u}_1, \mathbf{v}_1)$ is the following:

$$r_{\mathcal{S}_{sym}^{(g)}} (\mathbf{u}_1, \mathbf{v}_1) = \begin{bmatrix} 2 & 4 & 6 \\ 4 & 4 & 6 \\ 6 & 6 & 6 \end{bmatrix}.$$

Constructing $r_{\mathcal{S}_{sym}^{(g)}} (\mathbf{u}_2, \mathbf{v}_2)$ with $\mathbf{u}_2 = [2, 4, 6]^T$ and $\mathbf{v}_2 = [1, 1, 1]^T$ gives the following result:

$$r_{\mathcal{S}_{sym}^{(g)}} (\mathbf{u}_2, \mathbf{v}_2) = \begin{bmatrix} 2 & 4 & 6 \\ 4 & 4 & 6 \\ 6 & 6 & 6 \end{bmatrix}.$$

This means that the map

$$r_{\mathcal{S}_{sym}^{(g)}} : r_{\mathcal{S}_{sym}^{(g)}}^{-1} (\mathcal{S}_{sym}^{(g)}) = \mathbb{R}^n \times \mathbb{R}^n \to \mathcal{S}_{sym}^{(g)}$$

$$(\mathbf{u}, \mathbf{v}) \mapsto \mathrm{tril}(\mathbf{u}\mathbf{v}^T) + \mathrm{triu}(\mathbf{v}\mathbf{u}^T, 1)$$

is not bijective, where

$$r_{\mathcal{S}_{sym}^{(g)}}^{-1}\left(\mathcal{S}_{sym}^{(g)}\right) = \left\{(\mathbf{u}, \mathbf{v}) \in \mathbb{R}^n \times \mathbb{R}^n \middle| r_{\mathcal{S}^{(g)}}(\mathbf{u}, \mathbf{v}) \in \mathcal{S}_{sym}^{(g)}\right\}.$$

∎

Some kind of normalization is needed, such that every matrix will have a unique set of generators. The condition $u_1 = 1$ (u_1 is the first element of the vector $\mathbf{u}$) will often lead to a unique representation, but not always.

Example 2.4 Suppose the following matrix S is given:

$$S = \begin{bmatrix} 0 & 0 & 0 \\ 0 & 2 & 3 \\ 0 & 3 & 6 \end{bmatrix}.$$

The following generators both have the first element of the vector $\mathbf{u}$ equal to 1 but are not the same: $\mathbf{u}_1 = [1, 2, 3]^T, \mathbf{v}_1 = [0, 1, 2]^T$ and the couple $\mathbf{u}_2 = [1, 1, 3/2]^T, \mathbf{v}_2 = [0, 2, 4]^T$. ∎

Let us define a map which satisfies the conditions to be a possible choice for the map $s_{\mathcal{S}_{sym}^{(g)}}$. Suppose we have a nonzero (symmetric) matrix $S \in \mathcal{S}_{sym}^{(g)}$, $S = [K_1, K_2, \ldots, K_n]$, where K_i denotes the ith column of the matrix S. Suppose $1 \le l \le k \le n$ such that K_l is the first column in S different from zero, and K_k is the last column in S different from zero. The element $S(k, l)$ clearly has to be different from zero. Otherwise either the column K_k or K_l has to be zero. This is illustrated in the following figure:

$$\begin{array}{cc}
l & k \\
\downarrow & \downarrow
\end{array}$$

$$\begin{bmatrix}
0 & \vdots & 0 & \vdots & 0 \\
\cdots & \cdot & \cdots & S(k,l) & \cdots \\
0 & \vdots & \cdot & \vdots & 0 \\
\cdots & S(k,l) & \cdots & \cdot & \cdots \\
0 & \vdots & 0 & \vdots & 0
\end{bmatrix} \begin{array}{l} \\ \leftarrow l \\ \\ \leftarrow k \\ \\ \end{array}$$

Define $\mathbf{u} = K_l / S(k, l)$ and $\mathbf{v} = K_k$. Then we get a unique set $(\mathbf{u}, \mathbf{v})$ for every semiseparable matrix in $\mathcal{S}_{sym}^{(g)}$. Note that this construction is only suitable for symmetric semiseparable matrices. This allows us to define the following map as a possible choice for the function $s_{\mathcal{S}_{sym}^{(g)}}$.

Definition 2.5. *Suppose for each nonzero matrix $S \in \mathcal{S}_{sym}^{(g)}$ of dimension n (for a zero matrix $\mathbf{u}, \mathbf{v}$ are defined to be zero), K_l is the first column of S different from zero and K_k is the last column of S different from zero. Then we define the map*

$s_{\mathcal{S}_{sym}^{(g)}}$ *in the following way:*

$$s_{\mathcal{S}_{sym}^{(g)}} : \mathcal{S}_{sym}^{(g)} \to \mathbb{R}^n \times \mathbb{R}^n$$

$$S \mapsto \left[\frac{K_l}{S(k,l)}, K_k \right].$$

Because this defines the projection onto the vectors $\mathbf{u}$ and $\mathbf{v}$ in a unique way, we have that $r_{\mathcal{S}_{sym}^{(g)}}|_{s(\mathcal{S}_{sym}^{(g)})}$ is bijective[15]. We now have a unique representation for each element of the set $\mathcal{S}_{sym}^{(g)}$ but, as expected, we do not yet have a suitable representation for the complete class $\mathcal{S}_{sym}$ of symmetric semiseparable matrices.

To summarize: the representation based on the generators is suitable for representing symmetric generator representable semiseparable matrices, using thereby $2n$ parameters, stored in two vectors. If a well-defined normalization is used, one can use $2n - 1$ parameters. This conforms with the class of symmetric irreducible tridiagonal matrices, which are linked by inversion to the class $\mathcal{S}_{sym,inv}^{(g)}$, and which can also be represented by $2n - 1$ parameters.

Before searching for a suitable representation for the class of symmetric semiseparable matrices, we apply the representation above presented to the class of symmetric semiseparable plus diagonal and the class of symmetric quasiseparable matrices.

2.2.2   Application to other classes of matrices

The representation is only useful for representing a specific subclass of the class of semiseparable matrices; hence, we can use it solely to represent specific subclasses of the class of symmetric semiseparable plus diagonal $\mathcal{S}_{sym}^{(d)}$ and symmetric quasiseparable matrices $\mathcal{Q}_{sym}$.

For example, a straightforward use of this representation is to represent the class of symmetric generator representable semiseparable plus diagonal matrices, denoted as

$$\mathcal{S}_{sym}^{(g,d)} = \{S + D \mid S \in \mathcal{S}_{sym}^{(g)} \text{ and } D \in \mathcal{D}\}.$$

The map linked to this class of matrices is logically the following:

$$r_{\mathcal{S}_{sym}^{(g,d)}} : \mathbb{R}^n \times \mathbb{R}^n \times \mathbb{R}^n \to \mathcal{S}_{sym}^{(g)}$$

$$(\mathbf{u}, \mathbf{v}, \mathbf{d}) \mapsto \text{tril}(\mathbf{u}\mathbf{v}^T) + \text{triu}(\mathbf{v}\mathbf{u}^T, 1) + \text{diag}(\mathbf{d}).$$

With $\text{diag}(\mathbf{d})$, we denote a diagonal matrix with, as diagonal elements, the elements from the vector $\mathbf{d}$. This representation uses $3n$ parameters, or $3n - 1$, if a suitable

---

[15]The map considered here is not the only one to compute the generators of a generator representable semiseparable matrix. Moreover, there certainly exist more stable approaches (see the Givens-vector representation for more information). This map is merely included to show the problems and to present an example.

normalization is chosen. This class of matrices $\mathcal{S}_{sym}^{(g,d)}$ is used, e.g., in the publications [107, 279, 211, 104, 57, 54] (more information can be found in the notes and references).

In this way we can also define a special subclass of the quasiseparable matrices, namely the class of matrices, for which the strictly lower and the strictly upper triangular parts are representable with generators. Let us define this class of matrices as $\mathcal{Q}_{sym}^{(g)}$. The representation is of the following form:

$$r_{\mathcal{Q}_{sym}^{(g)}} : \mathbb{R}^{n-1} \times \mathbb{R}^{n-1} \times \mathbb{R}^n \to \mathcal{Q}_{sym}^{(g)}$$

$$(\mathbf{u}, \mathbf{v}, \mathbf{d}) \mapsto A,$$

where A is a matrix of the following form:

$$A = \begin{bmatrix} d_1 & u_1 v_1 & u_2 v_1 & u_3 v_1 & \cdots & & u_{n-1} v_1 \\ u_1 v_1 & d_2 & u_2 v_2 & u_3 v_2 & \cdots & & u_{n-1} v_2 \\ u_2 v_1 & u_2 v_2 & d_3 & u_3 v_2 & \cdots & & u_{n-1} v_3 \\ u_3 v_1 & u_3 v_2 & u_3 v_3 & \ddots & & & \vdots \\ \vdots & \vdots & \vdots & & d_{n-1} & u_{n-1} v_{n-1} \\ u_{n-1} v_1 & u_{n-1} v_2 & u_{n-1} v_3 & \cdots & u_{n-1} v_{n-1} & d_n \end{bmatrix}.$$

This representation uses $3n - 3$ parameters in case of a suitable normalization.

Taking a closer look at the class of matrices $\mathcal{S}_{sym}^{(g,d)}$ and $\mathcal{Q}_{sym}^{(g)}$ and Proposition 1.19, one sees that both these classes are in fact the same, as any matrix from $\mathcal{Q}_{sym}^{(g)}$, can be written as a symmetric semiseparable plus diagonal matrix, due to the specific structure imposed on the strictly lower and strictly upper triangular part. Readers familiar with semiseparable and related matrices will surely recognize the latter matrices. Often the representation of these matrices is just taken as their definition.

Notes and references

In the literature a lot of attention is being paid to the class of generator representable semiseparable matrices (see also the notes and references for Section 1.1 and Section 1.5), in particular, as they can be considered as the inverses of irreducible tridiagonal matrices.

☞ S. O. Asplund. Finite boundary value problems solved by Green's matrix. *Mathematica Scandinavica*, 7:49–56, 1959.

In this paper, S. O. Asplund, the father of E. Asplund, proves the same as Gantmacher and Kreĭn, by calculating the inverse via techniques for solving finite boundary value problems. A brief remark states that higher order band matrices have as inverses higher order Green's matrices. (The Green's matrix can be considered as a generator representable semiseparable matrix.) Higher order Green's (semiseparable) matrices are the subject of Part III. Also some theoretical results concerning inverses of nonsymmetric tridiagonal (not necessarily irreducible) matrices are included.

Also the closely related classes such as the generator representable semiseparable plus diagonal matrices have been studied.

☞ S. N. Roy, B. G. Greenberg, and A. E. Sarhan. Evaluation of determinants, characteristic equations and their roots for a class of patterned matrices. *Journal of the Royal Statistical Society. Series B. Statistical Methodology*, 22:348–359, 1960.

☞ S. N. Roy and A. E. Sarhan. On inverting a class of patterned matrices. *Biometrika*, 43:227–231, 1956.

In their 1956 paper, Roy and Sarhan invert very specific matrices arising in statistical applications, e.g., a lower triangular semiseparable matrix and several types of symmetric generator representable semiseparable (plus diagonal) matrices. As the statistical research in 1960 was interested in fast calculations for so-called patterned matrices, Roy, Greenberg and Sarhan designed an order n algorithm for calculating the determinant. The patterned matrices are of symmetric generator representable semiseparable and generator representable semiseparable plus diagonal form.

☞ F. Valvi. Explicit presentation of the inverses of some types of matrices. *Journal of the Institute of Mathematics and its Applications*, 19(1):107–117, 1977.

Valvi determines several explicit formulas for inverting specific patterned matrices arising in statistical applications. These matrices are specific types of symmetric generator representable semiseparable (plus diagonal matrices). Also one persymmetric example is considered.

☞ K. C. Mustafi. The inverse of a certain matrix with an application. *Annals of Mathematical Statistics*, 38:1289–1292, 1967.

Mustafi investigates the inverse of a symmetric generator representable plus identity matrix A of size $n \times n$, which he represents in the following way:

$$A = I + \sum_k = 1^n c_k W_k,$$

where W_k is an $n \times n$ matrix for which the upper left $k \times k$ block contains elements equal to 1, and the remaining elements are zero. Moreover, the author concludes that the inverse of this matrix is of the same form. The matrix results from a statistical problem.

☞ V. R. R. Uppuluri and J. A. Carpenter. An inversion method for band matrices. *Journal of Mathematical Analysis and Applications*, 31:554–558, 1970.

Uppuluri and Carpenter calculate the inverse of a tridiagonal Toeplitz matrix (not necessarily symmetric) via the associated difference equation. Moreover they formulate it in a theorem that, if a symmetric matrix is of generator representable form, its inverse necessarily needs to be tridiagonal. A little more information can be found in Section 14.1 in Chapter 14 on inverting banded Toeplitz matrices.

☞ J. Baranger and M. Duc-Jacquet. Matrices tridiagonales symétriques et matrices factorisables. *Revue Française d' Automatique, Informatique et de Recherce Opérationelle*, 5(R-3):61–66, 1971. (In French).

Baranger and Duc-Jacquet prove that the inverse of a symmetric generator representable semiseparable matrix (called "une matrice factorisable" in the paper) is a tridiagonal matrix. They explicitly calculate the inverse of a generator representable semiseparable matrix.

☞ B. Bukhberger and G. A. Emel'yanenko. Methods of inverting tridiagonal matrices. *Computational Mathematics and Mathematical Physics (translated from Zhurnal Vychislitel'noĭ Matematiki i Matematicheskoĭ Fiziki)*, 13:546–554, 1973.

In this paper Bukhberger and Emel'yanenko present a computational method for inverting symmetric tridiagonal matrices, for which all the elements are different from zero. The theoretical results are applicable in a physical application which studies the motion of charged particles in a particular environment. The results are identical to the ones in the book of Gantmacher and Kreĭn.

☞ D. Fasino and L. Gemignani. Direct and inverse eigenvalue problems, for diagonal-plus-semiseparable matrices. *Numerical Algorithms*, 34:313–324, 2003.

☞ M. Van Barel, D. Fasino, L. Gemignani, and N. Mastronardi. Orthogonal rational functions and diagonal plus semiseparable matrices. In F. T. Luk, editor, *Advanced Signal Processing Algorithms, Architectures, and Implementations XII*, volume 4791 of *Proceedings of SPIE*, pages 167–170, 2002.

In their paper Fasino and Gemignani (2003) study the direct and the inverse eigenvalue problem of diagonal plus semiseparable matrices. In the 2002 paper, the authors investigate the relation between orthogonal rational functions and symmetric (hermitian) generator representable semiseparable plus diagonal matrices. A specific orthonormal basis is searched for. The problem is solved by solving an inverse eigenvalue problem involving the symmetric (hermitian) generator representable semiseparable plus diagonal matrices. Mastronardi et al. present in [211] a divide-and-conquer algorithm to compute the eigendecomposition of symmetric generator representable semiseparable plus diagonal matrices. Chandrasekaran and Gu also present in [57] a divide-and-conquer method to calculate the eigendecomposition of a symmetric generator representable semiseparable plus a block diagonal matrix. For more information, see also Section 3.5 in Chapter 3.

In [104] Fasino proves that any hermitian matrix with pairwise distinct eigenvalues can be transformed into a similar hermitian generator representable semiseparable plus diagonal matrix. More information can be found in Section 3.5 in Chapter 3.

Bevilacqua and Del Corso in [26] investigate the existence and the uniqueness of a unitary similarity transformation of a symmetric matrix into a symmetric generator representable semiseparable matrix, based on [282]. See also Section 3.5 in Chapter 3.

In [54] Chandrasekaran and Gu present an algorithm to transform a generator representable quasiseparable matrix plus a band matrix into a similar tridiagonal matrix. This reduction is then used for computing the eigenvalues of the original matrix. More information can be found in Section 3.5 in Chapter 3.

2.3 The symmetric diagonal-subdiagonal representation

The next representation map we will investigate is based on the diagonal and subdiagonal of semiseparable matrices. Let $\mathbf{d}^{(s)}$ be the subdiagonal of a symmetric semiseparable matrix and $\mathbf{d}$ its diagonal. Which subclass can be represented by these two vectors? The elements up to the diagonal in rows of a semiseparable matrix are dependent on each other. Therefore, the diagonal and the subdiagonal elements contain all information about the semiseparable matrix in case all diagonal and subdiagonal elements are different from zero. More details about the information that is stored in the diagonal and subdiagonal can be found in [112, 118]. In the last paragraphs of this section we will briefly mention the most important properties connected with this representation. Firstly, we will focus attention on the semiseparable case, and secondly, towards the quasiseparables. We do not go into more detail as this representation is not widely used in the literature.

2.3.1   The representation for symmetric semiseparables

When all the $d_i^{(s)}$ and d_i are different from zero, we can construct a corresponding symmetric semiseparable matrix in the following way. Already known is the first 2×2 block. Note that because of symmetry we only construct the lower triangular semiseparable part. The upper left 2×2 block looks like

$$S(1:2, 1:2) = \left[\begin{array}{cc} d_1 & \\ d_1^{(s)} & d_2 \end{array} \right].$$

The upper 3×3 block is as follows

$$S(1:3, 1:3) = \left[\begin{array}{ccc} d_1 & & \\ d_1^{(s)} & d_2 & \\ s_{21} & d_2^{(s)} & d_3 \end{array} \right].$$

The issue now is how to compute the unknown element s_{21}. By definition, all submatrices in the lower triangular part of S have maximum rank 1. Hence, the following equation has to be satisfied:

$$\frac{s_{21}}{d_1^{(s)}} = \frac{d_2^{(s)}}{d_2}.$$

The equation is easily solved for s_{21}. Continuing this procedure gives us the following 4×4 matrix. Note, however, that it is essential that all diagonal and subdiagonal elements are different from zero.

$$S(1:4, 1:4) = \left[\begin{array}{cccc} d_1 & & & \\ d_1^{(s)} & d_2 & & \\ s_{21} & d_2^{(s)} & d_3 & \\ s_{31} & s_{32} & d_3^{(s)} & d_4 \end{array} \right].$$

Using the equations:

$$\frac{s_{31}}{s_{21}} = \frac{s_{32}}{d_2^{(s)}} = \frac{d_3^{(s)}}{d_3},$$

the unknown elements s_{31} and s_{32} can be found. This process can be repeated to complete the lower triangular part of the semiseparable matrix S. It is clear that the resulting symmetric semiseparable matrix is unique. The assumption that all the diagonal and subdiagonal elements have to be nonzero is quite strong, however, and cannot be guaranteed in general[16]. When zeros occur on the diagonal and/or the subdiagonal it is possible that there are different symmetric semiseparable matrices having the same diagonal and subdiagonal. Therefore, we have to make choices, such that our map will point to only one semiseparable matrix. We distinguish different cases.

- $d_i^{(s)} = 0$. Then all the elements s_{ij} with $1 \leq j < i$ can be chosen equal to zero such that the assumptions about the rank 1 blocks are still satisfied. (Later on we will show that there are also other possibilities.)

- $d_i = 0$ and $d_i^{(s)} \neq 0$. Because of this special situation one can check that the element $d_{i-1}^{(s)}$ has to be zero, and therefore, because of the assumption above, the complete row i equals zero. Because all the elements in the row $i + 1$ (except for the subdiagonal and diagonal) can take various values now, and still satisfy the semiseparable structure, we assume that all the elements in this row (except the subdiagonal and the diagonal) are zero.

Using the construction explained above, the following map can be defined:

$$r_{\mathcal{S}_{sym}^{(s)}} : \mathbb{R}^{n-1} \times \mathbb{R}^n \to \mathcal{S}_{sym}$$

$$(\mathbf{d}^{(s)}, \mathbf{d}) \mapsto S$$

with S the matrix as constructed above. We will now investigate if this map is a representation map. For the surjectivity of this map, one can expect problems, because in the preceding lines we already had to make distinctions between the different types of matrices. This can also be illustrated by the following example:

Example 2.6 For each of the following two symmetric semiseparable matrices there does not exist $\mathbf{d}^{(s)}$ and $\mathbf{d}$ such that $r_{\mathcal{S}_{sym}^{(s)}}(\mathbf{d}^{(s)}, \mathbf{d})$ equals the given matrix:

$$S_1 = \begin{bmatrix} 0 & 1 & 1 \\ 1 & 0 & 0 \\ 1 & 0 & 1 \end{bmatrix}, \qquad S_2 = \begin{bmatrix} 1 & 1 & 0 & 1 \\ 1 & 1 & 0 & 1 \\ 0 & 0 & 0 & 1 \\ 1 & 1 & 1 & 1 \end{bmatrix}.$$

The matrices with the same subdiagonal and diagonal that are constructed by applying $r_{\mathcal{S}_{sym}^{(s)}}(\mathbf{d}^{(s)}, \mathbf{d})$ are as follows:

$$S_1 = \begin{bmatrix} 0 & 1 & 0 \\ 1 & 0 & 0 \\ 0 & 0 & 1 \end{bmatrix}, \qquad S_2 = \begin{bmatrix} 1 & 1 & 0 & 0 \\ 1 & 1 & 0 & 0 \\ 0 & 0 & 0 & 1 \\ 0 & 0 & 1 & 1 \end{bmatrix}.$$

∎

---

[16]In Chapter 7 on inversion, however, it will be shown that this representation is useful for inverting, for example, lower semiseparable matrices.

Because of this nonsurjective behavior of the map, an adaptation of the target set is needed. Define the set of matrices $\mathcal{S}_{sym}^{(s)}$ [17] as

$$\mathcal{S}_{sym}^{(s)} = r_{\mathcal{S}_{sym}^{(s)}} (\mathbb{R}^{n-1} \times \mathbb{R}^n) = \{A \in \mathcal{S}|A \text{ can be represented by } \mathbf{d} \text{ and } \mathbf{d}^{(s)}\}.$$

Let us now redefine the map in the following sense:

$$r_{\mathcal{S}_{sym}^{(s)}} : \mathbb{R}^{n-1} \times \mathbb{R}^n \to \mathcal{S}_{sym}^{(s)}$$

$$(\mathbf{d}^{(s)}, \mathbf{d}) \mapsto S.$$

This map is a representation map for the new class, because surjectivity is now by construction satisfied, and the so-called inverse $s_{\mathcal{S}_{sym}^{(s)}}$ is defined in a natural way, by projecting the diagonal and the subdiagonal of the matrix S. It can clearly be seen that this map satisfies the wanted properties. One important question remains uninvestigated. Is the set $\mathcal{S}_{sym}^{(s)}$ closed for the pointwise convergence? Unfortunately this is not the case: consider for example the matrix (with $\epsilon \neq 0$)

$$\begin{bmatrix} 1 & 1 & \epsilon & 1 \\ 1 & 1 & \epsilon & 1 \\ \epsilon & \epsilon & \epsilon & 1 \\ 1 & 1 & 1 & 1 \end{bmatrix}.$$

This matrix can clearly be represented by the diagonal subdiagonal representation, but the limit of this matrix:

$$\lim_{\epsilon \to 0} \begin{bmatrix} 1 & 1 & \epsilon & 1 \\ 1 & 1 & \epsilon & 1 \\ \epsilon & \epsilon & \epsilon & 1 \\ 1 & 1 & 1 & 1 \end{bmatrix} = \begin{bmatrix} 1 & 1 & 0 & 1 \\ 1 & 1 & 0 & 1 \\ 0 & 0 & 0 & 1 \\ 1 & 1 & 1 & 1 \end{bmatrix}$$

cannot be represented by the diagonal and subdiagonal representation.

Some might find this choice of representation somewhat arbitrary, but the theorems in [112, 118, 133] provide theoretical results indicating that this class of diagonal subdiagonal representable matrices has very interesting properties. The main reason for including this type of representation in the book is the fact that this class has been the subject of a lot of research, and in Chapter 7 the results presented here will come in handy for inverting specific matrices.

As we will not use this type of representation extensively throughout the remainder of the book, we do not consider here possible applications to other types of classes of matrices. Only some important theorems of this representation are presented in their original formulation, with regard to quasiseparable matrices in the next section. The interested reader can find more information in the notes and references following this section.

---

[17]In fact this class is equally important as the class of generator representable semiseparable matrices, as they are defined as the image of a specific representation map.

2.3.2 The representation for symmetric quasiseparables

In this section we will present some interesting results covered in [112, 118, 133]. The matrices in these papers are called basic matrices or generalized Hessenberg matrices. A basic matrix is essentially a nonsymmetric quasiseparable matrix, whereas a generalized Hessenberg matrix is a matrix for which the lower triangular part is of quasiseparable form.

As the low rank part for quasiseparable matrices starts strictly below the diagonal it is logical that we do not store the diagonal and the subdiagonal but the subdiagonal and the second subdiagonal. As already indicated in the previous section, this representation is not capable of representing all matrices having a lower triangular part of quasiseparable form. In the manuscripts [112, 118, 133] a subclass is considered, called the class of complete basic matrices or the class of complete generalized Hessenberg matrices. The prefix 'complete' indicates that all the subdiagonal elements (also the superdiagonal in case of a basic matrix) are different from zero. In this way the structure of the lower (and upper for basic matrices) part of the matrices is uniquely determined.

In the remainder of this section we present some of the theoretical results. We only illustrate them for the lower triangular part.

Theorem 2.7. *Suppose A is a matrix, for which the lower triangular part is of quasiseparable form and the subdiagonal elements are different from zero. The strictly lower triangular part of A is uniquely determined by the first two subdiagonals. More explicitly, if $k - i > 2$, then*

$$a_{ki} = a_{k,k-2}\, a_{k-1,k-2}^{-1}\, a_{k-1,k-3}\, a_{k-2,k-3}^{-1} \cdots a_{i+2,i}. \tag{2.2}$$

Conversely, if the previous statement is true for a certain matrix A with the subdiagonal elements different from zero, then this matrix has the lower triangular structure of quasiseparable form.

Important to remark is the fact that, when recalculating an element in fact, a zigzag path is followed through the matrix as illustrated in the following example:

Example 2.8 Suppose we have a 6×6 matrix which has the lower triangular part of quasiseparable form. Then we have the following equations for the elements:

$$a_{51} = a_{53}\, a_{43}^{-1}\, a_{42}\, a_{32}^{-1}\, a_{31}$$
$$a_{62} = a_{64}\, a_{54}^{-1}\, a_{53}\, a_{43}^{-1}\, a_{42}$$
$$a_{61} = a_{64}\, a_{54}^{-1}\, a_{53}\, a_{43}^{-1}\, a_{42}\, a_{32}^{-1}\, a_{31}.$$

A zigzag path is followed through the matrix, starting at the lower right ❶, going through the elements ❷, ❸, ... and arriving at the upper left ❺ in order to compute the element $\otimes$ as can be seen in the figures below (for computing a_{51} and a_{61}):

$$
\begin{bmatrix}
\times & & & & & \\
\times & \times & & & & \\
\circledS & \circledA & \times & & & \\
\times & \circled3 & \circled2 & \times & & \\
\otimes & \times & \circled1 & \times & \times & \\
\times & \times & \times & \times & \times & \times
\end{bmatrix}
\qquad
\begin{bmatrix}
\times & & & & & \\
\times & \times & & & & \\
\times & \times & \times & & & \\
\times & \circledS & \circledA & \times & & \\
\times & \times & \circled3 & \circled2 & \times & \\
\times & \otimes & \times & \circled1 & \times & \times
\end{bmatrix}
$$

Also for computing a_{61} (marked with $\otimes$) a zigzag path is followed, starting at ❶ and ending at ❼:

$$
\begin{bmatrix}
\times & & & & & \\
\times & \times & & & & \\
❼ & ❻ & \times & & & \\
\times & ❺ & ❹ & \times & & \\
\times & \times & ❸ & ❷ & \times & \\
\otimes & \times & \times & ❶ & \times & \times
\end{bmatrix}
$$

It is clear from the previous example that for a certain element from the second subdiagonal $a_{k+1,k-1}$ equal to zero, the complete submatrix $A(N\backslash N_k; N_{k-1})$ will be zero for $N = \{1, 2, \ldots, n\}$ and $N_k = \{1, 2, \ldots, k\}$.

The next theorem gives an explicit inversion formula for matrices having their lower triangular part of quasiseparable form. This theorem can be very useful for designing inversion algorithms for this type of matrix. Moreover, when factorizing, for example, a quasiseparable matrix into the LU decomposition, both of the factors will be lower or upper triangular matrices having the quasiseparable structure.

Theorem 2.9 (Theorem 2.7 in [118]). *Suppose A is an $n \times n$ nonsingular matrix with the lower triangular part of quasiseparable form and subdiagonal elements different from zero. Then its inverse B will also have the lower triangular part of quasiseparable form with the subdiagonal entries different from zero*

$$
\begin{aligned}
b_{ii} &= a_{ii}^{-1}, & i &= 1, \ldots, n \\
b_{i+1,i} &= -a_{i+1,i+1}^{-1}\, a_{i+1,i}\, a_{ii}^{-1} & i &= 1, \ldots, n-1 \\
b_{i+1,i-1} &= a_{i+1,i+1}^{-1}\left(a_{i+1,i}\, a_{ii}^{-1}\, a_{i,i-1} - a_{i+1,i-1}\right) a_{i-1,i-1}^{-1} & i &= 2, \ldots, n-2
\end{aligned}
$$

The other elements can be calculated by using Theorem 2.7.

It is clear that the factor $\left(a_{i+1,i}\, a_{ii}^{-1}\, a_{i,i-1} - a_{i+1,i-1}\right)$ determines the zero in the position $b_{i+1,i-1}$ depending on the rank of the block $A(i : i+1, i-1 : i)$.

Theorem 2.10 (Corollary 2.2 in [112]). *Suppose A is an $n \times n$ nonsingular matrix with the lower triangular part of quasiseparable form and all the subdiagonal elements different from zero. If $A = LU$ is the LU-decomposition of A, the matrix L is also a matrix with the lower triangular part of quasiseparable form and the subdiagonal elements different from zero.*

Theorem 2.10 can easily be generalized to the upper triangular part of quasiseparable form, such that the U factor also inherits this structure and nonzero superdiagonal entries.

One can clearly see that a combination of Theorems 2.10 and 2.9 already presents a method for inverting a large class of matrices, for example, semiseparable, semiseparable plus diagonal and tridiagonal matrices, for which of course certain elements have to be different from zero.

Notes and references

As mentioned in the beginning of this section, the representation of structured rank matrices, based on the diagonal and the subdiagonal, is not so widely used. The representation is mainly investigated by Fiedler and coauthors. Important theorems related to the diagonal and subdiagonal representation can be found in the following manuscripts.

In the papers of [112, 2003] and [118, 2004], Fiedler investigates the properties of so-called basic matrices and complete basic matrices. Theorems as presented in this section concerning the representation, LU decompositions, factorizations of these matrices and inversion methods are presented. Also the class of extended basic matrices and generalized Hessenberg matrices is considered. A matrix has the extended structure, if some of the diagonal elements can be included in the low rank structure below the diagonal. This means in fact that for these diagonal elements the matrix has the semiseparable form. This is sort of a hybrid form between quasiseparable and semiseparable. Theorems are provided concerning the structure of the inverse. This representation is closely related to the LU-factorization, more precisely a product of elementary Gaussian transformation matrices gives rise to a kind of diagonal-subdiagonal representation. Hence more information on these matrices, with regard to the diagonal-subdiagonal representation and the link with the LU-factorization can be found in Section 4.6 in Chapter 4.

☞ M. Fiedler. Complementary basic matrices. *Linear Algebra and its Applications*, 384:199–206, 2004.

The class of complementary basic matrices (2004) can be considered as a special subclass of the class of basic matrices; these matrices satisfy a certain structure condition. The structure condition can be considered as 'antisymmetric' in some sense. This means that the sets of diagonal elements (from 2 to $n-1$) includable in the lower triangular structure are complementary to the diagonal elements includable in the upper triangular structure.

The manuscripts [111, 114], consider the nullity theorem and some properties of matrices, related to the diagonal-subdiagonal representation. (These manuscripts were already discussed in Chapter 1, Section 1.5.)

2.4   The symmetric Givens-vector representation

The Givens-vector representation for semiseparable matrices was recently proposed in [295]. We will show that this representation is capable of representing the complete class of semiseparable matrices by a sequence of $n-1$ Givens transformations and a vector. This section is divided in different subsections. Firstly, an illustration is presented, showing the construction of a semiseparable matrix, given the Givens-vector representation. Secondly, some examples illustrate the use of this representation. Thirdly, a way is presented to retrieve the Givens-vector representation in a numerically stable way and finally we briefly mention the two variants of the representation and possible applications to other classes of matrices.

2.4.1 The representation for symmetric semiseparables

We consider this representation for symmetric semiseparable matrices of dimension n. This representation consists of $n-1$ Givens transformations and a vector of length n.

A Givens transformation is an orthogonal 2×2 matrix of the following form

$$G = \begin{bmatrix} c & -s \\ s & c \end{bmatrix},$$

where $c^2 + s^2 = 1$. In fact c and s are, respectively, the cosine and sine of a certain angle. Suppose a vector $[x, y]^T$ is given, then there exists a Givens transformation such that

$$G \begin{bmatrix} x \\ y \end{bmatrix} = \begin{bmatrix} r \\ 0 \end{bmatrix},$$

with $r = \sqrt{x^2 + y^2}$. More information on how to compute Givens transformations can, e.g., be found in [152].

Let us denote the Givens transformations and the vector used in the representation of the semiseparable matrix as $G = [G_1, \ldots, G_{n-1}]$ and as $\mathbf{v}^T = [v_1, \ldots, v_n]$. It is clear that this representation needs $2n - 1$ parameters to reconstruct the complete semiseparable matrix, as a Givens transformation can be stored in one parameter as described in [152, Section 5.1.11]. Here however, we chose to store the cosine and the sine of the Givens transformation. As shown later on, this choice often reduces the computational complexity in algorithms.

The following figures denote how a semiseparable matrix can be reconstructed by using this information. How to retrieve the representation given a matrix is the subject of Section 2.4.3. The elements denoted by $\boxtimes$ make up the semiseparable part of the matrix. Initially one starts on the first 2 rows of the matrix. The element v_1 is placed in the upper left position, then the Givens transformation G_1 is applied, and finally to complete the first step, element v_2 is added in position $(2, 2)$. Only the first two columns and rows are shown here:

$$\begin{bmatrix} v_1 & 0 \\ 0 & 0 \end{bmatrix} \rightarrow G_1 \begin{bmatrix} v_1 & 0 \\ 0 & 0 \end{bmatrix} + \begin{bmatrix} 0 & 0 \\ 0 & v_2 \end{bmatrix} \rightarrow \begin{bmatrix} \boxtimes & 0 \\ \boxtimes & v_2 \end{bmatrix}.$$

The second step consists of applying the Givens transformation G_2 on the second and the third row, furthermore v_3 is added in position $(3, 3)$. Here only the first three columns are shown and the second and third row. This leads to:

$$\begin{bmatrix} \boxtimes & v_2 & 0 \\ 0 & 0 & 0 \end{bmatrix} \rightarrow G_2 \begin{bmatrix} \boxtimes & v_2 & 0 \\ 0 & 0 & 0 \end{bmatrix} + \begin{bmatrix} 0 & 0 & 0 \\ 0 & 0 & v_3 \end{bmatrix} \rightarrow \begin{bmatrix} \boxtimes & \boxtimes & 0 \\ \boxtimes & \boxtimes & v_3 \end{bmatrix}.$$

This process can be repeated by applying the Givens transformation G_3 on the third and the fourth row of the matrix, and afterwards adding the diagonal element v_4.

After applying all the Givens transformations and adding all the diagonal elements, the lower triangular part of a symmetric semiseparable matrix is constructed. (It is clear, by construction, that the constraints posed on a semiseparable matrix are naturally satisfied in this case.) Because of the symmetry the upper triangular part is also known.

When denoting a Givens transformation G_l as:

$$G_l = \left[\begin{array}{cc} c_l & -s_l \\ s_l & c_l \end{array} \right],$$

the elements $S(i, j) = s_{ij}$ are calculated in the following way:

$$\left\{ \begin{array}{ll} S(i, j) = s_{ij} = c_i s_{i-1} s_{i-2} \cdots s_j v_j & n > i \geq j, \\ S(n, j) = s_{nj} = s_{n-1} s_{n-2} \cdots s_j v_j & i = n. \end{array} \right.$$

The elements in the upper triangular part can be calculated similarly due the symmetry. The elements of the semiseparable matrix can therefore be calculated in a stable way based on the Givens-vector representation. This means that we have constructed the following map $r_{\mathcal{S}_{sym}}$

$$r_{\mathcal{S}_{sym}} : \mathbb{R}^{2 \times (n-1)} \times \mathbb{R}^n \to \mathcal{S}_{sym} \tag{2.3}$$

$$\left[\left[\begin{array}{ccc} c_1 & \cdots & c_{n-1} \\ s_1 & \cdots & s_{n-1} \end{array} \right], [v_1, \ldots, v_n]^T \right] \mapsto \left[\begin{array}{cccc} c_1 v_1 & & & \\ c_2 s_1 v_1 & c_2 v_2 & & \\ c_3 s_2 s_1 v_1 & c_3 s_2 v_2 & c_3 v_3 & \\ \vdots & \vdots & & \ddots \end{array} \right].$$

The storage costs $3n - 2$. We store the cosine and sine separately for numerical efficiency. Theoretically, only storing the cosine (or sine or tangent) would be enough leading to a storage cost of $2n - 1$. As can be seen from the structure of a matrix represented with this Givens-vector representation, the Givens transformations store the dependencies between the rows, while the vector contains a weight posed on each column. Hence this representation is also referred to as the Givens-weight representation [72].

Example 2.11 (Example 1.9 continued) The Givens-vector representation for the matrix of Example 1.9 is the following: (In the first row of G the elements $c_1, \ldots, c_4$ are stored and in the second row, the elements $s_1, \ldots, s_4$.)

$$G = \left[\begin{array}{cccc} 0.9534 & 0.9878 & 1.0000 & 1.0000 \\ -0.3017 & 0.1558 & 0.0008 & 0.0000 \end{array} \right]$$

and

$$\mathbf{v}^T = \left[\begin{array}{ccccc} 1.7049 & 1.4435 & 2.9485 & 9.9999 \cdot 10 & 1.0000 \cdot 10^4 \end{array} \right].$$

All the elements of the semiseparable matrix can be reconstructed now with high relative precision if the corresponding elements of G and $\mathbf{v}$ are known with high relative precision. In this example the maximum absolute error between the original semiseparable matrix and the semiseparable matrix represented with the Givens-vector representation is of the order 10^{-14}. For the generator representation we obtained elements which had only 2 significant digits left. ∎

In this section the construction of a symmetric semiseparable matrix given a sequence of Givens transformations and a vector was presented. Before developing a method for computing the representation given a symmetric semiseparable matrix, we will give some illustrative examples.

2.4.2   Examples

In the following examples the Givens-vector representation of several different symmetric semiseparable matrices is given. The construction of this representation is the subject of the next section.

$$
\begin{bmatrix} 1 & 1 & 1 \\ 1 & 1 & 1 \\ 1 & 1 & 1 \end{bmatrix} \leftrightarrow \begin{array}{l} G = \begin{bmatrix} \sqrt{3}/3 & \sqrt{2}/2 \\ \sqrt{6}/3 & \sqrt{2}/2 \end{bmatrix} \\ \mathbf{v} = \left[\sqrt{3}, \sqrt{2}, 1\right]^T \end{array}
$$

$$
\begin{bmatrix} 1 & 0 & 0 \\ 0 & 2 & 0 \\ 0 & 0 & 3 \end{bmatrix} \leftrightarrow \begin{array}{l} G = \begin{bmatrix} 1 & 1 \\ 0 & 0 \end{bmatrix} \\ \mathbf{v} = [1, 2, 3]^T \end{array}
$$

$$
\begin{bmatrix} 1 & 0 & 1 \\ 0 & 0 & 1 \\ 1 & 1 & 1 \end{bmatrix} \leftrightarrow \begin{array}{l} G = \begin{bmatrix} \sqrt{2}/2 & 0 \\ \sqrt{2}/2 & 1 \end{bmatrix} \\ \mathbf{v} = \left[\sqrt{2}, 1, 1\right]^T \end{array}
$$

$$
\begin{bmatrix} 1 & 1 & 1 \\ 1 & 0 & 0 \\ 1 & 0 & 1 \end{bmatrix} \leftrightarrow \begin{array}{l} G = \begin{bmatrix} \sqrt{3}/3 & \sqrt{2}/2 \\ \sqrt{6}/3 & \sqrt{2}/2 \end{bmatrix} \\ \mathbf{v} = \left[\sqrt{3}, 0, 1\right]^T \end{array}
$$

$$
\begin{bmatrix} 1 & 0 & 1 \\ 0 & 0 & 0 \\ 1 & 0 & 1 \end{bmatrix} \leftrightarrow \begin{array}{l} G = \begin{bmatrix} \sqrt{2}/2 & 0 \\ \sqrt{2}/2 & 1 \end{bmatrix} \\ \mathbf{v} = \left[\sqrt{2}, 0, 1\right]^T \end{array}
$$

$$
\begin{bmatrix} 0 & 1 & 0 \\ 1 & 0 & 0 \\ 0 & 0 & 1 \end{bmatrix} \leftrightarrow \begin{array}{l} G = \begin{bmatrix} 0 & 1 \\ 1 & 0 \end{bmatrix} \\ \mathbf{v} = [1, 0, 1]^T \end{array}
$$

$$
\begin{bmatrix} 1 & 1 & 0 & 1 & 1 \\ 1 & 0 & 0 & 0 & 0 \\ 0 & 0 & 0 & 1 & 1 \\ 1 & 0 & 1 & 1 & 1 \\ 1 & 0 & 1 & 1 & 1 \end{bmatrix} \leftrightarrow \begin{array}{l} G = \begin{bmatrix} 1/2 & \sqrt{3}/3 & 0 & \sqrt{2}/2 \\ \sqrt{3}/2 & \sqrt{6}/3 & 1 & \sqrt{2}/2 \end{bmatrix} \\ \mathbf{v} = \left[2, 0, \sqrt{2}, \sqrt{2}, 1\right]^T \end{array}
$$

In the following subsection we will derive a method for retrieving the Givens-vector representation in a stable manner.

2.4.3   Retrieving the Givens-vector representation

In the following example, we construct the Givens-vector representation of a symmetric semiseparable matrix based on the diagonal and subdiagonal elements. In fact we already know that this naïve procedure cannot work in all cases, because the diagonal-subdiagonal representable matrices do not cover the complete set of symmetric semiseparable matrices. The following example shows that the procedure just explained to determine the representation is not stable.

Example 2.12 Suppose we have a given symmetric semiseparable matrix,

$$S = \begin{bmatrix} 1 & 1 & 0 \\ 1 & 0 & 0 \\ 0 & 0 & 1 \end{bmatrix}$$

and we add random noise of the size of the machine precision 10^{-16} to this matrix. Numerically this matrix is still symmetric semiseparable. We will then construct the Givens-vector representation of this matrix, based on the diagonal and subdiagonal elements, and build up again the semiseparable matrix given the Givens-vector representation. We see that we get large errors, even though there has been only a small change in the elements of the order 10^{-16}. The disturbed matrix A has the following form:

$$\begin{bmatrix} 1.0000 & 1.0000 & 4.0570 \cdot 10^{-16} \\ 1.0000 & 7.3820 \cdot 10^{-16} & 9.3546 \cdot 10^{-16} \\ 7.9193 \cdot 10^{-16} & 1.7626 \cdot 10^{-16} & 1.0000 \end{bmatrix}.$$

Calculating the Givens-vector representation in the way presented above and then recalculating the semiseparable matrix gives the following result:

$$\begin{bmatrix} 1.0000 & 1.0000 & 2.3877 \cdot 10^{-1} \\ 1.0000 & 7.3820 \cdot 10^{-16} & 1.7626 \cdot 10^{-16} \\ 2.3877 \cdot 10^{-1} & 1.7626 \cdot 10^{-16} & 1.0000 \end{bmatrix}.$$

This matrix has the element in position $(3, 1)$ recovered in a very inaccurate way. In the next section a more robust and stable algorithm will be presented, and we will reconsider this example. ∎

Next, a method is proposed to retrieve the Givens-vector representation of a symmetric semiseparable matrix in a stable way.

In fact we search for the map:

$$ss_{\mathcal{S}_{sym}} : \mathcal{S}_{sym} \to \mathbb{R}^{2 \times (n-1)} \times \mathbb{R}^n$$
$$S \mapsto (G, \mathbf{v}).$$

Suppose we have a symmetric semiseparable matrix as in (2.3). The vector elements v_i can be retrieved rather easily from the matrix. In fact:

$$\|S(i:n,i)\|_2$$
$$= \sqrt{(c_i v_i)^2 + (c_{i+1} s_i v_i)^2 + \cdots + (c_{n-1} s_{n-2} \ldots s_i v_i)^2 + (s_{n-1} s_{n-2} \ldots s_i v_i)^2}$$
$$= \sqrt{(c_i v_i)^2 + (c_{i+1} s_i v_i)^2 + \cdots + (s_{n-1}^2 + c_{n-1}^2)(s_{n-2} \ldots s_i v_i)^2}$$
$$= \sqrt{(c_i^2 + s_i^2) v_i^2}$$
$$= |v_i|.$$

This means that the absolute value of v_i can be calculated by calculating the norm $\|S(i:n,i)\|_2$. All diagonal elements are determined now, except their signs, this is done at the same time as the calculation of the Givens transformations. To calculate the corresponding Givens transformations connected with the matrix in a robust way, we first map the matrix S towards another symmetric semiseparable matrix. This procedure is quite expensive, but results in a stable way to compute the Givens-vector representation.[18] The matrix S is mapped onto the following matrix of norms. Note that the choice of the norm does not play a role:

$$\hat{S} = \begin{bmatrix} \|S(1,1)\| & & & \\ \|S(2,1)\| & \|S(2,1:2)\| & & \\ \vdots & & \ddots & \\ \|S(n,1)\| & \|S(n,1:2)\| & \ldots \|S(n,1:n)\| \end{bmatrix}.$$

So in fact a new matrix $\hat{S}$ is created with elements $\hat{S}_{i,j} = \|S(i,1:j)\|$. Note that this matrix has the same dependencies between the rows as the matrix S (except for the signs). We start calculating the last Givens transformation G_{n-1} such that

$$G_{n-1}\left[r_{n-1} \ , \ 0\right]^T = \left[\hat{S}_{n-1,n-1} \ , \ \hat{S}_{n,n-1}\right]^T.$$

Before calculating the next Givens transformation we have to update the matrix $\hat{S}$ by applying the Givens transformation G_{n-1} to the rows $n-1$ and n. Denoting this new matrix as $\hat{S}^{(n-1)}$, the next Givens transformation G_{n-2} is calculated such that

$$G_{n-2}\left[r_{n-2} \ , \ 0\right]^T = \left[\hat{S}_{n-2,n-2} \ , \ \hat{S}^{(n-1)}_{n-1,n-2}\right]^T = \left[\hat{S}^{(n-1)}_{n-2,n-2} \ , \ \hat{S}^{(n-1)}_{n-1,n-2}\right]^T.$$

Updating again the matrix $\hat{S}^{(n-1)}$ by applying the Givens transformation G^T_{n-2} to the rows $n-2$ and $n-1$ we get the matrix $\hat{S}^{(n-2)}$ and we can calculate G_{n-3}. Consecutively, all the Givens transformations can be calculated, satisfying:

$$G_i\left[r_i \ , \ 0\right]^T = \left[\hat{S}_{i,i} \ , \ \hat{S}^{(i+1)}_{i+1,i}\right]^T = \left[\hat{S}^{(i+1)}_{i,i} \ , \ \hat{S}^{(i+1)}_{i+1,i}\right]^T.$$

This procedure gives us the Givens-vector representation, except for the signs of the diagonal elements, but the sign is determined rather easily, by looking at the signs of the elements in the original matrix. The Givens transformations are uniquely determined, because we take c_i always positive (in case $c_i = 0$, take $s_i > 0$), and when a Givens transformation of the form $G\left[0 \ , \ 0\right]^T = \left[0 \ , \ 0\right]$ has to be determined, we take G equal to the identity matrix. We illustrate the numerical stability with respect to the algorithm used in Example 2.12.

Example 2.13 (Example 2.12 continued) The same experiment is performed as in Example 2.12, but now with the newly designed algorithm: the output matrix

---

[18]The technique explored here is only one possible manner, other techniques based on computing partial singular value decompositions (SVD) do exist.

is the following one:

$$
\begin{bmatrix}
1.0000 & 1.0000 & 8.1131 \cdot 10^{-16} \\
1.0000 & 7.5895 \cdot 10^{-16} & 6.1575 \cdot 10^{-31} \\
8.1131 \cdot 10^{-16} & 6.1575 \cdot 10^{-31} & 1.0000
\end{bmatrix} .
$$

This matrix is much closer to the original one compared to the result in Example 2.12. ∎

This construction of the Givens-vector representation is of course quite expensive $\mathcal{O}(n^2)$ (faster, even $\mathcal{O}(n)$ algorithms can be constructed for arbitrary symmetric semiseparable matrices, but they are numerically unstable in general). In practical applications, however, faster $\mathcal{O}(n)$ and numerically stable methods can be designed to retrieve the representation, e.g., if the given symmetric semiseparable matrix of generator representable form is represented with two generators $\mathbf{u}$ and $\mathbf{v}$, one can derive the Givens-vector representation in $\mathcal{O}(n)$ computational time. These methods, however, are highly dependent of the application or problem one is trying to solve. Therefore we do not further investigate algorithms to retrieve the Givens-vector representation; in Subsection 2.13.2, however, some more methods are presented to go from one representation to another.

In [282] a reduction algorithm for reducing a symmetric matrix into a similar semiseparable one by orthogonal transformations was proposed. The output of the algorithm is already in the Givens-vector form. Also the QR-algorithm proposed in [294] uses as input this representation and gives as output the Givens-vector representation of the new symmetric semiseparable matrix after one iteration step. In both of these algorithms there is no need to apply this expensive procedure for calculating the Givens-vector representation from the elements of the symmetric semiseparable matrix.

Moreover, this representation reveals already the QR-factorization of the semiseparable matrix S: the Givens transformations appearing in the representation of the matrix are exactly the same as the Givens transformations appearing in the Q factor of the QR-factorization. (More information can be found in Chapter 5 in the discussion of methods for solving systems.)

2.4.4 Swapping the representation

Taking a closer look at the construction of a semiseparable matrix using the Givens-vector representation, we see that we keep, in fact, the dependency between the rows and a weight for each column. The matrix is constructed from top to bottom. In fact we can do completely the same by keeping the dependency between the columns and a weight for the rows. This corresponds to building the semiseparable matrix from the right towards the left.

These two possibilities are in fact very important, because as we will show later on in the implicit QR-algorithm, the choice between these two can make the implementation much more simple and cheaper in terms of operation cost. Therefore we present in this section an order $\mathcal{O}(n)$ algorithm to swap the representation from top to bottom into a representation from right to left.

We will show this construction by an example of order $n = 4$. We have in fact the following matrix:

$$\begin{bmatrix} c_1 v_1 & & & \\ c_2 s_1 v_1 & c_2 v_2 & & \\ c_3 s_2 s_1 v_1 & c_3 s_2 v_2 & c_3 v_3 & \\ s_3 s_2 s_1 v_1 & s_3 s_2 v_2 & s_3 v_3 & v_4 \end{bmatrix}.$$

A naive way to do it is to calculate the diagonal and subdiagonal elements and to reconstruct the representation in the other direction by using this information. But as prevosiously mentioned, not all semiseparable matrices can be represented with the diagonal subdiagonal approach and, moreover, this construction is numerically instable. We will start reconstructing the representation from right to left; this means that for $\hat{G}$ and $\hat{\mathbf{v}}$, we first calculate $\hat{v}_4$ and $\hat{G}_3$. One can calculate $\hat{v}_4$ immediately as $\hat{v}_4 = c_1 v_1$. The dependency between the first and second column has to be calculated from $[s_1 v_1, v_2]$, because it is possible that c_2 equals zero. Let us now use the following notation: $[s_1 v_1, v_2] \hat{G}_3^T = [0, r_2]$, where $\hat{G}_3$ is the Givens transformation annihilating $s_1 v_1$ and r_2 is the norm of $[s_1 v_1, d_2]$. Now we calculate $\hat{G}_2$ as the Givens transformation such that $[s_2 r_2, v_3] \hat{G}_2^T = [0, r_3]$, and one has to continue now with $[s_3 r_3, v_4] \hat{G}_1^T = [0, r_4]$. The reader can verify that with this construction we omit all the problems that can occur with zero structures. The new Givens transformations are determined by the Givens transformations $\hat{G}_i$ and the new vector elements are determined by the elements r_i, up to the sign.

From an implementational point of view this will be a very powerful tool. If, for example, one performs a row transformation on a semiseparable matrix presented with the Givens-vector representation from top to bottom, one can see that the Givens transformations do not contain the correct dependencies between the rows anymore, because the transformation changed them. If, however, the matrix was presented with the Givens-vector representation from right to left, performing a row transformation on the matrix would not change the dependencies between the columns; these dependencies are stored in the Givens transformations and are therefore not altered by this transformation, only the vector $\mathbf{v}$ needs to be updated.

2.4.5   Application to other classes of matrices

As the class of symmetric generator representable semiseparable matrices $\mathcal{S}_{sym}^{(g)}$ is a subclass of the class of symmetric semiseparable matrices, the use of the Givens-vector representation for representing them is straightforward. Also the use of this representation for symmetric semiseparable plus diagonal matrices is obvious.

As the class of quasiseparable matrices has the strictly lower triangular part of semiseparable form, we can use this representation to represent this part of the matrix using the Givens-vector representation, plus an extra diagonal. This gives us the following map for quasiseparable matrices using the Givens-vector representation.

$$r : \mathbb{R}^{2 \times (n-2)} \times \mathbb{R}^{n-1} \times \mathbb{R}^n \to \mathcal{Q}_{sym}$$
$$(G, \mathbf{v}, \mathbf{d}) \mapsto A,$$

where A is a matrix of the following form:

$$
\begin{bmatrix}
d_1 & c_1 v_1 & c_2 s_1 v_1 & c_3 s_2 s_1 v_1 & \cdots \\
c_1 v_1 & d_2 & c_2 v_2 & c_3 s_2 v_2 & \cdots \\
c_2 s_1 v_1 & c_2 v_2 & d_3 & c_3 v_3 & \cdots \\
c_3 s_2 s_1 v_1 & c_3 s_2 v_2 & c_3 v_3 & d_4 & \\
\vdots & \vdots & \vdots & & \ddots
\end{bmatrix}.
$$

This representation of the quasiseparable matrices uses essentially $3n-3$ parameters.

Notes and references

The Givens-vector representation was recently proposed by Vandebril, Van Barel and Mastronardi. The new representation originated from the lack of numerical stability for solving eigensystems using the generator representation. This new representation covering the complete set of symmetric semiseparable matrices solved this problem. The results presented in this section were based on the following manuscript.

☞ R. Vandebril, M. Van Barel, and N. Mastronardi. A note on the representation and definition of semiseparable matrices. *Numerical Linear Algebra with Applications*, 12(8):839–858, October 2005.

The authors used this representation for representing the symmetric semiseparable matrix, resulting from an orthogonal similarity reduction of a symmetric matrix to semiseparable form (see [282], and more information in Section 3.5, Chapter 3).

It was used for developing an implicit QR-algorithm for symmetric semiseparable matrices. It was also used for representing an upper triangular semiseparable matrix to compute its singular value decomposition (see [294, 292] and Section 3.5, Chapter 3). Finally the representation was also used for the implementation of the algorithms for solving generator representable semiseparable plus diagonal systems of equations in [284] (see also Section 5.5, Chapter 5). .

The Givens-vector representation as presented in this section has an extension towards higher order structured rank matrices in terms of the Givens-weight representation [72] (see also Section 8.5, Chapter 8).

2.5 The symmetric quasiseparable representation

In this section, we will present the quasiseparable representation. As already indicated, a representation is just a mathematical way to represent the low rank part in the matrices. Hence, we will define the quasiseparable representation for semiseparable matrices. The quasiseparable representation was defined in [32].

Firstly we will present the quasiseparable representation and prove that every symmetric semiseparable matrix can be represented in this way. Secondly we will illustrate that this type of representation is in fact the most general representation described in the book, as the Givens-vector and the generator representation can both be considered as special cases of this quasiseparable representation.

2.5.1   The representation for symmetric semiseparables

Let us define the quasiseparable representation in the following way.

$$r_{\mathcal{S}_{sym}^{(q)}} : \mathbb{R}^n \times \mathbb{R}^n \times \mathbb{R}^{n-1} \to \mathbb{R}^{n \times n} \tag{2.4}$$

$$(\mathbf{u}, \mathbf{v}, \mathbf{t}) \mapsto A,$$

where A is a symmetric matrix of the following form (only the lower triangular part is specified):

$$A = \begin{bmatrix} u_1 v_1 & & & & & \\ u_2 t_1 v_1 & u_2 v_2 & & & & \\ u_3 t_2 t_1 v_1 & u_3 t_2 v_2 & u_3 v_3 & & & \\ \vdots & & \ddots & \ddots & & \\ u_{n-1} t_{n-2} \cdots t_1 v_1 & \cdots & & & u_{n-1} v_{n-1} & \\ u_n t_{n-1} t_{n-2} \cdots t_1 v_1 & \cdots & & & u_n t_{n-1} v_{n-1} & u_n v_n \end{bmatrix}.$$

Reconsidering the previous representations for symmetric semiseparable matrices, it was always clear that the result of the representation map was (a subset of) the class of symmetric semiseparable matrices $\mathcal{S}_{sym}$. For the quasiseparable representation this is not straightforward, however. Therefore, our mapping has as a target set the complete class of $n \times n$ matrices. Let us prove now that the resulting matrices of this mapping are of semiseparable form.

Definition 2.14. *Suppose we have a class of matrices $\mathcal{S}'_{sym}$ such that $S \in \mathcal{S}'_{sym}$ has the following structure:*

$$S(i,j) = \begin{cases} u_i t_{i-1} t_{i-2} \dots t_j v_j & 1 \leq j < i \leq n \\ u_i v_i & 1 \leq i = j \leq n \\ u_j t_{j-1} t_{j-2} \dots t_i v_i & 1 \leq i < j \leq n \end{cases}$$

with $u_i, v_i \in \mathbb{R}$ for $1 \leq i \leq n$ and $t_i \in \mathbb{R}$ for $1 \leq i \leq n-1$.

This class of matrices is a special case of sequentially semiseparable matrices as defined in [53] and of quasiseparable matrices as defined in [93]. See also [83, 84, 96]. Note however that in the references mentioned above the diagonal is not incorporated into the structure while this is the case in Definition 2.14.

First we will prove that this class of matrices is equal to the class of semiseparable matrices.

Theorem 2.15. *The class of matrices $\mathcal{S}'_{sym}$ satisfying Definition 2.14 and the class of semiseparable matrices $\mathcal{S}_{sym}$ are equal.*

Proof. The proof is divided in two parts.

- $\mathcal{S}'_{sym} \subset \mathcal{S}_{sym}$: Let $S \in \mathcal{S}'_{sym}$. We have to prove that for all $1 \leq i \leq n$

$$\text{rank}\,(S(i:n,1:i)) \leq 1.$$

This corresponds to the demand that the determinant of every 2×2 submatrix of the lower triangular part of S equals 0. Let us consider the following 2×2 submatrix (with $i \geq k, l > i$ and $k > j$.)

$$\begin{bmatrix} S(i,j) & S(i,k) \\ S(l,j) & S(l,k) \end{bmatrix}$$

$$= \begin{bmatrix} u_i t_{i-1} \ldots t_k \ldots t_j v_j & u_i t_{i-1} \ldots t_k v_k \\ u_l t_{l-1} \ldots t_{i-1} \ldots t_k \ldots t_j v_j & u_l t_{l-1} \ldots t_{i-1} \ldots t_k v_k \end{bmatrix}.$$

This matrix can be written as the product of a diagonal and a rank 1 matrix:

$$\begin{bmatrix} u_i t_{i-1} \ldots t_k & 0 \\ 0 & u_l t_{l-1} \ldots t_k \end{bmatrix} \begin{bmatrix} t_{k+1} \ldots t_j v_j & v_k \\ t_{k+1} \ldots t_j v_j & v_k \end{bmatrix}.$$

Therefore, we have that the determinant of every 2×2 matrix of the lower triangular part of S equals zero and $S \in \mathcal{S}_{sym}$ which proves the first part of the theorem.

- $\mathcal{S}_{sym} \subset \mathcal{S}'_{sym}$: Let $S \in \mathcal{S}_{sym}$, then we know that S is a block diagonal matrix for which all the blocks are generator representable semiseparable matrices. Without loss of generality we assume there are only 2 blocks on the diagonal. Let us denote the generators of the first block with $\tilde{\mathbf{u}}$ and $\tilde{\mathbf{v}}$, both of length n_1, and denote the generators of the second block with $\hat{\mathbf{u}}$ and $\hat{\mathbf{v}}$ both of length n_2. If we define now the following vectors:

$$\mathbf{u} = \begin{bmatrix} \tilde{\mathbf{u}} \\ \hat{\mathbf{u}} \end{bmatrix} \qquad \mathbf{v} = \begin{bmatrix} \tilde{\mathbf{v}} \\ \hat{\mathbf{v}} \end{bmatrix}$$

and $t_1 = t_2 = \ldots = t_{n_1-1} = 1$, $t_{n_1} = 0$ and $t_{n_1+1} = t_{n_1+n_2-1} = 1$, then we have that the matrix constructed from Definition 2.14 with $\mathbf{u}, \mathbf{v}$ and $\mathbf{t}$ is equal to the matrix S. This means that the matrix $S \in \mathcal{S}'_{sym}$, which proves the other part of the theorem. $\qquad\qquad\square$

The theorem above proves in fact that the adjusted map $r_{\mathcal{S}^{(q)}_{sym}}$, with an adapted target set:

$$r_{\mathcal{S}^{(q)}_{sym}} : \mathbb{R}^n \times \mathbb{R}^n \times \mathbb{R}^{n-1} \to \mathcal{S}_{sym}$$

$$(\mathbf{u}, \mathbf{v}, \mathbf{t}) \mapsto S$$

for which S is defined as in Definition 2.14, is surjective.

The construction of the map $s_{\mathcal{S}^{(q)}_{sym}}$ is not so straightforward, however. The elements of the vector $\mathbf{t}$ are not defined uniquely and one should choose the elements $\mathbf{t}$ in such a way that the representation is stable. This choice however is not straightforward; see, for example, [32]. Moreover, the representation of the a symmetric semiseparable matrix as presented here needs $3n - 1$ parameters, whereas we know that it is also possible to represent this part essentially with $2n - 1$ parameters.

2.5.2 Application to other classes of matrices

Originally this representation comes from the class of quasiseparable matrices. As in the other sections, an adaptation of this type of representation towards the quasiseparable case is straightforward. Just represent the strictly lower triangular part of the quasiseparable matrix with the representation presented above.

In the remainder of this subsection, we will briefly illustrate that all previously shown representations are in fact special cases of this 'general representation'. Be aware, however, that the quasiseparable representation might perhaps be the most general one, but it also uses the largest number of parameters among all representations. It is also important to choose the parameters in such a way that they behave in a numerically stable way.

Considering u_i and t_i as the cosine and sine of a Givens transformation, the quasiseparable representation naturally reduces to the Givens-vector representation. This is clear comparing the representation map (2.3) for symmetric semiseparable matrices and the representation map (2.4) for the symmetric quasiseparable matrices.

Putting an extra condition on the elements: $t_i = 1$, we obtain the generator representation. Again this is clear by comparing the representation map (2.1) for symmetric generator representable semiseparable matrices and the representation map (2.4) for the symmetric quasiseparable matrices.

Notes and references

The name quasiseparable matrices was only introduced recently by Eidelman and Gohberg. Before several names were used such as matrices having low Hankel rank, weakly semiseparable. (See also the notes and references in Section 1.1.)

The authors Dewilde and van der Veen present [83] a new type of representation for sequentially semiseparable matrices or low Hankel rank matrices (quasiseparable matrices also belong to this class), as well as a QR-factorization and inversion formulas for this class of matrices. The class of matrices considered in the book is broader than the class of matrices considered in this section.

In [93], as already mentioned, the authors present a linear complexity inversion method for the general class of quasiseparable matrices. (See also Chapter 12.)

Bini, Gemignani and Pan derive in [32] an algorithm for performing a step of QR-algorithm on a generalized semiseparable matrix. The lower part of this generalized matrix is in fact lower triangular semiseparable and represented with the quasiseparable representation. See Section 3.5, Chapter 3 for more information.

The following papers deal with higher order quasiseparable and closely related matrices, such as sequentially semiseparable. As they are the subject of forthcoming investigations in Part III of this book, we will not yet describe them in detail [53, 84, 89].

2.6 Some examples

With this section we conclude the first part of this chapter, dedicated to symmetric representations. In the following section we will start investigating unsymmetric definitions. Here, however, some examples are first presented.

The reader can try to decide whether the following matrices are representable with the generator representation, the diagonal subdiagonal representation or the Givens-vector representation. We did not include the quasiseparable representation as this is the most general one, and the Givens-vector is a special case of it. With $\mathrm{diag}(\mathbf{v})$ we denote the diagonal matrix with as diagonal elements the elements coming from the vector $\mathbf{v}$.

$$D_1 = \mathrm{diag}([1,1,1,1,1]) \tag{2.5}$$

$$D_2 = \mathrm{diag}([0,1,1,1,1]) \tag{2.6}$$

$$S_1 = \begin{bmatrix} 1 & 0 & 0 \\ 0 & 1 & 1 \\ 0 & 1 & 0 \end{bmatrix} \tag{2.7}$$

$$S_2 = \begin{bmatrix} 0 & 0 & 0 & 0 \\ 0 & 1 & 0 & 0 \\ 0 & 0 & 1 & 1 \\ 0 & 0 & 1 & 0 \end{bmatrix} \tag{2.8}$$

$$S_3 = \begin{bmatrix} 0 & 1 & 0 \\ 1 & 0 & 0 \\ 0 & 0 & 1 \end{bmatrix} \tag{2.9}$$

$$S_4 = \begin{bmatrix} 0 & 1 & 1 \\ 1 & 0 & 0 \\ 1 & 0 & 1 \end{bmatrix} \tag{2.10}$$

$$S_5 = \begin{bmatrix} 1 & 0 & 1 & 1 \\ 0 & 0 & 0 & 0 \\ 1 & 0 & 1 & 1 \\ 1 & 0 & 1 & 0 \end{bmatrix} \tag{2.11}$$

The results can be found in the next table.

matrix	singular?	generator representation	diagonal-subdiagonal	Givens-vector representation
matrix 2.5	no	no	yes	yes
matrix 2.6	yes	no	yes	yes
matrix 2.7	no	no	yes	yes
matrix 2.8	yes	no	yes	yes
matrix 2.9	no	no	yes	yes
matrix 2.10	no	yes	no	yes
matrix 2.11	yes	yes	no	yes

One can clearly see that the Givens-vector representation is the most general one. Moreover, this representation is still an $\mathcal{O}(n)$ representation. Note also that the quasiseparable representation can represent all the matrices shown above.

2.7   The unsymmetric generator representation

In this section, we will present some possibilities for the generator representation for nonsymmetric generator representable semiseparable matrices. First, we will define the representation similarly to the way we defined the class of generator representable semiseparable matrices, by separating the representation of the upper and the lower triangular part. This representation uses approximately $4n$ parameters. However, inverting a nonsingular generator representable semiseparable matrix, we get an irreducible tridiagonal matrix which is essentially determined by $3n - 2$ parameters. Hence, the first presented representation of $4n$ parameters is not necessarily the most compact one. Therefore we present a way to adjust this representation for invertible generator representable semiseparable matrices.

2.7.1   The representation for semiseparables

Let us define now the generator representation for the class of unsymmetric generator representable semiseparable matrices $\mathcal{S}$. We already know from Section 2.2, that not all semiseparable matrices can be represented with generators. First of all we investigate if the map corresponding to the definition of generator representable semiseparable matrices (Definition 1.3) satisfies the properties of a representation map for $\mathcal{S}$. This map $r_{\mathcal{S}(g)}$ is defined in the following natural way:

$$r_{\mathcal{S}(g)} : \mathcal{R} \subset \mathbb{R}^n \times \mathbb{R}^n \times \mathbb{R}^n \times \mathbb{R}^n \to \mathcal{S}^{(g)} \tag{2.12}$$
$$(\mathbf{u}, \mathbf{v}, \mathbf{p}, \mathbf{q}) \mapsto \mathrm{tril}(\mathbf{u}\mathbf{v}^T) + \mathrm{triu}(\mathbf{p}\mathbf{q}^T, 1),$$

where $\mathcal{R}$ is the set of 4-tuples of the following form:

$$\mathcal{R} = \{(\mathbf{u}, \mathbf{v}, \mathbf{p}, \mathbf{q}) \in \mathbb{R}^n \times \mathbb{R}^n \times \mathbb{R}^n \times \mathbb{R}^n \mid u_i v_i = p_i q_i, \text{ for } i = 1, \dots, n\}.$$

The first condition on the map, $\dim(\mathbb{R}^n \times \mathbb{R}^n \times \mathbb{R}^n \times \mathbb{R}^n) \leq \dim(\mathbb{R}^{n \times n})$, is satisfied in a natural way. And also, due to the definition, we know that surjectivity is naturally satisfied. Hence, we can use this mapping to represent the general class of nonsymmetric generator representable semiseparable matrices. In fact this representation is often used in practice and has proved to be useful. See, for example, [222, 221, 105, 134] and the notes and references following this section.

The class of matrices $\mathcal{R}$ poses n constraints on its elements, however. This means that this class of matrices should be, in case of nontrivial constraints, representable with $3n$ parameters instead of $4n$. A general construction of these 3 vectors for representing this class of matrices is not straightforward, in case there are zero blocks or the matrices are singular. Therefore we will illustrate this with a construction for nonsingular generator representable semiseparable matrices.

The following representation finds it origin in [206]. Let us define the set of matrices $\mathcal{S}'$ as the following set:

$$\mathcal{S}' = \left\{ A \in \mathbb{R}^{n \times n} \mid \exists\, \mathbf{x}, \mathbf{y}, \mathbf{z} \in \mathbb{R}^n, \text{ such that } a_{ij} = \left\{ \begin{array}{ll} x_i y_j z_j, & i \geq j, \\ y_i x_j z_j, & i \leq j. \end{array} \right. \right\}.$$

This means that an invertible matrix A in $\mathcal{S}'_{inv}$ is of the following form:

$$
A = \begin{bmatrix}
x_1 y_1 z_1 & y_1 x_2 z_2 & y_1 x_3 z_3 & \cdots & y_1 x_n z_n \\
x_2 y_1 z_1 & x_2 y_2 z_2 & y_2 x_3 z_3 & \cdots & y_2 x_n z_n \\
x_3 y_1 z_1 & x_3 y_2 z_2 & x_3 y_3 z_3 & \cdots & y_3 x_n z_n \\
\vdots & \vdots & \vdots & \ddots & \vdots \\
x_n y_1 z_1 & x_n y_2 z_2 & x_n y_3 z_3 & \cdots & x_n y_n z_n
\end{bmatrix}.
\tag{2.13}
$$

Let us prove now that the class $\mathcal{S}'_{inv} = \mathcal{S}^{(g)}_{inv}$.

Theorem 2.16. *The class of matrices $\mathcal{S}'_{inv}$ as defined above is equal to the class of generator representable semiseparable matrices.*

Proof. All the matrices S in $\mathcal{S}^{(g)}_{inv}$ are of the form:

$$
S = \begin{bmatrix}
u_1 v_1 & q_2 p_1 & q_3 p_1 & \cdots & q_n p_1 \\
u_2 v_1 & u_2 v_2 & q_3 p_2 & \cdots & q_n p_2 \\
u_3 v_1 & u_3 v_2 & u_3 v_3 & \ddots & \vdots \\
\vdots & \vdots & \vdots & \ddots & q_n p_{n-1} \\
u_n v_1 & u_n v_2 & u_n v_3 & \cdots & u_n v_n
\end{bmatrix},
\tag{2.14}
$$

with $u_i v_i = p_i q_i$ for all $i = 1, \ldots, n$.

- $\mathcal{S}'_{inv} \subset \mathcal{S}^{(g)}_{inv}$. Suppose we have a matrix A in $\mathcal{S}'_{inv}$, where A is of the form (2.13). Define the vectors $\mathbf{u} = \mathbf{x}$, $\mathbf{v} = [y_1 z_1, y_2 z_2, \ldots, y_n z_n]^T$, $\mathbf{p} = \mathbf{y}$ and $\mathbf{q} = [x_1 z_1, x_2 z_2, \ldots, x_n z_n]^T$ and using these vectors in (2.14) gives us the desired inclusion, namely, $A \in \mathcal{S}^{(g)}_{inv}$.

- $\mathcal{S}^{(g)}_{inv} \subset \mathcal{S}'_{inv}$. Suppose a semiseparable matrix S is given of the form (2.14). Define the vectors $\mathbf{x} = \mathbf{u}$ and $\mathbf{y} = \mathbf{p}$. If all components in the generators $\mathbf{u}, \mathbf{v}, \mathbf{q}$ and $\mathbf{p}$ are different from zero, we can easily define for all j: $z_j = \frac{v_j}{p_j} = \frac{q_j}{u_j}$, which is well defined as $u_j v_j = p_j q_j$ for all j. An easy calculation reveals that indeed the matrix A from (2.13) with the vectors $\mathbf{x}, \mathbf{y}$ and $\mathbf{z}$ equals the matrix S.

Omit the nonzeroness condition on the generators $\mathbf{u}, \mathbf{v}, \mathbf{q}$ and $\mathbf{p}$. In fact the following two relations need to be satisfied for all j:

$$
u_i p_j z_j = u_i v_j \quad i \geq j,
\tag{2.15}
$$

$$
p_i u_j z_j = p_i q_j, \quad i \leq j.
\tag{2.16}
$$

Assume that not all elements are nonzero, we will illustrate then how to compute z_j. Fix a certain i, if $u_i = 0$ and $p_i \neq 0$, define $z_i = v_i / p_i$. Because $u_i v_i = p_i q_i$, we know that $q_i = 0$, and hence both Conditions (2.15) and (2.16) are satisfied. The case where $u_i \neq 0$ and $p_i = 0$ is similar. The case where both $u_i = p_i = 0$ corresponds to a singular matrix, which we excluded.

This concludes the proof. We have that $\mathcal{S}'_{inv} = \mathcal{S}^{(g)}_{inv}$.   □

Hence we can define a representation map for the invertible generator representable semiseparable matrices:

$$r_{\mathcal{S}^{(g)}_{inv}} : \mathbb{R}^n \times \mathbb{R}^n \times \mathbb{R}^n \to \mathcal{S}^{(g)}_{inv}$$

$$(\mathbf{x}, \mathbf{y}, \mathbf{z}) \mapsto A,$$

where the matrix A is of the form (2.13).

Even though the representation above is not suitable for representing all generator representable semiseparable matrices, it might come in handy when inverting, for example, an irreducible tridiagonal matrix (see [206]).

Example 2.17 The reader can verify that the new representation is not suitable for representing, for example, the following singular generator representable matrix.

$$\begin{bmatrix} 1 & 2 & 3 & 9 \\ 0 & 0 & 0 & 0 \\ 2 & 2 & 2 & 6 \\ 4 & 4 & 4 & 5 \end{bmatrix}$$

∎

The two investigated representations above (one with 3 and one with 4 vectors) are not the only ones, but the most frequently used ones. Several different forms also arise, such as the representation of a matrix S of the following form is given [24]:

$$A = \begin{bmatrix} x_1 y_1 & y_1 x_2 z_2 & y_1 x_3 z_3 & \cdots & y_1 x_n z_n \\ x_2 y_1 & x_2 y_2 & y_2 x_3 z_3 & \cdots & y_2 x_n z_n \\ x_3 y_1 & x_3 y_2 & x_3 y_3 & \cdots & y_3 x_n z_n \\ \vdots & \vdots & \vdots & \ddots & \vdots \\ x_n y_1 & x_n y_2 & x_n y_3 & \cdots & x_n y_n \end{bmatrix} . \tag{2.17}$$

Sometimes even the elements z_i are built up as products of elements of another vector. For more information see the notes and references.

The reader can verify that not all of these representations are suitable for representing a general generator representable semiseparable matrix. Most of the representations are only suitable for representing invertible generator representable semiseparable matrices. Therefore general investigations quite often stick to the first-mentioned representation, based on 4 vectors and the connection on the diagonal. Finally we present a last type of representation, also using 4 vectors which is also capable of representing all nonsymmetric generator representable semiseparable matrices.

Suppose 4 vectors are given, namely $\mathbf{w}$, $\mathbf{x}$, $\mathbf{y}$ and $\mathbf{z}$

$$A = \begin{bmatrix} x_1 y_1 z_1 w_1 & y_1 x_2 z_2 w_1 & y_1 x_3 z_3 w_1 & \cdots & y_1 x_n z_n w_1 \\ x_2 y_1 z_1 w_2 & x_2 y_2 z_2 w_2 & y_2 x_3 z_3 w_2 & \cdots & y_2 x_n z_n w_2 \\ x_3 y_1 z_1 w_3 & x_3 y_2 z_2 w_3 & x_3 y_3 z_3 w_3 & \cdots & y_3 x_n z_n w_3 \\ \vdots & \vdots & \vdots & \ddots & \vdots \\ x_n y_1 z_1 w_n & x_n y_2 z_2 w_n & x_n y_3 z_3 w_n & \cdots & x_n y_n z_n w_n \end{bmatrix}. \tag{2.18}$$

The verification that this type of matrix can represent the complete class is left to the reader.

2.7.2 Application to other classes of matrices

Even though the representation is only useful for representing generator representable semiseparable matrices, we can also use it to represent specific subclasses of the class of semiseparable plus diagonal and quasiseparable matrices.

For example, a straightforward use of this representation is to represent the class of generator representable semiseparable plus diagonal matrices, denoted as

$$\mathcal{S}^{(g,d)} = \{S + D \mid S \in \mathcal{S}^{(g)} \text{ and } D \in \mathcal{D}\}.$$

A representation map linked to this class of matrices, for example, is the following map:

$$r_{\mathcal{S}^{(g,d)}} : \mathcal{R} \times \mathbb{R}^n \to \mathcal{S}^{(g)}$$
$$(\mathbf{u}, \mathbf{v}, \mathbf{p}, \mathbf{q}, \mathbf{d}) \mapsto \text{tril}(\mathbf{u}\mathbf{v}^T) + \text{triu}(\mathbf{p}\mathbf{q}^T, 1) + \text{diag}(\mathbf{d}).$$

With diag $(\mathbf{d})$, we denote a diagonal matrix, with as diagonal elements, the elements from the vector $\mathbf{d}$. This class of matrices $\mathcal{S}^{(g,d)}$ is used in the publications [134, 105].

In this way we can also define a special subclass of the quasiseparable matrices, namely, this class of matrices for which the strictly lower and the strictly upper triangular part are representable with generators, namely $\mathcal{Q}^{(g)}$. This means, in fact, that the strictly lower and the strictly upper triangular part are coming from a rank 1 matrix[19]. A representation map can be of the following form:

$$r_{\mathcal{Q}^{(g,d)}} : \mathbb{R}^{n-1} \times \mathbb{R}^{n-1} \times \mathbb{R}^{n-1} \times \mathbb{R}^{n-1} \times \mathbb{R}^n \to \mathcal{Q}^{(g)}$$
$$(\mathbf{u}, \mathbf{v}, \mathbf{p}, \mathbf{q}, \mathbf{d}) \mapsto A,$$

where A is a matrix of the following form:

$$\begin{bmatrix} d_1 & q_1 p_1 & q_2 p_1 & q_3 p_1 & \cdots & q_{n-1} p_1 \\ u_1 v_1 & d_2 & q_2 p_2 & q_3 p_2 & \cdots & q_{n-1} p_2 \\ u_2 v_1 & u_2 v_2 & d_3 & q_3 p_2 & \cdots & q_{n-1} p_3 \\ u_3 v_1 & u_3 v_2 & u_3 v_3 & \ddots & & \vdots \\ \vdots & \vdots & \vdots & & d_{n-1} & q_{n-1} p_{n-1} \\ u_{n-1} v_1 & u_{n-1} v_2 & u_{n-1} v_3 & \cdots & u_{n-1} v_{n-1} & d_n \end{bmatrix}.$$

---

[19] As there is no relation posed between the upper and lower triangular part, we cannot expect to be able to reduce the complexity from $4n$ to $3n$ parameters.

Readers familiar with semiseparable and related matrices will surely recognize the latter matrix (see, e.g., [284]). Often this matrix is just taken as the definition of a semiseparable plus diagonal matrix, as shown in the first chapter of this book. This is however not a consistent way of defining these matrices; as in the nonsymmetric case, these classes are not the same.

To conclude this section, one might say that there are different ways to use the generator representation for nonsymmetric matrices. Even though theoretically $3n$ parameters should be enough, it seems that in practice (see notes and references), just the $4n$ representation is used quite often with the extra condition put on the diagonal. This was in fact the representation we used to define these matrices.

Notes and references

In the literature a lot of attention is being paid to the class of generator representable semiseparable matrices. Also the closely related classes such as the generator representable semiseparable plus diagonal matrix and the generator representable quasiseparable matrices have been studied.

☞ T. Torii. Inversion of tridiagonal matrices and the stability of tridiagonal systems of linear equations. *Information processing in Japan (Joho Shori)*, 6:41–46, 1966. (In Japanese).

☞ T. Torii. Inversion of tridiagonal matrices and the stability of tridiagonal systems of linear equations. *Technology Reports of the Osaka University*, 16:403–414, 1966.

The article of 1965 was published in a Japanese journal and translated afterwards to the article of 1966. The author provides explicit formulas for calculating the inverse of a nonsymmetric tridiagonal matrix. The method of inversion is based on solving the associated difference equation. The stability of solving a tridiagonal system of equations via inversion is investigated due to the explicit formulation of bounds between which the condition number needs to lie. The resulting semiseparable matrices are represented by $4n$ parameters. The upper triangular part and the lower triangular part have different generators linked on the diagonal, as in the representation map (2.12). As an example the formulas are applied on Toeplitz tridiagonal matrices. These results are afterwards expanded in [224], towards band matrices (see Section 14.3 in Chapter 14).

☞ M. Capovani. Sulla determinazione della inversa delle matrici tridiagonali e tridiagonali a blocchi. *Calcolo*, 7:295–303, 1970. (In Italian).

☞ M. Capovani. Su alcune proprietà delle matrici tridiagonali e pentadiagonali. *Calcolo*, 8:149–159, 1971. (In Italian).

In the paper of 1970, Capovani gives explicit formulas for the inverse of a nonsymmetric tridiagonal matrix (no demands are placed on the elements of the tridiagonal). The representation of the final generator representable semiseparable matrix is however based on $4n$ parameters. In the paper of 1971, Capovani proves that the inverse of a nonsymmetric tridiagonal matrix with the super and subdiagonal elements different from zero also has as inverse a generator representable matrix, where the upper and lower triangular part have different generators (in fact, this is the representation presented in the beginning of this section, and which was used as the definition for nonsymmetric generator representable semiseparable matrices). Using these results the author proves that a pentadiagonal matrix can be written as the product of two tridiagonal matrices, for which one of them is symmetric.

> ☞ R. Bevilacqua and M. Capovani. Proprietà delle matrici a banda ad elementi ed a blocchi. *Bolletino Unione Matematica Italiana*, 5(13-B):844–861, 1976. (In Italian).

Bevilacqua and Capovani extend in the manuscript of 1976 the results of papers [160] and [2, 3] to band matrices and to block band matrices (not necessarily symmetric). Formulas are presented for inverting band matrices whose elements on the extreme diagonals are different from zero. The proposed representation for a nonsymmetric generator representable matrix uses $3n$ parameters and presents a matrix of the following form (which is an adapted version of the matrix in Equation (2.17)). In fact this matrix also has connections with the quasiseparable structure (we show a 5×5 example):

$$A = \begin{bmatrix} x_1 & \alpha_2 x_2 & \alpha_2 \alpha_3 x_3 & \alpha_2 \alpha_3 \alpha_4 x_4 & \alpha_2 \alpha_3 \alpha_4 \alpha_5 x_5 \\ x_2 & x_2 y_1 & \alpha_3 x_3 y_1 & \alpha_3 \alpha_4 x_4 y_1 & \alpha_3 \alpha_4 \alpha_5 x_5 y_1 \\ x_3 & x_3 y_1 & x_3 y_2 & \alpha_4 x_4 y_2 & \alpha_4 \alpha_5 x_5 y_2 \\ x_4 & x_4 y_1 & x_4 y_2 & x_4 y_3 & \alpha_5 x_5 y_3 \\ x_5 & x_5 y_1 & x_5 y_2 & x_5 y_3 & x_5 y_4 \end{bmatrix}.$$

> ☞ T. Yamamoto and Y. Ikebe. Inversion of band matrices. *Linear Algebra and its Applications*, 24:105–111, 1979.

Yamamoto and Ikebe propose formulas for inverting band matrices with different bandwidths under the assumption that the elements on the extreme diagonals are different from zero. The formulas given for constructing the inverse suggest some kind of higher order generator representation. Application of their algorithm to a nonsymmetric tridiagonal matrix gives the result presented in the form (2.13).

Lewis provides in [206] an explicit formula which can be used to compute the inverse of tridiagonal matrices. The matrix does not necessarily need to be symmetric, nor do all elements have to be different from zero. Also interesting is a new kind of representation for nonsymmetric semiseparable matrices, which does not use 4 but three vectors x, y and z, where the elements of S are of the form (2.13). The explicit formula is a generalization of the theorem for symmetric matrices by Ukita [272]. More information on the inversion based on the method of Lewis can be found in Section 7.2 in Chapter 7 in which his method is depicted.

Nabben provides in [222] upper and lower bounds for the entries of the inverse of diagonally dominant tridiagonal matrices, where the inverse is represented by using 4 vectors like in the mapping (2.12). In the manuscript [221], decay rates for the inverse of special tridiagonal and band matrices are given. Again, the generator representable semiseparable matrix is represented with $4n$ parameters. More information and a more detailed analysis can be found in Section 7.5 in Chapter 7.

> ☞ D. Fasino and L. Gemignani. Structural and computational properties of possibly singular semiseparable matrices. *Linear Algebra and its Applications*, 340:183–198, 2001.

This paper by Fasino and Gemignani works with nonsymmetric generator representable semiseparable matrices defined as in mapping (2.12). They provide for this class of matrices, also singular ones, a sparse structured representation. More precisely, it is shown that a semiseparable matrix A always can be written as the inverse of a block tridiagonal matrix plus a sparse, low rank matrix Z.

☞ R. K. Mallik. The inverse of a tridiagonal matrix. *Linear Algebra and its Applications*, 325:109–139, 2001.

Mallik provides explicit inversion formulas for nonsymmetric irreducible tridiagonal matrices. The formulas are obtained by solving an associated boundary value problem. Also special cases such as cyclic tridiagonal matrices and matrices with a constant diagonal are considered. A cyclic tridiagonal matrix is a tridiagonal matrix with the upper right and the lower left element different from zero, hence a slightly perturbed tridiagonal matrix. Also a connection between the matrix inverse and orthogonal polynomials is established.

Mastronardi, Chandrasekaran and Van Huffel present in [209] an order $\mathcal{O}(n)$ algorithm to solve a system of equations where the coefficient matrix is in fact a quasiseparable matrix represented using the generator representation. (The authors refer to these matrices as diagonal plus semiseparable matrices, but in the context of the book, we name these matrices quasiseparable, represented using the Givens-vector representation.) The algorithm is suitable for an implementation on two processors. More information on this algorithm can be found in Section 5.6, Chapter 5.

Fasino and Gemignani describe in [134] an order $\mathcal{O}(n)$ solver for banded plus generator representable semiseparable systems of equations, based on the LU-decomposition. For more information see Section 4.5 in Chapter 4.

In manuscripts [108, 210] Mastronardi, et al. present different algorithms for reducing generator representable quasiseparable matrices to either tridiagonal or bidiagonal form. More information on these articles can be found in Section 3.5 in Chapter 3.

Chandrasekaran and Gu present in [56] a method for solving systems whose coefficient matrix is of quasiseparable form plus a band matrix. Moreover the parts below the band and above the band are completely decoupled and both represented by rank 1 matrices, hence the generator representation for representing a quasiseparable matrix plus a band matrix. This decoupling of the upper and lower triangular part is often used and is discussed in more detail in a forthcoming section (Section 2.10). The algorithm itself is presented in Section 5.6 of Chapter 5.

In the manuscripts [148, 150, 91, 90], the authors Eidelman and Gohberg consider different methods for inverting and solving systems of equations involving generator representable quasiseparable matrices. More information can be found in Section 14.5 in Chapter 3 and in the Chapter 12.

2.8   The unsymmetric Givens-vector representation

In this section we will provide an expansion of the Givens-vector representation for symmetric matrices, towards the nonsymmetric case. First we define this unsymmetric definition in terms of $4n$ parameters. Similarly, as in the generator representation case, it will be shown that it is not an easy task to design a parameterization of $3n$ variables for the general case.

2.8.1   The representation for semiseparables

The Givens-vector representation for unsymmetric semiseparable matrices admits the following natural representation map.

$$r_\mathcal{S} : \mathbb{R}^{2\times(n-1)} \times \mathbb{R}^n \times \mathbb{R}^{2\times(n-1)} \times \mathbb{R}^n \to \mathcal{S}$$
$$(G, \mathbf{v}, H, \mathbf{e}) \mapsto S.$$

Where we have that the matrices G and H contain the cosines and sines of Givens transformations and $\mathbf{v}$ and $\mathbf{e}$ are two vectors of dimension $n - 1$.

$$G = \begin{bmatrix} c_1 & c_2 & \cdots & c_{n-1} \\ s_1 & s_2 & \cdots & s_{n-1} \end{bmatrix}$$
$$\mathbf{v} = [v_1, v_2, \ldots, v_n]^T$$

$$H = \begin{bmatrix} r_1 & r_2 & \cdots & r_{n-2} \\ t_1 & t_2 & \cdots & t_{n-2} \end{bmatrix}$$
$$\mathbf{e} = [e_1, e_2, \ldots, e_{n-1}]^T.$$

The Givens transformations G_i and H_i are denoted as

$$G_i = \begin{bmatrix} c_i & -s_i \\ s_i & c_i \end{bmatrix} \text{ and } H_i = \begin{bmatrix} r_i & -t_i \\ t_i & r_i \end{bmatrix}.$$

Using this representation map the matrix S will be of the form:

$$S = \begin{bmatrix} c_1 v_1 & r_2 t_1 e_1 & r_3 t_2 t_1 e_1 & \cdots \\ c_2 s_1 d_1 & c_2 v_2 & r_3 t_2 e_2 & \cdots \\ c_3 s_2 s_1 d_1 & c_3 s_2 v_2 & c_3 v_3 & \\ \vdots & \vdots & & \ddots \end{bmatrix}. \tag{2.19}$$

And, for all $i = 1, \ldots, n - 1$ we have that $c_i v_i = r_i e_i$ and $v_n = e_n$.

The design of a general method using $3n$ parameters for representing a general semiseparable matrix is not trivial. As the representation above suits our needs, and as it was already illustrated in the previous section that this is not an easy task, we do not go into the details. We just provide one easy example.

If all the diagonal elements of the semiseparable matrix are nonzero, one can easily remove n parameters from the Givens-vector representation for nonsymmetric semiseparable matrices. In this case, the vector $\mathbf{e}$ is completely determined by the equalities $c_i v_i = r_i e_i$, for every $i = 1, \ldots, n - 1$ and $v_n = e_n$, and hence there is no need to use it in the representation. This will reduce the representation to $3n$ parameters.

2.8.2   Application to other classes of matrices

The use of the representation above for representing generator representable non-symmetric semiseparable plus diagonal matrices is obvious and hence not covered.

Using the Givens-vector representation for quasiseparable matrices is much easier than the construction above. Quasiseparable matrices have independent structures in the strictly upper triangular and the strictly lower triangular part, i.e., there is no connection on the diagonal, as is the case for semiseparable matrices. Hence the representation of the strictly upper triangular part is completely decoupled from the representation of the strictly lower triangular part. This means that two sequences of Givens transformations and two vectors are needed for representing a quasiseparable matrix. Hence, for representing a quasiseparable matrix, one always needs $4n + \mathcal{O}(1)$ for representing the strictly lower and the strictly upper triangular part plus an extra n parameters for the diagonal (in total $5n + \mathcal{O}(1)$).

Notes and references

To our knowledge this representation has not yet been used in the current form. The decoupled form, i.e., neglecting the connection on the diagonal, however, has already been used for actual implementations of the QR-factorization of a quasiseparable matrix (see [284]).

We would like to remark that unitary Hessenberg matrices, which are a special type of quasiseparable matrices, also admit a specific representation based on Givens transformations. This representation is the so-called Schur parametrization of the unitary Hessenberg matrix. More information concerning this representation can for example be found in the references in Section 1.3. Several algorithms for unitary Hessenberg matrices are based on this representation.

2.9   The unsymmetric quasiseparable representation

Throughout the remainder of the book, the quasiseparable representation is most often used for representing quasiseparable matrices; hence, we do not cover all details as in the previous sections anymore. First we will briefly, intuitively sketch the application of the quasiseparable representation towards the semiseparable case.

2.9.1   The representation for semiseparables

The quasiseparable representation uses approximately (up to a constant) $3n$ parameters for representing the semiseparable part in the matrix. Because the upper triangular part has the same structure, this means that $6n$ parameters are needed to represent a semiseparable matrix using this quasiseparable representation, with an extra condition, namely the upper triangular linked to the lower triangular part on the diagonal. Taking into consideration this extra information leads to a theoretical presentation of semiseparable matrices, using the quasiseparable representation, which uses $5n$ parameters. This representation is not used in this book, however, and therefore we do not cover it in detail.

2.9.2   Application to other classes of matrices

This representation can of course also be used for representing nonsymmetric semiseparable plus diagonal matrices, but for the same reasons as above, we do not go into the details of this representation.

The quasiseparable representation as presented in this section is the one we will often use for representing nonsymmetric quasiseparable matrices. The representation of the lower triangular part of the quasiseparable matrix A uses the vectors $\mathbf{u}, \mathbf{v}$ and $\mathbf{t}$, the diagonal is represented with the vector $\mathbf{d}$, and the upper triangular part is represented using the vectors $\mathbf{p}, \mathbf{q}$ and $\mathbf{r}$.

Let us introduce the new notation $t_{i:j}$ to denote the product $t_i t_{i-1} t_{i-2} \cdots t_j$. The quasiseparable matrix represented with this representation is of the following form:

$$
\begin{bmatrix}
d_1 & q_1 p_1 & q_2 r_1 p_1 & q_3 r_{2:1} p_1 & \cdots & q_{n-1} r_{n-2:1} p_1 \\
u_1 v_1 & d_2 & q_2 p_2 & q_3 r_2 p_2 & \cdots & q_{n-1} r_{n-2:2} p_2 \\
u_2 t_1 v_1 & u_2 v_2 & d_3 & q_3 p_3 & \cdots & q_{n-1} r_{n-2:3} p_3 \\
u_3 t_{2:1} v_1 & u_3 t_2 v_2 & u_3 v_3 & \ddots & & \vdots \\
\vdots & \vdots & \vdots & & d_{n-1} & q_{n-1} p_{n-1} \\
u_{n-1} t_{n-2:1} v_1 & u_{n-1} t_{n-2:2} v_2 & u_{n-1} t_{n-2:3} v_3 & \cdots & u_{n-1} v_{n-1} & d_n
\end{bmatrix}
\tag{2.20}
$$

Often (see papers on quasiseparable matrices) one also uses the following form, which in fact is the same, except for a slight change in the indices:

$$
\begin{bmatrix}
d_1 & q_2 p_1 & q_3 r_2 p_1 & q_4 r_{3:2} p_1 & \cdots & q_n r_{n-1:2} p_1 \\
u_2 v_1 & d_2 & q_3 p_2 & q_4 r_3 p_2 & \cdots & q_n r_{n-1:3} p_2 \\
u_3 t_2 v_1 & u_3 v_2 & d_3 & q_4 p_3 & \cdots & q_n r_{n-1:4} p_3 \\
u_4 t_{3:2} v_1 & u_4 t_3 v_2 & u_4 v_3 & \ddots & & \vdots \\
\vdots & \vdots & \vdots & & d_{n-1} & q_n p_{n-1} \\
u_n t_{n-1:2} v_1 & u_n t_{n-1:3} v_2 & u_n t_{n-1:4} v_3 & \cdots & u_n v_{n-1} & d_n
\end{bmatrix}.
\tag{2.21}
$$

Notes and references

For more references concerning quasiseparable matrices see Section 2.5, and also the higher order quasiseparable representation which will be presented in Chapter 8, Section 8.5.

The class of quasiseparable matrices is often defined with the representation from Equation (2.20), instead of using the rank definition as it was proposed in this book. The definition of quasiseparable matrices as well as this representation, in a more general form, can be found in [93]. More information on this representation and related algorithms can be found in Chapter 12.

Most of the references concerning quasiseparable matrices are considered with higher order quasiseparable matrices. These are covered in forthcoming chapters.

2.10 The decoupled representation for semiseparable matrices

The representations presented above for either nonsymmetric semiseparable or nonsymmetric generator representable semiseparable where representations take into consideration the inherent structure of the matrices. This means that the representations above tried to exploit the existing structure of the matrices as efficiently as possible, leading to the minimum number of parameters for representing these matrices. For example, to represent nonsymmetric invertible generator representable matrices it was shown that $3n$ parameters were sufficient, under some mild assumptions.

In applications, however, the connection between the lower and the upper triangular part is not necessarily exploited. Moreover, if one exploits the link between the upper and lower triangular part, this does not necessarily mean that the resulting algorithms will be faster or more accurate.

In this viewpoint, 'decoupled' representations are often used. This means that no assumptions are made concerning connections between the upper and lower triangular part in the semiseparable or generator representable semiseparable matrices. If the connection on the diagonal is lost, this means, of course, that one will be working with a slightly larger class of matrices, having as a subset the class of semiseparable matrices. The decoupled representation of matrices is often used for solving systems.

Note 2.18. *We like to draw attention to the class of quasiseparable matrices. As the class of quasiseparable matrices does not assume a connection between the strictly upper and strictly lower triangular part, this class in fact always has a decoupled representation. Therefore essentially the decoupled representation is only useful for semiseparable or generator representable semiseparable matrices.*

Let us illustrate what we mean with the decoupled representation in some examples.

2.10.1   The decoupled generator representation

The decoupled generator representation just neglects the connection between the upper and lower triangular part of a generator representable semiseparable matrix.

This leads to a mapping of the following form.

$$r : \mathbb{R}^n \times \mathbb{R}^n \times \mathbb{R}^{n-1} \times \mathbb{R}^{n-1} \to \mathbb{R}^{n \times n}$$
$$(\mathbf{u}, \mathbf{v}, \mathbf{p}, \mathbf{q}) \mapsto \text{tril}(\mathbf{u}\mathbf{v}^T) + \text{triu}(\mathbf{p}\mathbf{q}^T, 1).$$

Note that no conditions are put on the generators. We do not demand here that $u_i v_i = p_i q_i$, as required previously. Clearly the image of this mapping is a subset of $\mathbb{R}^{n \times n}$; the generator representable semiseparable matrices also belong to this class.

2.10.2   The decoupled Givens-vector representation

Also the decoupled Givens-vector representation no longer uses the connection between the upper triangular and the lower triangular part. Therefore, we represent the lower triangular part with the Givens-vector representation, and similarly we represent the strictly upper triangular part of the matrix with the Givens-vector representation.

Let us denote the sequence of Givens transformations and the vector of the Givens-vector representation for the lower triangular part as $G, \mathbf{v}$ and the ones for the strictly upper triangular part as $H, \mathbf{e}$. The matrices G, H and the vectors $\mathbf{e}$ and $\mathbf{v}$ are defined similarly as in Subsection 2.8.1.

Using the uncoupled representation, our resulting matrix A will be of the following form.

$$A = A_L + A_U, \qquad (2.22)$$

where

$$A_L = \begin{bmatrix} c_1 v_1 & 0 & 0 & \cdots & 0 & 0 \\ c_2 s_1 v_1 & c_2 v_2 & 0 & & \vdots & \vdots \\ c_3 s_2 s_1 v_1 & c_3 s_2 v_2 & c_3 v_3 & \ddots & & \\ \vdots & & \ddots & \ddots & 0 & 0 \\ c_{n-1} s_{n-2} \cdots s_1 v_1 & \cdots & & & c_{n-1} v_{n-1} & 0 \\ s_{n-1} s_{n-2} \cdots s_1 v_1 & \cdots & & & s_{n-1} v_{n-1} & v_n \end{bmatrix}$$

and

$$A_U = \begin{bmatrix} 0 & r_1 e_1 & r_2 t_1 e_1 & \cdots & r_{n-2} t_{n-3} \cdots t_1 e_1 & t_{n-2} t_{n-3} \cdots t_1 e_1 \\ 0 & 0 & r_2 e_2 & & \vdots & \vdots \\ 0 & 0 & 0 & \ddots & & \\ \vdots & & \ddots & \ddots & r_{n-2} e_{n-2} & t_{n-2} e_{n-2} \\ 0 & \cdots & & & 0 & e_{n-1} \\ 0 & \cdots & & & 0 & 0 \end{bmatrix}.$$

Clearly the class of matrices representable with this decoupled representation contains the class of semiseparable matrices.

Notes and references

For example, the decoupled representation, which represents in fact a more general class of matrices, is considered in the following publication [284]. Van Camp, Van Barel and Mastronardi provide two fast algorithms for solving a semiseparable plus diagonal matrix. The representation used for the semiseparable plus diagonal matrix is the decoupled generator representation. Hence the class of matrices considered is in fact the class of generator representable quasiseparable matrices. The solution method consists of an effective calculation of the QR-factorization of this type of matrices. Moreover, in the actual implementation they chose to use the decoupled Givens-vector representation. More information on the implementation using the decoupled representation can be found in Section 4.6 in Chapter 4.

2.11   Summary of the representations

In this section we will briefly summarize all the results presented in this chapter with regard to the representations. First, we will reconsider some special newly defined classes of matrices, due to the combinations of representations and definitions of matrices. Second, we will reconsider the number of parameters needed to represent the different classes of matrices with the different representations.

From Chapter 1, we already know the relations between all the following classes of matrices (symmetric as well as unsymmetric): $\mathcal{S}$, $\mathcal{S}^{(g)}$, $\mathcal{S}^{(d)}$, $\mathcal{Q}$ and $\mathcal{Q}^{(s)}$. In this chapter in fact three new classes were introduced: $\mathcal{Q}^{(g)}$, $\mathcal{S}^{(s)}$ and $\mathcal{S}^{(g,d)}$, which are, respectively, the class of generator representable quasiseparable matrices, the class of diagonal-subdiagonal representable semiseparable matrices and the class of generator representable semiseparable plus diagonal matrices. Especially classes $\mathcal{Q}^{(g)}$ and $\mathcal{S}^{(g,d)}$ are commonly used classes, with in fact each having an inherent representation. We know, due to the results of the first chapter, that $\mathcal{Q}^{(g)}_{sym} = \mathcal{S}^{(g,d)}$. But one should be aware that both these classes often use a different representation, whereas the class $\mathcal{Q}^{(g)}$ does not expand the generators to the diagonal; this is the case for the class $\mathcal{S}^{(g,d)}$. The classes $\mathcal{Q}^{(g)}$ and $\mathcal{S}^{(g,d)}$ in general are not the same classes. In the class $\mathcal{S}^{(g,d)}$ the semiseparable part is in general generator representable matrix, whereas (see the results in Chapter 1) it cannot be guaranteed that every nonsymmetric matrix from $\mathcal{Q}^{(g)}$ can be written in this way. If a coupling is required between the upper and lower triangular part of the generator representable semiseparable part of the semiseparable plus diagonal matrix, one has normally a matrix from $\mathcal{S}^{(g,d)}$, if no relation is required, in general the matrix belongs to $\mathcal{Q}^{(g)}$.

Below two tables will be given, showing on the left the different representations and on the top the different classes of matrices. The number of theoretically involved parameters is given on each intersection. The parameter count neglects the constants, as constants might sometimes change for specific problems, e.g., is normalization used for the generator representation or not? We also like to remark that the Givens-vector representation essentially uses $2n$ parameters but in practice, both the cosines and the sines (or even the completed Givens transformations) are often stored and hence we obtain $3n$ parameters. We denote however the essential number of parameters needed plus of course $\mathcal{O}(n)$.

The first table shows the number of parameters used for the symmetric matrices, and the symmetric representations. A $\times$ is inserted if the representation is not suitable to represent that class of matrices.

	$\mathcal{S}^{(g)}_{sym}$	$\mathcal{S}^{(g,d)}_{sym}$	$\mathcal{Q}^{(g)}_{sym}$	$\mathcal{S}_{sym}$	$\mathcal{S}^{(d)}_{sym}$	$\mathcal{Q}_{sym}$
Generator	$2n$	$3n$	$3n$	$\times$	$\times$	$\times$
Givens-Vector	$2n$	$3n$	$3n$	$2n$	$3n$	$3n$
Quasiseparable	$3n$	$4n$	$4n$	$3n$	$4n$	$4n$

The second table shows the number of parameters used for the nonsymmetric matrices and the nonsymmetric representations. The number of parameters, written in *(italics)*, are theoretical estimates which were not investigated in detail in the text. If the numbers are separated by a /, this means that in some cases the number of parameters can be reduced, e.g., by exploiting the connection in the diagonal if all diagonal elements are different from zero.

	$\mathcal{S}^{(g)}$	$\mathcal{S}^{(g,d)}$	$\mathcal{Q}^{(g)}$	$\mathcal{S}$	$\mathcal{S}^{(d)}$	$\mathcal{Q}$
Generator	$4n/3n$	$5n/4n$	$5n$	$\times$	$\times$	$\times$
Givens-Vector	$4n/3n$	$5n/4n$	$5n$	$4n/3n$	$5n/4n$	$5n$
Quasiseparable	$(6n/5n)$	$(7n/6n)$	$7n$	$(5n)$	$(6n)$	$7n$
Decoupled Generator	$4n$	$5n$	$5n$	$\times$	$\times$	$\times$
Decoupled Givens-Vector	$4n$	$5n$	$5n$	$4n$	$5n$	$5n$

2.12 Are there more representations?

The formal question which we would like to address in this section is: "Are there more representations for semiseparable? If yes, which ones?" We can easily state that there are many more representations, but it is impossible to say how many and hence equally difficult to say which ones.

Let us illustrate this with an example. Suppose we have a tridiagonal matrix T given to us in factored form. Assume $T = LU$, which is the LU-decomposition of the matrix T. Hence both L and U are bidiagonal matrices, respectively lower and upper bidiagonal. If we need to solve for example a system of equations for the matrix T, it would be preferrable to work with L and U, instead of explicitly forming the product T. Hence, we can conclude that L and U represent in some sense T. Moreover, one can check that, based on the L and U factors of the LU-decomposition of the tridiagonal matrix T, one can easily construct a representation map for the class of tridiagonal matrices.

We illustrated by the previous example that it is sometimes useful not to represent the matrix itself, but to represent lets say its factors if the matrix is in factored form.

The same is true for the semiseparable case. Even though it will be shown later on, when computing the QR-factorization of semiseparable matrices, we can already state that the Givens-vector representation is closely related to the QR-factorization of a semiseparable matrix. The Givens transformations correspond to the Q-factor of the QR-factorization of the semiseparable matrix.

In this viewpoint one can try to construct representations using the LU-decomposition by using partial QR-factorization, changing the QR-factorization with Givens to a factorization with Householders, using the essential factors in the Levinson algorithm, storing not the matrix but its inverse, . . . In fact there are numerous ways to represent matrices. And, as told before, the choice for a specific representation has to be made by the reader himself depending on problem-specific factors.

The representations more thoroughly discussed in this chapter are the most commonly used ones; therefore we chose to focus attention on these representations and to present a more detailed analysis of them. As mentioned previously the quasiseparable representation covered the generator representation and the Givens-vector representation. Moreover, if one represents the lower triangular part of a semiseparable matrix via its LU-decomposition (more information can be found in Chapter 4), we get also a special form of the quasiseparable representation.

To conclude one might say that there are infinitely many representations, all having their pros and cons.

2.13   Some algorithms related to representations

In this last section, we will provide some useful algorithms for dealing with the different representations. First, it will be shown that all structured rank matrices as presented in Chapter 1, with different representations as given in this chapter, admit order $\mathcal{O}(n)$ multiplications with a vector. Second, some algorithms will be given to interchange from one representation to another. Finally, to conclude this section and this chapter, a method is presented to calculate the determinant of a Givens-vector represented semiseparable matrix.

2.13.1   A fast matrix vector multiplication

We will derive three general schemes for order $\mathcal{O}(n)$ multiplication algorithms for multiplying a matrix A with a vector x: three schemes, because we will cover three representations, namely the generator, Givens-vector and the quasiseparable representation.

The matrix A will be a structured rank matrix of generator representable semiseparable, semiseparable, semiseparable plus diagonal or quasiseparable form. The matrix A can in fact always be written as the sum of a lower triangular, an upper triangular and a diagonal matrix

$$A = A_L + A_U + A_D.$$

For each type of matrix, these terms have a specific structured rank form.

Example 2.19 For example, if A is semiseparable, A can be written as the sum of a lower semiseparable matrix A_L and a strictly upper triangular matrix A_U, for which the strictly upper triangular part is also semiseparable. Or if A is quasiseparable, A can be written as the sum of a strict lower triangular, a strict upper triangular and a diagonal matrix. The strict lower (upper) triangular matrix has the strict lower (upper) triangular part of semiseparable form.

Please note that the upper and lower triangular matrices considered in this example are not yet subject to a specific representation.   ∎

Multiplication of the matrix A with a vector x can be decoupled now into a sum of three separate matrix vector multiplications:

$$A\mathbf{x} = A_L\mathbf{x} + A_U\mathbf{x} + A_D\mathbf{x}.$$

The multiplication of a lower and/or an upper triangular matrix of a specific representation with a vector are quite similar, and from an implementational point of view the multiplications with a strictly lower triangular and lower triangular are also comparable. Hence, we only include the algorithms for multiplication with a specific lower triangular matrix of a certain type of representation. Adapting this algorithm to a different case ((strictly)lower/(strictly)upper) is almost trivial. Below, algorithms are derived for a fast multiplication of the form $A_L\mathbf{x} = \mathbf{y}$.

The generator representation

The lower triangular part A_L is assumed to be generator representable and hence the equation $A_L \mathbf{x} = \mathbf{y}$ can be written as:

$$
\begin{bmatrix}
u_1 v_1 & & & & \\
u_2 v_1 & u_2 v_2 & & & \\
u_3 v_1 & u_3 v_2 & u_3 v_3 & & \\
\vdots & & & \ddots & \\
u_n v_1 & u_n v_2 & u_n v_3 & \cdots & u_n v_n
\end{bmatrix}
\begin{bmatrix}
x_1 \\ x_2 \\ x_3 \\ \vdots \\ x_n
\end{bmatrix}
=
\begin{bmatrix}
y_1 \\ y_2 \\ y_3 \\ \vdots \\ y_n
\end{bmatrix} .
$$

Expanding towards the rows leads to the following equations:

$$
\begin{aligned}
y_1 &= u_1 v_1 x_1 \\
y_2 &= u_2 v_1 x_1 + u_2 v_2 x_2 = u_2(v_1 x_1 + v_2 x_2) \\
y_3 &= u_3 v_1 x_1 + u_3 v_2 x_2 + u_3 v_3 x_3 = u_3(v_1 x_1 + v_2 x_2 + v_3 x_3) \\
&\vdots \\
y_n &= u_n(v_1 x_1 + v_2 x_2 + \ldots v_n x_n).
\end{aligned}
$$

Rewriting these formulas, one comes to a general formula

$$
y_i = u_i \left(\sum_{k=1}^{i} v_k x_k \right),
$$

leading to a complexity count of $3n-1$ operations for performing this multiplication.

The Givens-vector representation

Suppose the lower triangular part is represented with the Givens-vector representation.

$$
G = \begin{bmatrix}
c_1 & c_2 & \cdots & c_{n-1} \\
s_1 & s_2 & \cdots & s_{n-1}
\end{bmatrix}
$$

$$
\mathbf{v} = \begin{bmatrix} v_1 & v_2 & \cdots & v_n \end{bmatrix}^T
$$

This means that the multiplication with A_L is of the form:

$$
\begin{bmatrix}
c_1 v_1 & & & & \\
c_2 s_1 v_1 & c_2 v_2 & & & \\
c_3 s_2 s_1 v_1 & c_3 s_2 v_2 & c_3 v_3 & & \\
\vdots & & & \ddots & \ddots & \\
c_{n-1} s_{n-2} \cdots s_1 v_1 & \cdots & & & c_{n-1} v_{n-1} & \\
s_{n-1} s_{n-2} \cdots s_1 v_1 & \cdots & & & s_{n-1} v_{n-1} & v_n
\end{bmatrix}
\begin{bmatrix}
x_1 \\ x_2 \\ x_3 \\ \vdots \\ x_{n-1} \\ x_n
\end{bmatrix}
=
\begin{bmatrix}
y_1 \\ y_2 \\ y_3 \\ \vdots \\ y_{n-1} \\ y_n
\end{bmatrix}
$$

Expanding the equation above towards its rows gives us (only the first 4 components of y are denoted):

$$y_1 = c_1 v_1 x_1$$
$$y_2 = c_2 s_1 v_1 x_1 + c_2 v_2 x_2$$
$$y_3 = c_3 s_2 s_1 v_1 x_1 + c_3 s_2 v_2 x_2 + c_3 v_3 x_3$$
$$y_4 = c_4 s_3 s_2 s_1 v_1 x_1 + c_4 s_3 s_2 v_2 x_2 + c_4 s_3 v_3 x_3 + c_4 v_4 x_4.$$

We use some temporary variables called a_i. Rewriting the formulas reveals the order $\mathcal{O}(n)$ algorithm for the multiplication.

$$y_1 = c_1 \left(v_1 x_1 \right)$$
$$ = c_1 a_1$$
$$y_2 = c_2 \left(s_1 v_1 x_1 + v_2 x_2 \right)$$
$$ = c_2 \left(s_1 a_1 + v_2 x_2 \right)$$
$$ = c_2 a_2$$
$$y_3 = c_3 \left(s_2 \left(s_1 v_1 x_1 + v_2 x_2 \right) + v_3 x_3 \right)$$
$$ = c_3 \left(s_2 a_2 + v_3 x_3 \right)$$
$$ = c_3 a_3$$
$$y_4 = c_4 \left(s_3 \left(s_2 \left(s_1 v_1 x_1 + v_2 x_2 \right) + v_3 x_3 \right) x_3 + v_4 x_4 \right)$$
$$ = c_4 \left(s_3 a_3 + v_4 x_4 \right)$$
$$ = c_4 a_4.$$

More precisely, this involves $4n - 3$ operations.

The quasiseparable representation

Representing the matrix A_L in the quasiseparable form gives us structurally the same matrix as the matrix represented with the Givens-vector representation. A deduction of an multiplication hence is completely similar and therefore not included. The complexity is however (if implemented similar as for the Givens-vector multiplication) $4n - 2$. This involves one operation more, which becomes clear comparing the explicit structure of a quasiseparable and a Givens-vector represented lower semiseparable matrix.

Summary

Below you find a table, summarizing the complexities of multiplying a lower triangular semiseparable matrix, in a specific representation, with a vector. On top you find the representations, in the table the complexities.

	Generator	Givens-vector	Quasiseparable
Multiplication	$3n - 1$	$4n - 3$	$4n - 2$

2.13.2 Changing representations

In this section we will briefly illustrate some ways to change from one representation into another. First, we present some trivial ways for bringing the generator representation and the Givens-vector representation to the quasiseparable form. Second, methods are presented for going from the quasiseparable and the generator representation towards the Givens-vector form. Finally, we present some algorithms to go from the quasiseparable and the Givens-vector representation towards the generator representation. These final transformations into the generator representation can only be done if the matrix satisfies some additional constraints, as will be shown. The methods presented here are all of order $\mathcal{O}(n)$, and attention was also paid to the numerical reliability. It is only illustrated for changing the representation of a lower triangular rank structured matrix; other cases are similar. It was mentioned previously that the Givens-vector representation has two variants the one from top to bottom and the one from right to left. How to change between these representations was discussed in Section 2.4.4.

From Givens-vector to the quasiseparable representation

In the definition of the quasiseparable representation it has already been stated that this is the most general representation. But it is also not uniquely determined. Therefore, there are always different ways to represent a matrix using the quasiseparable representation. Here one possibility is presented, starting from the Givens-vector representation. As the Givens-vector representation is in fact a special quasiseparable representation, there are no essential changes necessary to go from the Givens-vector representation to the quasiseparable one. Comparing the structure of the parameters in both representations immediately reveals how the quasiseparable parameters should be chosen. Hence the complexity equals zero operations.

From generator to the quasiseparable representation

As already stated, the quasiseparable representation is not uniquely determined. This means that there are several quasiseparable representations for the generator representable semiseparable matrix. We illustrate two ways.

- The first possibility is straightforward. As the Givens-vector representation is in fact a quasiseparable representation, one might just transfer the generator represented semiseparable matrix into a Givens-vector represented one. The Givens-vector represented matrix is already in quasiseparable form. The procedure of transforming a generator represented matrix into a Givens-vector represented one is shown later in this section.

- The second possibility is an even easier approach. Assume we have a semiseparable matrix, which is represented by two vectors $\mathbf{u}$ and $\mathbf{v}$. Consider the structure of the quasiseparable representation from Section 2.5. One can clearly see that choosing as generators for the quasiseparable part the vectors $\mathbf{u}$, $\mathbf{v}$ and $\mathbf{t}$, where $t_i = 1$, for every i, that we get exactly the generator represented matrix. This costs zero operations.

From generator to Givens-vector representation

In this section we will explain a simple way to go from the generator representation towards the Givens-vector representation.

Assume we have the following 4×4 generator representable semiseparable matrix (only the lower triangular part is shown)

$$S(\mathbf{u}, \mathbf{v}) = \begin{bmatrix} u_1 v_1 \\ u_2 v_1 & u_2 v_2 \\ u_3 v_1 & u_3 v_2 & u_3 v_3 \\ u_4 v_1 & u_4 v_2 & u_4 v_3 & u_4 v_4 \end{bmatrix},$$

and we want to calculate the Givens-vector representation, where the Givens transformations are denoted by G_i and the vector is denoted by $\hat{v}$.

We construct the algorithm based on this simple example; the general case is similar. It can be clearly seen that the dependencies between the rows are only determined by the vector $\mathbf{u}$. So we can easily calculate the Givens transformations G_1, G_2 and G_3, because they satisfy the following equalities:

$$G_3 \begin{bmatrix} r_3 \\ 0 \end{bmatrix} = \begin{bmatrix} u_3 \\ u_4 \end{bmatrix}, \quad G_2 \begin{bmatrix} r_2 \\ 0 \end{bmatrix} = \begin{bmatrix} u_2 \\ r_3 \end{bmatrix} \text{ and } G_1 \begin{bmatrix} r_1 \\ 0 \end{bmatrix} = \begin{bmatrix} u_1 \\ r_2 \end{bmatrix}.$$

The elements of the vector $\hat{\mathbf{v}}$ for the Givens-vector representation can easily be retrieved computing the norm of each column, similarly as done in Section 2.4.3. The procedure on how to compute these vector elements in $\mathcal{O}(n)$ is clear, when looking at the following formulas.

$$\hat{v}_4 = \left(\sqrt{u_4^2} \right) v_4,$$

$$\hat{v}_3 = \left(\sqrt{u_4^2 + u_3^2} \right) v_3,$$

$$\hat{v}_2 = \left(\sqrt{u_4^2 + u_3^2 + u_2^2} \right) v_2,$$

$$\hat{v}_1 = \left(\sqrt{u_4^2 + u_3^2 + u_2^2 + u_1^2} \right) v_1.$$

Hence the complete procedure, to go from a generator represented semiseparable matrix to a Givens-vector represented one, can be done in $\mathcal{O}(n)$ operations. More precisely, denote the cost of computing the Givens transformation and the norm r_i of the involved vector by $\mathcal{C}_G$[20], then we get that we need: $4n - 1 + (n-1)\mathcal{C}_G$ operations.

---

[20] For example, in [152] an algorithm involving 6 operations is proposed for computing the Givens transformation and an extra 3 operations are needed for computing r (other variants exist).

From the quasiseparable to the Givens-vector representation

The lower triangular part of quasiseparable form means that the matrix A is of the following form (We consider again a 4×4 example, as this illustrates the general case.):

$$
A = \begin{bmatrix}
u_1 v_1 & & & \\
u_2 t_1 v_1 & u_2 v_2 & & \\
u_3 t_2 t_1 v_1 & u_3 t_2 v_2 & u_3 v_3 & \\
u_4 t_3 t_2 t_1 v_1 & u_4 t_3 t_2 v_2 & u_4 t_3 v_3 & u_4 v_4
\end{bmatrix}.
$$

The computation of the Givens-vector representation of this matrix proceeds quite similarly as for the generator representation. Let us illustrate the construction for this 4×4 example.

$$
G_3 \begin{bmatrix} r_3 \\ 0 \end{bmatrix} = \begin{bmatrix} u_3 \\ u_4 t_3 \end{bmatrix}, \quad G_2 \begin{bmatrix} r_2 \\ 0 \end{bmatrix} = \begin{bmatrix} u_2 \\ r_3 t_2 \end{bmatrix} \text{ and } G_1 \begin{bmatrix} r_1 \\ 0 \end{bmatrix} = \begin{bmatrix} u_1 \\ r_2 t_1 \end{bmatrix}.
$$

The linear computation of the vector $\hat{\mathbf{v}}$ is due to the following linked equations:

$$
\hat{v}_4 = \left(\sqrt{u_4^2} \right) v_4,
$$

$$
\hat{v}_3 = \left(\sqrt{u_4^2 t_3^2 + u_3^2} \right) v_3,
$$

$$
\hat{v}_2 = \left(\sqrt{(u_4^2 t_3^2 + u_3^2) t_2^2 + u_2^2} \right) v_2,
$$

$$
\hat{v}_1 = \left(\sqrt{((u_4^2 t_3^2 + u_3^2) t_2^2 + u_2^2) t_1^2 + u_1^2} \right) v_1.
$$

It is clear that this interchanging can be achieved in $\mathcal{O}(n)$ operations, namely $(7n - 4) + (n - 1)\mathcal{C}_G$, where $\mathcal{C}_G$ denotes the complexity of computing the Givens transformation.

From Givens-vector to the generator representation

As not every Givens-vector representable semiseparable matrix can be represented using the generator representation, we have to place a condition on these matrices. This condition is a direct consequence from Proposition 1.13; namely, we assume the bottom left element to be different from zero. This means there are no blocks below the diagonal.

This means that our lower triangular matrix A, Givens-vector represented is of the following form:

$$
\begin{bmatrix}
c_1 \hat{v}_1 & & & & \\
c_2 s_1 \hat{v}_1 & c_2 \hat{v}_2 & & & \\
c_3 s_2 s_1 \hat{v}_1 & c_3 s_2 \hat{v}_2 & c_3 \hat{v}_3 & & \\
\vdots & & \ddots & \ddots & \\
c_{n-1} s_{n-2} \cdots s_1 \hat{v}_1 & \cdots & & c_{n-1} \hat{v}_{n-1} & \\
s_{n-1} s_{n-2} \cdots s_1 \hat{v}_1 & \cdots & & s_{n-1} \hat{v}_{n-1} & \hat{v}_n
\end{bmatrix},
$$

with the lower left element different from zero. This means that all s_i values also need to be different from zero. Hence the following assignments are well defined:

$$\mathbf{v}^T = \left[\hat{v}_1, \frac{\hat{v}_2}{s_1}, \frac{\hat{v}_3}{s_1 s_2}, \ldots, \frac{v_n}{s_1 s_2 \cdots s_{n-1}}\right],$$

$$\mathbf{u}^T = [c_1, \; c_2 s_1, \; c_3 s_2 s_1, \; \ldots, \; s_{n-1} s_{n-2} \cdots s_1].$$

This costs $3n - 5$ operations.

From the quasiseparable to the generator representation

Because the structure of the quasiseparable representation is close to the Givens-vector representation we do not include these computations. They are a simple adaptation of the transformation from Givens-vector to generator representable which involves $3n - 4$ operations.

Summary

In the table below a summary of the computational complexity of changing between the representations is shown. On an intersection the complexity of going from the left representation to the top one is shown. If a complexity is put in *(italics)* this means that this transition is not always possible. As mentioned before $\mathcal{C}_G$ denotes the complexity of computing a Givens rotation and the norm of the involved vector.

	Generator	Givens-vector	Quasiseparable
Generator	$\times$	$4n - 1 + (n-1)\mathcal{C}_G$	0
Givens-vector	$(3n - 5)$	$\times$	0
Quasiseparable	$(3n - 4)$	$7n - 4 + (n-1)\mathcal{C}_G$	$\times$

2.13.3 Computing the determinant of a semiseparable matrix in $\mathcal{O}(n)$ flops

In this section, we will design an order $\mathcal{O}(n)$ algorithm for calculating the determinant of a semiseparable matrix represented with the Givens-vector representation.

We will use the fact that the Givens transformations for representing the matrix, in fact, contain all the information needed for the QR-factorization of the corresponding matrix. Using this information, we can very easily calculate the diagonal elements of the R factor of the semiseparable matrix. Multiplying these diagonal elements will give us the wanted determinant of the semiseparable matrix.

We have, because of the special structure of the representation, the Givens transformations $G_1, \ldots, G_{n-1}$ such that applying G_{n-1}^T on the last two rows will annihilate all the elements except for the diagonal element; applying G_{n-2}^T on the third last and second last row will annihilate all the elements in the second last row, except the last two elements (note that the Givens transformations by construction have the determinant equal to 1). In this fashion we can continue very easily to annihilate all the elements in the strictly lower triangular part. In fact we are only interested in the diagonal elements. Performing these transformations will change the diagonal elements in the following way. Denote with d_i the diagonal elements of the semiseparable matrix, and with $d_i^{(s)}$ the super diagonal elements. The diagonal elements $i = 2, \ldots, n$ change by performing:

$$G_i^T \left[\begin{array}{c} d_{i-1}^{(s)} \\ d_i \end{array} \right] = \left[\begin{array}{c} \hat{d}_{i-1}^{(s)} \\ \hat{d}_i \end{array} \right]$$

where the elements $\hat{d}_i$ denote the diagonal elements of the upper triangular matrix R, which are the only essential ones for computing the determinant. Using this information one can deduce an method involving $4n - 4$ operations for computing the determinant.

More information about the QR-factorization, from a computational point of view, can be found in Chapter 5 where the more general class of semiseparable plus diagonal matrices is considered. More information about the structure of the Q factor and the R factor when calculating the QR-factorization of a semiseparable plus diagonal matrix can also be found there.

Notes and references

As most of the manuscripts link a specific class of matrices to a specific type of representation, not many results are known about changing representations, stability issues, and so on.

Historically, however, attention was paid to calculations of the determinant of structured rank matrices. Statistical researchers were especially interested in these quantities.

The previously discussed book of Gantmacher and Kreĭn [131] and the paper of Roy, Greenberg, et al. [239] also cover interesting results concerning the determinants. Theoretical results as well as practical algorithms are discussed. More information on these references can be found in Section 2.2 in Chapter 2 for reference [239] and in Section 1.1, Chapter 1 for reference [131].

The book by Graybill [158] is concerned with statistics. Several specific types of semiseparable matrices and inversion algorithms are discussed, as well as ways to compute the determinant. More references on semiseparable matrices related to statistical applications are discussed in Section 3.2 in Chapter 3.

☞ W. D. Hoskins and M. C. Thurgur. Determinants and norms for the inverses of a set of band matrices. *Utilitas Mathematica*, 3:33–47, 1973.

Hoskins and Thurgur invert a specific type of band matrix (see [179]). The inversion formulas are obtained by calculating the LU-decomposition of the band matrix and formulas are given to compute the determinant and the infinity norm of this inverse. The considered matrices arise in the solution of a specific differential equation.

Historically a lot of matrix computations focused on computations with determinants. The following booklist includes several examples of this approach.

☞ R. Baltzer. *Theorie und Anwendung der Determinanten*. S. Hirzel, fifth edition, 1881. (In German).

☞ W. S. Burnside and A. W. Panton. *An introduction to determinants*, chapter from The theory of equations. Hodges, Figgis, & Co. and Longmans, Green & Co., 1899.

☞ C. E. Cullis. *Matrices and determinoids*, volume 1 of *Readership lectures.* Cambridge, University Press, 1913.

☞ W. L. Gifford. *A short course in the theory of determinants.* Macmillan and Co., 1893.

☞ W. L. Gifford. *Determinants*, volume 3 of *Mathematical monographs.* John Wiley and Sons, fourth edition, 1906.

☞ E. Pascal. *Die Determinanten.* Teubner, B.G., 1900. (In German).

☞ M. Reiss. *Beiträge zur Theorie der Determinanten.* B.G. Teubner, Leipzig, 1867. (In German).

☞ R. F. Scott. *Theory of Determinants and their applications.* Cambridge University Press, second edition, 1904.

2.14   Conclusions

In this chapter we first presented a definition for a representation of a class of matrices. Moreover, it was also clearly stated that a representation is just a manner of working with matrices from a specific class.

First, we started our investigations with the class of symmetric matrices for which four different types of representations were investigated: the generator representation, the diagonal-subdiagonal representation, the Givens-vector representation and the quasiseparable representation. For each of these representations attention was paid to several important issues, concerning the set of matrices to which this representation was applicable, the construction of the representation map and adaptation to different classes of matrices. It was shown that the quasiseparable representation was the most general one, covering as special cases the generator representation and the Givens-vector representation. The diagonal-subdiagonal representation was investigated from a theoretical viewpoint, but will become handy in inverting specific types of matrices.

Second, we started our analysis of the different representations by taking a closer look at the class of nonsymmetric matrices. We investigated the generator representation, the Givens-vector representation and the quasiseparable representation. The class of nonsymmetric generator representable semiseparable matrices were investigated in detail, because this class received a lot of attention in the literature. The nonsymmetric Givens-vector and quasiseparable representation for semiseparable matrices were not covered in detail as they are not so frequently used.

The last type of representations considered is the class of decoupled representations for the semiseparable and the generator representable semiseparable matrices. In fact these representations do not use the interaction between the upper triangular part and the lower triangular part. Therefore they use more parameters.

In a final section of this chapter we constructed some algorithms, working directly with some of the proposed representations in this chapter. First, we showed that all representations admit a linear multiplication algorithm with a vector. Second, different algorithms were proposed to interchange between the different representations. Finally, a specific example of an algorithm for computing the determinant of a Givens-vector represented matrix was given.

We are now familiar with the different classes of semiseparable matrices, as well as with the different ways of representing them. In the next and final chapter of this part we will show some traditional examples related to semiseparable matrices. We will also briefly discuss some other topics of interest, such as orthogonal rational functions related to semiseparable matrices and eigenvalue problems via semiseparable matrices. The chapter could be omitted, however, and the reader could immediately start with the part on system solving.

Chapter 3

Historical applications and other topics

In this chapter some historical applications and early appearances of semiseparable and related matrices are investigated. Some links to other closely related topics not covered in this book are also presented, e.g., the eigenvalue problem via semiseparable matrices. The matrices considered in this chapter are of the forms discussed in the first two chapters of this book. This chapter is not essential for a full understanding of the forthcoming chapters. It is merely intended to present some interesting applications in which semiseparable matrices appear. Also an overview of publications closely related to semiseparable matrices but not covered in this book is presented.

In the first section we examine the properties of oscillation matrices. After the definition and some properties of this type of matrices are given, some examples are considered. It is shown that Jacobi (i.e., irreducible tridiagonal matrices) and symmetric generator representable semiseparable matrices are oscillation matrices. Moreover, to our knowledge, the earliest proof that the inverse of an irreducible symmetric tridiagonal matrix is a generator representable semiseparable matrix is included. The proof is constructive, i.e., the inverse of the Jacobi matrix is explicitly calculated via determinantal formulas. After this proof a physical interpretation of such a generator representable matrix will be given. It is shown that the vibrational properties of a string are related to this matrix. To conclude the chapter it is shown that the properties of eigenvalues and eigenvectors of oscillation matrices are directly linked with the physical interpretation of such an oscillation system.

Semiseparable matrices do not only appear in physical applications, but also in the field of statistics. In Section 3.2 we investigate some of these matrices. We calculate the covariance matrix for a multinomial distribution. We will show that this matrix is a semiseparable plus diagonal matrix. Other multivariate distributions for which the covariance matrices have the semiseparable plus diagonal form are included. In some of the statistical applications the eigenvalues and/or eigenvectors of these covariance matrices are desired. Sometimes the complete inverse of these matrices is of interest.

In Section 3.3 an integral equation with a so-called Green's kernel is discretized via the trapezoidal rule. The resulting matrix is of semiseparable plus diagonal form.

In Section 3.4 the connection between orthogonal rational functions and semiseparable plus diagonal matrices is briefly investigated.

Section 3.5 presents some references related to semiseparable matrices with a brief explanation. The references are dedicated to items not covered in detail in the book. It deals with the eigenvalue problems related to semiseparable matrices and also some references to uncovered applications.

✎ *In this chapter some early results on semiseparable matrices are presented. The matrices are placed in their original settings such as oscillation matrices and covariance matrices. Also in brief some results related to orthogonal rational functions and references on eigenvalue problems are presented. These results are for the interested reader but do not really contribute towards a better understanding of the following chapters. We will however briefly indicate the main results of each section.*

Section 3.1 discusses oscillation matrices. The results presented here are most often referred to as being the first in which semiseparable matrices appeared. After an introduction to oscillation matrices the definition of an oscillation matrix is given in Definition 3.4. Following the definition, examples are given, including the example of a tridiagonal matrix. Based on the fact that the inverse of an oscillation matrix is again an oscillation matrix the inverse of this tridiagonal matrix is investigated. In Theorem 3.12 it is proved that the inverse of an irreducible tridiagonal matrix is a generator representable semiseparable matrix. The proof is based entirely on results related to determinants of matrices. An interesting example of oscillation matrices is given in Subsection 3.1.4.

Section 3.2 shows that several covariance matrices are of structured rank form. The example presented in Subsection 3.2.2 shows clearly the computations to obtain a semiseparable covariance matrix.

In Section 3.3 a discretization of an integral equation is presented leading to a system of equations involving a semiseparable matrix.

Section 3.5 contains an overview of references related to eigenvalue problems. For those interested in eigenvalue computations related to structured rank matrices, this list can point them directly to the interesting references.

3.1 Oscillation matrices

These matrices appeared for the first time in the manuscripts [126, 125] of Gantmacher and Kreĭn and combined in book form in [127, 128], which was published in Russia in (1940 first ed., 1950 second ed.). The book was written in Russian. Several translations were made of it, first into the German language [129, 1960] and afterwards [130, 1961] also into English. The complete book [131], as it appeared in 2002, is based on these three references. Because of the many translations, the original Russian name the authors gave to generator representable semiseparable matrices was translated in several papers and books into one-pair matrices, while other authors translated it into single-pair matrices. In our survey of this book we will use the name one-pair matrix. In fact these matrices are a special sort of semi-separable matrix, and as we will prove they have strong connections with "Jacobi" matrices. Jacobi matrices are nowadays more commonly known as irreducible tridiagonal matrices. Here, the origin of oscillation matrices in physics will be briefly mentioned. It is also shown that a symmetric generator representable semiseparable matrix, under some extra conditions, can be considered as an oscillation matrix. A proof based on properties of determinants is also included to show that the inverse of a one-pair matrix is a Jacobi matrix.

3.1.1 Introduction

Let us consider small transverse oscillations of a linear elastic continuum (e.g., a string or a rod), spread along the x-axis ranging from $x = a$ to $x = b$. The natural harmonic oscillation of the continuum is given by:

$$y(x, t) = \varphi(x) \sin(pt + \alpha).$$

The function $y(x, t)$ denotes the deflection at point x at time t, $\varphi(x)$ stands for the amplitude at point x, α for the phase and p for the frequency of the oscillation. A so-called segmental continuum has the following main oscillation properties (from [131, Introduction]):

1. All the frequencies p are simple. (This means that the amplitude function of a given frequency is uniquely determined up to a constant factor.)

2. At frequency p_i the oscillation has exactly i nodes. (Suppose x to be a node, this means that $\varphi(x) = 0$.)

3. The nodes of two successive overtones alternate.

4. When superposing natural oscillations with frequencies $p_k < p_l < \ldots < p_m$, the number of sign changes of the deflection fluctuates with time within the limits from k to m.

Now we will construct a so-called oscillation matrix, and in [131] it is shown that all the properties mentioned above are strictly connected to the properties of the oscillation matrix. Suppose our oscillation system consists of n masses $m_1, m_2, \ldots, m_n$, located at points $s_1, s_2, \ldots, s_n$. Note that the masses are not fixed but movable in the direction of the y-axis. With $K(x, s)$ we denote the so-called influence function. It denotes the deflection at point x under the influence of a unit force at point s. Denote $a_{ik} = K(s_i, s_k)$. Assume $y_1, y_2, \ldots, y_n$ to be the deflection of these masses, under the influence of a force:

$$-m_k \frac{d^2 y_k}{dt^2} \quad (k = 1, 2, \ldots, n).$$

We have that the deflection at point x at time t can be written as

$$y(x, t) = -\sum_{k=1}^{n} K(x, s_k) m_k \frac{d^2 y_k}{dt^2}.$$

For x equal to $s_1, s_2, \ldots, s_n$ we have

$$y_i = -\sum_{k=1}^{n} a_{ik} m_k \frac{d^2 y_k}{dt^2}.$$

When denoting the amplitude of the deflection as $u_i = \varphi(s_i)$ in the harmonic oscillation equation we get

$$y_i = u_i \sin(pt + \alpha),$$

which leads after differentiation of y_i to the following system of equations:

$$u_i = p^2 \sum_{k=1}^{n} a_{ik} m_k u_k \quad (k = 1, 2, \ldots, n)$$

which can be rewritten to obtain the following system of equations:

$$\begin{cases} (1 - p^2 a_{11} m_1) u_1 - p^2 a_{12} m_2 u_2 - \cdots - p^2 a_{1n} m_n u_n &=& 0 \\ -p^2 a_{21} m_1 u_1 + (1 - p^2 a_{22} m_2) u_2 - \cdots - p^2 a_{2n} m_n u_n &=& 0 \\ \vdots & & \vdots \\ -p^2 a_{n1} m_1 u_1 - p^2 a_{n2} m_2 u_2 - \cdots + (1 - p^2 a_{nn} m_n) u_n &=& 0. \end{cases}$$

Which has a solution if:

$$\det \begin{bmatrix} 1 - p^2 a_{11} m_1 & -p^2 a_{12} m_2 & \cdots & -p^2 a_{1n} m_n \\ -p^2 a_{21} m_1 & 1 - p^2 a_{22} m_2 & \cdots & p^2 a_{2n} m_n \\ \vdots & & \ddots & \vdots \\ -p^2 a_{n1} m_1 & -p^2 a_{n2} m_2 & \cdots & 1 - p^2 a_{nn} m_n \end{bmatrix} = 0$$

revealing thereby the possible frequencies p of the oscillation.

As already mentioned, in [203] Kreĭn observed that the oscillation properties given above are closely related to the influence coefficients a_{ik}. Even more, the oscillation properties are related to the fact that all minors (a minor is the determinant of a submatrix of the given matrix) of the matrix (a_{ik}) (of all orders) have to be nonnegative. The theory connected to these matrices is called the theory of oscillation matrices.

Definition 3.1. *An oscillation matrix A is a matrix such that all minors of all orders of this matrix (principal and nonprincipal) are nonnegative. A minor of A is the determinant of a square submatrix of A.*

Important properties of these oscillation matrices and their connections with the oscillation properties were investigated in the papers [124, 125, 126]. Other authors have also shown their interest in this field of matrices [103, 116, 117], in particular, the article [103] contains several references related to positive and nonnegative matrices. The choice of the term *oscillation matrix* by the authors of the book [131] comes from the following circumstance (citation from [131], Introduction, page 3):

As soon as for a finite system of points, the matrix of coefficients of influence of a given linear elastic continuum is an oscillation matrix (as it always is in the case for a string or a rod supported at the endpoints in the usual manner), this automatically implies the oscillation properties of the vibration of the continuum, for any distribution of masses at these points.

The theory of oscillation matrices (sometimes also other matrix theories) has an analogue in the theory of integral equations. For the oscillation matrices this corresponds to the theory of the following integral equation:

$$\varphi(x) = \lambda \int_a^b K(x,s)\varphi(s)d\sigma(s),$$

with an oscillation kernel $K(x,s)$.

Definition 3.2. *A kernel $K(x,s)$ is called an oscillation kernel if for every choice of $x_1 < x_2 < \ldots < x_n$ in the interval $[a,b]$, the matrix $K(x_i,x_k)$, with $(i,k \in \{1,2,\ldots,n\})$ is an oscillation matrix.*

It follows from the definition that an oscillation kernel is characterized by the following inequalities:

- $\det(K(x_i,s_k)) \geq 0$ for every choice of points $a < x_1 < x_2 < \ldots < x_n < b$ and $a < s_1 < s_2 < \ldots < s_n < b$ where the equality sign should be omitted when $x_i = s_i$;

- $K(x,s) > 0$ for $a < x < b$ and $a < s < b$.

This type of integral equation for $d\sigma = ds$ was investigated in [196, 197]. The more general case for a nonsymmetric kernel and an increasing function $\sigma(s)$ is studied for example in [123]. In Section 3.3 of this chapter we will show that discretization of this integral equation also results in solving a system of equations with a semiseparable matrix as coefficient matrix.

For the purpose of our book this introduction into the theory of oscillation matrices is enough. In the remainder of this section we will take a closer look at two types of oscillation matrices: Jacobi matrices and one-pair matrices. Jacobi matrices are a special type of tridiagonal matrices, namely irreducible ones, and one-pair matrices are a special type of semiseparable matrices, namely symmetric generator representable semiseparable matrices of semiseparability rank 1. Also attention will be paid to the most important properties of these matrices, which are of interest in the theory of oscillation matrices.

3.1.2 Definition and examples

First some new terms have to be defined:

Definition 3.3 (Chapter II, Definition 4 in [131]). *A matrix A is called totally nonnegative (or totally positive) if all its minors of any order are nonnegative (or positive).*

In [131], an oscillation matrix is defined in the following way (which is an extension of Definition 3.1):

Definition 3.4 (Chapter II, Definition 6 in [131]). *A matrix A is called an oscillation matrix if A is totally nonnegative and there exists a positive integer κ such that A^κ is totally positive.*

Some interesting properties of oscillation matrices:

Proposition 3.5. *If A is an oscillation matrix, then we have that*

- *A is nonsingular;*

- *also A^p with p an integer is an oscillation matrix;*

- *also $\left(A^{-1}\right)^T$ is an oscillation matrix.*

Some examples of totally positive and totally nonnegative matrices are given. (Note that a totally positive matrix is already an oscillation matrix). The corresponding proofs can be found in [131, Chapter II, p. 76].

Example 3.6 A generalized Vandermonde matrix:

$$A = (a_i^{\alpha_k})_{ik} = \begin{bmatrix} a_1^{\alpha_1} & a_1^{\alpha_2} & \dots & a_1^{\alpha_n} \\ a_2^{\alpha_1} & a_2^{\alpha_2} & \dots & a_2^{\alpha_n} \\ \vdots & & \ddots & \vdots \\ a_n^{\alpha_1} & a_n^{\alpha_2} & \dots & a_n^{\alpha_n} \end{bmatrix},$$

with $0 < a_1 < a_2 < \dots < a_n$ and $\alpha_1 < \alpha_2 < \dots < \alpha_n$, is totally positive and therefore an oscillation matrix.   ∎

Example 3.7 The Cauchy matrix:

$$A = \left(\frac{1}{x_i + y_k}\right)_{ik} = \begin{bmatrix} \frac{1}{x_1+y_1} & \frac{1}{x_1+y_2} & \dots & \frac{1}{x_1+y_n} \\ \frac{1}{x_2+y_1} & \frac{1}{x_2+y_2} & \dots & \frac{1}{x_2+y_n} \\ \vdots & & \ddots & \vdots \\ \frac{1}{x_n+y_1} & \frac{1}{x_n+y_2} & \dots & \frac{1}{x_n+y_n} \end{bmatrix},$$

with $0 < x_1 < x_2 < \dots < x_n$ and $0 < y_1 < y_2 < \dots < y_n$ is totally positive and therefore an oscillation matrix.   ∎

Example 3.8 A one-pair matrix S with elements:

$$s_{i,j} = \left\{ \begin{array}{ll} u_i v_j & (i \geq j) \\ u_j v_i & (i \leq j) \end{array} \right. ,$$

with all the elements u_i and v_j different from zero is totally nonnegative, if and only if all the numbers u_i and v_j have the same sign and:

$$\frac{v_1}{u_1} \leq \frac{v_2}{u_2} \leq \ldots \leq \frac{v_n}{u_n}. \tag{3.1}$$

Moreover the rank of the matrix S is equal to the number of "$<$" signs in (3.1) plus one. Note that the fact that the matrix is totally nonnegative does not imply that it is an oscillation matrix. ∎

Example 3.9 A Jacobi matrix J:

$$J = \begin{bmatrix} a_1 & c_1 & 0 & \ldots & & 0 \\ b_1 & a_2 & c_2 & & & \\ 0 & b_2 & a_3 & & & \\ \vdots & & \ddots & \ddots & & c_{n-1} \\ 0 & \ldots & 0 & b_{n-1} & a_n \end{bmatrix},$$

with all elements b_i and c_i different from zero, and the successive principal minors positive, is totally nonnegative. ∎

As already stated in the example of the one-pair matrices, the condition that the matrix is totally nonnegative is not sufficient to form an oscillation matrix. the following theorem can help us in the case of the one-pair and Jacobi matrices.

Theorem 3.10 (Chapter II, Theorem 10 in [131]). *It is necessary and sufficient for a totally nonnegative matrix A that:*

- *A is a nonsingular matrix;*

- *$a_{i,i+1} > 0$ and $a_{i+1,i} > 0$ for all $i = 1, 2, \ldots, n - 1$;*

in order to be an oscillation matrix.

With this theorem one can clearly see that one-pair matrices, plus one extra demand, are oscillation matrices and the only extra demand which has to be placed on the Jacobi matrix is that the sub and superdiagonal elements have to be positive. We know now what an oscillation matrix is and also that under certain circumstances a tridiagonal and a semiseparable matrix will be oscillation matrices. In the following subsections we will explain why the eigenvalues and eigenvectors of oscillation matrices are important, in connection with the physical interpretation. Also for the one-pair matrix we will construct the matrix directly from the physical application. Although we have already given a proof in Chapter 1, we will prove here in the old-fashioned way that the inverse of a one-pair matrix is a Jacobi matrix.

3.1.3   The inverse of a one-pair matrix

This section summarizes the results of [131, Chapter II]. The definitions and theorems are adapted such that they fit our notation style. Our main result will be to prove, as Gantmacher and Kreĭn did, that the inverse of a one-pair matrix is a Jacobi one and vice versa. Originally the one-pair matrices were defined as (equivalent to Example 3.8):

Definition 3.11.   *A one-pair matrix is a symmetric matrix S such that*

$$s_{ij} = \begin{cases} u_i v_j & (i \geq j) \\ u_j v_i & (i \leq j) \end{cases} \;,$$

where the elements u_i and v_j are chosen arbitrarily.

It is interesting to see that the proof is completely based on properties of determinants of matrices. Around 1900 several books were written, completely devoted to theoretical results concerning determinants of matrices, e.g., [8, 46, 63, 135, 136, 227, 233, 251] [21].

Theorem 3.12.   *Suppose A is a symmetric Jacobi matrix, i.e., a symmetric tridiagonal matrix of size n:*

$$\begin{bmatrix} a_1 & b_1 & & & & \\ b_1 & a_2 & b_2 & & & \\ & b_2 & a_3 & \ddots & & \\ & & \ddots & \ddots & b_{n-1} & \\ & & & b_{n-1} & a_n \end{bmatrix},$$

with all the b_i different from zero. The inverse of A will be a one-pair matrix.

The proofs of the different theorems and corollaries are mainly based on formulas for the determinants of matrices. First, some notation has to be introduced. Suppose we have an arbitrary $n \times n$ matrix A which is denoted as $A = (a_{i,j})_{i,j \in \{1,\dots,n\}}$.

Definition 3.13.   *Define the matrix $A(i_1, \dots, i_p; j_1, \dots, j_p)$ as the matrix:*

$$A(i_1, \dots, i_p; j_1, \dots, j_p) = (a_{i,j}),$$

with the indices belonging to the following sets:

$$i \in \{i_1, \dots, i_p\},$$
$$j \in \{j_1, \dots, j_p\}.$$

For the determinant of a matrix we will temporarily use the shorter notation:

$$\det(A) = |A|.$$

---

[21]Several of these books can be found in the online historical library at Cornell: http://historical.library.cornell.edu/math/.

Before we can prove the first important statement, a proposition is needed.

Proposition 3.14. *Suppose a Jacobi matrix (irreducible tridiagonal matrix) A of size n is given. If*

$$1 \le i_1 < i_2 < \ldots < i_p \le n,$$
$$1 \le j_1 < j_2 < \ldots < j_p \le n \tag{3.2}$$

and

$$i_1 = j_1, i_2 = j_2, \ldots, i_{\nu_1} = j_{\nu_1},$$
$$i_{\nu_1+1} \ne j_{\nu_1+1}, \ldots, i_{\nu_2} \ne j_{\nu_2},$$
$$i_{\nu_2+1} = j_{\nu_2+1}, \ldots, i_{\nu_3} = j_{\nu_3},$$
$$i_{\nu_3+1} \ne j_{\nu_3+1}, \ldots$$

then

$$|A(i_1, \ldots, i_p; j_1, \ldots, j_p)|$$
$$= |A(i_1, \ldots, i_{\nu_1}; j_1, \ldots, j_{\nu_1})| \cdot |A(i_{\nu_1+1}; j_{\nu_1+1})| \cdots$$
$$|A(i_{\nu_2}; j_{\nu_2})| \cdot |A(i_{\nu_2+1}, \ldots, i_{\nu_3}; j_{\nu_2+1}, \ldots, j_{\nu_3})| \cdots$$

Proof. We will prove that under the conditions (3.2), and $i_\nu \ne j_\nu$ the following equations hold:

$$|A(i_1, \ldots, i_p; j_1, \ldots, j_p)|$$
$$= |A(i_1, \ldots, i_\nu; j_1, \ldots, j_\nu)| \cdot |A(i_{\nu+1}, \ldots, i_p; j_{\nu+1}, \ldots, j_p)| \tag{3.3}$$
$$= |A(i_1, \ldots, i_{\nu-1}; j_1, \ldots, j_{\nu-1})| \cdot |A(i_\nu, \ldots, i_p; j_\nu, \ldots, j_p)|.$$

Proving the first of the two equations is enough to derive the complete desired result. If $i_\nu < j_\nu$ then we have, because the matrix A is tridiagonal,

$$a_{i_\lambda j_\mu} = 0 \quad (\lambda = 1, 2, \ldots, \nu; \mu = \nu + 1, \ldots, p);$$

otherwise, if $i_\nu > j_\nu$ would lead to:

$$a_{i_\lambda j_\mu} = 0 \quad (\lambda = \nu + 1, \ldots, p; \mu = 1, 2, \ldots, \nu).$$

These two last statements say that the matrix is either upper block triangular or lower block triangular. This proves (3.3).   □

We can now prove that the inverse of a symmetric Jacobi matrix is a so-called one-pair matrix.

Proof (Proof of Theorem 3.12). We prove the theorem by explicitly constructing the inverse of the symmetric Jacobi matrix A. This matrix will then appear to be a one-pair matrix. Suppose S is the inverse of the matrix A. This means that:

$$s_{i,j} = \frac{1}{\det(A)} (-1)^{i+j} |A(1, \ldots, i-1, i+1, \ldots, n; 1, \ldots, j-1, j+1, \ldots, n)|.$$

We distinguish between two cases now:

1. When $i \leq j$, we have

$$
\begin{aligned}
s_{i,j} &= \frac{1}{\det(A)}(-1)^{i+j}|A(1,\ldots,i-1;1,\ldots,i-1)| \cdot |A(i+1,i)| \cdots \\
&\quad |A(j,j-1)| \cdot |A(j+1,\ldots,n;j+1,\ldots,n)| \\
&= \frac{1}{\det(A)}(-1)^{i+j}|A(1,\ldots,i-1;1,\ldots,i-1)|\, b_i\, b_{i+1} \cdots \\
&\quad b_{j-1}\,|A(j+1,\ldots,n;j+1,\ldots,n)|.
\end{aligned}
$$

2. When $i \geq j$, we can do the same as above and one gets

$$
\begin{aligned}
s_{i,j} &= \frac{1}{\det(A)}(-1)^{i+j}|A(1,\ldots,j-1;1,\ldots,j-1)|\, b_j\, b_{j+1} \cdots \\
&\quad b_{i-1}\,|A(i+1,\ldots,n;i+1,\ldots,n)|.
\end{aligned}
$$

When writing u_i and v_j now as (under the assumption that all the $b_i \neq 0$):

$$
v_i = \frac{(-1)^i}{\det(A)}|A(1,\ldots,i-1;1,\ldots,i-1)|\, b_i\, b_{i+1}\ldots b_{n-1},
$$

$$
u_i = \frac{(-1)^i|A(i+1,\ldots,n;i+1,\ldots,n)|}{b_i\, b_{i+1}\ldots b_{n-1}},
$$

we get that

$$
s_{i,j} = \begin{cases} u_i v_j & (i \geq j) \\ u_j v_i & (i \leq j) \end{cases} .
$$

This proves the theorem.   □

The following properties are necessary to prove that the inverse of a one-pair matrix is an irreducible tridiagonal matrix.

Proposition 3.15. *Suppose we have a one-pair matrix S which is generated by the vectors u and v. If*

$$
1 \leq i_1, j_1 < i_2, j_2 < \ldots < i_p, j_p \leq n, \tag{3.4}
$$

then

$$
S(i_1,\ldots,i_p;j_1,\ldots,j_p) = v_{\alpha_1} \begin{vmatrix} u_{\beta_1} & u_{\alpha_2} \\ v_{\beta_1} & v_{\alpha_2} \end{vmatrix} \begin{vmatrix} u_{\beta_2} & u_{\alpha_3} \\ v_{\beta_2} & v_{\alpha_3} \end{vmatrix} \cdots \begin{vmatrix} u_{\beta_{p-1}} & u_{\alpha_p} \\ v_{\beta_{p-1}} & v_{\alpha_p} \end{vmatrix} u_{\beta_p},
$$

where

$$
\alpha_\nu = \min(i_\nu, j_\nu) \quad \beta_\nu = \max(i_\nu, j_\nu).
$$

Proof. Because the matrix S is a symmetric matrix, we can, without loss of generality, assume that $i_2 \leq j_2$. This means that $\alpha_2 = i_2$ and $\beta_2 = j_2$, leading to the following equality:

$$|S(i_1, \ldots, i_p; j_1, \ldots, j_p)|$$

$$= \det \begin{bmatrix} u_{\beta_1} v_{\alpha_1} & u_{j_2} v_{i_1} & u_{j_3} v_{i_1} & \cdots & u_{j_p} v_{i_1} \\ u_{i_2} v_{j_1} & u_{j_2} v_{i_2} & u_{j_3} v_{i_2} & \cdots & u_{j_p} v_{i_2} \\ \vdots & & \ddots & & \end{bmatrix}.$$

Subtracting from the first row, the second one multiplied with v_{i_1}/v_{i_2} gives the following equation:

$$|S(i_1, \ldots, i_p; j_1, \ldots, j_p)|$$

$$= \left(v_{\alpha_1} u_{\beta_1} - \frac{v_{j_1} u_{i_2} v_{i_1}}{v_{i_2}} \right) |S(i_2, \ldots, i_p; j_2, \ldots, j_p)|$$

$$= \frac{v_{\alpha_1}}{v_{\alpha_2}} \begin{vmatrix} u_{\beta_1} & u_{\alpha_2} \\ v_{\beta_1} & v_{\alpha_2} \end{vmatrix} |S(i_2, \ldots, i_p; j_2, \ldots, j_p)|.$$

Applying the equation above successively and using the fact that

$$S(i_p; j_p) = v_{\alpha_p} u_{\beta_p}$$

gives the desired result.						□

One more proposition about the minors of a one-pair matrix is needed.

Proposition 3.16. *Suppose we have a one-pair matrix S. If*

$$1 \le i_1 < i_2 < \ldots < i_p \le n,$$
$$1 \le j_1 < j_2 < \ldots < j_p \le n,$$

but Equation (3.4) is not satisfied then:

$$|S(i_1, \ldots, i_p; j_1, \ldots, j_p)| = 0.$$

Proof. Using Proposition 3.15 and assuming that

$$1 \le i_1, j_1 < i_2, j_2 < \ldots < i_r, j_r,$$

whereas for example $j_r > i_{r+1}$, we get that

$$|S(i_1, \ldots, i_p; j_1, \ldots, j_p)|$$

$$= \frac{u_{\alpha_1}}{u_{\alpha_2}} \begin{vmatrix} v_{\beta_1} & v_{\alpha_2} \\ u_{\beta_1} & u_{\alpha_2} \end{vmatrix} \cdots \begin{vmatrix} v_{\beta_{r-1}} & v_{\alpha_r} \\ u_{\beta_{r-1}} & u_{\alpha_r} \end{vmatrix} |S(i_r, \ldots, i_p; j_r, \ldots, j_p)|.$$

Because $j_r > i_{r+1}$, this means that the last determinant is zero.			□

Now the theorem stating that the inverse of a one-pair matrix is a Jacobi matrix can be stated.

Theorem 3.17. *Suppose S is a one-pair matrix with all the elements of the generators different from zero, then the inverse of S is a tridiagonal matrix with all the sub and superdiagonal elements different from zero.*

Proof. It is easily proved by using Propositions 3.15 and 3.16.   □

3.1.4   The example of the one-pair matrix

This subsection is based on [131, Section 7, Chapter III]. Suppose that we have a string fastened in the points $x = 0$ and $x = l$, with a tension T. We will now deduce the influence function $K(x, s)$ of this system.

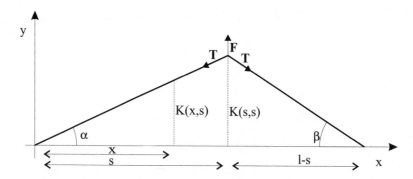

Figure 3.1. *Constructing the influence function of a string.*

Figure 3.1 should help to understand the following part. If we pull the string with a unit force F in the point s, we know that the tension must satisfy the following equality

$$T(\sin(\alpha) + \sin(\beta)) = 1.$$

Because the angles α and β are very small we can assume that

$$\sin(\alpha) = \frac{K(s, s)}{s},$$
$$\sin(\beta) = \frac{K(s, s)}{l - s}.$$

Using these equations combined with the equation in T we obtain

$$K(s, s) = \frac{(l - s)s}{Tl}.$$

When we want to calculate the deflection at an arbitrary point x now, we use the trigonometric equality:

$$\frac{K(x,s)}{x} = \frac{K(s,s)}{s},$$

which gives us the following equation:

$$K(x,s) = \begin{cases} \dfrac{x(l-s)}{Tl} & (x \leq s), \\ \dfrac{s(l-x)}{Tl} & (x \geq s). \end{cases}$$

Taking now a finite number of points on the string: $x_1 < x_2 < \ldots < x_n$, we have that:

$$K(x_i, x_k) = \begin{cases} u_i v_k & (i \geq k) \\ u_k v_i & (i \leq k) \end{cases},$$

with

$$v_i = \frac{x_i}{Tl}, \text{ and } u_k = l - x_k \text{ for } (i = 1, 2, \ldots, n),$$

which clearly is a one-pair matrix. Moreover, if one of the two ends of the string is not fixed, one gets again a one-pair matrix.

3.1.5 Some other interesting applications

This subsection is based on [131, Section 8-11, Chapter III]. Jacobi matrices are closely related to oscillations of Sturm systems. We will not go into the details of what exactly a Sturm system is but we will give some examples. For example an oscillating thread with beads is a Sturm system. Also a shaft on which a number of disks are fastened, all turning at the same speed, gives rise to a Sturm system when looking at the motion of the shaft, under an initial impulse. Also a pendulum, with not one mass, but different masses can be considered as a Sturm system. Such a pendulum consists of a thread attached at a fixed point, and hanging down, at certain places at the thread, masses are attached. The swinging movement of the system satisfies the conditions for a Sturm system.

When one investigates the influence function of a rod, one constructs an influence function which is in fact the sum of two one-pair matrices. This means that the upper triangular part can be considered as coming from a rank 2 and also the lower triangular part can be considered as coming from a rank 2 matrix.

3.1.6 The connection with eigenvalues and eigenvectors

To see the connection with the physical interpretation we have to take a closer look at the following theorem. First, we need to define what is meant by the number of sign changes. Suppose we have a sequence of numbers $u_1, u_2, \ldots, u_n$, the number of sign changes reading this sequence from the left to the right is of course dependent on the signs we give to the u_i's which are equal to zero. Nevertheless we can speak about the minimal number of sign changes and the maximum number of sign changes.

Theorem 3.18 (Theorem 6 in [131]). *Suppose A is an oscillation matrix, then this matrix has the following properties:*

1. *All the eigenvalues of A are positive and single.*

2. *If v_k is an eigenvector of A, corresponding to the eigenvalue λ_k, then for every sequence of coefficients $c_p, c_{p+1}, \ldots, c_q$ (with $1 \le p \le q \le n$ and not all c_i equal to zero) we have that the number of sign changes in the coordinates of the vector:*

$$u = c_p v_p + c_{p+1} v_{p+1} + \cdots c_q v_q$$

lies between $p-1$ and $q-1$. More precisely, for the vector v_k there are exactly $k-1$ sign changes.

3. *The nodes of two successive eigenvectors v_k and v_{k+1} alternate.*

One can fairly easily see the connection between the properties of the oscillations in Section 3.1.1 and Theorem 3.18. Property 1 of Theorem 3.18 corresponds to Property 1 in Section 3.1.1. Property 2 of the theorem corresponds with 2 and 4 of Section 3.1.1 and Property 3 corresponds with 3 in Section 3.1.1. This reveals the close connection between the eigenvalues and eigenvectors of oscillation matrices and the properties of the oscillations of a segmental continuum.

Notes and references

The references on which this section is based were presented previously in Section 1.1 of Chapter 1, in which a complete overview was given of all the different translations of the following book on oscillation matrices by Gantmacher and Kreĭn presented below.

☞ F. R. Gantmacher and M. G. Kreĭn. *Oscillation matrices and kernels and small vibrations of mechanical systems.* AMS Chelsea Publishing, Providence, Rhode Island, revised edition, 2002.

Often one is also concerned with the reconstruction of the physical parameters of a system, once its spectrum is known. These physical parameters correspond to the matrix. These problems are called Inverse Eigenvalue Problems (IEP). If one wants to reconstruct the parameters of, e.g., a string, this has strong relations with semiseparable matrices. More information on Inverse Eigenvalue Problems can be found in the book by M. Chu and G. Golub. The other references are related to semiseparable matrices.

☞ M. T. Chu and G. H. Golub. *Inverse Eigenvalue Problems: Theory, Algorithms and Applications.* Numerical Mathematics & Scientific Computations. Oxford University Press, 2006.

☞ G. M. L. Gladwell. The inverse problem for the vibrating beam. *Proceedings of the Royal Society of London A*, 393:277–295, 1984.

☞ G. M. L. Gladwell. The inverse mode problem for lumped mass systems. *The Quarterly Journal of Mechanics and Applied Mathematics*, 39(2):297–307, 1986.

☞ G. M. L. Gladwell and J. A. Gbadeyan. On the inverse problem of the vibrating string or rod. *The Quarterly Journal of Mechanics and Applied Mathematics*, 38(1):169–174, 1985.

☞ Y. M. Ram and G. M. L. Gladwell. Constructing a finite element model of a vibratory rod from eigendata. *Journal of Sound and Vibration*, 169(2):229–237, 1994.

More information on Inverse Eigenvalue Problems and more precisely on the computation of eigenvalues/singular values related to semiseparable matrices, will be published in Volume II of this book.

3.2 Semiseparable matrices as covariance matrices

This section is mainly inspired by the book of A. G. Franklin [158]. (More information about the connection between semiseparable and covariance matrices can be found in [14, 220, 121, 159, 160, 202, 237, 239, 247].) Several covariance matrices from multivariate distributions will be constructed. Here only these matrices will be considered which have very strong connections with semiseparable matrices.

In the theory of multivariate analysis, the joint distribution of random variables is considered. Suppose n random variables $X_1, \ldots, X_n$ are given. Statistical analysis is quite often concerned with the analysis of the covariance matrices V. These matrices have as diagonal elements the variances of the variables X_i and as off-diagonal elements v_{ij} the covariances between the variables X_i and X_j. These matrices are symmetric and they contain very important information concerning relations between the set of random variables. In particular, the inverse of these matrices is very important for statistical analysis (see, e.g., [220, 160, 230]).

In several examples, these matrices have structure, so-called patterned matrices in the book [158]. These structures are sometimes semiseparable, tridiagonal or semiseparable plus diagonal matrices, whose inverses can quite often be calculated in an easy way.

Here we will construct some of these matrices to show that semiseparable matrices also can be found in the field of statistics. The theorems proving how to calculate the inverses of these matrices are not explicitly given; they can be found in [158].

Before constructing covariance matrices a small recapitulation of basic statistics is given. This makes it possible for the reader to calculate some of the different covariance matrices mentioned here.

3.2.1 Covariance calculations

Suppose we have a statistical variable X for which the following distribution function $\mathrm{P}(X)$ is given. This function can either be discrete or continuous. For the discrete case this function is defined as:

$$\mathrm{P}(X = k) = p_k;$$

in the continuous case we have the following equality:

$$\mathrm{P}(X) = f(x).$$

Before we can define what is meant with covariances and variances, we have to declare what we mean with the expected value:

Definition 3.19. *Define the expected value of $g(X)$ (denoted as $\mathrm{E}(g(X))$) in the following way:*

$$\mathrm{E}(g(X)) = \sum_{k=-\infty}^{\infty} g(k)p_k \quad (discrete)$$

$$\mathrm{E}(g(X)) = \int_{-\infty}^{\infty} g(x)f(x)dx \quad (continuous).$$

Using the definition above we can define the mean and the variance of a variable X with a distribution function $\mathrm{P}(X)$.

Definition 3.20. *The mean and the variance of a variable X with a distribution function $\mathrm{P}(X)$ are defined in the following way: Mean: $\mu = \mathrm{E}(X)$ and the variance: $\mathrm{E}\left((X-\mu)^2\right)$.*

These definitions can be extended to the multivariable case in the following way. Suppose the following distribution function $\mathrm{P}(X_1, X_2, \ldots, X_k)$ is given. This function can be discrete or continuous (or a mixture of both).

Definition 3.21. *Define the expected value of $g(X_1, X_2, \ldots, X_k)$ (denoted as $\mathrm{E}(g(X_1, X_2, \ldots, X_k))$) in the following way:*

$$\mathrm{E}(g(X_1, X_2, \ldots, X_k))$$

$$= \sum_{j_1=-\infty}^{\infty} \cdots \sum_{j_k=-\infty}^{\infty} g(j_1, \ldots, j_k)\,\mathrm{P}(X_1 = j_1, \ldots, X_k = j_k) \quad (discrete)$$

$$\mathrm{E}(g(X_1, X_2, \ldots, X_k))$$

$$= \int_{-\infty}^{\infty} \cdots \int_{-\infty}^{\infty} g(x_1, \ldots, x_k)f(x_1, \ldots, x_k)dx_1 \ldots dx_2 \quad (continuous).$$

The mean and variance for multivariable distributions are defined in the same way as the monovariable case. The covariance for two distinct i and j is defined as:

$$\mathrm{Cov}(X_1, X_2) = \mathrm{E}\left((X_i - \mu_i)(X_j - \mu_j)\right).$$

Now that we have all the definitions we can start the calculations of the different covariance matrices.

3.2.2   The multinomial distribution

Before defining the multinomial distribution, the binomial distribution is discussed. These comments are based on the book [300]. The use of the distribution is explained with an example. In brief the binomial distribution describes the number of "success" outcomes when a Bernoulli experiment is repeated n times independently.

More precisely, the binomial distribution describes the behavior of a count variable X, which counts the number of successes in n observations, if the following conditions are satisfied:

- The number of observations n is fixed.

- Each observation is independent.

- Each observation represents one of two outcomes ("success" or "failure").

- The probability of "success" p is the same for each outcome.

The binomial distribution is denoted as $B(n, p)$, with n denoting the number of observations and p the chance of success. This distribution is used in several examples:

Example 3.22 Suppose we have a group of individuals with a certain gene. These people have a 0.60 probability of getting a certain disease. If a study is performed on 300 individuals with this specific gene, then the distribution of the variable describing the number of people who will get the disease has the following binomial distribution: $B(300, 0.6)$. ∎

Example 3.23 The number of sixes rolled by a single die in 30 rolls has a binomial distribution $B(30, 1/6)$. ∎

The distribution function for the binomial distribution $B(n, p)$ satisfies the following equation:

$$\mathrm{P}(X = k) = C(n, k)p^k(1 - p)^{n-k},$$

where

$$C(n, k) = \frac{n!}{k!(n - k)!}.$$

The mean and the variance of this distribution can be calculated using the formulas above and are (denote the mean as μ_X and the variance as σ_X^2):

$$\mu_X = np \qquad \sigma_X^2 = np(1 - p).$$

Before giving the distribution function we will try to explain what is meant with a multinomial distribution. A multinomial trials process is a sequence of independent, identically distributed random variables $U_1, U_2, \ldots$, where each random variable can take now k values. For the Bernoulli process, this corresponds to $k = 2$ (success and failure). Therefore this is a generalization of a Bernoulli trials process. We denote the outcomes by the integers $1, 2, \ldots, k$, for simplicity. This means that for a trial variable U_j we can write the distribution function in the following way:

$$p_i = \mathrm{P}(U_j = i) \text{ for } i = 1, 2, \ldots, k \text{ (and for any } j \text{)}.$$

Of course $p_i > 0$ for each i and $p_1 + p_2 + \cdots + p_k = 1$.

As with the binomial distribution, we are interested in the variables counting the number of times each outcome has occurred, where in the binomial case one variable for counting was enough, here we need $k - 1$. Thus, let (with $\#$ we denote the cardinality of the set)

$$Z_i = \#\{j \in \{1, 2, \ldots, n\} \text{ for which } U_j = i\} \text{ for } i = 1, 2, \ldots, k,$$

where n is the number of observations.

Note that

$$Z_1 + Z_2 + \cdots + Z_k = n.$$

So if we know the values of $k - 1$ of the counting variables, we can find the value of the remaining counting variable as mentioned previously. Generalizing the binomial distribution we get the following function (More information about these distributions can be found at `http://www.math.uah.edu/statold/bernoulli/`):

$$\mathrm{P}(Z_1 = j_1, Z_2 = j_2, \ldots, Z_n = j_k) = k p_1^{j_1} p_2^{j_2} \ldots p_k^{j_k},$$

with $j_1 + j_2 + \cdots + j_k = n$ and $p_1 + p_2 + \cdots + p_k = 1$. Before we start calculating the covariance matrix, we will give an example.

Example 3.24 It is very easy to see that the dice experiment also fits in here. For example if one rolls 10 dice, we can calculate the probability that 1 and 2 occur once, and the other occur all two times. To calculate this, one needs the multinomial distribution. ∎

For this distribution: $\mathrm{E}(Z_i) = np_i$, $\mathrm{Var}(Z_i) = np_i(1 - p_i)$ and $\mathrm{Cov}(Z_i, Z_j) = np_ip_j$. As an example we calculate the mean using the following binonium and multinonium formulas:

$$C(n, j) = \frac{n!}{(n-j)!j!},$$

$$(x + y)^n = \sum_{j=0}^{n} C(n, j) x^{n-j} y^j,$$

$$C(n; j_1, \ldots, j_k) = \frac{n!}{j_1! \ldots j_k!},$$

$$(x_1 + x_2 + \cdots + x_k)^n = \sum_{j_1 + j_2 + \cdots + j_k = n} C(n; j_1, \ldots, j_k) x_1^{j_1} \ldots x_k^{j_k}.$$

For this distribution we calculate the mean of the variable X_1, all the other variances and covariances can be calculated in a completely analogous way.

$$\mathrm{E}(X_1) = \sum_{j_1} \cdots \sum_{j_k} j_1 \, \mathrm{P}(X_1 = j_1, \ldots, X_k = j_k)$$

$$= \sum_{j_1} \cdots \sum_{j_k} j_1 C(n; j_1, \ldots, j_k) p_1^{j_1} \cdots p_k^{j_k}$$

$$= \sum_{j_1} j_1 p_1^{j_1} \left(\sum_{j_2} \cdots \sum_{j_k} \frac{n!}{j_1! \ldots j_k!} p_2^{j_2} \cdots p_k^{j_k} \right)$$

$$= \sum_{j_1} \frac{j_1}{j_1!} p_1^{j_1} \left(\sum_{j_2} \cdots \sum_{j_k} \frac{n!(n-j_1)!}{(n-j_1)! j_2! \ldots j_k!} p_2^{j_2} \cdots p_k^{j_k} \right)$$

$$= \sum_{j_1} \frac{j_1 n!}{j_1!(n-j_1)!} p_1^{j_1} (p_2 + \cdots + p_k)^{j_2 + \cdots + j_k}$$

$$= \sum_{j_1} j_1 \frac{n!}{j_1!(n-j_1)!} p_1^{j_1} (1 - p_1)^{n-j_1}$$

$$= \mathrm{E}(Z)$$

$$= np_1$$

where Z has a binomial distribution $B(n, p_1)$, which gives us the last equality. The covariance matrix of this distribution looks like

$$V = n \begin{bmatrix} p_1(1-p_1) & -p_1 p_2 & -p_1 p_3 & \cdots & -p_1 p_k \\ -p_1 p_2 & p_2(1-p_2) & -p_2 p_3 & \cdots & -p_2 p_k \\ -p_1 p_3 & -p_2 p_3 & p_3(1-p_3) & \cdots & -p_3 p_k \\ \vdots & \vdots & \vdots & \ddots & \vdots \\ -p_1 p_k & -p_2 p_k & -p_3 p_k & \cdots & p_k(1-p_k) \end{bmatrix}.$$

This matrix can be rewritten as a generator representable semiseparable plus diagonal matrix in the following way. Denote:

$$\mathbf{u} = [p_1, p_2, p_3, \ldots, p_n]^T,$$
$$\mathbf{v} = [-p_1, -p_2, -p_3, \ldots, -p_n]^T,$$
$$D = \mathrm{diag}([p_1, p_2, \ldots, p_n]),$$

then the matrix can be written in a more compact form as

$$V = D + \mathbf{u}\mathbf{v}^T.$$

This is a semiseparable plus diagonal matrix.

The main reason of writing the covariance matrices in this form is the simple expression of the inverses of this type of matrices. The book [158] states several theorems about the inverses of tridiagonal and semiseparable matrices, resulting in an inversion formula for this type of matrices, namely:

$$V^{-1} = D^{-1} + \gamma \hat{\mathbf{u}} \hat{\mathbf{v}}^T,$$

with $\hat{u}_i = u_i/d_i$, $\hat{v}_i = v_i/d_i$ and $\gamma = -\left(1 + \alpha \sum_{i=1}^n u_i v_i d_i^{-1}\right)^{-1}$, with d_i the diagonal elements of D. This leads to the following explicit structure of the matrix V^{-1},

$$V^{-1} = \begin{bmatrix} \frac{1}{p_1} + \gamma & \gamma & \gamma & \cdots & \gamma \\ \gamma & \frac{1}{p_2} + \gamma & \gamma & \cdots & \gamma \\ \gamma & \gamma & \frac{1}{p_3} + \gamma & & \vdots \\ \vdots & \vdots & & \ddots & \gamma \\ \gamma & \gamma & \cdots & \gamma & \frac{1}{p_n} + \gamma \end{bmatrix}.$$

This matrix is clearly again semiseparable plus diagonal, as we expected according to the theorems of Chapter 1. More precisely this is a rank 1 plus a diagonal matrix.

3.2.3 Some other matrices

To conclude this section we will briefly show two other matrices from the book [246]. No calculations are included anymore because they become too complicated. For a sample of size k of order statistics, with an exponential density, the covariance matrix looks like (see, e.g., also [160]):

$$\begin{bmatrix} \frac{1}{k^2} & \frac{1}{k^2} & \frac{1}{k^2} & \cdots & \frac{1}{k^2} \\ \frac{1}{k^2} & \frac{1}{k^2} + \frac{1}{(k-1)^2} & \frac{1}{k^2} + \frac{1}{(k-1)^2} & \cdots & \frac{1}{k^2} + \frac{1}{(k-1)^2} \\ \frac{1}{k^2} & \frac{1}{k^2} + \frac{1}{(k-1)^2} & \sum_{j=1}^3 \frac{1}{(k-j+1)^2} & \cdots & \sum_{j=1}^3 \frac{1}{(k-j+1)^2} \\ \vdots & \vdots & \vdots & & \vdots \\ \frac{1}{k^2} & \frac{1}{k^2} + \frac{1}{(k-1)^2} & \sum_{j=1}^3 \frac{1}{(k-j+1)^2} & \cdots & \sum_{j=1}^k \frac{1}{(k-j+1)^2} \end{bmatrix}.$$

This matrix is clearly a semiseparable matrix and the inverse can easily be calculated by the algorithms proposed in [158], and looks like

$$\begin{bmatrix} k^2 + (k-1)^2 & -(k-1)^2 & 0 & 0 & \cdots \\ -(k-1)^2 & (k-1)^2 + (k-2)^2 & -(k-2)^2 & 0 & \cdots \\ 0 & -(k-2)^2 & (k-2)^2 + (k-3)^2 & -(k-3)^2 & \\ 0 & 0 & -(k-3)^2 & \ddots & \ddots \\ \vdots & \vdots & & & \ddots \end{bmatrix}.$$

The following matrix arises in the statistical theory of ordered observations :

$$\begin{bmatrix} k & k-1 & k-2 & k-3 & \cdots & 1 \\ k-1 & 2(k-1) & 2(k-2) & 2(k-3) & \cdots & 2 \\ k-2 & 2(k-2) & 3(k-2) & 3(k-3) & \cdots & 3 \\ k-3 & 2(k-3) & 3(k-3) & 4(k-3) & \cdots & 4 \\ \vdots & \vdots & \vdots & \vdots & \ddots & \vdots \\ 1 & 2 & 3 & 4 & \cdots & k \end{bmatrix}.$$

This matrix is again symmetric and semiseparable. The last matrix arises in the theory of stationary time series:

$$\sigma^2 \begin{bmatrix} 1 & \rho & \rho^2 & \rho^3 & \ldots & \rho^{n-1} \\ \rho & 1 & \rho & \rho^2 & \ldots & \rho^{n-2} \\ \rho^2 & \rho & 1 & \rho & \ldots & \rho^{n-3} \\ \rho^3 & \rho^2 & \rho & 1 & \ldots & \rho^{n-4} \\ \vdots & \vdots & \vdots & \vdots & \ddots & \vdots \\ \rho^{n-1} & \rho^{n-2} & \rho^{n-3} & \rho^{n-4} & \ldots & 1 \end{bmatrix}.$$

Also this matrix is symmetric and semiseparable. The details of the origin of these matrices are not included; this would lead us too far away from the basic subject of the book. An explicit formula for the inverses of these matrices can also be calculated rather easily. The inverses and the corresponding theorems can be found in [158].

We will end this section with a quote ([158, Introduction])

> The theory of multivariate analysis often centers around an analysis of a covariance matrix V. When this is the case, it may be necessary to find the determinant of V, the characteristic roots of V, the inverse of V if it exists, and perhaps to determine these and other quantities for certain submatrices of V.

This quote shows that it can be important to calculate the inverse, the determinant or even the eigenvalues of these covariance matrices. In this book we focus on system solving and the computation of the inverses. The second volume will pay attention to specific techniques for computing the eigendecomposition of these matrices.

Notes and references

We briefly provide some information concerning the references given in the beginning of this section. These papers were related to statistics in which certain types of semiseparable or related matrices appeared.

☞ L. Guttman. A generalized simplex for factor analysis. *Psychometrika*, 20:173–195, 1955.

In this manuscript Gutmann inverses a very specific covariance matrix, whose inverse will have the following form:

$$T = \begin{bmatrix} a_1 & -a_1 & & \\ -a_1 & a_1 + a_2 & -a_2 & \\ & -a_2 & a_3 + a_2 & a_3 \\ & & \ddots & \ddots & \ddots \end{bmatrix}.$$

Necessary conditions are provided on the original matrix, such that this matrix is invertible (these relations can also be found in [272]).

☞ F. A. Graybill. *Matrices with applications in statistics*. Wadsworth international group, Belmont, California, 1983.

The author Graybill presents in the book above several specific matrices, which are of semiseparable or related form, and inversion formulas for these matrices are given.

☞ W. W. Barrett and P. J. Feinsilver. Gaussian families and a theorem on patterned matrices. *Journal of Applied Probability*, 15:514–522, 1978.

Barrett and Feinsilver provide a probabilistic proof, stating that the inverse of a symmetric semiseparable matrix (with diagonal elements different from zero) is a symmetric tridiagonal matrix. Note that the matrices are characterized via the rank 1 assumptions in the lower and upper triangular part. The results are restricted to positive definite and symmetric matrices. Some special examples of patterned matrices are given and inverted. (See also the manuscripts [13, 15], by the same authors, discussed in Section 1.5 in Chapter 1.)

The author Mustafi inverts in [220] a symmetric generator representable semiseparable plus identity matrix, coming from a statistical problem. More information can be found in Section 1.5 in Chapter 1.

☞ B. G. Greenberg and A. E. Sarhan. Matrix inversion, its interest and application in analysis of data. *Journal of the American Statistical Association*, 54:755–766, 1959.

In this paper by Greenberg and Sarhan, the papers [272, 240] are generalized and applied to several types of matrices arising in statistical applications. A relation is introduced which needs to be satisfied, such that the inverse of the matrix is a diagonal matrix of type r (these diagonal matrices of type r correspond to band matrices of width r). Different semiseparable matrices are given and the relation is investigated for r equal to $1, 2$ and 3, thereby proving that these matrices have a banded inverse. The sufficient condition that needs to be posed on the matrix, such that it becomes of band form, corresponds with the demand that the matrix is of generator representable semiseparable form.

The author Kounias provides in [202] explicit formulas for inverting some patterned matrices. For example, a tridiagonal Toeplitz matrix and a matrix completely filled with one element plus a diagonal are inverted. More information on this manuscript can be found in Section 2.2 in Chapter 2, and related to the Toeplitz examples in Section 14.1 in Chapter 14.

The authors Roy, Sarhan and Greenberg investigate in the manuscripts [239, 240], several patterned matrices, arising from statistical applications. Methods for inverting and computing the determinant are presented. (See Section 2.2 in Chapter 2 for more information.)

☞ V. R. R. Uppuluri and J. A. Carpenter. The inverse of a matrix occurring in first-order moving-average models. *Sankhyā The Indian Journal of Statistics Series A*, 31:79–82, 1969.

Uppuluri and Carpenter present an exact formula to compute the inverse of a specific covariance matrix, which is a symmetric tridiagonal Toeplitz matrix.

In [278] Valvi determines explicit formulas for inverting specific patterned matrices. For more information see Section 2.2 in Chapter 2.

3.3 Discretization of integral equations

In this section a discretization of a special integral equation is made. The discretization will lead to a semiseparable system of equations. The resulting matrix will satisfy the semiseparable plus diagonal structure. More information about this topic can be found for example in [228, 193].

We discretize the following integral equation:

$$y(x) - \int_0^1 k(x,t)y(t)dt = a(x). \tag{3.5}$$

We want to compute the function $y(x)$ while $a(x)$ and $k(x,t)$ are known. The kernel $k(x,t)$ is called a Green's kernel, satisfying the following properties:

$$k(x,t) = \begin{cases} G(x)F(t) & x \le t \\ F(x)G(t) & x > t \end{cases}.$$

More information about discretization techniques can be found in one of the following books [45, 62, 132]. We use the following discretization scheme with the trapezoidal rule [132, p. 154]. Suppose a function $f(x)$ is given, and we want to integrate it in the interval $[a,b]$. Divide the interval in equally spaced smaller intervals of length h by using the points x_i. Denote $f(x_i) = f_i$, then we get the following discretization scheme:

$$\int_a^b f(x)dx \approx \frac{h}{3}\left(f_0 + 4f_1 + 2f_2 + \cdots + 4f_{n-1} + f_n\right).$$

We will use this scheme to discretize the integral Equation (3.5). Slicing the interval $[0,1]$ with a distance h between the successive points t_i, denoting $G(x_i) = G_i$, $F(x_i) = F_i$ and $a(x_i) = a_i$, we get:

$$a(x) \approx y(x) - \frac{h}{3}\Big(k(x,t_0)y(t_0) + 4k(x,t_1)y(t_1) + 2k(x,t_2)y(t_2) + \ldots$$

$$+ 4k(x,t_{n-1})y(t_{n-1}) + k(x,t_n)y(t_n)\Big).$$

Substituting for x the different values of $x_i = t_i$, we get the following system of equations equal to $[a_0, a_1, a_2, \ldots, a_n]^T$

$$\left(-\frac{h}{3}\right)\begin{bmatrix} G_0F_0 - \frac{3}{h} & 4G_0F_1 & 2G_0F_2 & \ldots & 4G_0F_{n-1} & G_0F_n \\ G_0F_1 & 4G_1F_1 - \frac{3}{h} & 2G_1F_2 & \ldots & 4G_1F_{n-1} & G_1F_n \\ G_0F_2 & 4G_1F_2 & \ldots & & & \\ \vdots & & & \ddots & & \\ G_0F_n & & & & & G_nF_n - \frac{3}{h} \end{bmatrix}\begin{bmatrix} y_0 \\ y_1 \\ y_2 \\ \vdots \\ y_n \end{bmatrix}$$

$$= \left(-\frac{h}{3}\right)\begin{bmatrix} G_0F_0 - \frac{3}{h} & G_0F_1 & G_0F_2 & \ldots & G_0F_{n-1} & G_0F_n \\ G_0F_1 & G_1F_1 - \frac{3}{4h} & G_1F_2 & \ldots & G_1F_{n-1} & G_1F_n \\ G_0F_2 & G_1F_2 & \ldots & & & \\ \vdots & & & \ddots & & \\ G_0F_n & & & & & G_nF_n - \frac{3}{h} \end{bmatrix}\begin{bmatrix} y_0 \\ 4y_1 \\ 2y_2 \\ \vdots \\ y_n \end{bmatrix}.$$

The last system of equations is a semiseparable plus diagonal matrix. Later on in this book, we will solve this system of equations in a stable and fast way.

When one wants to solve two point boundary value problems as in [256], one can sometimes translate these problems into a second order integral equation with a Green's kernel. In general this kernel is not semiseparable anymore, but so-called recursively semiseparable or sequentially semiseparable. This means that the matrices have the form as given in Figure 3.2, for which all the white blocks have the same rank. Here, we will not go into the details of this type of structure, but the

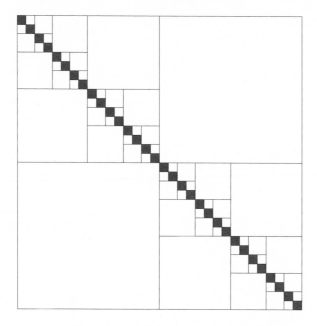

Figure 3.2. *A recursively semiseparable matrix.*

reader should know that this type of matrix appears in several types of applications, and there are already several techniques for solving this type of matrices, e.g., [53, 55, 83, 93]. More information on how to deal with these matrices will be presented Part III.

Notes and references

More information on integral operators with this type of kernel can be found in the following references.

☞ R. A. Gonzales, J. Eisert, I. Koltracht, M. Neumann, and G. Rawitscher. Integral equation method for the continuous spectrum radial Schrödinger equation. *Journal of Computational Physics*, 134:134–149, 1997.

☞ S.-Y. Kang, I. Koltracht, and G. Rawitscher. High accuracy method for integral equations with discontinuous kernels, 1999. http://arxiv.org/abs/math.NA/9909006.

☞ J. Petersen and A. C. M. Ran. *LU-* versus *UL*-factorization of integral operators with semiseparable kernel. *Integral Equations and Operator Theory*, 50:549–558, 2004.

3.4 Orthogonal rational functions

This section will give a brief example in which way orthogonal rational functions are connected to semiseparable matrices. This section is not expanded as a thorough study of the relations between orthogonal rational functions and semiseparable matrices will be given in Volume II of this book [297]. This section is mainly based on the paper [279], but the papers [43, 44, 106, 107, 280] are also closely related to this problem.

Suppose we have a vector space V_n of all proper rational functions having possible poles in $y_1, y_2, \ldots, y_n$:

$$V_n = \text{Span}\left\{1, \frac{1}{z-y_1}, \frac{1}{z-y_2}, \ldots, \frac{1}{z-y_n}\right\},$$

with real numbers $y_1, y_2, \ldots, y_n$ all different from zero. On this vector space one can define the following bilinear form:

Definition 3.25. *Given the real numbers $z_0, z_1, \ldots, z_n$ which are, together with the y_i, all different from each other, and the weights w_i which are different from zero, one can define for two elements ϕ, φ of V_n the following inner product:*

$$\langle \phi, \varphi \rangle = \sum_{i=0}^{n} w_i^2 \phi(z_i)\varphi(z_i).$$

The aim is to compute an orthonormal basis $\alpha_0(z), \alpha_1(z), \ldots, \alpha_n(z)$ such that

$$\alpha_j \in V_j \backslash V_{j-1} \quad (V_{-1} = \emptyset), \quad \langle \alpha_i, \alpha_j \rangle = \delta_{i,j}.$$

Define $\mathbf{w} = [w_1, w_2, \ldots, w_n]^T$, $\mathbf{e}_1 = [1, 0, \ldots, 0]^T$, $D_y = \text{diag}([y_0, y_1, \ldots, y_n])$ with y_0 arbitrary and $D_z = \text{diag}([z_0, z_1, \ldots, z_n])$. It is proved in [279] that if one solves an inverse eigenvalue problem of the following form:

$$Q^T \mathbf{w} = \mathbf{e}_1 \|\mathbf{w}\|,$$
$$Q^T D_z Q = S + D_y,$$

where S is a symmetric generator representable semiseparable matrix, then the columns of the orthogonal Q satisfy the following recurrence relation:

$$\left(D_z - y_{i+1}I\right)b_j Q_{j+1} = \left(I - (D_z - y_j I)a_j\right)Q_j - \left(D_z - y_{j-1}I\right)b_{j-1}Q_{j-1},$$

with

$$j = 0, 1, \ldots, n-1,$$
$$Q_0 = \mathbf{w}/\|\mathbf{w}\|$$
$$Q_{-1} = 0.$$

and the elements a_i and b_i are, respectively, the diagonal and subdiagonal elements of the matrix $T = S^{-1}$ which is a tridiagonal matrix.

Even more, if we define now the functions $\alpha_j(z)$ using the following recursion relation

$$\left(z - y_{j+1}\right)b_j\alpha_{j+1}(z) = \alpha_j(z)\left(1 - (z - y_j)a_j\right) - \alpha_{j-1}(z)\left(z - y_{j-1}\right)b_{j-1},$$

with $\alpha_0(z) = 1/\sqrt{\sum w_i^2}, \alpha_{-1}(z) = 0$. Then we have the following relation between the Q_j's and the α_j's:

$$Q_j = \operatorname{diag}(w)(\alpha_j(z_i))_{i=0}^n$$

and

$$\alpha_j \in \mathcal{V}_j \backslash \mathcal{V}_{j-1} \quad (\mathcal{V}_{-1} = \emptyset), \quad \langle \alpha_i, \alpha_j \rangle = \delta_{i,j}.$$

This means that we can calculate the desired orthonormal basis by solving an inverse eigenvalue problem for semiseparable matrices.

Notes and references

We remark that these inverse eigenvalue problems, as well as general eigenvalue problems for semiseparable matrices will be discussed in Volume II of this book.

The following papers are closely related to the subject presented in this section. In the paper [279], Van Barel, Fasino, Gemignani and Mastronardi investigate the relation between orthogonal rational functions and generator representable semiseparable plus diagonal matrices. (For more information see Section 2.2 in Chapter 2.)

Fasino and Gemignani, study in the manuscript [107] the direct and the inverse eigenvalue problem of generator representable symmetric diagonal plus semiseparable matrices.

More information on the relation between structured rank matrices and orthogonal rational functions can be found in the manuscript [280] (discussed in Section 1.3 of Chapter 1) and the following manuscripts.

    ☞ A. Bultheel, M. Van Barel, and P. Van gucht. Orthogonal basis functions in discrete least-squares rational approximation. *Journal of Computational and Applied Mathematics*, 164-165:175–194, 2004.

    ☞ S. Delvaux and M. Van Barel. Orthonormal rational function vectors. *Numerische Mathematik*, 100(3):409–440, May 2005.

3.5   Some comments

In this section some references to papers and books closely related to semiseparable matrices, whose content is not covered in this book are presented. We will briefly discuss the name 'semiseparable' matrix, eigenvalue problems via semiseparable matrices and also some links to other applications.

3.5.1   The name "semiseparable" matrix

The name semiseparable matrix finds its origin in the discretization of kernels. A separable kernel $k(x, y)$ means that the kernel is of the following form

$$k(x, y) = g(x)f(y).$$

Associating a matrix with this kernel gives a rank 1 matrix. This matrix can be written as the outer product of two vectors. A semiseparable kernel satisfies the following properties (sometimes also called a Green's kernel [194, p. 110]):

$$k(x, y) = \begin{cases} g(x)f(y) & \text{if} \quad x \leq y \\ g(y)f(x) & \text{if} \quad y \leq x \end{cases}.$$

Associating a matrix with this semiseparable kernel gives us a generator representable semiseparable matrix. This might seem illogical at first sight, but we chose to define a semiseparable matrix as generically coming from the inverse of a tridiagonal matrix. This definition also admits easy generalizations of semiseparable matrices to higher order, as we will see later on.

3.5.2   Eigenvalue problems

A traditional algorithm to compute the eigenvalues of a symmetric matrix is by first reducing it to tridiagonal form, by means of orthogonal similarity transformations and then applying a QR-method on these tridiagonal matrices. The following references provide similar tools for computing the eigenvalues of symmetric matrices via intermediate semiseparable matrices instead of tridiagonal ones. Also different algorithms are included to compute the eigenvalues of semiseparable matrices itself, via, e.g., divide and conquer methods. A more elaborate study of eigenvalue problems via semiseparable matrices can be found in Volume II of this book.

☞ Kh. D. Ikramov and L. Elsner. On matrices that admit unitary reduction to band form. *Mathematical Notes*, 64(6):753–760, 1998.

The authors Elsner and Ikramov prove in this manuscript that low rank perturbations of hermitian matrices admit a unitary reduction to band form

☞ S. Chandrasekaran and M. Gu. A divide and conquer algorithm for the eigendecomposition of symmetric block-diagonal plus semi-separable matrices. *Numerische Mathematik*, 96(4):723–731, February 2004.

☞ N. Mastronardi, M. Van Barel, and E. Van Camp. Divide and conquer algorithms for computing the eigendecomposition of symmetric diagonal-plus-semiseparable matrices. *Numerical Algorithms*, 39(4):379–398, 2005.

Chandrasekaran and Gu present a divide and conquer method to calculate the eigendecomposition of a symmetric generator representable semiseparable plus a block diagonal matrix. Mastronardi, Van Camp and Van Barel present a different divide and conquer algorithm to compute the eigendecomposition of a generator representable semiseparable plus diagonal matrix.

☞ M. Van Barel, R. Vandebril, and N. Mastronardi. An orthogonal similarity
reduction of a matrix into semiseparable form. *SIAM Journal on Matrix
Analysis and its Applications*, 27(1):176–197, 2005.

☞ R. Vandebril and M. Van Barel. Necessary and sufficient conditions for
orthogonal similarity transformations to obtain the Arnoldi(Lanczos)-Ritz
values. *Linear Algebra and its Applications*, 414:435–444, 2006.

☞ R. Vandebril, E. Van Camp, M. Van Barel, and N. Mastronardi. On
the convergence properties of the orthogonal similarity transformations to
tridiagonal and semiseparable (plus diagonal) form. *Numerische Mathematik*, 104:205–239, 2006.

☞ R. Vandebril, E. Van Camp, M. Van Barel, and N. Mastronardi. Orthogonal similarity transformation of a symmetric matrix into a diagonal-plus-
semiseparable one with free choice of the diagonal. *Numerische Mathematik*, 102:709–726, 2006.

These papers are concerned with two types of reduction algorithms: namely the
orthogonal similarity reduction of a symmetric matrix into a similar semiseparable
or a similar semiseparable plus diagonal one. These two reductions can be seen as
the first step for the eigenvalue algorithm based on semiseparable matrices. The
proposed reduction algorithms to semiseparable and semiseparable plus diagonal
have in common with the reduction to tridiagonal form that intermediate matrices
have as eigenvalues the Lanczos-Ritz values, which is proved in the manuscript
of 2006 in a general form. The two presented reduction schemes, however, have
an extra convergence behavior with regard to the reduction to tridiagonal form,
namely a kind of nested multishift iteration, which is examined in the most recent
publication.

☞ M. Van Barel, E. Van Camp, and N. Mastronardi. Orthogonal similarity
transformation into block-semiseparable matrices of semiseparability rank
k. *Numerical Linear Algebra with Applications*, 12:981–1000, 2005.

Van Barel, Van Camp and Mastronardi provide an orthogonal similarity transformation to reduce symmetric matrices into a similar semiseparable one of rank k.
This is an extension of one of the articles above.

☞ D. Fasino. Rational Krylov matrices and QR-steps on Hermitian diagonal-
plus-semiseparable matrices. *Numerical Linear Algebra with Applications*,
12(8):743–754, October 2005.

Fasino proves that any hermitian matrix, with pairwise distinct eigenvalues can
be transformed into a similar hermitian generator representable semiseparable plus
diagonal matrix, with a prescribed diagonal. Moreover, it is proved that the unitary transformation in this similarity reduction is the unitary matrix in the QR-
factorization of a suitably defined Krylov subspace matrix.

☞ R. Bevilacqua and G. M. Del Corso. Structural properties of matrix unitary
reduction to semiseparable form. *Calcolo*, 41(4):177–202, 2004.

Bevilacqua and Del Corso investigate in this report the existence and the uniqueness of a unitary similarity transformation of a symmetric matrix into semiseparable form. Also the implementation of a QR-step on a semiseparable matrix without shift is investigated.

QR-algorithms for semiseparable matrices can be found in the following articles.

☞ D. A. Bini, L. Gemignani, and V. Y. Pan. QR-like algorithms for generalized semiseparable matrices. Technical Report 1470, Department of Mathematics, University of Pisa, Largo Bruno Pontecorvo 5, 56127 Pisa, Italy, 2004.

☞ E. Van Camp, M. Van Barel, R. Vandebril, and N. Mastronardi. An implicit QR-algorithm for symmetric diagonal-plus-semiseparable matrices. Technical Report TW419, Department of Computer Science, Katholieke Universiteit Leuven, Celestijnenlaan 200A, 3000 Leuven (Heverlee), Belgium, March 2005.

☞ R. Vandebril, M. Van Barel, and N. Mastronardi. An implicit QR-algorithm for symmetric semiseparable matrices. *Numerical Linear Algebra with Applications*, 12(7):625–658, 2005.

Bini, Gemignani and Pan derive an algorithm for performing a step of the QR-method on a generalized semiseparable matrix, the lower part of this generalized matrix is in fact lower triangular semiseparable. The authors present an alternative representation consisting of 3 vectors to represent this part (this is in fact the quasiseparable representation). The two manuscripts by Van Barel et al. derive an implicit QR-algorithm for semiseparable and semiseparable plus diagonal matrices.

☞ S. Delvaux and M. Van Barel. Rank structures preserved by the QR-algorithm: the singular case. *Journal of Computational and Applied Mathematics*, 189:157–178, 2006.

☞ S. Delvaux and M. Van Barel. Structures preserved by the QR-algorithm. *Journal of Computational and Applied Mathematics*, 187(1):29–40, March 2006.

In these manuscripts by Delvaux and Van Barel, the preservation of a rank structure under a step of the QR-iteration is examined. The authors provide general results, including, e.g., quasiseparable, semiseparable plus diagonal, semiseparable matrices, as well as their higher order rank variants.

☞ R. Vandebril, M. Van Barel, and N. Mastronardi. An implicit Q theorem for Hessenberg-like matrices. *Mediterranean Journal of Mathematics*, 2:59–275, 2005.

An implicit Q-theorem, like for tridiagonal matrices, was proposed in this manuscript.

☞ R. Vandebril, M. Van Barel, and N. Mastronardi. A QR-method for computing the singular values via semiseparable matrices. *Numerische Mathematik*, 99:163–195, November 2004.

The authors present in this paper a translation of the traditional algorithm for computing the singular values via bidiagonal matrices towards the semiseparable case. The intermediate matrices are now of upper triangular semiseparable form. This means that first a reduction to an upper triangular semiseparable matrix is proposed, represented with the Givens-vector representation. Then this upper triangular matrix is used in an implicit QR-algorithm for computing the singular values.

☞ S. Chandrasekaran and M. Gu. Fast and stable eigendecomposition of symmetric banded plus semi-separable matrices. *Linear Algebra and its Applications*, 313:107–114, 2000.

This article by Chandrasekaran and Gu uses a different approach to compute the eigenvalues of a symmetric generator representable semiseparable plus band matrix. The considered matrix is brought to tridiagonal form via orthogonal similarity transformations. First the low rank parts above and below the diagonal are removed via orthogonal transformations. This results in a band matrix, for which the bandwidth is dependent of the original bandwidth and the semiseparability rank of the parts below and above the diagonal (higher order semiseparability ranks are discussed in Part III). This band matrix is then reduced to tridiagonal form. Standard techniques for computing the eigenvalues of this tridiagonal matrix can then be applied.

☞ D. Fasino, N. Mastronardi, and M. Van Barel. Fast and stable algorithms for reducing diagonal plus semiseparable matrices to tridiagonal and bidiagonal form. *Contemporary Mathematics*, 323:105–118, 2003.

☞ N. Mastronardi, S. Chandrasekaran, and S. Van Huffel. Fast and stable algorithms for reducing diagonal plus semiseparable matrices to tridiagonal and bidiagonal form. *BIT*, 41(1):149–157, 2003.

Mastronardi, Chandrasekaran and Van Huffel provide an algorithm to transform a symmetric generator representable quasiseparable matrix into a similar tridiagonal one. Also a second algorithm to reduce an unsymmetric generator representable quasiseparable matrix to a bidiagonal one by means of orthogonal transformations is included. In the paper of 2003, Fasino et al. propose two new algorithms for transforming the same matrices from above to either tridiagonal or bidiagonal form. In the manuscripts above the matrices are referred to as semiseparable plus diagonal matrices. We know that due to the theoretical considerations in the previous chapter, but in the context of this book, they are considered as quasiseparable matrices. These algorithms can be used as a preprocessing step for computing for example the eigenvalues and/or singular values of the original matrices.

Several variants of QR-algorithms for structured rank matrices are proposed by the authors Bini,Gemignani and Daddi.

☞ D. A. Bini, F. Daddi, and L. Gemignani. On the shifted QR iteration applied to companion matrices. *Electronic Transactions on Numerical Analysis*, 18:137–152, 2004.

☞ D. A. Bini, Y. Eidelman, L. Gemignani, and I. C. Gohberg. Fast QR eigenvalue algorithms for Hessenberg matrices which are rank-one perturbations of unitary matrices. *SIAM Journal on Matrix Analysis and its Applications*, 29(2):566–585, 2007.

☞ D. A. Bini, L. Gemignani, and V. Y. Pan. Fast and stable QR eigenvalue algorithms for generalized companion matrices and secular equations. *Numerische Mathematik*, 100(3):373–408, 2005.

☞ L. Gemignani. A unitary Hessenberg QR-based algorithm via semiseparable matrices. *Journal of Computational and Applied Mathematics*, 184:505–517, 2005.

The manuscripts above are all concerned with the development of efficient QR-algorithms for structured rank matrices. Special attention is being paid to the class of companion matrices. Companion matrices are linked to polynomials in the standard basis in such a manner that the roots of the polynomial are the eigenvalues of the companion matrix. In this way the development of a QR-algorithm for companion matrices contributes to the fast computation of roots of polynomials. The authors propose in these manuscripts algorithms for performing a QR-step on companion matrices and they provide theoretical results concerning the structure of the factors, rank reduction and so on.

☞ R. Bevilacqua, E. Bozzo, and G. M. Del Corso. Transformations to rank structures by unitary similarity. Technical Report TR-04-19, Università di Pisa, F. Buonarroti 2, 56127 Pisa, Italy, November 2004.

The authors prove in this manuscript, via Krylov methods, the existence of unitary similarity transformations of square matrices to structured rank matrices. The matrices are characterized by low rank blocks and can hence be of band, semiseparable, or of semiseparable plus band form.

☞ Y. Eidelman, I. C. Gohberg, and V. Olshevsky. The QR iteration method for Hermitian quasiseparable matrices of an arbitrary order. *Linear Algebra and its Applications*, 404:305–324, July 2005.

The authors Eidelman, Gohberg and Olshevsky develop in this manuscript a QR-method for hermitian quasiseparable matrices, represented in using the standard quasiseparable representation. The method is in some sense general, as it covers the semiseparable as well as the tridiagonal case.

☞ Y. Eidelman, I. C. Gohberg, and V. Olshevsky. Eigenstructure of order-one-quasiseparable matrices. three-term and two-term recurrence relations. *Linear Algebra and its Applications*, 405:1–40, 2005.

In this manuscript, Eidelman, Gohberg and Olshevsky present a three-term recurrence relation for the characteristic polynomial of quasiseparable matrices of order 1. Based on this recurrence a bisection method and explicit formulas for the structured of eigenvectors are derived.

Recently also a new representation for higher order structured rank matrices was introduced by Delvaux and Van Barel [72]. Different algorithms are introduced

by the same authors for computing the QR-factorization, performing QR-steps and
so on. More information about this representation can be found in Section 8.5 in
Chapter 8; the QR-factorization is briefly discussed in Chapter 9. In the following
manuscripts the authors introduce a similarity reduction to Hessenberg form of a
higher order structured rank matrix and the explicit QR-algorithm for structured
rank matrices, all represented with the Givens-weight representation.

> ☞ S. Delvaux and M. Van Barel. The explicit QR-algorithm for rank struc-
> tured matrices. Technical Report TW459, Department of Computer Sci-
> ence, Katholieke Universiteit Leuven, Celestijnenlaan 200A, 3000 Leuven
> (Heverlee), Belgium, May 2006.

> ☞ S. Delvaux and M. Van Barel. A Hessenberg reduction algorithm for rank
> structured matrices. Technical Report TW460, Department of Computer
> Science, Katholieke Universiteit Leuven, Celestijnenlaan 200A, 3000 Leu-
> ven (Heverlee), Belgium, May 2006.

3.5.3   References to applications

In the remaining part of this section we will briefly mention some classes of matrices
closely related to semiseparable matrices. Some of these classes are examined more
thoroughly further in the text (namely in Part III). Currently the class of sequentially semiseparable matrices is investigated thoroughly. (This class is sometimes
also called quasiseparable, or recursively semiseparable.) Interesting algorithms and
the definition of this class of matrices can be found in [53, 55, 83, 84, 89, 93, 96].
Recently a new class of matrices $\mathcal{H}$-matrices is introduced. This class is also closely
related to the class of semiseparable and sequentially semiseparable matrices (see
[38, 169, 170] and the references therein).

Semiseparable matrices also appear in various types of applications, e.g., the
field of integral equations [154, 228, 193], operator theory [95, 97, 166], boundary
value problems [154, 161, 204], in the theory of Gauss-Markov processes [195], time
varying linear systems [83, 145], in statistics [158], acoustic and electromagnetic
scattering theory [60], signal processing [218, 219], numerical integration and differentiation [256], the following articles by Usmani and Moskovitz also contain explicit
inversion formulas.

> ☞ D. Moskovitz. The numerical solution of Laplace's and Poisson's equation.
> *Quarterly of Applied Mathematics*, 2:148–163, 1944.

> ☞ R. A. Usmani. Explicit inverse of a band matrix with application to com-
> puting eigenvalues of a Sturm-Liouville system. *Linear Algebra and its
> Applications*, 86:189–198, 1987.

These matrices also appear in inverse problems related to oscillation and vibration
analysis [139, 140, 141, 142, 143, 144, 231] and rational interpolation [279, 44, 280,
70]. Also in the biological field, applications exist directly resulting in semiseparable
matrices [102].

3.6 Conclusions

In this chapter we showed that the interest in the class of semiseparable matrices is not only of a theoretical nature. We briefly mentioned different fields in which semiseparable and related matrices arise. We pointed out that semiseparable matrices, under some assumptions, can be seen as oscillation matrices. Some covariance matrices are also semiseparable or semiseparable plus diagonal matrices. Finally we indicated the role these matrices play in the field of integral equations, statistics and orthogonal rational functions.

In the last section we provided some references related to applications, and to the eigenvalue problem for semiseparable matrices, which will be covered in a forthcoming book.

Part II

Linear systems with semiseparable and related matrices

The second part, combined with the first part of the book, can be seen as the most technical piece of this book. Even though we only cover the easy classes in this first and second part, we hand out different ideas and we will actually present details for implementing the methods. The third part is less algorithmically oriented than this part. In some sense this is logical as this part covers structured rank matrices in general. But the class of structured rank matrices is quite large, and instead of limiting ourselves to a specific class such as quasiseparable or semiseparable, we choose to cover all classes at a higher level. This means Part III will focus more directly to the structure of the matrix, instead of on directly applicable algorithms.

The goal of this part is to present some ideas and methods for solving systems of equations involving the easy classes of structured rank matrices. Not all algorithms are explored in deep detail, but ingredients and ideas which can be combined to successfully create algorithms are presented.

The first chapter of this part will discuss the LU-decomposition. Instead of immediately starting to describe how to efficiently compute the LU-decomposition, we first pay attention to the structure of the factors in this decomposition. A generalization of the nullity theorem suitable for the LU-decomposition is given. This theorem states that the structure of the matrix $A = LU$ propagates in some sense into the factors L and U. Some basic building blocks for computing the LU-decomposition of semiseparable and quasiseparable matrices are deduced. The elementary Gaussian transformation matrices seem to be especially important, as they can be used for representing a lower triangular semiseparable matrix in a natural way. Moreover, this representation presents a direct manner for computing the inverse of this matrix, which is a bidiagonal matrix. Using these elementary Gaussian transformation matrices we deduce the LU-factorization for both quasiseparable and semiseparable matrices. For the semiseparable matrices, we also discuss the case in which the matrix is not strongly nonsingular. This implies that, in general, pivoting will be necessary to compute the LU-decomposition. This is not an easy problem and has a strong effect on the structured rank of the factors L and U in the factorization. We conclude the chapter by investigating the LU-decomposition for quasiseparable matrices, hereby restricted to the strongly nonsingular case.

Chapter 5 will focus attention to the computation of a QR-factorization. First we take a closer look at this factorization, from a theoretical viewpoint. Using a generalization of the nullity theorem, just as we could predict the structure of the L and U factors in the LU-decomposition, we can predict the structure of the orthogonal matrix Q. Unfortunately this theorem cannot provide us with information about the upper triangular matrix R. The rank structure of this upper triangular matrix will in general be higher than 1. More on the structure of the matrix R can be found in the forthcoming chapters on the QR-factorization of higher order structured rank matrices. The QR-factorization as it will be discussed in this chapter will be based on at most two sequences of Givens transformations. One sequence of transformations will start at the bottom and go up to annihilate the rank structure. The second sequence of transformations is necessary in case there are nonzero subdiagonal elements remaining, e.g., in the case of a quasiseparable and a semiseparable plus diagonal matrix. In addition the implementation of the

QR-factorization for a semiseparable plus diagonal matrix will be discussed in more detail.

A Levinson-like solver for semiseparable and quasiseparable matrices will be discussed in Chapter 6. The Levinson idea is applied on semiseparable plus diagonal matrices to show the general idea of the method. After having proposed this method, a more general scheme is presented. This framework is suitable for so-called Levinson conform matrices. These Levinson conform matrices need to obey some specific conditions with regard to a block factorization of the matrix. It turns out that this framework is rather general in the sense that quasiseparable, semiseparable, arrowhead matrices, and other classes of easy structured rank matrices all are of Levinson conform form. Moreover it is also indicated how one can overcome possible singular submatrices by a look-ahead scheme and also a factorization in terms of triangular factors of the inverse matrix is presented.

The last chapter of this part presents some methods for inverting these simple classes of structured rank matrices. Some inversion methods based on factorizations of the original matrix are discussed briefly. This is followed by some direct inversion methods for either semiseparable and tridiagonal matrices. Also some general formulas for inverting matrices are presented. A small section is devoted to the scaling of semiseparable matrices. Some information is provided with regard to the decay of the elements of the inverse of a tridiagonal matrix. To conclude, some results on the inverses of tridiagonal Toeplitz matrices are presented.

✎ *Being familiar with the basic types of structured rank matrices such as semiseparable, semiseparable plus diagonal, generator representable semiseparable and so forth, algorithms for these matrices are presented here. Results in this chapter will be presented for the LU-decomposition (a Gaussian elimination-like solver), the QR-factorization, a Levinson and Schur-like solver and finally on the inversion of these matrices. These methods are all applicable for solving systems of equations involving these structured rank matrices. Depending on the structure some methods can be advantageous above others.*

Chapter 4 discusses system solving via the LU-decomposition. To construct such an algorithm the following issues are discussed: the structured rank properties of the involved matrices, the inversion of lower and upper semiseparable matrices, solving an upper triangular structured rank system via backward substitution. More detailed pointers to the key ideas and theorems are found in the introduction connected to the chapter.

The second chapter of this part focuses on the development of a QR-based solver. The key ideas/issues discussed in this chapter are: again, the structural properties of the matrices involved in the factorization, an orthogonalization procedure of a semiseparable matrix based on a sequence of Givens transformations from bottom to top, explicit computations on a quasiseparable represented semiseparable matrix, the QR-factorization and implementation of a quasiseparable matrix and finally some alternative factorizations such as the URV and the RQ-decomposition are discussed.

Chapter 6 discusses Schur and Levinson-type algorithms for solving structured rank systems of equations. After an initial introduction to the topic, in which a

Levinson-like algorithm is created for generator representable semiseparable matrices the general theory is developed. The most important ideas in the method are a characterization of matrices for which the idea is applicable and the effective fast method for these matrices. Several examples are included. Following the Levinson method the Schur-type solver is discussed. Important is the discussion on the preservation of the rank structure when considering the Schur complement. This leads immediately to a Schur-like algorithm.

The final chapter of this part discusses inversion methods for the different types of matrices. Methods based on factorizations (LU and QR) as well as direct inversion techniques are discussed. Also interesting are the results on the boundedness of the sizes of the elements of structured rank matrices when inverting sparse matrices.

Chapter 4

Gaussian elimination

In this chapter we will investigate an efficient Gaussian elimination pattern for some classes of structured rank matrices. Gaussian elimination with partial row pivoting computes a factorization of the matrix A of the form $PA = LU$, where L and U are, respectively, a lower and an upper triangular matrix and P is a permutation matrix. In case of a decomposition without row pivoting of the form $A = LU$, the matrices L and U will have a specific rank structure depending on the rank structure of the original matrix A. The rank structure of these factors L and U needs to be exploited for efficiently using the Gaussian elimination pattern for solving rank structured systems of equations.

The chapter is divided in several sections. Before being able to use the LU-decomposition as presented in this chapter as a solver, some elementary tools have to be developed. First of all we need to be able to solve upper triangular structured rank systems of equations. This can be done effectively, for example, by applying backward substitution, which will be investigated in Section 4.2. Inversion of an upper triangular semiseparable matrix can also be used to solve this system. It will be shown that the matrix L in the LU-decomposition of a strongly nonsingular semiseparable matrix is lower triangular semiseparable. In the algorithm, however, one does not construct the matrix L, but its inverse. Hence we investigate in Section 4.3 the inverse of lower semiseparable matrices to quickly compute the matrix L and also to possibly use it for solving the upper triangular system of equations.

Before starting the construction of an algorithm to compute effectively the LU-decomposition of semiseparable and/or quasiseparable matrices, we investigate the structural properties of this decomposition. In Section 4.4, we derive, based on the nullity theorem, a theorem which predicts the rank structure in the L and U-factor of the nonpivoted LU-decomposition of a general structured rank matrix. Before investigating the LU-decomposition of a general quasiseparable matrix, we study in Section 4.5, the easy class of semiseparable matrices. A general elimi-

nation pattern based on Gaussian elementary transformations[1] is first described
for strongly nonsingular semiseparable matrices. Strongly nonsingular matrices are
considered first, as they admit a nonpivoted LU-decomposition. Combined with
the knowledge gained in Section 4.4, we know that the L and U factors will both
have a lower and upper triangular semiseparable form, respectively. Second the
more general case is investigated, and using the special elimination pattern as described in that section we will be able to prove also that this pivoted decomposition
will output factors having a specific rank structure; more precisely, they will be
of quasiseparable form. The final section of this chapter investigates the class of
quasiseparable matrices. These matrices have less structure than the semiseparable
matrices, and hence more operations are needed to compute the LU-decomposition
of the quasiseparable matrix. A first example shows possible problems that might
occur if we extend directly the algorithm developed for semiseparable matrices. The
problem involves the pivoting, which in this case will propagate into the matrix L,
and which we cannot get rid of. In case the matrix is strongly nonsingular, as
proved before, a decomposition without pivoting always exists. Therefore we focus
in this part on the nonpivoted decomposition and a way to compute it. For the last
two sections attention is also paid to an effective computation of the factors L and
U and to the application of system solving.

✎ *This chapter discusses system solving via the LU-decomposition. In our
opinion, Theorem 4.14 contains the most important results for the development of
an LU-factorization. This theorem (and the proposition following it) reveals the
structure of the factors L and U and provides therefore essential information for
the construction of the LU-factorization. Example 4.17 clearly illustrates the impact of this theorem. How to compute the LU-decomposition of a semiseparable
matrix is presented in Section 4.5. The algorithm is developed for quasiseparable
represented semiseparable matrices. In order to solve a system of equations via this
factorization some pointers are given to earlier sections in which hints on solving
an upper triangular system via backward substitution and methods for inverting
upper triangular matrices are discussed. More precisely, the inversion of bidiagonal
and triangular semiseparable matrices is discussed in Section 4.3 with main Theorems 4.2 and 4.4 for bidiagonal matrices. The remaining theorems are similar but
for triangular semiseparable matrices. The algorithm for performing backward substitution with a quasiseparable represented upper triangular semiseparable matrix
is discussed in Section 4.2.*

*Information on the LU-decomposition of quasiseparable matrices is presented
in Section 4.6. The last two sections discuss the factorization of a strongly nonsingular quasiseparable matrix with and without pivoting. More information on how
to compute the LU-decomposition in general is presented in Chapter 10. This is
because, in general, pivoting will be needed (for numerical stability) when computing LU-factorizations. The standard example in Section 4.7.1 shows that numerical
instability can also cause problems in the structured rank case.*

---

[1]For brevity sometimes called Gauss transforms (see [152]).

4.1 About Gaussian elimination and the *LU*-factorization

This first section briefly introduces the Gaussian elimination algorithm and its link with the *LU*-factorization of matrices. We do not intend to deeply analyze Gaussian elimination; only a brief discussion is presented.

The building blocks for Gaussian elimination are the Gaussian elementary transformation matrices, sometimes referred to as Gauss transforms. Suppose we have a vector $\mathbf{x} = [x_1, x_2, \ldots, x_n]^T$ given and we define the vector $\boldsymbol{\tau}_k$ of length n as

$$\boldsymbol{\tau}_k = \left[0, \ldots, 0, \frac{x_{k+1}}{x_k}, \ldots, \frac{x_n}{x_k}\right]^T.$$

This means that the component $(\boldsymbol{\tau}_k)_i$ is zero for $i \leq k$ and $(\boldsymbol{\tau}_k)_i = x_i/x_k$ for $k < i \leq n$, assuming x_k to be different from zero. The element x_k is called the pivot. A Gauss transform is now defined as the matrix M_k:

$$M_k = I_n - \boldsymbol{\tau}_k \mathbf{e}_k^T,$$

where $\mathbf{e}_k$ stands for the kth standard basis vector. This transformation matrix M_k acts on the vector $\mathbf{x}$ as follows:

$$M_k \mathbf{x} = \hat{\mathbf{x}},$$

where

$$\hat{\mathbf{x}} = [x_1, \ldots, x_k, 0, \ldots, 0]^T.$$

Hence this transformation M_k annihilated the last $n - k$ elements of the vector $\mathbf{x}$.

Applying successive Gauss transforms on a matrix A can now be used efficiently for bringing the matrix A to upper triangular form. Assume that all involved Gauss transforms are well defined (this means that the pivots are different from zeros.), with M_k acting on column k and defined as above. This leads to the following equation:

$$M_{n-1}M_{n-2}\cdots M_1 A = U,$$

where U is an upper triangular matrix. As the inverse of the matrix $M_k = I_n - \boldsymbol{\tau}_k \mathbf{e}_k^T$ equals $M_k^{-1} = I_n + \boldsymbol{\tau}_k \mathbf{e}_k^T$, we get that

$$\begin{aligned} A &= M_1^{-1} \cdots M_{n-1}^{-1} U \\ &= LU. \end{aligned}$$

Because all M_k^{-1} are unit lower triangular, L will also be unit lower triangular. This factorization in a unit lower triangular matrix and an upper triangular matrix U is called the *LU*-decomposition.

One can formulate and prove the following theorems related to the unicity of the *LU*-decomposition of a matrix A, where we take L unit lower triangular.

Theorem 4.1. *Suppose the matrix A is strongly nonsingular, then its LU-decomposition exists and is unique.*

If a matrix is nonsingular and has an LU-decomposition, then this decomposition will be unique.

Proof. Can be found in [152].   □

In practice, the LU-decomposition does not always exist, if, for example there are pivots equal to zero. Moreover, the method described above is not numerically robust for computing this LU-decomposition; hence pivoting is applied. Row pivoting means that one interchanges rows within the matrix A, such that the pivot is the maximal element in absolute value one can take as a pivot. Also column pivoting exists. More details on pivoting and references related to this subject can be found in [152, 265, 258].

Row pivoting leads to a more stable approach for computing the pivoted LU-decomposition. This leads to a factorization of the form:

$$PA = LU,$$

where P is a permutation matrix, L is unit lower triangular and U is upper triangular.

4.2   Backward substitution

Consider an upper triangular semiseparable matrix $S^{(u)}$, represented using the quasiseparable representation, with vectors $\mathbf{p}$, $\mathbf{q}$ and $\mathbf{r}$:

$$
S^{(u)} =
\begin{bmatrix}
q_1\,p_1 & q_2\,r_1\,p_1 & q_3\,r_{2:1}\,p_1 & \cdots & q_n\,r_{n-1:1}\,p_1 \\
         & q_2\,p_2   & q_3 r_2 p_2   & \cdots & q_n\,r_{n-1:2}\,p_2 \\
         &   & q_3 p_3   & \cdots & q_n\,r_{n-1:3}\,p_3 \\
         &   &   & \ddots & \vdots \\
         &   &   &   & q_n\,p_n
\end{bmatrix}.
$$

We cover here the quasiseparable representation, as this includes both the generator and the Givens-vector representation as special cases. Remember that $r_{i:j} = r_i r_{i-1} \cdots r_j$.

In this section we will compute the solution of

$$S^{(u)}\mathbf{x} = \mathbf{b},$$

with $\mathbf{x} = [x_1, \ldots, x_n]^T$ and $\mathbf{b} = [b_1, \ldots, b_n]^T$ via backward substitution.

Let us write down the equations corresponding to the last three rows of the system:

$$
\begin{aligned}
b_n &= q_n p_n x_n, \\
b_{n-1} &= q_{n-1} p_{n-1} x_{n-1} + q_n r_{n-1} p_{n-1} x_n, \\
b_{n-2} &= q_{n-2} p_{n-2} x_{n-2} + q_{n-1} r_{n-2} p_{n-2} x_{n-1} + q_n r_{n-1} r_{n-2} p_{n-2} x_n.
\end{aligned}
$$

Rewriting these equations in terms of the products $q_i x_i$ gives us:

$$q_n x_n = \frac{b_n}{p_n},$$

$$q_{n-1} x_{n-1} = \frac{b_{n-1}}{p_{n-1}} - r_{n-1} \left(q_n x_n \right),$$

$$q_{n-2} x_{n-2} = \frac{b_{n-2}}{p_{n-2}} - r_{n-2} \left(q_{n-1} x_{n-1} + r_{n-1} \left(q_n x_n \right) \right),$$

which are all well-defined as $p_i \neq 0$, otherwise the system would be singular. These last equations can be put, in a straightforward way, in a for-loop for computing the products $q_i x_i$ and hence also x_i. Using an auxiliary variable for storing the products $r_{n-1} \left(q_n x_n \right)$, then $r_{n-2} \left(q_{n-1} x_{n-1} + r_{n-1} \left(q_n x_n \right) \right)$, and so on leads naturally to an order n method. More precisely in every step of the loop, the updating of the auxiliary value takes 2 operations: a summation and a multiplication. The computation of $x_i q_i$ using thereby this auxiliary variable costs 1 subtraction and 1 division. And finally the computation of x_i from $x_i q_i$ costs 1 more division. This gives us an algorithm of $5n$ operations, plus lower order terms; more precisely $5n - 4$ operations are involved.

The changes needed in the method above for computing the solution of an upper triangular quasiseparable or semiseparable plus diagonal system of equations are few. Hence this is not included but left to the reader.

Notes and references

Backward substitution for solving upper triangular systems of equations is often used. Books in which the backward substitution is investigated are legion. For example the following general books cover the backward and/or the forward substitution technique [152, 258, 265].

As we will show later on, the solution of systems via the LU-decomposition or the solution of systems via the QR- or URV-factorization always requires the solution of an upper triangular structured rank system of equations in the last step. For example the manuscript [284] uses backward substitution for the practical implementation of the QR-algorithm; see Section 5.5 in Chapter 5.

Another application of upper triangular system solving of semiseparable matrices can be found in [292]. In this manuscript the authors compute the singular value decomposition of matrices via intermediate upper triangular semiseparable matrices. The above-defined technique can be used, for efficiently computing the singular vectors of the intermediate used semiseparable matrix, via, e.g., inverse iteration.

4.3 Inversion of triangular semiseparable matrices

In this section we will present a method for inverting a lower bidiagonal and a lower semiseparable matrix. Firstly, we will start with the inversion procedure of the lower bidiagonal matrix, as this is the easiest case. Secondly, we will derive in a similar way a method for inverting a lower semiseparable matrix. Numerical examples are presented after the theoretical investigations. It is clear that a combination of the techniques provided in this section, with the existence of a fast matrix vector multiplication for either bidiagonal and lower triangular semiseparable matrices, leads to another type of solver for bidiagonal or lower triangular semiseparable systems of equations.

4.3.1   The inverse of a bidiagonal

In this section we will present a simple method to invert a bidiagonal matrix and to obtain the generators of the inverse of this bidiagonal matrix. We know from Chapter 1, Section 1.5, that the inverse of a bidiagonal matrix will be a lower triangular semiseparable matrix. In this section we will represent this lower triangular semiseparable matrix by using the quasiseparable representation, as this results naturally from our inversion procedure. It is important to remark that in this subsection we do not pose any constraints on the subdiagonal elements of the bidiagonal matrix. Often the considered bidiagonal matrices are required to have a nonzero off-diagonal, as the inverse will then be of generator representable semiseparable form. Of course the splitting of a bidiagonal matrix into a block diagonal matrix, for which the blocks on the diagonal have nonzero off-diagonal elements, is trivial, but it costs computing time[2]. Therefore, we omit this constraint in this section.

A unit bidiagonal matrix can easily be decomposed as the product of Gaussian elementary transformation matrices[3].

Theorem 4.2. *Suppose B to be a unit bidiagonal*[4] *matrix*

$$
B = \begin{bmatrix}
1 & & & & \\
\beta_1 & 1 & & & \\
0 & \beta_2 & 1 & & \\
\vdots & \ddots & \ddots & & \\
0 & \cdots & 0 & \beta_{n-1} & 1
\end{bmatrix},
$$

then B can be written as $M_1 M_2 \cdots M_{n-1}$ where M_i is a Gauss transform of the form $M_i = I + \beta_i \mathbf{e}_{i+1} \mathbf{e}_i^T$, where $\mathbf{e}_i$ denotes the ith vector of the standard basis of $\mathbb{R}^n$.

Proof. Trivial.   □

Let us illustrate this with an example.

Example 4.3 The following bidiagonal matrix can naturally be written as the

---

[2]We did not even mention the rounding errors which might occur when neglecting some elements.

[3]More information concerning Gauss transforms can be found, e.g., in [152].

[4]With a unit bidiagonal matrix, a bidiagonal matrix with ones on the diagonal is meant.

product of Gauss elementary transformation matrices:

$$\begin{bmatrix} 1 & & \\ 2 & 1 & \\ & 3 & 1 \end{bmatrix} = \begin{bmatrix} 1 & & \\ 2 & 1 & \\ & & 1 \end{bmatrix} \begin{bmatrix} 1 & & \\ & 1 & \\ & 3 & 1 \end{bmatrix}.$$

∎

The inverse of such a unit bidiagonal matrix can be factorized in a similar fashion.

Theorem 4.4. *The inverse of a unit bidiagonal matrix B is a lower triangular semiseparable matrix, and can be factorized as $B^{-1} = M_{n-1} \cdots M_2 M_1$, where each M_i is a Gauss transform of the form $M_i = I - \beta_i \mathbf{e}_{i+1} \mathbf{e}_i^T$.*

Proof. We know that $B = \hat{M}_1 \hat{M}_1 \cdots \hat{M}_{n-1}$, with $\hat{M}_i = I + \beta_i \mathbf{e}_{i+1} \mathbf{e}_i^T$. As the inverse of $\hat{M}_i$ equals $\hat{M}_i^{-1} = M_i = I - \beta_i \mathbf{e}_{i+1} \mathbf{e}_i^T$, we find that the inverse of B is of the form:

$$B^{-1} = M_{n-1} \cdots M_2 M_1,$$

which proves our assertion. □

Taking a closer look at the product of the matrices M_i from the previous proof, we see that B^{-1} is of the following form:

$$\begin{bmatrix} 1 & & & & & \\ (-\beta_1) & 1 & & & & \\ (-\beta_1)(-\beta_2) & (-\beta_2) & 1 & & & \\ (-\beta_1)(-\beta_2)(-\beta_3) & (-\beta_2)(-\beta_3) & (-\beta_3) & 1 & & \\ \vdots & \vdots & \vdots & & \ddots & \\ (-\beta_1)\cdots(-\beta_{n-1}) & (-\beta_2)\cdots(-\beta_{n-1}) & (-\beta_3)\cdots(-\beta_{n-1}) & \cdots & (-\beta_{n-1}) & 1 \end{bmatrix}$$

This matrix is clearly a unit lower semiseparable matrix. Hence, one can easily derive a quasiseparable representation of this matrix. In correspondence with the notation used in Section 2.5 for the quasiseparable representation we have $\mathbf{u} = \mathbf{v} = [1, 1, \ldots, 1]^T$ and $\mathbf{t} = -\boldsymbol{\beta} = [-\beta_1, -\beta_2, \ldots, -\beta_{n-1}]^T$.

Calculating the inverse of a general bidiagonal matrix is now straightforward by scaling first the diagonal elements with a post- or premultiplication with a diagonal matrix.

Theorem 4.5. *Suppose an invertible bidiagonal matrix B is given, with diagonal elements $a_1, a_2, \ldots, a_n$ and lower bidiagonal elements $b_1, b_2, \ldots, b_{n-1}$. The inverse of B admits a quasiseparable representation, with $\mathbf{u} = [a_1^{-1}, a_2^{-1}, \ldots, a_n^{-1}]^T$, $\mathbf{v} = [1, 1, \ldots, 1]^T$ and $\mathbf{t} = -\boldsymbol{\beta} = [-\beta_1, -\beta_2, \ldots, -\beta_{n-1}]^T$, where $\beta_i = b_i a_i^{-1}$.*

Proof. The bidiagonal matrix B can be written as $B = \hat{B}D$, where D is a diagonal matrix, $D = \text{diag}([a_1, a_2, \ldots, a_n])$ and the matrix $\hat{B}$ is unit bidiagonal with subdiagonal elements $\boldsymbol{\beta} = [b_1 a_1^{-1}, b_2 a_2^{-1}, \ldots, b_{n-1} a_{n-1}^{-1}]^T$. Hence we have that the inverse of B equals $D^{-1} \hat{B}^{-1}$, where $\hat{B}^{-1}$ is a unit lower semiseparable matrix. Hence as a representation for the resulting lower semiseparable matrix, we get $\mathbf{u} = [a_1^{-1}, a_2^{-1}, \ldots, a_n^{-1}]^T$, $\mathbf{v} = [1, 1, \ldots, 1]^T$ and $\mathbf{t} = -\boldsymbol{\beta} = [-\beta_1, -\beta_2, \ldots, -\beta_{n-1}]^T$.

$\square$

Note 4.6. *In the previous proof, the diagonal elements of the matrix B were all made equal to one by a scaling of the columns. The same can be done by scaling the rows and by premultiplying with a diagonal matrix.*

The inversion method as presented in this section is clearly an order $\mathcal{O}(n)$ method. More precisely, computing the inverse of a unit bidiagonal matrix is for free, as a sign change is not counted as an operation. For inverting a general bidiagonal matrix, it costs n divisions for computing the elements a_i^{-1}, which define the generator $\mathbf{u}$ and $n - 1$ more multiplications to define the vector $\mathbf{t}$. So in total, inverting a bidiagonal matrix costs $2n - 1$ operations, plus lower order terms.

4.3.2   The inverse of lower semiseparable matrices

In this section we will derive an algorithm for inverting a lower semiseparable matrix in an analogous way as in the previous section. We will factorize the lower semi-separable matrix as a product of Gauss transforms (see also factorization schemes presented in [112, 118]).

Let us factorize the matrix $S^{(l)}$ in a similar way as we factorized the bidiagonal matrix in the previous section.

Theorem 4.7. *Suppose $S^{(l)}$ is a unit lower semiseparable matrix. Then $S^{(l)}$ can be written as $S^{(l)} = M_{n-1} \cdots M_2 M_1$, where each M_i is a Gauss transform of the form $M_i = I + \beta_i \mathbf{e}_{i+1} \mathbf{e}_i^T$. The elements β_i denote the subdiagonal elements, $\beta_i = S^{(l)}(i+1, i)$.*

Proof. As $S^{(l)}$ is a unit lower semiseparable matrix, with subdiagonal elements β_i, with $i = 1, \ldots, n-1$, we know, due to the semiseparable structure, that every element $S^{(l)}(i, j)$, with $i > j$ is of the form $S^{(l)}(i, j) = \beta_j \beta_{j+1} \cdots \beta_{i-2} \beta_{i-1}$. All the elements $S^{(l)}(i, i)$ are equal to 1, with $i \in \{1, \ldots, n\}$. As this is the structure of the matrix $S^{(l)}$, a straightforward calculation reveals that

$$S^{(l)} = M_{n-1} \cdots M_2 M_1,$$

where $M_i = I + \beta_i \mathbf{e}_{i+1} \mathbf{e}_i^T$. $\square$

Note 4.8. *For representing unit lower semiseparable matrices, one only needs the subdiagonal elements. This corresponds to the diagonal-subdiagonal representation as investigated in Chapter 2, Section 2.3.*

Note 4.9. *We briefly note that, for the inversion of this lower semiseparable matrix, the type of representation does not play any role. One only needs to be able to calculate the subdiagonal elements efficiently.*

As one has this representation, computing the inverse of a unit lower semiseparable as well as the inverse of a lower semiseparable becomes straightforward.

Theorem 4.10. *The inverse of a unit lower semiseparable matrix $S^{(l)}$ is a lower unit bidiagonal matrix, with subdiagonal elements $-\beta_i$, where the β_i are the subdiagonal elements of the matrix $S^{(l)}$.*

Proof. Invert the factorization of the matrix $S^{(l)}$ from Theorem 4.7. $\quad\square$

Theorem 4.11. *Suppose a lower semiseparable matrix $S^{(l)}$ is given, with diagonal elements $a_1, a_2, \ldots, a_n$ and subdiagonal elements $b_1, b_2, \ldots, b_{n-1}$. The inverse of $S^{(l)}$ is a bidiagonal matrix, with diagonal elements $a_1^{-1}, a_2^{-1}, \ldots, a_n^{-1}$ and subdiagonal elements $-\beta_1 a_2^{-1}, -\beta_2 a_3^{-1}, \ldots, -\beta_{n-1} a_n^{-1}$, where $\beta_i = b_i a_i^{-1}$.*

Proof. Similar to the proof of Theorem 4.5. $\quad\square$

The results in this section will come in very handy when computing the LU-factorization of semiseparable and related matrices, as we compute the LU-decomposition by Gaussian elementary transformation matrices. We will now derive the LU-decomposition of some structured rank matrices, as these results will be used for deriving general inversion methods for more general structured rank matrices, such as quasiseparable and semiseparable plus diagonal matrices.

4.3.3 Examples

Let us illustrate the theoretical considerations on inversion of the previous subsections with some examples. We present some examples of unit lower triangular matrices.

Example 4.12 In the examples below some matrices with their inverses are given.

$$B_1 = \begin{bmatrix} 1 & & \\ 1 & 1 & \\ 0 & 1 & 1 \end{bmatrix}, \quad B_1^{-1} = \begin{bmatrix} 1 & & \\ -1 & 1 & \\ 1 & -1 & 1 \end{bmatrix},$$

$$B_2 = \begin{bmatrix} 1 & & \\ 2 & 1 & \\ 0 & 2 & 1 \end{bmatrix}, \quad B_2^{-1} = \begin{bmatrix} 1 & & \\ -2 & 1 & \\ 4 & -2 & 1 \end{bmatrix},$$

$$B_3 = \begin{bmatrix} 1 & & \\ 2 & 1 & \\ 0 & 3 & 1 \end{bmatrix}, \quad B_3^{-1} = \begin{bmatrix} 1 & & \\ -2 & 1 & \\ 6 & -3 & 1 \end{bmatrix}.$$

Let us consider a larger matrix

$$
B_4 = \begin{bmatrix} 1 & & & & \\ 1 & 1 & & & \\ & 2 & 1 & & \\ & & 3 & 1 & \\ & & & 4 & 1 \end{bmatrix}, \quad B_4^{-1} = \begin{bmatrix} 1 & & & & \\ 1 & 1 & & & \\ 2 & -2 & 1 & & \\ -6 & 6 & -3 & 1 & \\ 24 & -24 & 12 & -4 & 1 \end{bmatrix}.
$$

∎

To conclude we present some examples of nonunit lower triangular matrices.

Example 4.13

$$
B_1 = \begin{bmatrix} 2 & & \\ 1 & 2 & \\ 0 & 1 & 2 \end{bmatrix}, \quad B_1^{-1} = \begin{bmatrix} 1/2 & & \\ -1/4 & 1/2 & \\ 1/8 & -1/4 & 1/2 \end{bmatrix},
$$

$$
B_2 = \begin{bmatrix} 2 & & \\ 4 & 2 & \\ 0 & 4 & 2 \end{bmatrix}, \quad B_2^{-1} = \begin{bmatrix} 1/2 & & \\ -1 & 1/2 & \\ 2 & -1 & 1/2 \end{bmatrix},
$$

$$
B_3 = \begin{bmatrix} 1 & & \\ 1 & 2 & \\ 0 & 1 & 3 \end{bmatrix}, \quad B_3^{-1} = \begin{bmatrix} 1 & & \\ -1/2 & 1/2 & \\ 1/6 & -1/6 & 1/3 \end{bmatrix},
$$

$$
B_3 = \begin{bmatrix} 1 & & \\ 2 & 2 & \\ 0 & 6 & 3 \end{bmatrix}, \quad B_3^{-1} = \begin{bmatrix} 1 & & \\ -1 & 1/2 & \\ 2 & -1 & 1/3 \end{bmatrix}.
$$

∎

Notes and references

The decomposition of the bidiagonal matrix as a product of Gauss elementary transformation matrices is in fact a direct consequence from the LU-decomposition of the bidiagonal matrix, where the matrix U will then be the identity matrix. Gauss transformation matrices are the basic tools for computing the LU-factorization of matrices. They can be found in general textbooks on numerical linear algebra.

In the manuscript [240], the authors invert a lower triangular generator representable semiseparable matrix S, and compute its inverse B. More information on this manuscript can be found in Section 2.2 of Chapter 2.

☞ T. Ting. A method of solving a system of linear equations whose coefficients form a tridiagonal matrix. *Quarterly of Applied Mathematics*, 22(2):105–106, 1964.

The author Ting describes, based on the LU-decomposition of a tridiagonal matrix T (not necessarily symmetric), a special decomposition of T^{-1} as a product of inverses of bidiagonal matrices, multiplied on the left with a diagonal matrix. This can be used for solving various systems of equations with the same tridiagonal matrix T.

In the following paper, the author gives explicit formulas for inverting a lower block bidiagonal matrix. The blocks on the diagonal have to be invertible.

☞ V. N. Singh. The inverse of a certain block matrix. *Bulletin of the Australian Mathematical Society*, 20:161–163, 1979.

As mentioned in a note in this section, the diagonal-subdiagonal representation is closely related to the decomposition in terms of Gauss transforms. Hence, closely related results can be found in the publications in the notes and references part of Section 2.3 in Chapter 2.

4.4 Theoretical considerations of the LU-decomposition

The algorithmic description on how to solve a system of equations with a structured rank coefficient matrix by using the LU-decomposition is the subject of the next section. We will first investigate theoretically the structure of the factors of an LU-decomposition of a structured rank matrix. Knowing structural properties of the involved matrices might help constructing the algorithm in an efficient way.

Based on the proof of the nullity theorem by Fiedler (see Chapter 1 and [114]) it is very easy to generalize the nullity theorem and apply it to decompositions of structured rank matrices.

Theorem 4.14. *Suppose we have an invertible matrix A, with an LU-factorization of the following form:*

$$A = LU.$$

Suppose A is partitioned in the following form

$$A = \left[\begin{array}{cc} A_{11} & A_{12} \\ A_{21} & A_{22} \end{array} \right],$$

with A_{11} of dimension $p \times q$. When the inverse B of U is partitioned as

$$U^{-1} = B = \left[\begin{array}{cc} B_{11} & B_{12} \\ B_{21} & B_{22} \end{array} \right],$$

with B_{11} of dimension $q \times p$, then the nullities $\mathrm{n}(A_{12})$ and $\mathrm{n}(B_{12})$ are equal.

We remark that, due to the free choice of q and p, the block B_{21} is not necessarily equal to zero. The block is, however, always zero if $p \leq q$. The proof is very similar to the one of the nullity theorem.

Proof. First, we will prove that $\mathrm{n}(A_{12}) \geq \mathrm{n}(B_{12})$, by using the relation $AU^{-1} = L$. Suppose that the nullity of B_{12} equals c. Then there exists a matrix F with c linearly independent columns such that $B_{12}F = 0$. Partitioning L in the following way

$$L = \left[\begin{array}{cc} L_{11} & 0 \\ L_{21} & L_{22} \end{array} \right],$$

with L_{11} of dimension $p \times p$, we can write down the following equations, which we multiply with F:

$$A_{11}B_{12} + A_{12}B_{22} = 0,$$
$$A_{12}B_{22}F = 0$$

and

$$A_{21}B_{12} + A_{22}B_{22} = L_{22},$$
$$A_{22}B_{22}F = L_{22}F.$$

Therefore rank $(B_{22}F) \geq c$, because L_{22} is of full rank. This leads us to the result:

$$\mathrm{n}\,(A_{12}) \geq \mathrm{rank}\,(B_{22}F) \geq c = \mathrm{n}\,(B_{12})\,.$$

This proves already one direction. For the other direction, $\mathrm{n}\,(A_{12}) \leq \mathrm{n}\,(B_{12})$, we use a partitioning for the inverse of A and the matrix U, such that the upper left block of $C = A^{-1}$, denoted as C_{11} has size $q \times p$ and the upper left block of U denoted as U_{11} has size $p \times q$. Using the equation $UA^{-1} = L^{-1}$ we can prove in a similar way as above that

$$\mathrm{n}\,(U_{12}) \geq \mathrm{n}\,(C_{12})\,.$$

Using the nullity theorem and the remarks following it, gives us

$$\mathrm{n}\,(B_{12}) = \mathrm{n}\,(U_{12}) \geq \mathrm{n}\,(C_{12}) = \mathrm{n}\,(A_{12})\,.$$

This proves the theorem.   □

Let us combine this with the nullity theorem.

Theorem 4.15. *Suppose we have an invertible matrix A, with an LU-factorization of the following form:*
$$A = LU.$$

Suppose A to be partitioned in the following form

$$A = \left[\begin{array}{cc} A_{11} & A_{12} \\ A_{21} & A_{22} \end{array}\right]$$

with A_{11} of dimension $p \times q$, while the matrices U and L are partitioned as

$$U = \left[\begin{array}{cc} U_{11} & U_{12} \\ U_{21} & U_{22} \end{array}\right] \text{ and } L = \left[\begin{array}{cc} L_{11} & L_{12} \\ L_{21} & L_{22} \end{array}\right],$$

with U_{11} and L_{11} of dimension $p \times q$. Then the nullities $\mathrm{n}(A_{12})$ and $\mathrm{n}(U_{12})$ are equal as well as the nullities $\mathrm{n}(A_{21})$ and $\mathrm{n}(L_{21})$.

We remark once more that only for specific choices of p and q do the blocks U_{21} and L_{12} become zero. The matrix U is upper triangular, however, and L is lower triangular.

Proof. Combining the results of Theorem 4.14 with the nullity theorem (see Theorem 1.33) leads directly to a proof. □

The theorem above can easily be translated towards the rank structure of these blocks.

Proposition 4.16. *Using the same notation as from Theorem 4.15, we have the following relations for the rank:*

$$\text{rank}\,(A_{12}) = \text{rank}\,(U_{12}),$$
$$\text{rank}\,(A_{21}) = \text{rank}\,(L_{21}).$$

Proof. A combination of the fact that the nullity of matrix A of dimension $p \times q$ satisfies

$$\text{n}(A) = q - \text{rank}\,(A),$$

combined with the results of the previous theorem, leads directly to this observation. □

Let us illustrate the power of this theorem with some examples.

Example 4.17 We assume that all the following matrices A are nonsingular and are decomposed as $A = LU$.

- If the matrix A is semiseparable, the matrices L and U are, respectively, lower triangular semiseparable and upper triangular semiseparable.

- If the matrix A is quasiseparable, the matrices L and U are, respectively, lower triangular quasiseparable, and upper triangular quasiseparable.

- If the matrix A is a tridiagonal matrix, the matrices L and U are, respectively, lower bidiagonal and upper bidiagonal matrices.

- If the matrix A is a band matrix, the matrices L and U are lower triangular and upper triangular band matrices, with the bandwidths corresponding to the bandwidths in the matrix A.

- Using the results above we can also start predicting the structures of the LU-decomposition of band plus semiseparable matrices etc.

∎

Note that for the more general LU-factorization

$$PA = LU,$$

with P a nontrivial permutation matrix the two factors L and U are not necessarily semiseparable. Take, for example, the matrix

$$A = \begin{bmatrix} 0 & 1 & 1 \\ 1 & 0 & 0 \\ 1 & 0 & 1 \end{bmatrix}.$$

Then, $PA = LU$ with

$$P = \begin{bmatrix} 0 & 1 & 0 \\ 1 & 0 & 0 \\ 0 & 0 & 1 \end{bmatrix}, \quad L = \begin{bmatrix} 1 & 0 & 0 \\ 0 & 1 & 0 \\ 1 & 0 & 1 \end{bmatrix}, \quad U = \begin{bmatrix} 1 & 0 & 0 \\ 0 & 1 & 1 \\ 0 & 0 & 1 \end{bmatrix}.$$

Hence, it is clear that L is not semiseparable. Similar examples can be constructed for band matrices. If pivoting is also applied the band structure of the factors can get lost (see, e.g., [152]).

The solution to the pivoted LU-decomposition is in fact straightforward.

Theorem 4.18.   *Suppose we have an invertible matrix A with a pivoted LU-factorization of the following form:*

$$PA = LU.$$

Suppose PA is partitioned in the following form

$$PA = \begin{bmatrix} (PA)_{11} & (PA)_{12} \\ (PA)_{21} & (PA)_{22} \end{bmatrix}$$

with $(PA)_{11}$ of dimension $p \times q$. The matrices U and L are partitioned as

$$U = \begin{bmatrix} U_{11} & U_{12} \\ U_{21} & U_{22} \end{bmatrix} \text{ and } L = \begin{bmatrix} L_{11} & L_{12} \\ L_{21} & L_{22} \end{bmatrix},$$

with U_{11} and L_{11} of dimension $p \times q$. Then the nullities $\mathrm{n}((PA)_{12})$ and $\mathrm{n}(U_{12})$ are equal as well as the nullities $\mathrm{n}((PA)_{21})$ and $\mathrm{n}(L_{21})$.

The proof is a direct application of Theorem 1.33.

Notes and references

The results in this section are theoretical results mainly based on the nullity theorem by Fiedler and on the results provided in the paper by Vandebril, Van Barel and Mastronardi (see [291, 114]). Vandebril, Van Barel and Mastronardi investigate in more detail the nullity theorem and present some extensions of it, towards the LU- and QR-factorization. These manuscripts were discussed previously in Section 1.5 in Chapter 1.

A weaker formulation of the nullity theorem translated to the LU-decomposition can be found in [111], discussed previously in Section 1.5 in Chapter 1. Results in this paper are straightforward consequences from the nullity theorem. Fiedler investigates in the manuscript [112] the properties of so-called basic matrices; these matrices correspond, in fact, to the class of quasiseparable matrices. Results concerning the LU-factorization and the inverses of these matrices are presented. It is proved that the factors in the LU-decomposition have the same structure as the original matrix, and also that the inverse remains of quasiseparable form. In the manuscript [118], so-called generalized Hessenberg matrices are investigated. These are matrices for which the lower triangular part is of quasiseparable form. Results concerning the LU-factorization and inversion are obtained. More information on the contents of these matrices with regard to their representation can be found in Section 2.3 in Chapter 2. More information on the structure of these matrices can be found in Section 4.6 in of this chapter.

4.5 The LU-decomposition for semiseparable matrices

First, we will present a method for computing the LU-decomposition of a semiseparable matrix. We derive the theoretical derivation of the decomposition, and also an algorithm to compute the representation of the L and the U factor. In a later section, we derive this algorithm for matrices for which the diagonal is not incorporated in the structure of the lower triangular part, such as quasiseparable and semiseparable plus diagonal matrices.

4.5.1 Strongly nonsingular matrices

We will first derive the decomposition for strongly nonsingular matrices, as these matrices (see [152, p. 97]), have a unique nonpivoted LU-decomposition. In the next subsection, a more thorough investigation will be made concerning matrices that are not necessarily strongly nonsingular.

The matrix we are working with is a semiseparable matrix, represented using the quasiseparable representation, using the vectors $\mathbf{u}$, $\mathbf{v}$ and $\mathbf{t}$. Let us consider initially only the lower triangular part:

$$
S = \begin{bmatrix}
u_1 v_1 & & & & & \\
u_2 t_1 v_1 & u_2 v_2 & & & & \\
u_3 t_2 t_1 v_1 & u_3 t_2 v_2 & u_3 v_3 & & & \\
\vdots & & \ddots & \ddots & & \\
u_{n-1} t_{n-2} \cdots t_1 v_1 & \cdots & & & u_{n-1} v_{n-1} & \\
u_n t_{n-1} t_{n-2} \cdots t_1 v_1 & \cdots & & & u_n t_{n-1} v_{n-1} & u_n v_n
\end{bmatrix}.
$$

We will construct now factors M_{n-1} up to M_1 such that $M_1 \ldots M_{n-1} S = U$, where U is upper triangular. Consider the transformation matrices $\breve{M}_i$ (with $i = 1, \ldots, n-1$), which are unit lower triangular:

$$
\breve{M}_i = \begin{bmatrix} 1 & \\ -\tau_i & 1 \end{bmatrix},
$$

where τ_i is defined as $u_{i+1} t_i / u_i$. A division of a nonzero element by zero is not possible, because this implies $u_i = 0$, which implies in turn that the matrix was not strongly nonsingular.

Applying the transformation matrix $\check{M}_{n-1}$ onto the rows $n-1$ and n up to the diagonal gives us (due to the semiseparable structure):

$$\check{M}_{n-1}\left[\begin{array}{c} S(n-1,1:n-1) \\ S(n,1:n-1) \end{array}\right] = \left[\begin{array}{c} S(n-1,1:n-1) \\ -\tau_n S(n-1,1:n-1)+S(n,1:n-1) \end{array}\right]$$
$$= \left[\begin{array}{c} S(n-1,1:n-1) \\ 0 \end{array}\right].$$

For notational convenience, we denote a transformation matrix working on part of the matrix as $\check{M}_i$. The embedding of this matrix in the identity matrix is denoted as M_i. This means that $M_i = I - \tau_i \mathbf{e}_{i+1} \mathbf{e}_i^T$, where $\mathbf{e}_i$ denotes the ith vector of the standard basis of $\mathbb{R}^n$. For this reason, $M_{n-1}S$ will have the last row equal to zero, except for the diagonal element.

Sequentially applying the transformations M_{n-1} up to M_1 gives us

$$M_1 \ldots M_{n-1}S = U,$$

where clearly U is an upper triangular matrix. Moreover, as all the matrices M_i are unit lower triangular, therefore also the matrix $L = M_{n-1}^{-1} \ldots M_1^{-1}$ will be unit lower triangular. Hence we get an LU-factorization of the semiseparable matrix $S = LU$. We know from the previous section that these matrices L and U inherit the structure from the matrix S. Hence they need to be lower and upper triangular semiseparable matrices, respectively. Let us take a closer look at the structure of these matrices.

The factor L

The computation of the factor L becomes very easy due to the special structure of the factors M_i (see Section 4.3). The matrix $M = M_1 \cdots M_{n-1} = L^{-1}$ will be unit bidiagonal, with subdiagonal elements $-\tau_1, -\tau_2, \ldots, -\tau_{n-1}$. We know from Section 4.3 that the inverse of this bidiagonal matrix is a lower semiseparable matrix L, in correspondence with Theorem 4.14.

The inverse of a matrix $M_i = I - \tau_i \mathbf{e}_{i+1} \mathbf{e}_i^T$ equals the matrix $M_i^{-1} = I + \tau_i \mathbf{e}_{i+1} \mathbf{e}_i^T$. As our matrix L is of the form $L = M_{n-1}^{-1} \ldots M_1^{-1}$ we get:

$$L = \left[\begin{array}{cccccc} 1 & & & & & \\ \tau_1 & 1 & & & & \\ \tau_1\tau_2 & \tau_2 & 1 & & & \\ \tau_1\tau_2\tau_3 & \tau_2\tau_3 & \tau_3 & 1 & & \\ \vdots & & & & \ddots & \\ \tau_1\cdots\tau_{n-1} & \tau_2\cdots\tau_{n-1} & \tau_3\cdots\tau_{n-1} & \cdots & \tau_{n-1} & 1 \end{array}\right],$$

which clearly has a semiseparable structure in the lower triangular part. The representation of this lower triangular part in terms of the quasiseparable representation is $\mathbf{u} = [1,\ldots,1]^T$, $\mathbf{v} = [1,\ldots,1]^T$, and $\mathbf{t} = [\tau_1, \tau_2, \ldots, \tau_{n-1}]^T$. Hence the computation of the factor L is almost trivial if the original matrix is semiseparable and strongly nonsingular.

The factor U

Let us consider first how the generators of the upper triangular part change in the case of a nonpivoted LU-decomposition. The structure of the upper triangular part of S is of the following form:

$$
\begin{bmatrix}
q_1\,p_1 & q_2\,r_1\,p_1 & q_3\,r_{2:1}\,p_1 & \cdots & q_n\,r_{n-1:1}\,p_1 \\
 & q_2\,p_2 & q_3 r_2 p_2 & \cdots & q_n\,r_{n-1:2}\,p_2 \\
 & & q_3 p_3 & \cdots & q_n\,r_{n-1:3}\,p_3 \\
 & & & \ddots & \vdots \\
 & & & & q_n\,p_n
\end{bmatrix},
$$

which is a standard quasiseparable representation with vectors $\mathbf{q}$, $\mathbf{p}$ and $\mathbf{r}$, where we have that $p_i q_i = u_i v_i$ for all i.

The application of the transformations M_{n-1} up to M_1 results in an upper triangular semiseparable matrix. As every row is changed into a sum of its own and the preceding row, the generators of the new upper triangular semiseparable matrix are calculated easily. The quasiseparable generators of the upper triangular part remain the same, except for the vector $\mathbf{p}$. The new vector $\hat{\mathbf{p}}$ is of the following form

$$
\hat{\mathbf{p}}^T = \left[\, p_1\,,\ (p_2 - \tau_1 r_1 p_1)\,,\ (p_3 - \tau_2 r_2 p_2)\,,\ \ldots\,,\ (p_n - \tau_{n-1} r_{n-1} p_{n-1})\,\right]^T.
$$

System solving

The LU-decomposition, discussed above, can be used in a straightforward manner to solve systems involving semiseparable coefficient matrices. Consider the system $S\mathbf{x} = \mathbf{b}$. Application of the transformations M_{n-1} up to M_1 on the left of the system gives:

$$
U\mathbf{x} = (M_1 \cdots M_{n-1} S)\,\mathbf{x} = M_1 \cdots M_{n-1}\mathbf{b} = \hat{\mathbf{b}},
$$

where $\hat{\mathbf{b}} = [\, b_1\,,\ b_2 - \tau_1 b_1\,,\ \ldots\,,\ b_n - \tau_{n-1} b_{n-1}]^T$. The solution of the remaining system $U\mathbf{x} = \hat{\mathbf{b}}$, can be done efficiently using the backward substitution algorithm of Section 4.2, or one can explicitly compute the inverse and then make a multiplication with this inverse via the techniques in Section 4.3.

4.5.2 General semiseparable matrices

In the previous section it was assumed that the considered semiseparable matrix was strongly nonsingular. This implied that the values τ_i were always well-defined. Let us loosen this condition in this section, and see what changes in the algorithm.

The aim of each successive transformation M_i is to make the row i zero up to the diagonal. The only problem might occur, as u_i is zero, while $u_{i+1} t_i$ is different from zero. In this case, we cannot determine the factor τ_i, such that multiplication of the matrix S on the left with M_i makes row i zero.

Example 4.19 The next matrix is of this special form:

$$A = \begin{bmatrix} 1 & 0 & 0 \\ 0 & 0 & 1 \\ 0 & 1 & 0 \end{bmatrix}.$$

It is clear that no Gauss transformation matrix, acting on the last two rows of this matrix, can make the last row zero up to the diagonal. ∎

To overcome this problem, we apply row pivoting. We swap row i and row $i-1$. This is done by applying an elementary row operation. The next equation clearly shows that this will create the desired zero row.

$$\breve{M}_i \begin{bmatrix} 0 \\ S(i,1:i-1) \end{bmatrix} = \begin{bmatrix} 0 & 1 \\ 1 & 0 \end{bmatrix} \begin{bmatrix} 0 \\ S(i,1:i-1) \end{bmatrix}$$
$$= \begin{bmatrix} S(i,1:i-1) \\ 0 \end{bmatrix}.$$

The matrix $\breve{M}_i$ is the 2×2 counteridentity matrix. Embedding the matrix $\breve{M}_i$ into the matrix M_i gives us a transformation, which makes all elements of the row i up to and including the diagonal zero.

Unfortunately, pivoting destroys in general the structure of the corresponding semiseparable matrix. We will analyze here in more detail what will happen with the semiseparable structure in case pivoting is applied.

In the end, we would like to obtain a factorization of the form $PA = LU$, where P is a permutation matrix and L and U are, respectively, lower and upper triangular. The permutation matrices M_i are located however in the sequence of Gauss transforms; hence, there is the need to interchange the order of the operations applied on the matrix A, such that we obtain $L^{-1}PA = U$, instead of a mixture of Gauss transforms and pivoting matrices.

Suppose all matrices M_i are calculated, we obtain a matrix factorization of the form:

$$M_1 \cdots M_{n-1} S = U. \tag{4.1}$$

Assume the matrices $\breve{M}_i$ with $i \in I = \{i_1, \ldots, i_\alpha\} \subset N = \{1, \ldots, n-1\}$ are 2×2 counteridentity matrices, with $i_1 < i_2 < \ldots < i_\alpha$. Define $J = N \backslash I = \{j_1, \ldots, j_\beta\}$, with $j_1 < j_2 < \ldots < j_\beta$ and $\alpha + \beta = n - 1$. It is easy to see that the matrices M_{j_l} and M_{i_k} commute, i.e., $M_{i_k} M_{j_l} = M_{j_l} M_{i_k}$ if $j_l < i_k - 1$. This is done in order to shift the permutation matrices to the right and the Gauss transforms to the left.

There can be problems however if $j_l = i_k - 1$. Then we have the following result, when dragging through the permutation matrix:

$$\begin{bmatrix} 0 & 1 & 0 \\ 1 & 0 & 0 \\ 0 & 0 & 1 \end{bmatrix} \begin{bmatrix} 1 & 0 & 0 \\ 0 & 1 & 0 \\ 0 & -\tau_{j_l} & 1 \end{bmatrix} = \begin{bmatrix} 1 & 0 & 0 \\ 0 & 1 & 0 \\ -\tau_{j_l} & 0 & 1 \end{bmatrix} \begin{bmatrix} 0 & 1 & 0 \\ 1 & 0 & 0 \\ 0 & 0 & 1 \end{bmatrix}. \tag{4.2}$$

It is clear that we still have a Gauss transform, but a slightly changed one. Let us denote in general these changed Gauss transforms as $\tilde{M}_{j_l}$. This means that we can rewrite (4.1) as

$$\left(\tilde{M}_{j_1} \tilde{M}_{j_2} \cdots \tilde{M}_{j_\beta} \right) \left(M_{i_1} M_{i_2} \cdots M_{i_\alpha} \right) S = U.$$

Define $\tilde{M} = \tilde{M}_{j_1} \tilde{M}_{j_2} \cdots \tilde{M}_{j_\beta} = L^{-1}$ and $P = M_{i_1} M_{i_2} \cdots M_{i_\alpha}$, then we have in fact a pivoted LU-decomposition of the matrix S, of the following form:

$$PS = LU.$$

Even though we applied now permutations to the matrix S, we will see that both the matrix L and U will still have some particular rank properties. Unfortunately the matrices do not necessarily inherit the structure of the semiseparable matrix S anymore, as illustrated in the following example.

Example 4.20 Suppose we have a semiseparable matrix S of the following form:

$$S = \begin{bmatrix} 1 & 0 & 1 \\ 0 & 0 & 1 \\ 0 & 1 & 1 \end{bmatrix}.$$

A row-pivoted LU-decomposition of this matrix can be of the following form:

$$\begin{bmatrix} 1 & 0 & 0 \\ 0 & 0 & 1 \\ 0 & 1 & 0 \end{bmatrix} \begin{bmatrix} 1 & 0 & 1 \\ 0 & 0 & 1 \\ 0 & 1 & 1 \end{bmatrix} = \begin{bmatrix} 1 & 0 & 0 \\ 0 & 1 & 0 \\ 0 & 0 & 1 \end{bmatrix} \begin{bmatrix} 1 & 0 & 1 \\ 0 & 1 & 1 \\ 0 & 0 & 1 \end{bmatrix},$$
$$PA = LU.$$

It is clear that the resulting upper triangular matrix U is not of upper triangular semiseparable form. ∎

Even though the lower triangular matrix in the previous example is still of semiseparable form, this is not always the case, as illustrated in the following example.

Example 4.21 The reader can try to compute the pivoted LU-decomposition of the following matrix, involving only one permutation matrix:

$$S = \begin{bmatrix} \times & 0 & \times & \times \\ 0 & 0 & \times & \times \\ \times & \times & \times & \times \\ \times & \times & \times & \times \end{bmatrix}.$$

Writing this matrix factorization in the form $PA = LU$, the reader will come across the interchange of one permutation matrix and one Gauss transform as in Equation (4.2). Finally the permuted matrix PA will not have the lower triangular part of semiseparable form anymore, but of quasiseparable form. Hence, and this is clear in the matrix L, the matrix L will also not be semiseparable anymore, but quasiseparable in general. ∎

We will now prove that the matrix L will be lower quasiseparable, and the matrix U will be upper quasiseparable, i.e., in general, when pivoting is used, the diagonal will not be includable in the structure of the matrix U or L.

The structure of the factors L and U

As already indicated, the only problem occurs when a certain u_i equals zero. It is easy to verify that it is not possible to have two successive values u_i and u_{i+1} both equal to zero, as the matrix has to be singular otherwise (due to the semiseparable structure of the upper triangular part). Hence let us assume that the involved pivoting matrix P only interchanges two successive rows, whereas one of the interchanged rows was completely zero up to (and including) the diagonal.

Assume we have now a pivoted LU-decomposition of the semiseparable matrix A of the form $PA = LU$, where P is of this specific form. Partition the semiseparable matrix A in the following form:

$$ PA = \begin{bmatrix} (PA)_{11} & (PA)_{12} \\ (PA)_{21} & (PA)_{22} \end{bmatrix}, $$

with $(PA)_{11}$ of dimension $p \times q$. The matrix L is partitioned as

$$ L = \begin{bmatrix} L_{11} & L_{12} \\ L_{21} & L_{22} \end{bmatrix}. $$

Then we have that $\operatorname{rank}((PA)_{21})$ and $\operatorname{rank}(L_{21})$ are equal. Moreover, due to the structure of the permutation matrix P (interchanging only two successive rows), we get that the involved transformation can destroy the diagonal element. Hence, in general, the diagonal element will not be includable anymore into the semiseparable structure and our matrix L will become of quasiseparable form.

One can write down similar relations for the upper triangular part that will lead to the same observation. Namely, the involved permutation matrix will destroy the structure of the diagonal. This causes the matrix U not be of upper semiseparable form but to become of quasiseparable form.

The factor L

The structure of the factor L can be derived in a completely similar way as in the previous section. The matrix $\tilde{M} = L^{-1} = \tilde{M}_{i_1} \tilde{M}_{i_2} \cdots \tilde{M}_{i_\alpha}$, where all the matrices $\tilde{M}_{i_l}$ are Gauss transforms. This means that the matrix M will have at most two subdiagonals with elements on either the first or second subdiagonal $-\tau_i$. Similarly as in Section 4.3, we can compute the structure of its inverse, which will be a lower quasiseparable matrix. Moreover this quasiseparable matrix is neatly presented in factored form.

The factor U

Here, we will derive the quasiseparable representation of the strictly upper triangular matrix U. It will be proven by induction that the matrix U is upper triangular quasiseparable. As we know from the first part of the book this means that the strictly upper triangular part will be of semiseparable form.

Theorem 4.22. *Suppose a semiseparable matrix S is given, with a factorization of the form $M_1 \cdots M_{n-1} S = U$, where the matrices M_i are defined as before. Then we have that the matrix U is upper triangular quasiseparable.*

Proof. We prove the theorem by induction. We start with a semiseparable matrix S and assume its upper triangular part is represented with the quasiseparable representation:

$$S^{(u)} = \begin{bmatrix} q_1\,p_1 & q_2\,r_1\,p_1 & q_3\,r_{2:1}\,p_1 & \cdots & q_n\,r_{n-1:1}\,p_1 \\ & q_2\,p_2 & q_3 r_2 p_2 & \cdots & q_n\,r_{n-1:2}\,p_2 \\ & & q_3 p_3 & \cdots & q_n\,r_{n-1:3}\,p_3 \\ & & & \ddots & \vdots \\ & & & & q_n\,p_n \end{bmatrix}.$$

By induction, we will construct the representation of the upper triangular part, which will represent an upper triangular quasiseparable matrix. Suppose the upper triangular part of the matrix is represented with vectors $\mathbf{p}, \mathbf{q}$ and $\mathbf{r}$. We will construct a new representation for the matrix in which we will represent the strictly upper triangular part with the vectors $\hat{\mathbf{p}}$, $\hat{\mathbf{q}}$ and $\hat{\mathbf{r}}$, the diagonal elements of the upper triangular matrix will be stored in the matrix $\hat{\mathbf{d}}$.

- The first step, namely performing M_{n-1}. After performing transformation M_{n-1}, we store diagonal element $\hat{\mathbf{d}}_n$, as the last row will not change anymore.

- Assume we have already successfully created the representation of the strictly upper triangular part of U from row $l+1$ to row n. Also the diagonal elements $\hat{d}_{l+1}, \ldots, \hat{d}_n$ are known. Now we have to distinguish between two cases. Either we have to perform a Gauss transform following a permutation (as we cannot have two successive permutations) or we have to perform a permutation following a Gauss transform. The case of two successive Gauss transforms is almost identical to the construction of U in the previous chapter, and hence omitted.

  - Suppose we have to perform now a permutation involving row $l-1$ and row l, and the previous step was a Gauss transform. The two involved rows are of the following form with diagonal elements 0 and $q_l^{(l)} p_l^{(l)}$:

$$\begin{bmatrix} 0 & \cdots & 0 & q_l^{(l)} r_{l-1}^{(l)} p_{l-1}^{(l)} & q_{l+1}^{(l)} r_l^{(l)} r_{l-1}^{(l)} p_{l-1}^{(l)} & \cdots & \times \\ \times & \cdots & u_l t_{l-1} v_{l-1} & q_l^{(l)} p_l^{(l)} & q_{l+1}^{(l)} r_l^{(l)} p_l^{(l)} & \cdots & \times \end{bmatrix}.$$

    Interchanging these rows immediately reveals how to update the representation below row l to obtain a representation also covering row l. The new diagonal element equals $q_l^{(l)} r_{l-1}^{(l)} p_{l-1}^{(l)}$, and because the strictly upper triangular part of the matrix remains of semiseparable form during the complete procedure, the new elements in the generators are computed in a straightforward way. We leave the details to the reader as the index shuffling makes it too obfuscated.

— The second case arises when the previous transform was a permutation and now we need to apply a Gauss transform. Remark once more that it is not possible to have two successive Gauss transforms as this implies the singularity of the original semiseparable matrix. In this case we have the following situation, with diagonal elements $u_{l+1}t_l v_l$ and $\boxtimes$:

$$
\begin{bmatrix}
\boxtimes & \cdots & \boxtimes & q_l^{(l)} r_{l-1}^{(l)} p_{l-1}^{(l)} & q_{l+1}^{(l)} r_l^{(l)} r_{l-1}^{(l)} p_{l-1}^{(l)} & \cdots & \times \\
\boxtimes & \cdots & \boxtimes & u_{l+1}t_l v_l & q_{l+1}^{(l+1)} p_{l+1}^{(l+1)} & \cdots & \times
\end{bmatrix}.
$$

Again we can see that the strictly upper triangular part remains of semiseparable plus diagonal form. The application of a Gauss transform does not destroy this structure, the new diagonal element in the representation of U can easily be obtained as $u_{l+1}t_l v_l - \tau_l q_l^{(l)} r_{l-1}^{(l)} p_{l-1}^{(l)}$, the computation of the new generators involves a similar computation as in the previous section, but with slightly changed indices. It is clear in this case that the diagonal is not includable in the upper triangular structure.

We proved that for each type of matrix M_l the strictly upper triangular part of U remains of semiseparable form, whereas the diagonal is not necessarily includable in the structure anymore. $\square$

System solving

System solving using this technique is completely similar to the method explained in the previous section. Apply all transformations $M_1 \cdots M_{n-1}$ directly to the right-hand side **b**. Due to the previous theorem, the explicit structure of the upper triangular part U is known and we can use backward substitution to solve the final upper triangular semiseparable system of equations.

Notes and references

Fasino and Gemignani describe an order $\mathcal{O}(n)$ solver for banded plus semiseparable systems of equations. The algorithm exploits the structure of the inverse of the semiseparable matrix, which is a tridiagonal matrix. The LU-decomposition of the tridiagonal matrix is used to factorize the semiseparable part in a kind of UL-decomposition. Moreover in case of small diagonal elements in the semiseparable matrix a technique is provided for adding a correction to this matrix to make the method more numerically stable.

   ☞ L. Gemignani and D. Fasino. Fast and stable solution of banded-plus-semiseparable linear systems. *Calcolo*, 39(4):201–217, 2002.

Neuman provides in [223] (see Section 7.1 in Chapter 7) algorithms for inverting tri- and pentadiagonal matrices via the UL decomposition.

In the manuscript [229] the authors develop a Cholesky based iterative method for computing the eigenvalues of positive definite semiseparable plus diagonal matrices. See Section 3.5 in Chapter 3.

4.6 The LU-decomposition for quasiseparable matrices

Let us present in this section the LU-decomposition for quasiseparable matrices. These matrices have less structure than the semiseparable matrices, and hence we have to perform more operations to derive the LU-decomposition.

Note that in the previous section a pivoted LU-decomposition was proposed, such that a maximal of structure in the factors L and U was preserved. It was shown before by a counterexample that, in general, applying pivoting does not maintain the structure of both these factors. The same is observed, for example, for band matrices (see [152]). In this section we will show that maintaining of the structure of the factors L and U in the quasiseparable case is not always as straightforward as for the semiseparable case. We will derive a method for computing the LU-factorization of a strongly nonsingular quasiseparable matrix.

4.6.1 A first naive factorization scheme

In this subsection we will, based on the previous section, derive a general scheme, which in most of the cases will compute effectively an LU-decomposition of the quasiseparable matrix. Moreover in most cases also the factors L and U will inherit the quasiseparable structure. We will however indicate that in fact we do not even always compute an LU-decomposition, nor does the rank structure of the matrix A propagate always into the factors L and U.

The first step in the computation of this LU-decomposition is just the application of the Gauss transforms defined in the previous section. Assume we are working with a quasiseparable matrix A, which can be written as the sum of a strictly lower semiseparable matrix $S^{(l)}$ plus an upper triangular matrix R: $A = S^{(l)} + R$. Assume the matrix $S^{(l)}$ is represented as follows, with quasiseparable generators $\mathbf{u}, \mathbf{v}$ and $\mathbf{t}$:

$$S^{(l)} = \begin{bmatrix} 0 & & & & & \\ u_1 v_1 & 0 & & & & \\ u_2 t_1 v_1 & u_2 v_2 & 0 & & & \\ u_3 t_2 t_1 v_1 & u_3 t_2 v_2 & u_3 v_3 & 0 & & \\ \vdots & & \ddots & \ddots & & \\ u_{n-1} t_{n-2} \cdots t_1 v_1 & \cdots & & & u_{n-1} v_{n-1} & 0 \end{bmatrix}.$$

Determine the first $n - 2$ transforms similarly as we defined them in the previous section. This means that these transformation matrices are either Gauss transforms or pivoting matrices. We remark that these $n - 2$ transformations, are completely determined by the strictly lower triangular part of the matrix $S^{(l)}$, which coincides with the strictly lower triangular part of the matrix A. Hence applying these $n - 2$ transformations transforms the matrix $S^{(l)}$ into upper Hessenberg form. Hence this is an upper Hessenberg matrix H_1. These transforms also make the matrix R of upper Hessenberg form H_2. So we get:

$$M_2 \cdots M_{n-1} A = M_2 \cdots M_{n-1} \left(S^{(l)} + R \right)$$
$$= H_1 + H_2 = H,$$

for which H is an upper Hessenberg matrix.

This upper Hessenberg matrix H can now in turn be made upper triangular by applying $n - 1$ transformation matrices, either Gauss transforms or pivoting matrices, from top to bottom.

$$\hat{M}_{n-1} \cdots \hat{M}_1 \left(M_2 \cdots M_{n-1} A \right) = \hat{M}_{n-1} \cdots \hat{M}_1 H$$
$$= U,$$

where U is an upper triangular matrix. It is clear that, in general, this procedure will always reduce the quasiseparable matrix to upper triangular form U. This reduction can therefore already be used to solve systems of equations involving quasiseparable matrices. Applying all the elementary transformations to the right-hand side and then solving an upper triangular system of equations gives us the desired solution.

Moreover, this reduction will look very familiar to the QR-decomposition of semiseparable matrices, which will be discussed in Chapter 5, where we will replace the Gauss transforms and the pivoting matrices by orthogonal transformations, more precisely Givens transformations.

We did not say anything concerning the structure of the matrix L and, in general, this matrix will also not necessarily be lower triangular, as we will show in the remainder of this subsection.

Assume now that while reducing the Hessenberg matrix to upper triangular form, that we encounter a matrix of the following structure:

$$\begin{bmatrix} \times & \times & \times & \times & \times \\ 0 & \times & \times & \times & \times \\ & 0 & 0 & \times & \times \\ & & \times & \times & \times \\ & & & \times & \times \end{bmatrix},$$

where this matrix is the result of applying a first transformation $\hat{M}_1$ for annihilating the element in position $(2, 1)$ and a second transformation $\hat{M}_2$ for annihilating the element in position $(3, 2)$. The next transformation $\hat{M}_3$ we want to apply needs to annihilate the element in position $(4, 3)$, but there is no elementary Gauss transform that can do this. Hence we need to apply a pivoting matrix. As a result of the total procedure for this 5×5 example we get:

$$\hat{M}_4 \hat{M}_3 \hat{M}_2 \hat{M}_1 M_2 M_3 M_4 A = U.$$

The right part U is clearly upper triangular. The matrix $\hat{M}_3$ is a pivoting matrix, and, in general, this matrix cannot be shifted through to the right of all the applied transformations. Moreover, if we compute the product

$$M = \hat{M}_4 \hat{M}_3 \hat{M}_2 \hat{M}_1 M_2 M_3 M_4,$$

it can be seen that the matrix is not necessarily lower triangular. This is shown in the following example.

Example 4.23 In this example we will calculate the *LU*-decomposition of the following quasiseparable matrix A. The algorithm we use is based on the ideas explained above. Suppose we have a quasiseparable matrix A of the following form:

$$A = \begin{bmatrix} 1 & 2 & 2 & 3 & 4 \\ 2 & 2 & 2 & 3 & 4 \\ 3 & 4 & 4 & 3 & 4 \\ 3 & 4 & 6 & 6 & 8 \\ 6 & 8 & 12 & 3 & 4 \end{bmatrix}.$$

Applying the following transformations M_4, M_3 and M_2 to the matrix A, where the matrices $\breve{M}_i$ are embedded in the matrices M_i as described before,

$$\breve{M}_4 = \begin{bmatrix} 1 & \\ -2 & 1 \end{bmatrix}, \quad \breve{M}_3 = \begin{bmatrix} 1 & \\ -1 & 1 \end{bmatrix} \text{ and } \breve{M}_2 = \begin{bmatrix} 1 & \\ -\frac{3}{2} & 1 \end{bmatrix},$$

gives us the following transformed matrix:

$$\hat{A} = M_2 M_3 M_4 A = \begin{bmatrix} 1 & 2 & 2 & 3 & 4 \\ 2 & 2 & 2 & 3 & 4 \\ & & 1 & 1 & -\frac{3}{2} & -2 \\ & & & 2 & 3 & 4 \\ & & & -9 & -12 \end{bmatrix}.$$

Let us apply now the following transformations $\breve{M}_1$ and $\breve{M}_2$ to this matrix, where:

$$\breve{M}_1 = \begin{bmatrix} 1 & \\ -2 & 1 \end{bmatrix} \text{ and } \breve{M}_2 = \begin{bmatrix} 1 & \\ \frac{1}{2} & 1 \end{bmatrix}.$$

This gives us:

$$\hat{M}_2 \hat{M}_1 \hat{A} = \begin{bmatrix} 1 & 2 & 2 & 3 & 4 \\ & -2 & -2 & -3 & -4 \\ & & 0 & -3 & -4 \\ & & 2 & 3 & 4 \\ & & & -9 & -12 \end{bmatrix}.$$

It is clear that only a permutation of rows can now be applied to make this matrix upper triangular. This means that

$$\breve{M}_3 = \begin{bmatrix} 0 & 1 \\ 1 & 0 \end{bmatrix} \text{ and } \breve{M}_4 = \begin{bmatrix} 1 & \\ -3 & 1 \end{bmatrix}.$$

We get the following matrix:

$$U = \hat{M}_4 \hat{M}_3 \hat{M}_2 \hat{M}_1 \hat{A} = \begin{bmatrix} 1 & 2 & 2 & 3 & 4 \\ & -2 & -2 & -3 & -4 \\ & & 2 & 3 & 4 \\ & & & -3 & 4 \\ & & & & 0 \end{bmatrix}.$$

Computing now the lower triangular matrix L gives us the following matrix:

$$L^{-1} = \hat{M}_4 \hat{M}_3 \hat{M}_2 \hat{M}_1 M_2 M_3 M_4$$

$$= \begin{bmatrix} 1 & & & & \\ -2 & 1 & & & \\ 0 & 0 & -1 & 1 & \\ -1 & -1 & 1 & 0 & 0 \\ 3 & 3 & -3 & -2 & 1 \end{bmatrix}.$$

It is clear that also extra pivoting, to obtain a factorization $PA = LU$, cannot bring the matrix L to upper triangular form. Hence, another type of factorization is needed. ∎

The example above clearly shows that the presented factorization scheme does not always provide the desired result. The scheme is useful, however, for solving for example systems of equations. We know, however, that strongly nonsingular quasiseparable matrices always permit a decomposition in terms of quasiseparable factors L and U. Let us consider therefore the factorization for the strongly nonsingular case. This case is split into two parts: the first part poses an extra constraint on the matrix, such that the algorithm of this section is directly applicable, and the second part considers the more general case without constraints.

4.6.2   The strongly nonsingular case, without pivoting

In this subsection we will briefly illustrate the construction of the L and U factors of the LU-decomposition of a strongly nonsingular quasiseparable matrix A, where in the first step no pivoting is needed. In fact, because the matrix is strongly nonsingular, there is always a decomposition without pivoting! For simplicity, however, we assume in this subsection that no pivoting is needed to make the matrix upper Hessenberg, i.e., that in the quasiseparable representation of the lower triangular part all elements u_i are different from zero.

We know from a previous section that the structure of the matrix $A = LU$ propagates into the factors L and U if the matrix A is strongly nonsingular. Suppose we have a problematic upper Hessenberg form. Let us work in the reverse direction and examine what the structural constraints are on the original matrix. We have a matrix of the following form:

$$\begin{bmatrix} \times & \times & \times & \times & \times \\ 0 & \times & \times & \times & \times \\ 0 & 0 & 0 & \times & \times \\ 0 & 0 & \times & \times & \times \\ 0 & 0 & 0 & \times & \times \end{bmatrix}. \tag{4.3}$$

We know from the previous section that these matrices create problems for our final structure of the lower triangular part, as they require a permutation of the rows. Let us reverse the order of the applied transformations and see what the structure of the initial matrix was. The matrix above is the result of applying the following transformations onto an initial matrix A:

$$\hat{M}_2\hat{M}_1M_2M_3M_4A = \begin{bmatrix} \times & \times & \times & \times & \times \\ 0 & \times & \times & \times & \times \\ 0 & 0 & 0 & \times & \times \\ 0 & 0 & \times & \times & \times \\ 0 & 0 & 0 & \times & \times \end{bmatrix}.$$

It is clear that this matrix is not strongly nonsingular. Moreover, one can see that, due to the application of only Gaussian elementary transformation matrices to obtain this matrix, the original matrix A also was not strongly nonsingular. Hence we can conclude that if a matrix A is strongly nonsingular, and no pivoting is applied in the first annihilating steps, the factorization as described in the previous section always exists. We do not prove this statement, but the analysis above clearly illustrates a way to prove this theorem, by contradiction.

Before considering the case in which pivoting is applied, we take a look at the structure of the factors L and U. Both factors should have a quasiseparable structure.

The factor L

The application of the first sequence of transformations is identical to transformations needed for the semiseparable case. The matrix $M_{n-1}^{-1}\cdots M_2^{-1}\hat{M}_1^{-1}$ is of the following form:

$$M_{n-1}^{-1}\cdots M_2^{-1}\hat{M}_1^{-1} = \begin{bmatrix} 1 & & & & & \\ \hat{\tau}_1 & 1 & & & & \\ \hat{\tau}_1\tau_2 & \tau_2 & 1 & & & \\ \hat{\tau}_1\tau_2\tau_3 & \tau_2\tau_3 & \tau_3 & 1 & & \\ \vdots & & & & \ddots & \\ \hat{\tau}_1\tau_2\cdots\tau_{n-1} & \tau_2\cdots\tau_{n-1} & \tau_3\cdots\tau_{n-1} & \cdots & \tau_{n-1} & 1 \end{bmatrix}.$$

Hence, without applying the last $n-2$ transformations $\hat{M}_2^{-1},\ldots,\hat{M}_{n-1}^{-1}$, we get a lower triangular semiseparable matrix. We know that the result will be a lower triangular quasiseparable matrix, so let us represent this matrix already in the quasiseparable representation. The diagonal $\mathbf{d} = [1,1,\ldots,1]$, $\mathbf{u} = [1,\ldots,1]^T$, $\mathbf{v} = [\hat{\tau}_1,\tau_2,\ldots,\tau_{n-1}]^T$, and $\mathbf{t} = [\tau_2,\ldots,\tau_{n-1}]^T$. In the following inductive procedure, we briefly illustrate the updating procedure of the quasiseparable generators of the lower triangular part of the matrix. Let us denote the matrix $M_{n-1}^{-1}\cdots M_2^{-1}\hat{M}_1^{-1}$ as $L^{(1)}$ and the corresponding generators of this matrix as $\mathbf{d}^{(1)} = \mathbf{d}$, $\mathbf{v}^{(1)} = \mathbf{v}$, $\mathbf{t}^{(1)} = \mathbf{t}$ and $\mathbf{u}^{(1)} = \mathbf{u}$.

Let us start with the procedure. Applying the next transformation $\hat{M}_2^{-1}$ to the right of the matrix $L^{(1)}$ gives us the matrix $L^{(2)} = L^{(1)}\hat{M}_2^{-1}$, which has the second column of $L^{(1)}$ replaced by a sum of the second and the third column. This leads to the following generators for the matrix $L^{(2)}$: The diagonal $\mathbf{d}^{(2)} = [1,1,\ldots,1]$, $\mathbf{u}^{(2)} = [1,\ldots,1]^T$, $\mathbf{v}^{(2)} = [\hat{\tau}_1,\hat{\tau}_2+\tau_2,\ldots,\tau_{n-1}]^T$, and $\mathbf{t}^{(3)} = [\tau_2,\ldots,\tau_{n-1}]^T$.

The multiplication on the right with the matrix $\hat{M}_3$ proceeds completely similarly. The quasiseparable generators of the matrix $L^{(3)}$ are the following ones: $\mathbf{d}^{(3)} = [1, 1, \ldots, 1]$, $\mathbf{u}^{(3)} = [1, \ldots, 1]^T$, $\mathbf{v}^{(3)} = [\hat{\tau}_1, \hat{\tau}_2 + \tau_2, \hat{\tau}_3 + \tau_3, \ldots, \tau_{n-1}]^T$, and $\mathbf{t}^{(3)} = [\tau_2, \ldots, \tau_{n-1}]^T$.

It is straightforward that the generators of the complete lower triangular matrix L will be the following ones: $\mathbf{d}^{(n-1)} = [1, 1, \ldots, 1]$, $\mathbf{u}^{(n-1)} = [1, \ldots, 1]^T$, $\mathbf{v}^{(n-1)} = [\hat{\tau}_1, \hat{\tau}_2 + \tau_2, \ldots, \hat{\tau}_{n-1} + \tau_{n-1}]^T$, and $\mathbf{t}^{(n-1)} = [\tau_2, \ldots, \tau_{n-1}]^T$. These generators clearly represent a lower quasiseparable matrix.

The factor U

The computation of the factor U proceeds in in a similar fashion as the computation of the factor L. First one can apply the first $n - 1$ transformations to obtain an upper triangular quasiseparable matrix. The second sequence of transformations will be applied on this quasiseparable matrix and will again, similarly as for the factor L, lead to a quasiseparable matrix. The construction is rather similar and hence not included.

System solving

System solving is straightforward by applying all the transformations to the right-hand side $\mathbf{b}$ and afterwards solving the upper triangular system of equations.

4.6.3   The strongly nonsingular case

In this section, we loosen the condition of nonpivoting for the first sequence of transformations. We know that for a strongly nonsingular matrix, however, there always exists a decomposition in terms of the factors L and U, without pivoting. This means that we need to apply another technique for computing the L and U factors of this matrix. We know also that for a strongly nonsingular matrix, the L and U-factors in the decomposition need to be of quasiseparable form.

We illustrate this technique by an example. This illustrates that it is possible to find a nonpivoted LU-decomposition, for which both factors inherit the structure of the original matrix A. We do not go further in detail on how to construct this factorization, because a detailed examination of this technique would lead us too far, and the example illustrates the general case. Moreover, there are other easier techniques to compute the LU-factorization of the matrix (see Chapter 6).

In the following example an illustration is given on how to compute the LU-decomposition in the case of a special quasiseparable matrix.

Example 4.24 In this example we will calculate the LU-decomposition of the following quasiseparable matrix A. Suppose the quasiseparable matrix A is of the following form:

$$A = \begin{bmatrix} 1 & 2 & 2 & 3 & 4 \\ 2 & 2 & 2 & 3 & 4 \\ 0 & 0 & 4 & 3 & 4 \\ 3 & 4 & 4 & 3 & 8 \\ 6 & 8 & 8 & 3 & 4 \end{bmatrix}.$$

It is an easy exercise to compute the decomposition of the matrix using the standard techniques described in the beginning of this section. This means first making the matrix upper Hessenberg, which involves one permutation. And, second, making the upper Hessenberg matrix upper triangular involves another permutation. Clearly this technique does not give us the desired lower triangular matrix L. The problem arises when applying the first permutation. This will interchange row 4 with row 3, and create the problem rank 1 block as illustrated in the beginning of this section (see Equation 4.3). Therefore it is a bad idea to interchange these two rows, because this interchanging will generate the problematic rank structure. Instead of this interchanging, we will use now this specific rank structure to compute our LU-factorization. In fact row 3 is already in upper triangular form and so we will leave it untouched.

To compute the LU-decomposition without pivoting, as mentioned above, we do not involve row 3. First, starting at the bottom, we calculate the transformation M_4 traditionally. Second, we annihilate the first elements of row 4 by subtracting a multiple of row 2 from it; hence we omit row 3. In this way we do not get a Hessenberg structure! Then we continue and subtract from row 2 a multiple of row 1, such that the two first rows become upper triangular. Then we subtract again from row 4 a multiple of row 2, but, due to rank 1 parts present in the structure, we obtain immediately the desired upper triangular part in the first 4 rows (if the matrix is not yet upper triangular one, more Gauss transform should be performed on rows 3 and 4 to annihilate the remaining element in row 4). Finally one more row operation, acting on rows 5 and 4, is needed. We leave the actual calculations to the reader, but present the resulting L and U factors, which are clearly of the desired quasiseparable structure.

$$L = \begin{bmatrix} 1 & & & & \\ 2 & 1 & & & \\ 0 & 0 & 1 & & \\ 3 & 1 & 0 & 1 & \\ 6 & 2 & 0 & 3 & 1 \end{bmatrix} \text{ and } U = \begin{bmatrix} 1 & 2 & 2 & 3 & 4 \\ & -2 & -2 & -3 & -4 \\ & & 4 & 3 & 4 \\ & & & -3 & 0 \\ & & & & -12 \end{bmatrix}.$$

We do not further consider the general LU-factorization of a not necessarily strongly nonsingular quasiseparable matrix A, because one cannot guarantee in general that the factors are also of quasiseparable form, and this would lead us again too far. Moreover there are different and more reliable techniques to solve systems of equations with structured rank matrices.

Notes and references

In this notes and references part we would like to discuss briefly a class of matrices, considered by Fiedler, named the class of extended complete basic matrices.

☞ M. Fiedler. Basic matrices. *Linear Algebra and its Applications*, 373:143–151, 2003.

☞ M. Fiedler and Z. Vavřín. Generalized Hessenberg matrices. *Linear Algebra and its Applications*, 380:95–105, 2004.

A basic matrix can be considered as a quasiseparable matrix. This basic matrix is called complete if all entries in the subdiagonal and superdiagonal are different from zero. Note that these matrices can be efficiently represented using the diagonal subdiagonal representation as presented in Chapter 1 of the book. Finally the name extended indicates that, if an element $a_{k+1,k-1}$ is nonzero in the complete basic matrix A, that the diagonal element $a_{k,k}$ can be included in the semiseparable structure of the strictly lower triangular part. The same needs to be valid for the upper triangular part.

For these matrices Fiedler derived an efficient LU-decomposition consisting of $n-1$ elementary Gauss transformation matrices of the form $M_i = I + \tau_i \mathbf{e}_{i+1} \mathbf{e}_i^T$. In these manuscripts, these matrices are referred to as elementary bidiagonal matrices. In the manuscript it is shown that there exists a factorization of an extended complete basic matrix of the form $A = LDU$, where L can be written as the product of $n-1$ elementary lower bidiagonal matrices, and U can be written as the product of $n-1$ upper bidiagonal matrices. For example the following matrix is of the correct form. We only depict the lower triangular part in schematic form; the elements $\boxtimes$ are includable in the structure.

$$
A = \begin{bmatrix}
\times & & & & \\
\boxtimes & \boxtimes & & & \\
\boxtimes & \boxtimes & \times & & \\
0 & 0 & \times & \times & \\
0 & 0 & 0 & \times & \times
\end{bmatrix}
$$

It is easy to see that for this specific structure only $n-1$ Gauss transformation matrices are needed.

These results are examined in a similar fashion for extended complete generalized Hessenberg matrices, which are matrices for which only the lower triangular part satisfies this specific structure.

4.7   Some comments

The Gaussian elimination pattern as described in this section has an interesting theoretical viewpoint, namely that the structure of the original matrix is transferred, under some constraints, to the factors L and U. Let us, in this section, briefly comment on the numerical stability of this factorization and on the use of this factorization as a possible representation.

4.7.1   Numerical stability

Gaussian elimination is widely spread and often used for solving systems of equations. Even though worst case scenarios exist, with regard to numerial stability, in

practice they seldom occur, if pivoting is used. In our case, however, it was shown
that the application of pivoting often leads to a loss of structure in the factors L and
U. This structural loss leads to an increase in complexity. Hence for the structured
rank matrices there is a trade-off between numerical accuracy and speed.

The following traditional example is a matrix with a quasiseparable structure
in the lower triangular part (see [265]):

$$
A = \begin{bmatrix}
1 & & & & 1 \\
-1 & 1 & & & 1 \\
-1 & -1 & 1 & & 1 \\
-1 & -1 & -1 & 1 & 1 \\
-1 & -1 & -1 & -1 & 1
\end{bmatrix}.
$$

The LU-factorization of this quasiseparable matrix, computed with the techniques
of this section, leads to a matrix U, with fast growing elements in the last column. This can lead to severe numerical instabilities. The algorithm as proposed in
this section is even more sensitive to numerical instabilities due to the lack of real
pivoting possibilities.

In the next chapter a decomposition is presented based on orthogonal transformations. In general this approach will be more stable than the approach based
on the LU-decomposition.

4.7.2   A representation?

In the first part of the book, it was briefly mentioned that also the Gaussian elimination pattern can be used for representing structured rank matrices efficiently. In
fact it is clear that if there are are only Gaussian elementary transformation matrices involved, and no pivoting is needed, that one can represent the lower triangular
part of a semiseparable matrix with Gaussian elementary transformation matrices
and a vector. In fact this representation was already used in the beginning of this
chapter for representing lower triangular semiseparable matrices. Let us name this
representation the Gauss-vector representation. A similar remark can be found in
[72, Remark 12]; it is explained there that instead of Givens transformations, for
the general Givens-weight representation, one can also use Gauss transforms. This
leads to a Gauss-weight representation.

4.8   Conclusions

In this chapter we discussed the computation of the LU-decomposition of semiseparable and quasiseparable matrices. It was proved, by an extension of the nullity
theorem, that a nonpivoted LU-decomposition of a matrix A passes the rank structure of the matrix A onto its L and U factors. Using this knowledge, we developed
a constructive way to compute the LU-decomposition of a semiseparable matrix.
Formulas were presented proving once more that the factors L and U of this decomposition are also in the semiseparable or quasiseparable form. In a final section
the class of quasiseparable matrices was investigated. This class was not so easy

to deal with, and a general formulation like for the semiseparable matrices was not possible. Hence we restricted ourselves to the case of strongly nonsingular matrices. For strongly nonsingular matrices having a special nonzero structure, it was proved that the algorithm for semiseparable matrices could easily be extended. For the general strongly nonsingular case it was illustrated how to compute this decomposition. We did not go into detail as a forthcoming chapter (see Chapter 6) will present another, more elegant way to compute the LU-factorization of strongly nonsingular matrices. Before examining this other method, we will reconsider the techniques for computing the LU-decomposition and adapt them so that we obtain a QR-factorization of the involved matrix.

Chapter 5

The QR-factorization

In this chapter we will consider two fast algorithms for solving linear systems involving semiseparable, quasiseparable and semiseparable plus diagonal matrices. Exploiting the structure of such matrices, algorithms with $\mathcal{O}(n)$ computational complexity can be developed, where n is the size of the considered problems; see, e.g., the LU-decomposition as discussed in the previous chapter.

The two algorithms we propose are based on two decompositions of the coefficient matrix of the linear system, by using orthogonal transformations. The first method computes the QR-factorization of the involved matrix. The second method computes in fact a URV-decomposition.

The first method involving the QR-factorization is explored for three main classes of matrices, namely semiseparable, quasiseparable, and semiseparable plus diagonal matrices. Before starting the initial description on how to compute these factorizations we present a theoretical result. A theorem will be presented, similar to the nullity theorem, which will predict the structure of the Q factor in the decomposition of the structured rank matrices. Following these theoretical results we will propose a method for computing the QR-factorization of a semiseparable matrix. The factorization will consist of applying successively $n - 1$ Givens transformations from the bottom to the top. After a description of the method using figures, we go into more detail and investigate the behavior of the generators in case the semiseparable matrix is represented with the quasiseparable representation.

After having deduced the decomposition of the semiseparable matrix, it is time to derive a decomposition scheme for a larger set of matrices, namely the set of quasiseparable matrices. Again, the method itself is depicted with figures. Instead of now using $n - 1$ Givens transformations from bottom to top, an extra sequence of transformations will be necessary. The first sequence of transformations will transform the quasiseparable matrix into an upper Hessenberg one; the second sequence of transformations will be necessary to bring the upper triangular Hessenberg matrix back to upper triangular form. This approach resembles strongly the LU-decomposition of the previous chapter.

To conclude the QR-part of this chapter a detailed example of an implementation of the presented method is given for solving systems of equations.

In Section 5.6.1, we present briefly the URV-decomposition of quasiseparable matrices. The first step in this decomposition is exactly the same as for the computation of the QR-factorization. First, $n-1$ Givens transformations are performed on the left to reduce the quasiseparable matrix to upper triangular form. Second, $n-1$ Givens transformations are needed to reduce the upper Hessenberg matrix to upper triangular form. These transformations, however, will be performed on the right side of the matrix, leading to orthogonal operations performed both on the right and the left side at the same time.

✎ *The focus of this chapter is the computation of the QR-factorization of some structured rank matrices. To obtain this factorization we will use Givens transformations. Again, a generalization of the nullity theorem is important predicting now the structure of the factor Q in the QR-factorization of a structured rank matrix. Proposition 5.5 contains a direct translation of these results towards a relation between the rank structure of the matrix A and Q. The example following the proposition clearly illustrates the power of this theorem.*

Subsection 5.3.1 contains the essential step in the computation of the QR-factorization, namely the application of a sequence of Givens transformations from bottom to top in order to annihilate the rank structure in the lower triangular part of a semiseparable matrix. If one is interested in implementing this method, Subsection 5.3.1 provides the details for a quasiseparable representation.

Section 5.4 applies tricks similar to the previous section for computing the QR-factorization for a quasiseparable matrix. Two sequences of Givens transformations are necessary: one sequence from bottom to top annihilating the rank structure and one sequence for annihilating the remaining subdiagonal elements. The following section discusses a similar implementation for the Givens-vector representation.

In the final section two other types of factorizations are discussed. Theorems 5.9, 5.12 and 5.14 discuss, respectively, the URV, RQ, LQ and QL decompositions, all small variations on the QR-decomposition and computed, based on similar ideas.

5.1 About the QR-decomposition

Before starting theoretical derivations on the structure of the Q and the R factor in the QR-decomposition, or investigating an algorithmic way to compute the QR-decomposition, we will first briefly mention some general properties of the QR-decomposition of matrices.

Suppose we have a matrix A and a factorization of the form:

$$A = QR,$$

with Q a matrix having orthonormal columns and R an upper triangular matrix. If Q is a square matrix, and hence orthogonal, this is called the full QR-factorization (in this case R is rectangular). Otherwise, if Q has the same number of columns as A this is called the thin (or reduced) QR-factorization (in this case R is square).

In fact this means that the first k columns of A span the same space as the first k columns of Q. Computation of the QR-factorization of a matrix can be done in theory by applying for example the Gram-Schmidt orthogonalization procedure to the columns of the matrix A.

One can formulate and prove the following two theorems related to the QR-decomposition of a matrix A.

Theorem 5.1. *For every matrix A there exists a full (and hence also a reduced) QR-factorization.*

Concerning unicity of the decomposition we have the following result.

Theorem 5.2. *Each matrix A of full rank has an essentially[5] unique reduced QR-factorization.*

Proofs of both of the theorems above can be found, for example, in [265, 82, 258, 152]. The last theorem implies that a square matrix of full rank has an essentially unique QR-factorization. The QR-factorization can, as already mentioned above, be computed by the classical Gram-Schmidt procedure. However, the modified Gram-Schmidt procedure gives better numerical results. Techniques used mostly involve Householder/Givens transformations for making a matrix upper triangular. This technique is also applied in this chapter. The QR-factorization is a basic tool and is often used for solving systems of equations and least-squares problems. Moreover the QR-factorization is also the basic tool in the QR-algorithm for computing the eigenvalue decomposition of matrices. More information concerning this factorization can be found in the following references.

Notes and references

The QR-factorization is said to be a basic linear algebra tool, having its impact in different topics of linear algebra. The factorization is therefore already covered in many basic linear algebra textbooks; see, e.g., [265, 82, 258, 152]. These books discuss the interpretation via Gram-Schmidt and derive the modified Gram-Schmidt procedure. Finally other triangularization methods, via Householder and/or Givens transformations, are also discussed.

5.2 Theoretical considerations of the QR-decomposition

Similar theorems as Theorem 4.14 can be deduced for other types of decompositions. We prove here a similar theorem for the QR-decomposition, which will give us information about the structured rank of the factor Q. In general, however, we will also have information on the factor L, because the matrix Q can be decomposed, as, for example, a sequence of Givens transformations from bottom to the top. Taking

---

[5]By essentially we mean that up to the signs, the columns are identical.

a closer look at this decomposition of the matrix Q reveals information on the rank structure of the matrix L. Let us deduce information concerning rank structure of the matrix Q based on the following extension of the nullity theorem.

Theorem 5.3. *Suppose we have an invertible matrix A with a QR-factorization $A = QR$. Suppose A is partitioned in the following form*

$$A = \left[\begin{array}{cc} A_{11} & A_{12} \\ A_{21} & A_{22} \end{array} \right],$$

with A_{11} of dimension $p \times q$. The inverse of Q, the matrix B, is partitioned as

$$Q^{-1} = Q^T = B = \left[\begin{array}{cc} B_{11} & B_{12} \\ B_{21} & B_{22} \end{array} \right],$$

with B_{11} of dimension $q \times p$. Then the nullities $\mathrm{n}(A_{21})$ and $\mathrm{n}(B_{21})$ are equal.

Proof. Similar to the proof from Theorem 4.14.   □

Let us combine the theorem above, with the nullity theorem to get direct structural information concerning the matrix Q, instead of its inverse $Q^{-1} = Q^T$.

Theorem 5.4. *Suppose we have an invertible matrix A with a QR-factorization of the following form:*

$$A = QR.$$

Suppose A is partitioned in the following form

$$A = \left[\begin{array}{cc} A_{11} & A_{12} \\ A_{21} & A_{22} \end{array} \right],$$

with A_{11} of dimension $p \times q$. The matrix Q is partitioned as

$$Q = \left[\begin{array}{cc} Q_{11} & Q_{12} \\ Q_{21} & Q_{22} \end{array} \right],$$

with Q_{11} of dimension $p \times q$. Then the nullities $\mathrm{n}(A_{21})$ and $\mathrm{n}(Q_{21})$ are equal to each other.

Proof. Combining the results of Theorem 5.3, with the nullity theorem (see Theorem 1.33), leads directly to a proof.   □

The theorem above can easily be translated towards the rank structure of these blocks.

Proposition 5.5. *Using the same notation as from Theorem 5.3, we obtain the following relations, according to the rank:*

$$\mathrm{rank}\,(A_{21}) = \mathrm{rank}\,(Q_{21}).$$

Proof. As the nullity of a matrix $A \in \mathbb{R}^{p \times q}$ satisfies

$$\mathrm{n}(A) = q - \mathrm{rank}\,(A)\,,$$

the result of the previous theorem leads directly to this observation. $\Box$

Using this theorem one can deduce the structure of the lower triangular part of the orthogonal matrix Q. Let us illustrate with some examples the power of this simple theorem.

Example 5.6 We do not include real examples, but the readers can easily generate matrices in MATLAB and use the built-in function for computing the QR-decomposition to check that the following statements are true. All considered matrices are assumed to be nonsingular.

- Consider a nonsingular tridiagonal matrix T. Its QR-factorization will generate a unitary Hessenberg matrix Q, which clearly has the same rank structure in the lower triangular part as the matrix T. Moreover when some of the subdiagonal elements in the matrix T are equal to zero, these zeros propagate simply into the matrix Q.

- Taking the QR-factorization of a Hessenberg matrix H gives us again a unitary Hessenberg matrix Q.

- The QR-factorization of band matrix B has the same number of nonempty subdiagonals in Q as in B.

- If A is a semiseparable matrix, the lower triangular part of the matrix Q will also be of semiseparable form.

- If A is quasiseparable matrix, the lower triangular part of the matrix Q will also be of quasiseparable form.

- The results above can be extended in a straightforward way to more general rank structures, such as rank k matrix plus a band matrix, or even to sequentially or hierarchically semiseparable matrices.

$\blacksquare$

The next example illustrates that the demand of invertibility of the matrix A is really essential.

Example 5.7 Consider the singular matrix A.

$$A = \begin{bmatrix} 0 & 0 & 1 \\ 0 & 0 & 0 \\ 0 & 0 & 1 \end{bmatrix}$$

This matrix has a simple (nonunique), QR-decomposition of the following form:

$$A = QR = \begin{bmatrix} 0 & 0 & 1 \\ 0 & 1 & 0 \\ 1 & 0 & 0 \end{bmatrix} \begin{bmatrix} 0 & 0 & 1 \\ 0 & 0 & 0 \\ 0 & 0 & 1 \end{bmatrix}.$$

It is clear that the structure of the matrix A below the diagonal does not propagate into the matrix Q.   ∎

We know now in which manner the structure of the lower triangular part propagates into the matrix Q. Let us search now for an effective way to compute the QR-decomposition of semiseparable and quasiseparable matrices.

Notes and references

These theoretical results are based on the nullity theorem by Fiedler and some of closely related articles; see notes and references of Section 1.5.1. The results concerning the structure of this QR-decomposition are based on notes from [291]. (See Section 1.5 in Chapter 1 for more information.)

5.3   A QR-factorization of semiseparable matrices

In the preceding chapter, concerning the LU-decomposition of semiseparable matrices, Gaussian transformation matrices were applied on semiseparable matrices in order to get upper triangular matrices. The combination of all the Gaussian transformation matrices led to a pivoted LU-decomposition of semiseparable matrices. In this chapter, we want to compute however a QR-factorization of a semiseparable matrix. In order to obtain this factorization we will apply orthogonal transformations to the left of the semiseparable matrix S in order to make this matrix upper triangular. Whereas it was not possible, in general (also for singular matrices), for the LU-decomposition to obtain a specific structure for the L and U matrix, we will derive in this chapter a method which will always, even for singular matrices, give us a specific structure for the factors Q and R. This means that we will design a specific QR-factorization, because nonsingular matrices do not necessarily have a unique QR-factorization.

Let us start first by deriving an algorithm to compute the QR-decomposition of a semiseparable matrix.

5.3.1   The factorization using Givens transformations

In this subsection we will use embedded Givens transformations, which are orthogonal transformations, to make the semiseparable matrix upper triangular. As already mentioned, a Givens transformation, is a 2×2 matrix of the following form

$$G = \begin{bmatrix} c & -s \\ s & c \end{bmatrix},$$

where $c^2 + s^2 = 1$. In fact c and s are, respectively, the sine and cosine of a certain angle. Suppose a vector $[x, y]^T$ is given, then there exists a Givens transformation such that

$$G \begin{bmatrix} x \\ y \end{bmatrix} = \begin{bmatrix} r \\ 0 \end{bmatrix},$$

with $r = \sqrt{x^2 + y^2}$. Givens transformations are often used in different applications for creating zeros in matrices. A study on how to compute Givens transformations accurately can be found in [29], for example. The Givens transformations we will be working with in this section are chosen to annihilate elements in a vector, not necessarily of length 2. Hence we need to embed our Givens transformation (see, e.g., [152]), in a larger matrix. The notation we use in this section is the following: $\check{G}$ denotes the 2×2 Givens transformation matrix, consisting of all the sines and cosines. And the matrix G is the matrix acting on the complete vector; in fact, the matrix $\check{G}$ is embedded in G as indicated below.

Before starting the algoritmic derivation of the method, we will describe by figures what happens to the semiseparable matrix. Suppose we have a 5×5 semiseparable matrix. Similarly as before, the elements marked as $\boxtimes$ depict the part of the matrix having semiseparable structure. Notice that we also only present the lower triangular part of the matrix, as this is the part we would like to annihilate. This means that we would like to annihilate the complete lower triangular part of the following matrix:

$$A = \begin{bmatrix} \boxtimes & & & & \\ \boxtimes & \boxtimes & & & \\ \boxtimes & \boxtimes & \boxtimes & & \\ \boxtimes & \boxtimes & \boxtimes & \boxtimes & \\ \boxtimes & \boxtimes & \boxtimes & \boxtimes & \boxtimes \end{bmatrix}.$$

Let $\check{G}_i^T$ be a 2×2 Givens transformation, acting on the rows i and $i - 1$, such that $\check{G}_i^T$ annihilates elements in row i. As mentioned before G_i^T is the Givens transformation acting on the complete matrix annihilating the desired element.

Take the Givens transformation G_5^T in order to annihilate a nonzero element in the last row of A^6, contained in the first 4 columns. Due to the semiseparable structure, this means that the last two rows, up to and including column 4, are dependent from each other. Hence annihilating one nonzero element in the last row will make all nonzero elements in the last row equal to zero. Graphically depicted this means:

---

[6]In practical implementations the choice of the element can be very important for numerical stability.

$$\begin{bmatrix} \boxtimes & & & & \\ \boxtimes & \boxtimes & & & \\ \boxtimes & \boxtimes & \boxtimes & & \\ \boxtimes & \boxtimes & \boxtimes & \boxtimes & \\ \boxtimes & \boxtimes & \boxtimes & \boxtimes & \boxtimes \end{bmatrix} \longrightarrow \begin{bmatrix} \boxtimes & & & & \\ \boxtimes & \boxtimes & & & \\ \boxtimes & \boxtimes & \boxtimes & & \\ \boxtimes & \boxtimes & \boxtimes & \boxtimes & \\ 0 & 0 & 0 & 0 & \boxtimes \end{bmatrix}$$

$$\Updownarrow$$

$$A \longrightarrow G_5^T A.$$

Remark that the performed Givens transformation G_5^T on the matrix A did not destroy the existing dependencies between rows 3 and 4, the first 3 elements of both these rows are still dependent of each other.

Take now G_4^T in order to annihilate a nonzero element, in row 4, out of column 1 to 3, from the matrix $G_5^T A$. Hence, we get G_4^T acting only on rows 3 and 4:

$$\begin{bmatrix} \boxtimes & & & & \\ \boxtimes & \boxtimes & & & \\ \boxtimes & \boxtimes & \boxtimes & & \\ \boxtimes & \boxtimes & \boxtimes & \boxtimes & \\ 0 & 0 & 0 & 0 & \boxtimes \end{bmatrix} \longrightarrow \begin{bmatrix} \boxtimes & & & & \\ \boxtimes & \boxtimes & & & \\ \boxtimes & \boxtimes & \boxtimes & & \\ 0 & 0 & 0 & \boxtimes & \\ 0 & 0 & 0 & 0 & \boxtimes \end{bmatrix}$$

$$\Updownarrow$$

$$G_5^T A \longrightarrow G_4^T G_5^T A.$$

Continuing this process gives us

$$\begin{bmatrix} \boxtimes & & & & \\ \boxtimes & \boxtimes & & & \\ \boxtimes & \boxtimes & \boxtimes & & \\ 0 & 0 & 0 & \boxtimes & \\ 0 & 0 & 0 & 0 & \boxtimes \end{bmatrix} \longrightarrow \begin{bmatrix} \boxtimes & & & & \\ \boxtimes & \boxtimes & & & \\ 0 & 0 & \boxtimes & & \\ 0 & 0 & 0 & \boxtimes & \\ 0 & 0 & 0 & 0 & \boxtimes \end{bmatrix}$$

$$\Updownarrow$$

$$G_4^T G_5^T A \longrightarrow G_3^T G_4^T G_5^T A.$$

And the last step

$$\begin{bmatrix} \boxtimes & & & & \\ \boxtimes & \boxtimes & & & \\ 0 & 0 & \boxtimes & & \\ 0 & 0 & 0 & \boxtimes & \\ 0 & 0 & 0 & 0 & \boxtimes \end{bmatrix} \longrightarrow \begin{bmatrix} \boxtimes & & & & \\ 0 & \boxtimes & & & \\ 0 & 0 & \boxtimes & & \\ 0 & 0 & 0 & \boxtimes & \\ 0 & 0 & 0 & 0 & \boxtimes \end{bmatrix}$$

$$\Updownarrow$$

$$G_3^T G_4^T G_5^T A \longrightarrow G_2^T G_3^T G_4^T G_5^T A.$$

It is clear that there are, in general, $n - 1$ Givens transformations necessary for reducing an $n \times n$ semiseparable matrix A to upper triangular form.

$$A = G_n G_{n-1} \cdots G_3 G_2 R$$
$$= QR.$$

This gives us the QR-factorization of this semiseparable matrix as:

$$G_2^T G_3^T \cdots G_{n-1}^T G_n^T A = R.$$

5.3.2 The generators of the factors

The matrices we will be working with in this section are semiseparable matrices, represented using the quasiseparable representation. We derive here the method for the quasiseparable representation, but later on in this section one will see that for computing the QR-factorization there are other, much more efficient representations. This subsection is divided into two parts: a first part for constructing the orthogonal matrix Q and a second part for an efficient representation of the matrix R.

The orthogonal matrix Q

The Q-factor is clearly already factored as a product of Givens transformations. In fact it is a lower unitary Hessenberg matrix, which all admit a factorization of this specific form (see, e.g., [137]). The only remaining problem in the computation of the orthogonal matrix Q is the efficient and accurate computation of the involved Givens transformations for making the original matrix A of upper triangular form.

As already mentioned, our matrix is represented with the quasiseparable representation. As illustrated in the previous section, the Givens transformations are determined by the lower triangular part, which is, using the quasiseparable representation of the following form, with generators $\mathbf{u}, \mathbf{v}$ and $\mathbf{t}$ (the upper triangular part is not specified, as it is not needed in this analysis):

$$A = \begin{bmatrix} u_1 v_1 & & & & & \\ u_2 t_1 v_1 & u_2 v_2 & & & & \\ u_3 t_2 t_1 v_1 & u_3 t_2 v_2 & u_3 v_3 & & & \\ \vdots & & \ddots & \ddots & & \\ u_{n-1} t_{n-2} \cdots t_1 v_1 & \cdots & & & u_{n-1} v_{n-1} & \\ u_n t_{n-1} t_{n-2} \cdots t_1 v_1 & \cdots & & & u_n t_{n-1} v_{n-1} & u_n v_n \end{bmatrix}. \quad (5.1)$$

Compute first the Givens transformation G_n^T, which acts on the last two rows and would annihilate the final row, up to the diagonal. The last two rows, without the last column, are of the following form:

$$\begin{bmatrix} u_{n-1} t_{n-2} \cdots t_1 v_1 & \cdots & u_{n-1} v_{n-1} \\ u_n t_{n-1} t_{n-2} \cdots t_1 v_1 & \cdots & u_n t_{n-1} v_{n-1} \end{bmatrix}.$$

It is clear that a Givens transformation $\breve{G}_n^T$ satisfying the following equation will do the job:

$$\breve{G}_n^T \begin{bmatrix} u_{n-1} \\ u_n t_{n-1} \end{bmatrix} = \begin{bmatrix} c_n & -s_n \\ s_n & c_n \end{bmatrix} \begin{bmatrix} u_{n-1} \\ u_n t_{n-1} \end{bmatrix}$$

$$= \begin{bmatrix} c_n u_{n-1} - s_n u_n t_{n-1} \\ s_n u_{n-1} + c_n u_n t_{n-1} \end{bmatrix} = \begin{bmatrix} \times \\ 0 \end{bmatrix}.$$

For reasons of simplicity we will slightly change the notation. The generators of the matrix A's lower triangular part are $\mathbf{u}^{(1)}$, $\mathbf{v}^{(1)}$ and $\mathbf{t}^{(1)}$. After performing a Givens transformation to the left of the matrix A, we again obtain a lower triangular part of semiseparable form. The generators for this lower triangular part are denoted as $\mathbf{u}^{(2)}$, $\mathbf{v}^{(2)}$ and $\mathbf{t}^{(2)}$, and so on.

Applying now the Givens transformation G_n^T to the left of the matrix A, we obtain the following matrix $G_n^T A$.

$$
\begin{bmatrix}
u_1^{(1)}v_1^{(1)} & & & & \\
u_2^{(1)}t_1^{(1)}v_1^{(1)} & & & & \\
\vdots & & & \ddots & \\
(c_n^{(1)}u_{n-1}^{(1)} - s_n^{(1)}u_n^{(1)}t_{n-1}^{(1)})t_{n-2}^{(1)}\cdots t_1^{(1)}v_1^{(1)} & \cdots & (c_n^{(1)}u_{n-1}^{(1)} - s_n^{(1)}u_n^{(1)}t_{n-1}^{(1)})v_{n-1}^{(1)} & & \\
0 & & \cdots & 0 & u_n^{(1)}v_n^{(1)}
\end{bmatrix}
$$

The lower triangular part of this matrix can then be represented with $\mathbf{u}^{(2)}$, $\mathbf{v}^{(2)}$ and $\mathbf{t}^{(2)}$, where $\mathbf{v}^{(2)} = \mathbf{v}^{(1)}$, $\mathbf{t}^{(2)} = \mathbf{t}^{(1)}$, only the vector $\mathbf{u}^{(2)}$ is slightly different from $\mathbf{u}^{(1)}$. Namely the first $n-2$ components remain untouched: $u_i^{(2)} = u_i^{(1)}$, for $1 \leq i \leq n-2$, $u_{n-1}^{(2)} = c_n^{(1)}u_{n-1}^{(1)} - s^{(1)}$ and $u_n^{(2)} = 0$. Using these generators for $G_n^T A$, we can easily compute G_{n-1}^T as the Givens transformation, satisfying:

$$
\breve{G}_{n-1}^T \begin{bmatrix} u_{n-2}^{(2)} \\ u_{n-1}^{(2)}t_{n-2}^{(2)} \end{bmatrix} = \begin{bmatrix} c_{n-1} & -s_{n-1} \\ s_{n-1} & c_{n-1} \end{bmatrix} \begin{bmatrix} u_{n-2}^{(2)} \\ u_{n-1}^{(2)}t_{n-2}^{(2)} \end{bmatrix}
$$

$$
= \begin{bmatrix} c_{n-1}u_{n-2}^{(2)} - s_{n-1}u_{n-1}^{(2)}t_{n-2}^{(2)} \\ s_{n-1}u_{n-2}^{(2)} + c_{n-1}u_{n-1}^{(2)}t_{n-2}^{(2)} \end{bmatrix} = \begin{bmatrix} \times \\ 0 \end{bmatrix}.
$$

This presents us with a simple algorithm for computing the Givens transformations in the decomposition of the unitary Hessenberg matrix Q, the so-called Schur parametrization of the unitary Hessenberg matrix [137].

The upper triangular matrix R

In this section we will find an efficient representation of the upper triangular part R of the QR-factorization of the semiseparable matrix A. In fact we will prove here, by construction, that the upper triangular matrix R will consist of the sum of two rank 1 quasiseparable matrices. In fact there are numerous ways of constructing the representation of the upper triangular part of the matrix R, e.g., in [284] a method is presented which works on generator representable semiseparable matrices.

Here we will use the decoupled quasiseparable representation. This means that the lower triangular part is represented in quasiseparable form, and also the strictly upper triangular part is represented in the quasiseparable form. We will use the following decomposition of the quasiseparable matrix $A = A_L + A_U$, where A_L denotes the lower triangular part of the matrix A and A_U denotes the strictly upper triangular part of the matrix A. It will be shown that for $R = Q^T A = Q^T A_L + Q^T A_U$, the matrices QA_L and QA_U will be of upper semiseparable, respectively, quasiseparable form.

The lower triangular part A_L is of the form (5.1), whereas the strictly upper triangular part A_U is represented with the quasiseparable representation, involving the vectors $\mathbf{q}^{(1)}$, $\mathbf{p}^{(1)}$ and the vector $\mathbf{r}^{(1)}$.

$$
A_U = \begin{bmatrix} 0 & q_1^{(1)} p_1^{(1)} & q_2^{(1)} r_1^{(1)} p_1^{(1)} & q_3^{(1)} r_{2:1}^{(1)} p_1^{(1)} & \cdots & q_{n-1}^{(1)} r_{n-2:1}^{(1)} p_1^{(1)} \\ & 0 & q_2^{(1)} p_2^{(1)} & q_3^{(1)} r_2^{(1)} p_2^{(1)} & \cdots & q_{n-1}^{(1)} r_{n-2:2}^{(1)} p_2^{(1)} \\ & & 0 & q_3^{(1)} p_3^{(1)} & \cdots & q_{n-1}^{(1)} r_{n-2:3}^{(1)} p_3^{(1)} \\ & & & \ddots & \ddots & \vdots \\ & & & & 0 & q_{n-1}^{(1)} p_{n-1}^{(1)} \\ & & & & & 0 \end{bmatrix}.
$$

We will prove by construction that the upper triangular matrix R will be the sum of two upper triangular semiseparable matrices. We denote the vectors with a $^{(1)}$ indicating that transformations are not yet applied on the matrix.

The only thing left to do is to compute now the products $Q^T A_U$ and $Q^T A_L$ by applying the successive Givens transformations from bottom to top on these two matrices. The details are left to the reader, but an easy computation reveals that the matrix A_U will gradually, by performing more transformations, fill up the upper triangular part with a quasiseparable matrix. Further, the matrix A_L will become an upper triangular matrix for which the upper triangular part is of semiseparable from.

For the readers planning to implement the method above, or for those wanting to get more structure, in fact, the sum of both transformed matrices can be written as the sum of two upper semiseparable matrices. The key to seeing this is to combine both transformation methods described above. Suppose one applies a first Givens transformation on the matrix A_U and on the matrix A_L. This creates fillin in the bottom right element of the matrix A_L and it changes the element in the last column on position $n-1$. One should change now the diagonal element in the matrix $G_n^T A_U$ in position $n-1$, such that it can be included in the semiseparable structure of the strictly upper triangular part. This change results in subtracting something from the $(n-1)$th diagonal element of the matrix $G_n^T A_L$; however, this will not compromise the structure of this matrix. This procedure should be continued then by successively changing the $(n-2)$nd element, after applying transformation G_{n-1}^T and so on. These are details, however, which only slightly change the complexity of the proposed method.

All details of this implementation are not presented simply because the next section contains information on a specific type of implementation for a more general class of structured rank matrices.

5.3.3 The Givens-vector representation

Suppose our matrix A is represented in the lower triangular part with the Givens-vector representation. An easy calculation reveals that the Givens transformations used in the representation are exactly the same up to the transpose as the Givens transformations needed for computing the QR-factorization of the quasiseparable matrix A.

Hence we might conclude that the Givens-vector representation of the lower triangular part finds in some sense its origin in the QR-decomposition of the lower triangular semiseparable part. The right left variant of the Givens-vector representation corresponds in some sense with the RQ-decomposition of the lower triangular part. One might now wonder if this representation has its extension towards the higher order case? The answer is affirmative; see, e.g., [72].

5.3.4   Solving systems of equations

In a straightforward way one can expand the procedure above to solve systems of equations via this QR-factorization. Suppose a system of equations with a matrix S of semiseparable form is presented, which one wants to solve: $S\mathbf{x} = \mathbf{b}$. Applying the successive Givens transformations to both sides of the equations gives us

$$Q^T S \mathbf{x} = Q^T \mathbf{b}$$
$$R \mathbf{x} = Q^T \mathbf{b} = \hat{\mathbf{b}}.$$

The latter system of equations can be solved rather easily by exploiting the structure of the upper triangular matrix R: the matrix is the sum of two upper semiseparable matrices. Information on how to solve this system of equations can be found in Section 4.2 of Chapter 4 and in Section 5.5.

Notes and references

The results in this section are based mainly on the results presented in the manuscript [284] and in the Ph.D. thesis [283]. See also Section 4.6 in Chapter 4. In this article the authors develop the QR-factorization for generator representable semiseparable plus diagonal matrices. Direct computations are given for the class of generator representable semiseparable matrices. A more detailed description of the implementation of the QR-factorization for semiseparable plus diagonal matrices can be found in [286]. This implementation is discussed in more detail in Section 4.6 in Chapter 4.

5.4   A QR-factorization of quasiseparable matrices

In this section, we show that the R-factor and the Q-factor of a QR-factorization of a quasiseparable matrix have some special structure, namely, the upper triangular part will be of semiseparability rank 2, while the Q factor will again be a quasiseparable matrix. Let us start with a description of the algorithm.

The factorization of the quasiseparable matrix as we will present it here is in fact a straightforward expansion of the techniques described in the previous section. Suppose we have a quasiseparable matrix A, it is known that the strictly lower triangular part is of semiseparable form. Hence we can easily annihilate all the elements strictly below the subdiagonal, with one sequence of Givens transformations going from the bottom to the top.

We will depict what happens with figures, as in the previous section. Suppose we have a 5×5 quasiseparable matrix. The elements marked with $\boxtimes$ depict the part of the matrix of semiseparable structure. Notice that we also only present the lower triangular part of the matrix, as this is the part we would like to annihilate.

$$A = \begin{bmatrix} \times & & & & \\ \boxtimes & \times & & & \\ \boxtimes & \boxtimes & \times & & \\ \boxtimes & \boxtimes & \boxtimes & \times & \\ \boxtimes & \boxtimes & \boxtimes & \boxtimes & \times \end{bmatrix}.$$

Denote again with $\breve{G}_i^T$ a 2×2 Givens transformation, acting on rows i and $i-1$, such that $\breve{G}_i^T$ annihilates elements in row i.

Annihilate now, by using the Givens transformation $\breve{G}_n$, the elements in the last row of this matrix, except for the last two:

$$\begin{bmatrix} \times & & & & \\ \boxtimes & \times & & & \\ \boxtimes & \boxtimes & \times & & \\ \boxtimes & \boxtimes & \boxtimes & \times & \\ \boxtimes & \boxtimes & \boxtimes & \boxtimes & \times \end{bmatrix} \longrightarrow \begin{bmatrix} \times & & & & \\ \boxtimes & \times & & & \\ \boxtimes & \boxtimes & \times & & \\ \boxtimes & \boxtimes & \boxtimes & \times & \\ 0 & 0 & 0 & \times & \times \end{bmatrix}$$

$$\Updownarrow$$

$$A \longrightarrow G_5^T A.$$

Construct now G_4^T in order to annihilate a nonzero element, in row 4, out of column 1 to 2, from the matrix $G_5^T A$. Hence, as G_4^T only acts on rows 3 and 4 we get:

$$\begin{bmatrix} \times & & & & \\ \boxtimes & \times & & & \\ \boxtimes & \boxtimes & \times & & \\ \boxtimes & \boxtimes & \boxtimes & \times & \\ 0 & 0 & 0 & \times & \times \end{bmatrix} \longrightarrow \begin{bmatrix} \times & & & & \\ \boxtimes & \times & & & \\ \boxtimes & \boxtimes & \times & & \\ 0 & 0 & \times & \times & \\ 0 & 0 & 0 & \times & \times \end{bmatrix}$$

$$\Updownarrow$$

$$G_5^T A \longrightarrow G_4^T G_5^T A.$$

Continuing this process gives us

$$\begin{bmatrix} \boxtimes & & & & \\ \boxtimes & \boxtimes & & & \\ \boxtimes & \boxtimes & \times & & \\ 0 & 0 & \times & \times & \\ 0 & 0 & 0 & \times & \times \end{bmatrix} \longrightarrow \begin{bmatrix} \times & & & & \\ \times & \times & & & \\ 0 & \times & \times & & \\ 0 & 0 & \times & \times & \\ 0 & 0 & 0 & \times & \times \end{bmatrix}$$

$$\Updownarrow$$

$$G_4^T G_5^T A \longrightarrow G_3^T G_4^T G_5^T A.$$

It is clear that $n-2$ Givens transformations are necessary for reducing the lower triangular part of the quasiseparable matrix A to upper Hessenberg form:

$$G_3^T \cdots G_{n-1}^T G_n^T A = H.$$

The reduction of an upper Hessenberg matrix into an upper triangular one is a standard procedure and it can be performed by $n-1$ Givens transformations from top to bottom. Let us denote these Givens transformations $\hat{G}_1^T, \dots \hat{G}_{n-1}^T$. Hence we get an overall factorization of the following form:

$$\hat{G}_{n-1}^T \dots \hat{G}_1^T G_3^T \cdots G_{n-1}^T G_n^T A = R.$$

A detailed calculation of the structure of the upper triangular matrix after these transformations reveals that this matrix R will have the strictly upper triangular part of semiseparable form of rank 2; this means that the strictly upper triangular part can be written as the sum of two semiseparable matrices. A detailed analysis with formulas of this statement can be found in [284]. By construction we will illustrate in the next section that this is true.

Notes and references

The algorithm to compute the QR-factorization as presented in this chapter is not the only existing one. Eidelman and Gohberg present in [96] a QR-factorization method for solving a system of equations with a quasiseparable matrix based on a modification of the algorithms presented in [83, 84]. Also inversion and factorization formulas are presented. More information on this article can be found in Chapter 12

5.5 Implementing the QR-factorization

Here we will describe an $\mathcal{O}(n)$ implementation to solve a system of equations where the coefficient matrix is a semiseparable plus diagonal matrix. The implementation we present computes first the QR-factorization of the semiseparable plus diagonal matrix and will be used to solve the system via backward substitution. The semiseparable matrices considered in this section are represented with the decoupled Givens-vector representation.

For the implementation the semiseparable plus diagonal matrix is not split up in two parts, as in the theoretical overview, but into three separate parts, namely the lower triangular part of the semiseparable matrix, the strictly upper triangular part of the semiseparable matrix and a diagonal. In fact we have the following decomposition

$$S + D$$
$$= S_L + S_U + D$$

$$
= \begin{bmatrix}
c_1 v_1 & 0 & 0 & \cdots & 0 & 0 \\
c_2 s_1 v_1 & c_2 v_2 & 0 & & \vdots & \vdots \\
c_3 s_2 s_1 v_1 & c_3 s_2 v_2 & c_3 v_3 & \ddots & & \\
\vdots & & \ddots & \ddots & 0 & 0 \\
c_{n-1} s_{n-2} \cdots s_1 v_1 & \cdots & & & c_{n-1} v_{n-1} & 0 \\
s_{n-1} s_{n-2} \cdots s_1 v_1 & \cdots & & & s_{n-1} v_{n-1} & v_n
\end{bmatrix}
$$

$$
+ \begin{bmatrix}
0 & r_1 e_1 & r_2 t_1 e_1 & \cdots & r_{n-2} t_{n-3} \cdots t_1 e_1 & t_{n-2} t_{n-3} \cdots t_1 e_1 \\
0 & 0 & r_2 e_2 & & \vdots & \vdots \\
0 & 0 & 0 & \ddots & & \\
\vdots & & \ddots & \ddots & r_{n-2} e_{n-2} & t_{n-2} e_{n-2} \\
0 & \cdots & & & 0 & e_{n-1} \\
0 & \cdots & & & 0 & 0
\end{bmatrix}
$$

$$
+ \begin{bmatrix}
d_1 & & & \\
& d_2 & & \\
& & \ddots & \\
& & & d_n
\end{bmatrix}.
$$

In the previous splitting of the semiseparable plus diagonal matrix, we already assumed that our semiseparable parts are represented with the Givens-vector representation. For the lower triangular semiseparable matrix we denote the representation with G and $\mathbf{v}$, for the strictly upper triangular part we use H and $\mathbf{e}$:

$$G = \begin{bmatrix} c_1 & c_2 & \cdots & c_{n-1} \\ s_1 & s_2 & \cdots & s_{n-1} \end{bmatrix},$$

$$\mathbf{v} = [v_1, v_2, \ldots, v_n]^T,$$

$$H = \begin{bmatrix} r_1 & r_2 & \cdots & r_{n-2} \\ t_1 & t_2 & \cdots & t_{n-2} \end{bmatrix},$$

$$\mathbf{e} = [e_1, e_2, \ldots, e_{n-1}]^T.$$

The Givens transformations G_i and H_i are denoted as

$$G_i = \begin{bmatrix} c_i & -s_i \\ s_i & c_i \end{bmatrix} \text{ and } H_i = \begin{bmatrix} r_i & -t_i \\ t_i & r_i \end{bmatrix}.$$

Let us denote with d_i the diagonal elements of the matrix D. As we already used G_i for the Givens transformations involved in the Givens-vector representation of the matrix we denote with $\tilde{G}_i$ the Givens transformations from the QR-factorization. These Givens transformations satisfy $\tilde{G}_n = G_{n-1}, \ldots, \tilde{G}_2 = G_1$.

Applying $\tilde{G}_n^T, \ldots, \tilde{G}_2^T$ to the lower triangular semiseparable matrix from bottom to top will transform this matrix into an upper triangular semiseparable matrix. We take a closer look at the matrix $S_L + D$ after applying the Givens transformation $\tilde{G}_n^T$ to the left. We get the following structure for the lower triangular matrix S_L:

$$
\begin{bmatrix}
c_1 v_1 & 0 & 0 & \cdots & & 0 & 0 \\
c_2 s_1 v_1 & c_2 v_2 & 0 & & & \vdots & \vdots \\
c_3 s_2 s_1 v_1 & c_3 s_2 v_2 & c_3 v_3 & \ddots & & & \\
\vdots & & \ddots & \ddots & & 0 & 0 \\
c_{n-2} s_{n-3} \cdots s_1 v_1 & \cdots & & & c_{n-2} v_{n-2} & 0 & 0 \\
s_{n-2} s_{n-3} \cdots s_1 v_1 & \cdots & & & s_{n-2} v_{n-2} & v_{n-1} & s_{n-1} v_n \\
0 & \cdots & & & & 0 & c_{n-1} v_n
\end{bmatrix}
$$

The diagonal matrix D will change in the following way:

$$
\begin{bmatrix}
d_1 & & & & \\
& d_2 & & & \\
& & \ddots & & \\
& & & c_{n-1} d_{n-1} & s_{n-1} d_n \\
& & & -s_{n-1} d_{n-1} & c_{n-1} d_n
\end{bmatrix}.
$$

It can be seen now that the upper triangular zero part is slowly filled with elements depending on each other. In fact a semiseparable part is created from bottom to top, for which the Givens transformations in the Givens-vector representation are the Givens transformations $G_{n-1}, \ldots, G_1$. One more Givens transformation $\tilde{G}_{n-1}^T$ is applied such that we can also see the pattern in the expansion for the elements in the vector of the new representation. After applying the transformation $\tilde{G}_{n-1}^T$ on rows $n-1$ and $n-2$ we get the following matrices:

$$
\begin{bmatrix}
c_1 v_1 & 0 & 0 & \cdots & & 0 & 0 \\
c_2 s_1 v_1 & c_2 v_2 & 0 & & & \vdots & \vdots \\
c_3 s_2 s_1 v_1 & c_3 s_2 v_2 & c_3 v_3 & \ddots & & & \\
\vdots & & \ddots & \ddots & & 0 & 0 \\
s_{n-3} s_{n-4} \cdots s_1 v_1 & \cdots & & & v_{n-2} & s_{n-2} v_{n-1} & s_{n-2} s_{n-1} v_n \\
0 & \cdots & & & 0 & c_{n-2} v_{n-1} & c_{n-2} s_{n-1} v_n \\
0 & \cdots & & & 0 & & c_{n-1} v_n
\end{bmatrix}
$$

and

$$
\begin{bmatrix}
d_1 & & & & & \\
& d_2 & & & & \\
& & \ddots & & & \\
& & & c_{n-2} d_{n-2} & s_{n-2}\left(c_{n-1} d_{n-1}\right) & s_{n-2} s_{n-1} d_n \\
& & & -s_{n-2} d_{n-2} & c_{n-2}\left(c_{n-1} d_{n-1}\right) & c_{n-2} s_{n-1} d_n \\
& & & & -s_{n-1} d_{n-1} & c_{n-1} d_n
\end{bmatrix}.
$$

After all the Givens transformations $\tilde{G}_i$ are applied on the matrix $S_L + D$, the subdiagonal elements from top to bottom are the following:

$$[-s_1 d_1, \ldots, -s_{n-2} d_{n-2}, -s_{n-1} d_{n-1}].$$

The new vector in the Givens-vector representation from bottom to top has the following form:

$$\tilde{\mathbf{v}} = [v_n + d_n, v_{n-1} + c_{n-1} d_{n-1}, v_{n-2} + c_{n-2} d_{n-2}, \ldots, v_2 + c_2 d_2, v_1 + c_1 d_1]^T.$$

The procedure described above is an $\mathcal{O}(n)$ procedure, when implementing everything with the Givens-vector representation. However, we did not yet took into consideration the strictly upper triangular part S_U. We know that this matrix has the representation from left to right. This is an advantage in the next procedure, because the Givens transformations are performed on the left, they will not change the dependencies between the columns. The rough idea will be to apply a Givens transformation, which will clearly disturb the semiseparable structure because of fill-in in the diagonal. However, because a following sequence of Givens transformations is still needed , we can take out a term in the corresponding diagonal element of $\tilde{\mathbf{v}}$ and add it to our transformed matrix S_U, such that the dependency between the columns is maintained. In this way, we will create an upper triangular semiseparable matrix for which the dependencies between the columns did not change; only the new diagonal elements need to be calculated and this is illustrated in the following equations.

Applying the Givens transformation $\tilde{G}_n^T = G_{n-1}^T$ on the left of the matrix S_U we get the following structure:

$$\begin{bmatrix} 0 & r_1 e_1 & r_2 t_1 e_1 & \cdots & r_{n-2} t_{n-3} \cdots t_1 e_1 & t_{n-2} t_{n-3} \cdots t_1 e_1 \\ 0 & 0 & r_2 e_2 & & \vdots & \vdots \\ 0 & 0 & 0 & \ddots & & \\ \vdots & & \ddots & \ddots & r_{n-2} e_{n-2} & t_{n-2} e_{n-2} \\ 0 & \cdots & & & 0 & c_{n-1} e_{n-1} \\ 0 & \cdots & & & 0 & -s_{n-1} e_{n-1} \end{bmatrix}$$

To create an upper triangular semiseparable matrix, an element x_{n-1} has to be added in position $(n-1, n-1)$ such that the last two columns will satisfy again the semiseparable structure. This means in fact that an element x_{n-1} has to be added such that the following block will be of rank 1:

$$\begin{bmatrix} r_{n-2} e_{n-2} & t_{n-2} e_{n-2} \\ x_{n-1} & c_{n-1} e_{n-1} \end{bmatrix}.$$

When doing so, the corresponding element of $\tilde{\mathbf{v}}$, namely $\tilde{v}_{n-1}$, has to be diminished with this term x_{n-1}. (Note that this can be done without any problem, because the Givens transformations applied to the left are the same for S_L and S_U.)

We can now continue and perform the next Givens transformation $\tilde{G}_{n-1}$ on the matrix in which we added the element x_{n-1} (only the last four rows and columns are shown):

$$
\begin{bmatrix}
0 & r_{n-3}e_{n-3} & r_{n-2}t_{n-3}e_{n-3} & t_{n-2}t_{n-3}e_{n-3} \\
0 & 0 & c_{n-2}r_{n-2}e_{n-2} + s_{n-2}x_{n-1} & c_{n-2}t_{n-2}e_{n-2} - s_{n-2}c_{n-1}e_{n-1} \\
0 & 0 & c_{n-2}x_{n-1} - s_{n-2}r_{n-2}e_{n-2} & c_{n-2}c_{n-1}e_{n-1} - s_{n-2}t_{n-2}e_{n-2} \\
0 & 0 & 0 & -s_{n-1}e_{n-1}
\end{bmatrix}.
$$

We can continue this process and calculate x_{n-2}, thereby diminishing element $\tilde{v}_{n-2}$ with x_{n-2}. Using this procedure, one can design an $\mathcal{O}(n)$ calculation transforming S_U into an upper triangular semiseparable part having the following representation from left to right. For the Givens transformations we have:

$$
\tilde{H} = \begin{bmatrix}
0 & r_1 & r_2 & \cdots & r_{n-2} \\
1 & t_1 & t_2 & \cdots & t_{n-2}
\end{bmatrix}
$$

and the new diagonal elements, denoted with $\tilde{\mathbf{e}}$, are:

$$
\begin{bmatrix}
\dfrac{c_1 r_1 e_1 + s_1 x_2}{r_1}, & \dfrac{-s_1 r_1 e_1 + c_1 x_2}{r_1}, & \ldots, & \dfrac{-s_{n-2}r_{n-2}e_{n-2} + c_{n-2}x_{n-1}}{r_{n-2}}, & -s_{n-1}e_{n-1}
\end{bmatrix}^T.
$$

One should not forget to subtract from the elements of the vector $\tilde{\mathbf{v}}$ the corresponding elements x_i.

Note 5.8. *The implementation as designed here presumes that we are in the ideal case, i.e., there are no divisions by zero, all the elements are different from zero, etc. The practical implementation is changed a little bit, such that the not-ideal situations are also dealt with in a correct way.*

The final step consists of transforming the Hessenberg matrix into an upper triangular matrix. We will not describe this step in detail, as in the previous ones, because it is a straightforward generalization of the Givens transformations applied to S_U. First we need to swap the representation of G and $\tilde{\mathbf{v}}$ towards a representation from left to right. In this way, both the representations of the upper triangular part of the Hessenberg matrix are from left to right. The application of the second sequence of Givens transformations to annihilate the subdiagonal elements of the Hessenberg matrix will therefore not change the dependencies between the columns, and only the diagonal will be disturbed. Therefore we exclude the diagonal and calculate the new Givens-vector representation of the strictly upper triangular part of the matrix R. In this way we get a diagonal and two Givens-vector representations which will represent the strictly upper triangular part.

Solving a linear system of equations with the QR-algorithm for semiseparable plus diagonal matrices is an easy application of the decomposition above. First we calculate the QR-factorization of $S + D = QR$. Suppose we want to solve the system $(S + D)\mathbf{x} = \mathbf{b}$. This is equivalent to solving $R\mathbf{x} = Q^T\mathbf{b}$ and then solving the final part via backward substitution. This final backward substitution can also be implemented in $\mathcal{O}(n)$. To do so the same techniques as described in Chapter 4 have to be applied.

This algorithm can be used, for example, to calculate eigenvectors corresponding to particular eigenvalues via inverse iteration.

Notes and references

A more detailed and profound analysis of the QR-decomposition as depicted in this section can be found, together with a URV-decomposition in the following manuscripts.

☞ E. Van Camp. *Diagonal-plus-semiseparable matrices and their use in numerical linear algebra.* Ph.D. thesis, Department of Computer Science, Katholieke Universiteit Leuven, Celestijnenlaan 200A, 3000 Leuven (Heverlee), Belgium, May 2005.

☞ E. Van Camp, N. Mastronardi, and M. Van Barel. Two fast algorithms for solving diagonal-plus-semiseparable linear systems. *Journal of Computational and Applied Mathematics*, 164-165:731–747, 2004.

☞ R. Vandebril. *Semiseparable matrices and the symmetric eigenvalue problem.* Ph.D. thesis, Dept. of Computer Science, K.U.Leuven, Celestijnenlaan 200A, 3000 Leuven, May 2004.

The Ph.D. thesis of Vandebril describes the implementation as presented here in detail.

5.6 Other decompositions

In this section we will discuss some other decompositions. These decompositions are in some sense straightforward extensions of the QR-decomposition as presented in this chapter.

5.6.1 The URV-decomposition

In this section we will show that a URV-decomposition of a quasiseparable matrix can be easily derived in analogy to the QR decomposition.[7]

Theorem 5.9. *Suppose A is a quasiseparable matrix. Then A admits a factorization of the form*

$$A = URV^T,$$

with the matrices U^T and V^T unitary Hessenberg matrices and R an upper triangular matrix.

Proof. The proof is constructive. A constructive way to compute the decomposition and the Schur parametrization of the matrices U^T and V^T will be presented. Performing a first sequence of $n-2$ Givens transformations, chosen similarly as for the QR-decomposition of the matrix A, transforms this quasiseparable matrix as

---

[7]We call this the URV-decomposition, but in fact we compute here, for consistency reasons, a URV^T-decomposition.

follows:

$$G_3^T G_4^T \cdots G_n^T A = H,$$

with the matrix H an upper Hessenberg matrix. Defining the matrix U as $U^T = G_3^T G_4^T \cdots G_n^T$ gives us the following factorization:

$$U^T A = H.$$

It is clear that U^T is a unitary Hessenberg matrix.

The matrix H is now transformed into an upper triangular semiseparable matrix R by performing $n-1$ Givens transformations on the right of H. The Givens transformations are performed in an order to annihilate the subdiagonal elements, starting from the right and moving to the left. Denoting these Givens transformations as V_i, where V_i is the Givens transformation annihilating the subdiagonal element in row i and column $i-1$, leads to the following factorization:

$$\begin{aligned}
G_3^T G_4^T \cdots G_n^T A V_n V_{n-1} \cdots V_2 &= H V_n V_{n-1} \cdots V_2 \\
&= R.
\end{aligned}$$

Let $V = V_n V_{n-1} \cdots V_2$. This gives us:

$$A = U R V^T,$$

with U^T and V^T two unitary Hessenberg matrices for which the factorization is known.   □

Basically, there are two equivalent algorithms to compute the URV-decomposition of a quasiseparable matrix. The first algorithm is derived in a straightforward way from Theorem 5.9.

Algorithm 5.10.

  1. *Reduce the quasiseparable matrix A to upper Hessenberg form by means of the Givens transformations G_i, $i = n, n-1, \ldots, 2$:*

$$G_3^T G_4^T \cdots G_n^T A = H.$$

  2. *Compute the upper triangular matrix R from H applying to the right the $n-1$ Givens transformations V_i, $i = n, n-1, \ldots, 2$:*

$$H V_n V_{n-1} \cdots V_2 = R.$$

Hence, the URV-decomposition is obtained defining $U = G_n G_{n-1} \cdots G_3 G_2$ (with $G_2 = I$) and $V = V_n V_{n-1} \cdots V_2$.

The second algorithm computes the URV-decomposition applying simultaneously at step i, $i = n, n-1, \ldots, 2$, the Givens matrices G_i^T and V_i.

Algorithm 5.11.

1. $R^{(1)} = A$.

2. For $i = n, \ldots, 2$ do:

   (a) $R^{(n-i+2)} = G_i^T R^{(n-i+1)} V_i$,

   endfor;

Hence, the URV-decomposition of A is obtained defining $U = G_n G_{n-1} \cdots G_2$, $V = V_n V_{n-1} \cdots V_2$ and $R = R^{(n)}$. Since each orthogonal transformation G_i and V_i, $i = n, n-1, \ldots, 2$, modifies the rows and columns $i-1$ and i in the matrix $R^{(n-i+1)}$, it turns out that $R^{(n-i+2)}(i:n, i:n) = R(i:n, i:n)$ and $R^{(n-i+2)}(i:n, 1:i-1) = 0$, i.e., the submatrices $R^{(n-i+2)}(i:n, i:n)$, $i = n, n-1, \ldots, 2$, are upper triangular and the strictly upper part is semiseparable of semiseparability rank 2.

An algorithm for solving linear systems $A\mathbf{x} = \mathbf{b}$, where A is a quasiseparable matrix, based on an implicit URV-factorization, was proposed in [56]. (In this manuscript the authors considered diagonal plus semiseparable matrices, but the approach is also valid for quasiseparable matrices.) Here we describe the basic idea of the algorithm.

Let A be a quasiseparable matrix and let G_i and V_i, $i = n, n-1, \ldots, 2$, be the Givens transformations introduced in Theorem 5.9 to compute its URV-decomposition $A = URV^T$.

Let $R^{(1)} = A$, $\mathbf{x}^{(1)} = \mathbf{x}$ and $\mathbf{b}^{(1)} = \mathbf{b}$. First, the Givens transformation G_n^T is applied to both the left- and right-hand side of the system and V_n is applied to the right of the matrix. As a consequence the vector of unknowns is modified by the multiplication to the left by V_n:

$$G_n^T R^{(1)} \mathbf{x}^{(1)} = G_n^T \mathbf{b}^{(1)},$$
$$G_n^T R^{(1)} V_n V_n^T \mathbf{x}^{(1)} = G_n^T \mathbf{b}^{(1)}.$$

This can be rewritten as

$$R^{(2)} \mathbf{x}^{(2)} = \mathbf{b}^{(2)}, \tag{5.2}$$

with $R^{(2)} = G_n^T R^{(1)} V_n$, $\mathbf{x}^{(2)} = V_n^T \mathbf{x}^{(1)}$ and $\mathbf{b}^{(2)} = G_n^T \mathbf{b}^{(1)}$.

Since we have that $R^{(2)}(n, 1:n-1) = 0$, we can reduce our system of equations as follows:

$$R^{(2)}(1:n-1, 1:n-1)\mathbf{x}^{(2)}(1:n-1) = \mathbf{b}^{(2)}(1:n-1) - R^{(2)}(1:n-1, n)\mathbf{x}^{(2)}(n).$$

This is possible as we can easily compute $\mathbf{x}^{(2)}(n)$. Hence one is left with a smaller system of equations.

Repeating the same machinery with, $R^{(3)} = G_{n-1}^T R^{(2)} V_{n-1}$, $\mathbf{x}^{(3)} = V_{n-1}^T \mathbf{x}^{(2)}$ and $\mathbf{b}^{(3)} = G_{n-1}^T \mathbf{b}^{(2)}$. We obtain the next step, since $R^{(3)}(n-1:n, 1:n-2) = 0$:

$$R^{(3)}(1:n-2,1:n-2)\mathbf{x}^{(3)}$$
$$= \mathbf{b}^{(3)}(1:n-2) - R^{(3)}(1:n-2,n-1:n)\mathbf{x}^{(3)}(n-1:n)$$
$$= \mathbf{b}^{(3)} - R^{(3)}(1:n-2,n-1)\mathbf{x}^{(3)}(n-1) - R^{(3)}(1:n-2,n)\mathbf{x}^{(3)}(n)$$
$$= \mathbf{b}^{(3)} - R^{(3)}(1:n-2,n)\mathbf{x}^{(2)}(n) - R^{(3)}(1:n-2,n-1)\mathbf{x}^{(3)}(n-1)$$
$$= G_{n-1}^T \left(\mathbf{b}^{(2)} - R^{(2)}(1:n-2,n)\mathbf{x}^{(2)}(n) \right)$$
$$-R^{(3)}(1:n-2,n-1)\mathbf{x}^{(3)}(n-1).$$

Due to this recursive structure one can easily compute the components of the solution vector $\mathbf{x}^{(n)}$ from bottom to top.

This means that the original system has been transformed into one with an upper triangular coefficient matrix having the strict upper triangular part semiseparable of rank 2. Moreover, $\mathbf{x}^{(n)} = V^T\mathbf{x}$, where $V = V_n V_{n-1} \cdots V_2$ is a unitary Hessenberg semiseparable matrix. Therefore, the solution of the original linear system requires one more step, namely the solution of the system $V^T\mathbf{x} = \mathbf{x}^{(n)}$. This latter step is accomplished with $\mathcal{O}(n)$ flops because we know the factorization of the matrix V.

5.6.2   Some other orthogonal decompositions

In this section we will briefly indicate that the techniques presented in this chapter can be slightly modified to obtain different factorizations of the involved semiseparable or quasiseparable matrix. We will briefly cover the RQ- the QL- and the LQ-decomposition for specific structured rank matrices. Details on the rank structure of the factors in the proposed decompositions are left to the reader. Only indications are given on how to construct the factorizations in an efficient manner.

Based on a combination of the techniques presented in the proof of the Theorem 5.9, and on our knowledge of the QR-decomposition, we can formulate the following theorem.

Theorem 5.12. *Suppose a quasiseparable matrix A is given. Then there exists a factorization*

$$A = RQ,$$

with R an upper triangular matrix and Q an orthogonal matrix. Where R has the strictly upper triangular of semiseparable form of rank 2 and Q can be written as the product of $2n - 3$ Givens transformations.

Proof. The first step in the computation of the RQ-factorization of a matrix, is bringing the matrix A into upper Hessenberg form, by performing $n - 2$ Givens transformations on the right of A. This can be achieved by annihilating the semiseparable part in the lower triangular part of A, starting with the first column and moving to the right. Denote with G_i^T the transformation acting on column i and $i + 1$, such that the element $i + 2$ to element n in column i are annihilated. This gives us

$$AG_1^T \cdots G_{n-2}^T = H.$$

Similarly as in the proof of Theorem 5.9, the remaining upper triangular matrix H can be brought to upper triangular form. Hence a RQ-factorization, as requested, is obtained. The matrix R will be of a specific structured rank form. The proof is similar as in the previous chapters. The matrix Q is the product of $2n - 3$ Givens transformations, thereby making the lower and upper triangular part of the matrix of a very specific rank form. We leave the details to the reader. $\square$

Note 5.13. *For these kinds of factorizations one can easily adapt the nullity theorem of the QR-factorization as proved in a previous section. This nullity theorem can then easily predict the structure of the factor Q in, e.g., the RQ-factorization and the coming QL- and LQ-factorizations.*

The structure of a quasiseparable matrix is maintained under transposition. This leads to the following decompositions involving lower triangular matrices instead of upper triangular ones.

Theorem 5.14. *Suppose A is a quasiseparable matrix. Then there exist the following two factorizations of this matrix A:*

$$A = QL,$$
$$A = LQ.$$

For both of these factorizations Q represents an orthogonal matrix and L a lower triangular one, having the strictly lower triangular part of semiseparable form of rank 2. Also the factor Q, consisting of $2n - 3$ Givens transformations, is of a very specific rank structured form.

Proof. The proofs are a straightforward consequence if one computes the QR- and RQ-factorization of the matrix A^T. Taking the transpose of this resulting decomposition leads immediately to the factorizations with lower triangular matrices. $\square$

Note 5.15. *One can derive similar theorems for an ULV-decomposition.*

Notes and references

The URV-decomposition was first introduced by G.W. Stewart as a rank-revealing factorization of a matrix.

☞ G. W. Stewart. An updating algorithm for subspace tracking. *IEEE Transactions on Signal Processing*, 40:1535–1541, 1992.

The algorithm for solving a system of equations using the URV-decomposition as proposed in this subsection was based on the results in the following manuscript.

☞ S. Chandrasekaran and M. Gu. Fast and stable algorithms for banded plus semiseparable systems of linear equations. *SIAM Journal on Matrix Analysis and its Applications*, 25(2):373–384, 2003.

An extension of the latter algorithm has been proposed in the following article.

☞ N. Mastronardi, S. Chandrasekaran, and S. Van Huffel. Fast and stable two-way algorithm for diagonal plus semi-separable systems of linear equations. *Numerical Linear Algebra with Applications*, 8(1):7–12, 2001.

The reduction of the coefficient matrix into triangular form is accomplished simultaneously from the top-left and the bottom-right corners of the involved matrix, yielding an algorithm suitable for an efficient implementation on a parallel computer.

The presented solvers are also suitable for semiseparable plus band matrices. The upper and lower triangular parts of the semiseparable matrix are generator representable, but of higher order. In the following part of the book, more attention will be paid to higher order structured rank matrices.

Also the paper of Mastronardi, Van Camp and Van Barel provides a URV-decomposition for semiseparable plus diagonal matrices [284]. (See Section 4.6 in Chapter 4.)

5.7   Conclusions

The rank structure of matrices can be exploited when computing a QR-decomposition of structured rank matrices. Moreover the factors in these decompositions will also be of a specific rank structured form. In this chapter we presented a theoretical consideration about the rank structure of these factors. Finally we concluded this chapter by proposing some different orthogonal factorizations (and a solver), which are straightforward extensions of the factorizations proposed in the previous sections.

In a later chapter in this book we will derive methods for computing the QR-factorization of more general structured rank matrices.

Chapter 6

A Levinson-like and Schur-like solver

Different algorithms for solving systems of equations with semiseparable plus diagonal coefficient matrices have been proposed in the previous chapters. The method proposed in this chapter is based on the underlying idea of the Durbin and the Levinson algorithm for solving Toeplitz systems of equations. Assume we would like to solve the system $Ax = b$ with A of dimension n. The Levinson idea is to solve n systems of equations of increasing size. At step k of the Levinson algorithm a system of dimension $k \times k$ is solved with as coefficient matrix the upper left $k \times k$ block of the matrix A, and right-hand side the first k elements of the vector b. In the next step, step $k + 1$, we use the solution of step k to construct the solution of step $k + 1$.

The Levinson algorithm for Toeplitz matrices is widespread and described, for example, in [152, 191, 205]. It takes $\mathcal{O}(k)$ operations in each step of the loop. The overall complexity to solve the Toeplitz equations using the Levinson algorithm is therefore $\mathcal{O}(n^2)$. In this chapter, we will derive a similar algorithm for solving systems of equations involving structured rank matrices. Indeed, examples will show that the method is suitable for different types of structured rank matrices, including, for example, band, higher order semiseparable, band plus higher order semiseparable and different other classes. The scope of this second part of the book, however, is limited to the easy classes of structured rank matrices such as semiseparable, quasiseparable, semiseparable plus diagonal and so on. Further on in the text, starting in Part III, higher order structured rank matrices will also be considered.

The part on Levinson in this chapter consists of 4 sections. The first section is concerned with a description of the Levinson algorithm for Toeplitz systems. The Yule-Walker equation, the Durbin and the Levinson method are briefly revisited to understand better the following sections involving the structured rank matrices. Section 6.2 covers a Levinson-like method for solving generator representable semiseparable plus diagonal matrices. The considered matrices are symmetric and positive definite. This easy example illustrates the basic ideas of the general method, which is constructed in the third section. This section constructs a Levinson-like solver framework for general matrices. Examples of matrices for which this framework is applicable are provided in the last section. As told earlier, only the simplest classes are discussed. In Chapter 11 we will reconsider this solver and illustrate its power by applying the general framework to different types of structured rank matrices.

The last section in this chapter is dedicated to the design of a Schur-like algorithm for generator representable quasiseparable matrices. After the description of the algorithm a more general theorem is proven for the preservation of the rank under Schur complementation. This theorem proves, for example, that the Schur complement of a semiseparable and a quasiseparable matrix inherits the rank structure of the original matrix. Due to this theorem we know the existence of Schur-like algorithms for these classes of matrices.

✎ *This chapter discusses a Levinson-like algorithm for solving systems of equations. Section 6.2 proposes a Levinson-like method for generator representable semiseparable plus diagonal matrices. This is a very specific example, but it helps in creating the general picture; however, without loss of generality, one can immediately switch to the general case. The Levinson framework as discussed in Section 6.3 contains the algorithm to which this chapter is dedicated. Definition 6.3 and Definition 6.4 contain an abstract formulation of a matrix block decomposition. Matrices satisfying this decomposition will admit a Levinson-like solver. Section 6.3.2 contains the derivation of the algorithm. The first two subtitles, namely the Yule-Walker-like system and the Levinson-like solver present the main ideas. A first method is presented, but this approach is further tuned, reducing the computational complexity. Algorithm 6.9 proposes then the real Levinson-like algorithm, with reduced complexity. Section 6.3.3 discusses the connection with an upper triangular factorization of the involved matrix, which might come in handy later on when inverting structured rank matrices. Numerical stability issues are discussed in Subsection 6.3.4. Section 6.4 contains examples related to different types of structured rank matrices to see how these matrices fit in. Examples are given for Givens-vector represented semiseparable matrices, quasiseparable matrices, tridiagonal matrices, arrowhead matrices and so forth. Depending on the reader's interest one or more of these subsections can be taken into consideration.*

A second important algorithm of this chapter is discussed in Section 6.5. Important for the development of a Schur algorithm is the preservation of specific structures. The Schur reduction presented in Section 6.5.2 is translated in its subsections (Subsection 6.5.3 and Subsection 6.5.4) towards generator representable quasiseparable matrices. To show the applicability for other classes of structured rank matrices Section 6.5.5 provides some theorems and a list of matrices on which the Schur algorithm can be applied.

6.1   About the Levinson algorithm

Before deriving a Levinson-like algorithm, for semiseparable and related matrices, we will briefly discuss the Levinson algorithm for Toeplitz matrices, as it was historically developed. We will not go into the details of look-ahead, and possible numerical issues. Just the basic idea on how to solve a Toeplitz system, via the Levinson method, is explored.

Suppose a Toeplitz matrix T

$$T = \begin{bmatrix} t_0 & t_1 & \cdots & t_{n-1} \\ t_{-1} & t_0 & \ddots & \vdots \\ \vdots & \ddots & \ddots & t_1 \\ t_{-n+1} & \cdots & t_{-1} & t_0 \end{bmatrix}$$

is given. A Toeplitz matrix is characterized by the fact that all entries appearing on any of the sub- or superdiagonals have the same value. We will consider here only the symmetric case:

$$T = \begin{bmatrix} t_0 & t_1 & \cdots & t_{n-1} \\ t_1 & t_0 & \ddots & \vdots \\ \vdots & \ddots & \ddots & t_1 \\ t_{n-1} & \cdots & t_1 & t_0 \end{bmatrix}.$$

Toeplitz matrices arise, for example, in signal processing.

The purpose of this section is the development of a solver involving Toeplitz matrices

$$T\mathbf{x} = \mathbf{b}.$$

We will achieve this goal by solving systems of equations, where the coefficient Toeplitz matrices increase in size at every step. One has to solve special types of equations, the so-called Yule-Walker problem, followed by the inductive procedure to compute the final solution.

6.1.1   The Yule-Walker problem and the Durbin algorithm

The introduction in this section is based on [152], but is covered in many textbooks (see notes and references).

Suppose we have a symmetric positive definite Toeplitz matrix T of order n. Denote with T_k the principal leading $k \times k$ submatrix of T. The Yule-Walker problem of dimension k is the following one:

$$T_k \mathbf{y}_k = -[t_1, \ldots, t_k]^T = -\mathbf{t}_k.$$

To solve this system for $k = 1, \ldots, n$, one can use the Durbin algorithm (see [87]), which works by induction. Let us illustrate this. Assume by induction, that we know the solution of the kth Yule-Walker problem, and we would like to solve now the $(k+1)$th one.

Our Toeplitz matrix can be split up in blocks, leading to the following equations:

$$T_{k+1}\mathbf{y}_{k+1} = -\mathbf{t}_{k+1}$$

$$\begin{bmatrix} T_k & E_k\mathbf{t}_k \\ \mathbf{t}_k^T E_k & t_0 \end{bmatrix} \begin{bmatrix} \mathbf{z}_{k+1} \\ \alpha_{k+1} \end{bmatrix} = - \begin{bmatrix} \mathbf{t}_k \\ t_{k+1} \end{bmatrix},$$

where E_k is the kth order exchange matrix, sometimes called the counteridentity matrix. This matrix has zeros everywhere except for ones on the antidiagonal. This equation expands into two equalities which can be rewritten as:

$$\mathbf{z}_{k+1} = \mathbf{y}_k - \alpha_{k+1} T_k^{-1} E_k \mathbf{t}_k,$$

$$\alpha_{k+1} = \frac{1}{t_0} \left(-t_{k+1} - \mathbf{t}_k^T E_k \mathbf{z}_{k+1} \right).$$

Since T_k^{-1} is a persymmetric[8] matrix, the matrices T_k^{-1} and E_k commute, leading to

$$\mathbf{z}_{k+1} = \mathbf{y}_k + \alpha_{k+1} E_k \mathbf{y}_k.$$

Substituting the latter equation in the equation for α_k leads to the following value for α_k:

$$\alpha_{k+1} = \frac{- \left(t_{k+1} + \mathbf{t}_k^T E_k \mathbf{y}_k \right)}{t_0 \left(1 + \mathbf{t}_k^T \mathbf{y}_k \right)}. \tag{6.1}$$

Substituting the value computed for α_{k+1} into the equation for computing $\mathbf{z}_{k+1}$ gives us the solution of the $(k+1)$th order Yule-Walker system of equations. This algorithm requires $3n^2$ flops for computing the final solution $\mathbf{y} = \mathbf{y}_n$. However, one can compute the denominator in Equation (6.1) recursively and hence reduce the complexity to $2n^2$. The algorithm for computing this solution is the so-called Durbin algorithm. This method can now be expanded in a similar way to compute the solution of a system of equations with an arbitrary right-hand side.

6.1.2   The Levinson algorithm

The Levinson algorithm solves in fact a general system of equations $T\mathbf{x} = \mathbf{b}$, with an arbitrary right-hand side $\mathbf{b}$. The Levinson algorithm also works inductively, increasing the dimension of the coefficient matrix in each step, using thereby the solutions of the kth order Yule-Walker system of equations (see [205]). Let us illustrate this algorithm.

Assume the following two problems (including the kth order Yule-Walker problem) are solved:

$$T_k \mathbf{y}_k = -[t_1, \dots, t_k]^T = -\mathbf{t}_k,$$
$$T_k \mathbf{x}_k = [b_1, \dots, b_k]^T = \mathbf{b}_k.$$

Based on these two solved systems of equations, one can now compute the solution of

$$T_{k+1} \mathbf{x}_{k+1} = \mathbf{b}_{k+1} \tag{6.2}$$

$$\begin{bmatrix} T_k & E_k \mathbf{t}_k \\ \mathbf{t}_k^T E_k & t_0 \end{bmatrix} \begin{bmatrix} \mathbf{w}_{k+1} \\ \mu_{k+1} \end{bmatrix} = \begin{bmatrix} \mathbf{b}_k \\ b_{k+1} \end{bmatrix},$$

---

[8]Persymmetric means symmetric with regard to the antidiagonal.

in a similar fashion as in the previous subsection. A straightforward calculation, using thereby the solution of the kth order Yule-Walker problem, leads to the following equations for μ_{k+1} and $\mathbf{w}_{k+1}$:

$$\mu_{k+1} = \frac{(b_{k+1} - \mathbf{t}_k E_k \mathbf{x}_k)}{(1 + \mathbf{t}_k^T \mathbf{y}_k)},$$

$$\mathbf{w}_{k+1} = \mathbf{x}_{k+1} + \mu_{k+1} E_k \mathbf{t}_k^T.$$

This results in the solution of system (6.2). Using this in a recursive fashion leads to an overall method for computing the solution of the system of equations involving the Toeplitz matrix.

6.1.3 An upper triangular factorization of the inverse

The Durbin algorithm, as explained above, computes in fact an upper triangular factorization of the inverse of the Toeplitz matrix.

The kth order Yule-Walker equation is of the following form

$$T_k \mathbf{y}_k = -\mathbf{t}_k. \tag{6.3}$$

This equation can be rewritten towards the matrix T_{k+1} as

$$T_{k+1} \begin{bmatrix} \mathbf{y}_k \\ 1 \end{bmatrix} = \left[\begin{array}{c|c} T_k & \mathbf{t}_k \\ \hline \mathbf{t}_k^T & t_0 \end{array} \right] \begin{bmatrix} \mathbf{y}_k \\ 1 \end{bmatrix} = \begin{bmatrix} 0 \\ \sigma_{k+1} \end{bmatrix}.$$

Putting the different equations for every k in an upper triangular matrix we get (let us denote with $y_{k,j}$ the components in the vector $\mathbf{y}_k$):

$$T \begin{bmatrix} 1 & y_{1,1} & y_{2,1} & \cdots & y_{n-1,1} \\ 0 & 1 & y_{2,2} & \cdots & y_{n-1,2} \\ 0 & 0 & 1 & & y_{n-1,3} \\ \vdots & \vdots & \ddots & \ddots & \vdots \\ 0 & 0 & \cdots & 0 & 1 \end{bmatrix} = \begin{bmatrix} \sigma_1 & 0 & 0 & \cdots & 0 \\ \times & \sigma_2 & 0 & \cdots & 0 \\ \vdots & \times & \sigma_3 & & 0 \\ \vdots & \vdots & \ddots & \ddots & \vdots \\ \times & \times & \cdots & \times & \sigma_n \end{bmatrix}.$$

The $\times$ denote arbitrary elements in the matrix. The equations above can be rewritten and, defining the upper triangular matrix to the right of T as U and the right-hand side as L, we get the following equations ($\hat{L}$ is another lower triangular matrix):

$$TU = L$$

$$TU = U^{-T}\hat{L}$$

$$U^T T U = \hat{L}.$$

Because $\hat{L}$ is lower triangular and because of symmetry, the matrix $\hat{L}$ is a diagonal matrix Λ. Inverting the equation above yields

$$U^{-1}T^{-1}U^{-T} = \Lambda^{-1}$$

$$T^{-1} = U\Lambda^{-1}U^T,$$

which is a triangular factorization of the inverse of the Toeplitz matrix. In fact there are more direct methods based on the Levinson algorithm for computing the inverse, but we will not go into the details. They can be found, e.g., in [152].

Notes and references

The presentation of this section is based on the book [152] by Golub and Van Loan and on the following book.

☞ T. Kailath and A. H. Sayed, editors. *Fast reliable algorithms for matrices with structure.* SIAM, Philadelphia, PA, USA, May 1999.

The scope of these books is way beyond the presentation given here. They cover the same material, plus look-ahead schemes, inversion using this method (Trench algorithm) and nonsymmetric Toeplitz systems. The book by Kailath and Sayed is also focused on displacement operators, in which Toeplitz matrices play a very important role. The book by Golub and Van Loan covers an extensive list of references dedicated to the Levinson algorithm, including references for the inversion, stability and so on. We will briefly give the two original references related to the algorithms proposed in this section.

☞ J. Durbin. The fitting of time series in models. *Review of the International Statistical Institute*, 28:233–243, 1960.

☞ N. Levinson. The Wiener RMS error criterion in filter desing and prediction. *Journal of Mathematical Physics*, 25:261–278, 1947.

6.2   Generator representable semiseparable plus diagonal matrices

The next section of this chapter is concerned with a general Levinson framework. This means that a general theoretical framework will be constructed, and different types of matrices will fit nicely into that framework. Before presenting the general formulation, we will derive a Levinson-like solver for the class of symmetric positive definite generator representable semiseparable plus diagonal matrices. We do so to make the reader already familiar with the upcoming ideas.

This section is, as all sections are in this chapter, organized as follows. In the following introduction section, we briefly reconsider the class of matrices we are working with and provide some comments. In Subsections 6.2.2 and 6.2.3, we derive the analogue Durbin and Levinson algorithms, respectively, for this class of matrices. The relation with an upper triangular factorization of the inverse of the semiseparable plus diagonal matrix is investigated in Subsection 6.2.4.

6.2.1   The class of matrices

Before starting the algoritmic description of the Levinson and Durbin-like algorithms for semiseparable matrices, we will briefly reconsider the class of matrices we will be working with. Throughout the entire section we consider symmetric and positive definite generator representable semiseparable plus diagonal matrices, even though we might not always mention it.

Suppose three vectors $\mathbf{u}, \mathbf{v}$ and $\mathbf{d}$ of length n are given. The matrices considered here are of the following form, as defined in the beginning of the book:

$$S(\mathbf{u}, \mathbf{v}) + D = S + D = \begin{bmatrix} u_1 v_1 & u_2 v_1 & \cdots & u_n v_1 \\ u_2 v_1 & u_2 v_2 & & u_n v_2 \\ \vdots & & \ddots & \vdots \\ u_n v_1 & \cdots & & u_n v_n \end{bmatrix} + \begin{bmatrix} d_1 & & & \\ & d_2 & & \\ & & \ddots & \\ & & & d_n \end{bmatrix}.$$

We remark once more that the class of generator representable semiseparable matrices is less general than the class of semiseparable matrices, but because every symmetric semiseparable matrix as defined in Section 1.1 can be seen as a block diagonal matrix with the block diagonal matrices of the generator representable form, there is no loss in generality here. In this section we solve systems, and solving a system with a block diagonal matrix reduces in a natural way to the solution of the separate blocks. In the next section, however, we will loosen these conditions and come to a similar method for not necessarily symmetric semiseparable matrices. Let us call the vectors $\mathbf{u}$ and $\mathbf{v}$ the generators of the matrix S and the vector $\mathbf{d}$ contains the diagonal elements of the matrix D.

6.2.2 A Yule-Walker-like problem

In this first subsection, we will solve a Yule-Walker-like problem for semiseparable plus diagonal matrices.

The Yule-Walker-like equation we would like to solve is the following one:

$$(S + D)\mathbf{y} = -\mathbf{v},$$

where S is a semiseparable matrix generated by $\mathbf{u}$ and $\mathbf{v}$, and the diagonal elements of D are stored in the vector $\mathbf{d}$. We will solve this problem in a way analogous to solving the Toeplitz case. First, we will develop a completely similar $\mathcal{O}(n^2)$ method. Then, a slight adaptation will lead to an $\mathcal{O}(n)$ solver.

The method

We solve the system inductively. Let us denote with S_k the upper left $k \times k$ submatrix of S; a similar notation is used for D.

Let us assume by induction that we have solved the kth order Yule-Walker-like problem:

$$(S_k + D_k)\mathbf{y}_k = -\mathbf{r}_k,$$

where $\mathbf{r}_k = [v_1, v_2, \ldots, v_k]^T$. The $(k+1)$th order Yule-Walker-like system of equations

$$(S_{k+1} + D_{k+1})\mathbf{y}_{k+1} = -\mathbf{r}_{k+1}$$

$$\left[\begin{array}{c|c} S_k + D_k & u_{k+1}\mathbf{r}_k \\ \hline u_{k+1}\mathbf{r}_k^T & u_{k+1}v_{k+1} + d_{k+1} \end{array} \right] \left[\begin{array}{c} \mathbf{z}_{k+1} \\ \alpha_{k+1} \end{array} \right] = - \left[\begin{array}{c} \mathbf{r}_k \\ v_{k+1} \end{array} \right], \tag{6.4}$$

will be solved in $\mathcal{O}(k)$ flops. Using Equation (6.4), first we observe that

$$(S_k + D_k)\,\mathbf{z}_{k+1} + u_{k+1}\,\mathbf{r}_k\,\alpha_{k+1} = -\mathbf{r}_k \qquad (6.5)$$

and, secondly,

$$u_{k+1}\,\mathbf{r}_k^T \mathbf{z}_{k+1} + \alpha_{k+1}\,(u_{k+1}v_{k+1} + d_{k+1}) = -v_{k+1}. \qquad (6.6)$$

For Equation (6.5) this gives us:

$$\begin{aligned}
\mathbf{z}_{k+1} &= (1 + u_{k+1}\,\alpha_{k+1})(S_k + D_k)^{-1}(-\mathbf{r}_k) \\
&= (1 + u_{k+1}\,\alpha_{k+1})\mathbf{y}_k. \qquad (6.7)
\end{aligned}$$

Substituting the latter equation in (6.6) leads to:

$$u_{k+1}\,\mathbf{r}_k^T\,(1 + u_{k+1}\alpha_{k+1})\,\mathbf{y}_k + \alpha_{k+1}\,(u_{k+1}v_{k+1} + d_{k+1}) = -v_{k+1},$$

which can be rewritten towards α_{k+1} as:

$$\alpha_{k+1} = \frac{(-1)(v_{k+1} + u_{k+1}\mathbf{r}_k^T\mathbf{y}_k)}{(u_{k+1}^2\mathbf{r}_k^T\mathbf{y}_k + u_{k+1}v_{k+1} + d_{k+1})}. \qquad (6.8)$$

Using α_{k+1} in Equation (6.7), we can compute the vector $\mathbf{z}_{k+1}$.

The denominator in (6.8) is always different from zero, because we assumed our initial matrix $S + D$ to be positive definite. As $S + D$ is positive definite, all the leading principal submatrices are also positive definite and hence also $S_{k+1} + D_{k+1}$. This means that $\mathbf{w}^T(S_{k+1} + D_{k+1})\mathbf{w} > 0$ for every vector $\mathbf{w}$ different from zero. Taking $\mathbf{w}^T = [u_{k+1}\mathbf{y}_k^T\,, 1]$ gives us the following relations:

$$\begin{aligned}
\mathbf{w}^T\,(S_{k+1} + D_{k+1})\,\mathbf{w} &= [u_{k+1}\,\mathbf{y}_k^T\,, 1]\,(S_{k+1} + D_{k+1})\begin{bmatrix} u_{k+1}\,\mathbf{y}_k^T \\ 1 \end{bmatrix} \\
&= [u_{k+1}\,\mathbf{y}_k^T\,, 1]\left[\begin{array}{c|c} S_k + D_k & u_{k+1}\mathbf{r}_k \\ \hline u_{k+1}\mathbf{r}_k^T & u_{k+1}v_{k+1} + d_{k+1} \end{array}\right]\begin{bmatrix} u_{k+1}\mathbf{y}_k^T \\ 1 \end{bmatrix} \\
&= u_{k+1}^2\,\mathbf{y}_k^T\,(S_k + D_k)\,\mathbf{y}_k + u_{k+1}^2\,\mathbf{y}_k^T\,\mathbf{r}_k \\
&\quad + u_{k+1}^2\,\mathbf{r}_k^T\,\mathbf{y}_k + u_{k+1}\,v_{k+1} + d_{k+1} \\
&> 0.
\end{aligned}$$

Using now the relation $\mathbf{y}_k^T\,(S_k + D_k)\,\mathbf{y}_k = -\mathbf{y}_k^T\,\mathbf{r}_k$ in the equation above, we find that

$$u_{k+1}^2\,\mathbf{r}_k^T\,\mathbf{y}_k + u_{k+1}\,v_{k+1} + d_{k+1} > 0,$$

and hence the formula for calculating α_{k+1} is well defined.

Clearly, the computation of the inner product $\mathbf{r}_k^T\mathbf{y}_k$ at each step and the computation of $\mathbf{z}_k$ via the multiplication of a vector with a scalar leads to an order k complexity for solving the $(k+1)$th order Yule-Walker-like equation. Overall this leads to an order $\mathcal{O}(n^2)$ solver.

How to achieve an $\mathcal{O}(n)$ complexity

Two bottlenecks arise when solving the Yule-Walker-like equation of order $(k+1)$. These are the computation of $\mathbf{r}_k^T\mathbf{y}_k$ and the computation of $\mathbf{z}_k$ in each step. We will show now how to change the algorithm from the previous section in order to come to an iteration step with a fixed number of flops.

Suppose we just solved the $(k+1)$th Yule-Walker-like system of equations and we start now solving the $(k+2)$nd system. Let us denote the solution and the right-hand side of the $(k+1)$th system as follows: $(S_{k+1}+D_{k+1})\mathbf{y}_{k+1} = -\mathbf{r}_{k+1}$. For the new $(k+2)$nd system we need the inner product $\mathbf{r}_{k+1}^T\mathbf{y}_{k+1}$. This corresponds to the inner product (using hereby the notation from Section 6.2.2):

$$\mathbf{r}_{k+1}^T\mathbf{y}_{k+1} = [\mathbf{r}_k^T, v_{k+1}] \begin{bmatrix} \mathbf{z}_{k+1} \\ \alpha_{k+1} \end{bmatrix}.$$

Rewriting the equation above, using the expression for $\mathbf{z}_{k+1} = (1 + u_{k+1}\alpha_{k+1})\mathbf{y}_k$, leads to the following recursive formula for the inner product:

$$\mathbf{r}_{k+1}^T\mathbf{y}_{k+1}^T = (1 + u_{k+1}\alpha_{k+1})\mathbf{r}_k^T\mathbf{y}_k + v_{k+1}\alpha_{k+1}.$$

It is clear that the inner product $\mathbf{r}_{k+1}^T\mathbf{y}_{k+1}$ for solving the Yule-Walker-like equation of order $(k+2)$ can already be computed at the end of step $(k+1)$ in a constant number of operations if $\mathbf{r}_k^T\mathbf{y}_k$ is known. Incorporating this inductive construction of the inner product in the algorithm, we remove the $\mathcal{O}(k)$ flops in each step connected to the calculation of the inner product.

The only remaining slow step is the computation of the vector $\mathbf{z}_k$, and hence $\mathbf{y}_k$, in each step of the algorithm. Let us denote the solution of the kth Yule-Walker-like type of equations with $\mathbf{y}_k^T = [\mathbf{z}_k^T \ \alpha_k]$. In fact, we calculate consecutively the following solutions $\mathbf{y}_i$ (where the f_i's denote different factors $f_i = (1 + u_i\alpha_i)$):

$$\mathbf{y}_1 = \alpha_1,$$
$$\mathbf{y}_2^T = [\mathbf{z}_2, \ \alpha_2] = [f_2\mathbf{y}_1, \ \alpha_2] = [f_2\alpha_1, \alpha_2],$$
$$\mathbf{y}_3^T = [\mathbf{z}_3^T, \ \alpha_3] = [f_3\mathbf{y}_2^T, \ \alpha_3] = [f_3 f_2\alpha_1, f_3\alpha_2, \alpha_3],$$
$$\vdots$$
$$\mathbf{y}_n^T = [\mathbf{z}_n^T, \ \alpha_3] = [f_n\mathbf{y}_{n-1}^T, \ \alpha_n] = [f_n f_{n-1} \dots f_2\alpha_1, f_n \dots f_3\alpha_2, \dots, \alpha_n].$$

Instead of calculating now in each step the vector $\mathbf{y}_i$ and $\mathbf{z}_i$, we will store the factors f_i and the numbers α_i. Storing only these two numbers at each step of the algorithm, we achieve a constant complexity for solving a kth order Yule-Walker-like equations. The only remaining problem is the computation in order $\mathcal{O}(n)$ of the solution vector $\mathbf{y}_n$. This is done rather easily by starting the construction of the vector $\mathbf{y}_n$ from bottom to top and by updating in each step a certain factor. This is a rather straightforward computation and is therefore not included. The complete algorithm is given in the next subsection.

The algorithm

Two separate loops are needed for solving the Yule-Walker-like problem. One loop "solves" the kth order Yule-Walker-like equations. We say "solves" because, in fact, the solution is not calculated; only the factors determining the solution are calculated. A second loop will construct the solution, given the two sequences of numbers constructed by the first loop.

Given the generators $\mathbf{u}, \mathbf{v}$ and the diagonal $\mathbf{d}$ of the symmetric positive definite semiseparable matrix $S + D$, the following algorithm computes the solution $\mathbf{y}$ such that $(S+D)\mathbf{y} = -\mathbf{v}$. Two extra vectors are needed for the algorithm. In the vector $\boldsymbol{\alpha}$ we will store the separate values of the parameter α_k, and in the vector $\mathbf{f}$, the consecutive factors f_i.

Algorithm 6.1. *Initialize*

$$\alpha_1 = -v_1/(d_1 + u_1 v_1)$$
$$f_1 = 1$$
$$\mathbf{r}_1^T \mathbf{y}_1 = \alpha_1 v_1$$

For $k = 2, \ldots, n$ do:

1. *Compute and store the following variable:*

$$t = u_k \mathbf{r}_{k-1}^T \mathbf{y}_{k-1} + v_k$$

2. $\alpha_k = -t/(u_k t + d_k)$

3. $f_k = 1 + u_k \alpha_k$

4. $\mathbf{r}_k^T \mathbf{y}_k = f_k \mathbf{r}_{k-1}^T \mathbf{y}_{k-1} + v_k \alpha_k$

endfor;

Compute now the solution

$$y_n = \alpha_n$$
$$h = f_n$$

For $k = n - 1, \ldots, 1$ do:

1. $y_k = h \alpha_k$;

2. $h = h f_k$;

endfor;

So the overall cost is $12n - 8$ operations[9] for solving the Yule-Walker-like system of equations for positive definite symmetric semiseparable plus diagonal matrices.

---

[9] We did not count the changing of a sign as an operation.

6.2.3 A Levinson-type algorithm

Based on the results of the previous section, a Levinson-like solver for solving semi-separable plus diagonal systems of equations will be constructed here. We would like to solve the system in a way similar to the previous subsection:

$$(S + D)\mathbf{x} = \mathbf{b},$$

where $\mathbf{b}$ is an arbitrary right-hand side.

The method

Let us assume that we know the solutions of the $(k + 1)$th order Yule-Walker-like equations. Use the same notations as in Section 6.2.2. Moreover assume that we also know the solution of the system:

$$(S_k + D_k)\mathbf{x}_k = \mathbf{b}_k,$$

where $\mathbf{b}_k^T = [b_1, \ldots, b_k]$.

The system we would like to solve now in $\mathcal{O}(k)$ operations (in the next subsection we reduce this cost to a constant amount of flops) is the following one:

$$(S_{k+1} + D_{k+1})\,\mathbf{x}_{k+1} = \mathbf{b}_{k+1}$$

$$\left[\begin{array}{c|c} S_k + D_k & u_{k+1}\mathbf{r}_k \\ \hline u_{k+1}\mathbf{r}_k^T & u_{k+1}v_{k+1} + d_{k+1} \end{array}\right] \left[\begin{array}{c} \mathbf{w}_{k+1} \\ \mu_{k+1} \end{array}\right] = \left[\begin{array}{c} \mathbf{b}_k \\ b_{k+1} \end{array}\right]. \tag{6.9}$$

The idea is identical to the one in the previous section. Expanding Formula (6.9) leads to the following two equations

$$(S_k + D_k)\,\mathbf{w}_{k+1} + \mu_{k+1}\,u_{k+1}\,\mathbf{r}_k = \mathbf{b}_k \tag{6.10}$$

and

$$u_{k+1}\,\mathbf{r}_k^T\,\mathbf{w}_{k+1} + \mu_{k+1}\,(u_{k+1}\,v_{k+1} + d_{k+1}) = b_{k+1}. \tag{6.11}$$

Equation (6.10) can be rewritten towards $\mathbf{w}_{k+1}$:

$$\mathbf{w}_{k+1} = \mathbf{x}_k + \mu_{k+1}\,u_{k+1}\,\mathbf{y}_k.$$

Substituting the equation for $\mathbf{w}_{k+1}$ into Equation (6.11) and rewriting this equation leads to the following expression for μ_{k+1}:

$$\mu_{k+1} = \frac{b_{k+1} - u_{k+1}\mathbf{r}_k^T\mathbf{x}_k}{(u_{k+1}^2\mathbf{r}_k^T\mathbf{y}_k + u_{k+1}v_{k+1} + d_{k+1})}.$$

The denominator is always positive (see Section 6.2.2). Again one can clearly see that there are two computations in every step which will lead to a complexity of $\mathcal{O}(k)$, namely the computation of the vector $\mathbf{w}_{k+1}$ and the computation of the inner product $\mathbf{r}_k^T\mathbf{x}_k$. The heavy computational cost of computing $\mathbf{r}_k^T\mathbf{y}_k$ was solved in the previous subsection. Again we can simplify the calculation of the inner product, and we can postpone the calculation of the solution vector up to the very end.

Reduction of the complexity

We will clarify how we can reduce the complexity in the previous step to come to a step which takes a constant amount of arithmetic work.

Firstly we will derive a recursion for calculating the inner product $\mathbf{r}_k^T \mathbf{x}_k$. Suppose we just solved the $(k+1)$th Levinson system of equations, and we start now solving the $(k+2)$nd system. Let us denote the solution and the right-hand side of the $(k+1)$th system as follows: $(S_{k+1} + D_{k+1})\mathbf{x}_{k+1} = \mathbf{b}_{k+1}$, with $\mathbf{r}_{k+1}^T = [\mathbf{r}_k^T, v_{k+1}]$. For the new $(k+2)$nd system we need the inner product $\mathbf{r}_{k+1}^T \mathbf{x}_{k+1}$. This corresponds to the inner product:

$$\mathbf{r}_{k+1}^T \mathbf{x}_{k+1} = [\mathbf{r}_k^T, v_{k+1}] \begin{bmatrix} \mathbf{w}_{k+1} \\ \mu_{k+1} \end{bmatrix}.$$

As in the previous section we can rewrite this in terms of the previous inner product $\mathbf{r}_k^T \mathbf{x}_k$ plus the inner product $\mathbf{r}_k^T \mathbf{y}_k$:

$$\mathbf{r}_{k+1}^T \mathbf{x}_{k+1}^T = \mathbf{r}_k^T \mathbf{x}_k + \mu_{k+1}(u_{k+1} \mathbf{r}_k^T \mathbf{y}_k + v_{k+1}).$$

Using this recursive formula already decreases the amount of work in one step.

Instead of constructing the intermediate solutions of all the kth order Levinson equations, one can construct the overall solution in $\mathcal{O}(n)$ operations at the end of the algorithm. This construction is similar to the one described for solving the Yule-Walker-like equations. We will briefly illustrate it. Let us denote with $\mathbf{x}_k$ the solution of the kth order Levinson equation; similarly we denote with μ_k the last component of $\mathbf{x}_k$. The g_i's denote factors. We illustrate the computation for a matrix of dimension 4×4. We get the following equations with $g_i = \mu_i u_i$:

$$\mathbf{x}_1 = \mu_1,$$
$$\mathbf{x}_2^T = [\mathbf{x}_1 + g_2 \mathbf{y}_1, \mu_2],$$
$$\mathbf{x}_3^T = [\mathbf{x}_2^T + g_3 \mathbf{y}_2^T, \mu_3],$$
$$\mathbf{x}_4^T = [\mathbf{x}_3^T + g_4 \mathbf{y}_3^T, \mu_4].$$

Expanding everything now towards vector x_4 gives us the following vector:

$$\mathbf{x}_4 = \begin{bmatrix} \mu_1 + (g_2 + f_2(g_3 + f_3 g_4))\alpha_1 \\ \mu_2 + (g_3 + f_3 g_4)\alpha_2 \\ \mu_3 + g_4 \alpha_3 \\ \mu_4 \end{bmatrix}.$$

The latter equation gives a clear clue on how to compute the solution in linear time. This means that for a fast construction of the solution we need to store 4 vectors, namely 2 vectors which store the values of α_i and μ_i and two more vectors for storing the values of the factors f_i and g_i.

The algorithm

Let us denote with $\mathbf{u}$ and $\mathbf{v}$ the generators of the semiseparable matrix. The vector $\mathbf{d}$ contains the diagonal elements and $\mathbf{b}$ is an arbitrary right-hand side. The following algorithm solves the following system of equations $(S + D)\mathbf{x} = \mathbf{b}$, using thereby the Yule-Walker-like results as developed in the previous section. We have to store now 4 extra vectors for recalculating the solution, namely $\boldsymbol{\alpha}$ and $\boldsymbol{\mu}$ for storing the values of α_i and μ_i in each step and $\mathbf{f}$ and $\mathbf{g}$ for storing the values of f_i and g_i in each step.

Algorithm 6.2. *Initialization:*

$$denominator = d_1 + u_1 v_1$$
$$\alpha_1 = -v_1 / denominator$$
$$\mu_1 = b_1 / denominator$$
$$f_1 = 1$$
$$g_1 = 1$$
$$\mathbf{r}_1^T \mathbf{y}_1 = \alpha_1 v_1$$
$$\mathbf{r}_1^T \mathbf{x}_1 = \mu_1 v_1$$

For $k = 2, \ldots, n$ do:

1. *Compute and store the following variables:*

$$t = u_k \mathbf{r}_{k-1}^T \mathbf{y}_{k-1} + v_k$$
$$denominator = (u_k t + d_k)$$

2. $\alpha_k = -t / denominator$

3. $f_k = (1 + u_k a_k)$

4. $\mu_k = (b_k - u_k \mathbf{r}_{k-1}^T \mathbf{x}_{k-1}) / denominator$

5. $g_k = \mu_k u_k$

6. $\mathbf{r}_k^T \mathbf{x}_k = \mathbf{r}_{k-1}^T \mathbf{x}_{k-1} + \mu_k t$

7. $\mathbf{r}_k^T \mathbf{y}_k = a_k t + \mathbf{r}_{k-1}^T \mathbf{y}_{k-1}$

endfor;
Compute now the solution:

$$x_n = \mu_n$$
$$h = g_n$$

For $k = n - 1, \ldots, 1$ do:

1. $x_k = \mu_k + h a_k$

2. $h = g_k + f_k h$

end

An easy calculation reveals that the complexity of this solver is $19n - 13$ operations.

6.2.4 An upper triangular factorization of the inverse

Similarly to the Toeplitz case, the Durbin algorithm presented here is closely related to an upper triangular factorization of the inverse of the semiseparable plus diagonal matrix.

Let us denote the solution of the kth order Yule-Walker-like equation as

$$(S_k + D_k)\mathbf{y}_k = -\mathbf{r}_k. \tag{6.12}$$

This equation can be rewritten towards the matrix $S_{k+1} + D_{k+1}$, by bringing the factor r_k to the other side. The system (6.12) is similar to

$$(S_{k+1} + D_{k+1}) \begin{bmatrix} \mathbf{y}_k \\ 1/u_{k+1} \end{bmatrix} = \left[\begin{array}{c|c} S_k + D_k & u_{k+1}\mathbf{r}_k \\ \hline u_{k+1}\mathbf{r}_k^T & u_{k+1}v_{k+1} + d_{k+1} \end{array} \right] \begin{bmatrix} \mathbf{y}_k \\ 1/u_{k+1} \end{bmatrix}$$

$$= \begin{bmatrix} 0 \\ \sigma_{k+1} \end{bmatrix}.$$

If we put the successive equations above in an upper triangular matrix we get (let us denote with $y_{k,j}$ the successive components in the vector $\mathbf{y}_k$):

$$(S_n + D_n) \begin{bmatrix} 1/u_1 & y_{1,1} & y_{2,1} & \cdots & y_{n-1,1} \\ 0 & 1/u_2 & y_{2,2} & \cdots & y_{n-1,2} \\ 0 & 0 & 1/u_3 & & y_{n-1,3} \\ \vdots & \vdots & \ddots & \ddots & \vdots \\ 0 & 0 & \cdots & 0 & 1/u_n \end{bmatrix} = \begin{bmatrix} \sigma_1 & 0 & 0 & \cdots & 0 \\ \times & \sigma_2 & 0 & \cdots & 0 \\ \vdots & \times & \sigma_3 & & 0 \\ \vdots & \vdots & & \ddots & \vdots \\ \times & \times & \cdots & \times & \sigma_n \end{bmatrix}.$$

The $\times$ denote arbitrary elements in the matrix. Rewriting the equations above and defining the upper triangular matrix to the right of $(S_n + D_n)$ as U and the right-hand side as L, we get the following equations ($\hat{L}$ is another lower triangular matrix):

$$(S + D)U = L$$
$$(S + D)U = U^{-T}\hat{L}$$
$$U^T(S + D)U = \hat{L}.$$

Because $\hat{L}$ is lower triangular, and because of symmetry, the matrix $\hat{L}$ is a diagonal matrix Λ. Inverting the equation above gives:

$$U^{-1}(S + D)^{-1}U^{-T} = \Lambda^{-1}$$
$$(S + D)^{-1} = U\Lambda^{-1}U^T$$

which gives us a triangular factorization of the inverse of the semiseparable plus diagonal matrix.

6.2.5   Some general remarks

The results as presented in this section are formulated for positive definite symmetric semiseparable plus diagonal matrices. The implementations as given here are based on the generator representation. This representation is suitable here for solving systems of equations. If, however, one wants to use this upper triangular decomposition as presented here for an iterative algorithm (e.g., for computing the spectrum), one should be aware of numerical instabilities that can occur. More information on possible problems with this representation were mentioned in Chapter 2.

Notes and references

This section was based mainly on the article.

> ☞ N. Mastronardi, M. Van Barel, and R. Vandebril. A Levinson-like algorithm for symmetric positive definite semiseparable plus diagonal matrices. Technical Report TW423, Department of Computer Science, Katholieke Universiteit Leuven, Celestijnenlaan 200A, 3000 Leuven (Heverlee), Belgium, March 2005.

This paper includes the Levinson algorithm for generator representable semiseparable plus diagonal matrices, as well as numerical experiments related to the software proposed in this section.

6.3   A Levinson framework

The method presented in this section is also based on the Levinson algorithm. A class of matrices called $\{p_1, p_2\}$-Levinson conform is defined, which will admit a Levinson-like algorithm. In this section we focus to a specific subclass, called simple $\{p_1, p_2\}$-Levinson conform. This class is called simple, because we will prove that these matrices admit a solver of complexity $\mathcal{O}(p_1 p_2 n)$, which is in fact linear in time. Matrices, such as Toeplitz or Hankel, are not incorporated in the class of simple $\{p_1, p_2\}$-Levinson conform matrices, merely because the focus of the book is dedicated to rank structured matrices. However, as will be shown in this chapter, several classes of matrices do belong to this class and hence admit an order $\mathcal{O}(p_1 p_2 n)$ solver. For example, the matrices considered in [212, 288] fit in this framework and are given as an example.

   The algorithmic idea is exactly the same as for solving Toeplitz systems via the Levinson algorithm. First n systems with a special right-hand side, called the Yule-Walker-like equations, need to be solved. Based on these solutions, we can then solve the linear system with an arbitrary right-hand side.

   This section is organized as follows. In Subsection 6.3.1 the class of simple $\{p_1, p_2\}$-Levinson conform matrices is defined. In fact this is just based on a special block decomposition of the matrix involved. In Subsection 6.3.2, the algorithm for solving simple $\{p_1, p_2\}$-Levinson matrices is presented. First a direct way to solve systems in this manner is given having complexity $\mathcal{O}(p_1 p_2 n^2)$. It will be shown however how one can further reduce this complexity to come to a method which costs $\mathcal{O}(p_1 p_2 n)$ operations. It is therefore, important to choose the factors p_1 and p_2 as small as possible. In Subsection 6.3.3, we investigate the upper triangular factorization related to this method.

6.3.1   The matrix block decomposition

In this section we will develop a Levinson-like solver for structured systems of equations. In order to develop such a solver, our coefficient matrix should admit a specific partitioning of the matrix, making it able to derive the recursive algorithm.

Let us therefore introduce the notion of (simple) $\{p_1\}$-Levinson and (simple) $\{p_1, p_2\}$-Levinson conform matrices.

Definition 6.3 ($\{p_1\}$-Levinson conform matrices). *Given a matrix $A = (a_{i,j})$, for $i, j = 1, \ldots, n$, and denote with $A_k = (a_{i,j})$, for $i, j = 1, \ldots, k$ the upper $k \times k$ submatrix of A. The matrix A is said to be $\{p_1\}$-Levinson conform if the matrix can be decomposed in the following way:*

1. *For every $1 \leq k \leq n - 1$, there is a splitting in blocks of the matrix A_{k+1} of the following form:*

$$A_{k+1} = \left[\begin{array}{c|c} A_k & E_k R_k \mathbf{c}_{k+1} \\ \hline \mathbf{d}_{k+1}^T S_k^T F_k^T & a_{k+1,k+1} \end{array} \right]$$

   where $\mathbf{c}_{k+1} \in \mathbb{R}^{p_1}$, $\mathbf{d}_{k+1} \in \mathbb{R}^{p_2}$, $R_k \in \mathbb{R}^{k \times p_1}$, $F_k, E_k \in \mathbb{R}^{k \times k}$ $S_k \in \mathbb{R}^{k \times p_2}$ and $A_k \in \mathbb{R}^{k \times k}$.

2. *The following relation for the matrices R_{k+1} (with $1 \leq k \leq n-1$) needs to be satisfied:*

$$R_{k+1} = \left[\begin{array}{c} R_k P_k \\ \hline \boldsymbol{\xi}_{k+1}^T \end{array} \right],$$

   where P_k is a matrix of dimension $p_1 \times p_1$ and $\boldsymbol{\xi}_{k+1}$ is a column vector of length p_1.

We call the matrix simple $\{p_1\}$-Levinson conform if the matrix E_k equals the identity matrix of order k and the multiplication of a vector with the matrix P_k can be done in linear time (i.e., only $\mathcal{O}(p_1)$ operations are involved).

No conditions were placed on the matrix S_k. If we put similar conditions on S_k as on the matrix R_k we call the matrix $\{p_1, p_2\}$-Levinson conform.

Definition 6.4 ($\{p_1, p_2\}$-Levinson conform matrices). *A matrix A is called $\{p_1, p_2\}$-Levinson conform, if the matrix is $\{p_1\}$-Levinson conform, i.e., that Conditions (1) and (2) from Definition 6.3 are fulfilled and the following condition for the matrices S_k is satisfied:*

3. *The matrices S_{k+1} (with $1 \leq k \leq n-1$) can be decomposed as:*

$$S_{k+1} = \left[\frac{S_k Q_k}{\boldsymbol{\eta}_{k+1}^T} \right],$$

where Q_k is a matrix of dimension $p_2 \times p_2$ and $\boldsymbol{\eta}_{k+1}$ is a column vector of length p_2.

We call a matrix simple $\{p_1, p_2\}$-Levinson conform, if both the matrices E_k and F_k are equal to the identity matrix of order k and the multiplication of a vector with the matrices P_k and Q_k can be performed in $\mathcal{O}(p_1)$ and $\mathcal{O}(p_2)$ operations, respectively.

In fact, every matrix is simple Levinson conform.

Lemma 6.5. *Suppose an arbitrary matrix $A \in \mathbb{R}^{n \times n}$ is given; the matrix A is simple $\{n-1, n-1\}$-Levinson conform.*

Proof. The proof is straightforward. Define the $k \times (n-1)$ matrices R_k and S_k as follows:

$$R_k = S_k = [I_k, 0]$$

and assume for every k the matrices E_k, F_k, P_k and Q_k to be equal to the identity matrix. Defining

$$\mathbf{d}_k^T = [a_{k,1}, a_{k,2}, \dots, a_{k,k-1}, 0, \dots, 0],$$
$$\mathbf{c}_k^T = [a_{1,k}, a_{2,k}, \dots, a_{k-1,k}, 0, \dots, 0],$$

with both vectors of length $n-1$. One can easily check that the conditions of Definition 6.4 are satisfied. $\qquad\square$

The lemma also shows that the choice of $\mathbf{c}_k, \mathbf{d}_k, R_k$ and Q_k is not always unique. But, the notion of simple $\{p_1, p_2\}$-Levinson conformity is strongly related to the complexity of the method we will deduce in this section. The overall solver will have a complexity $\mathcal{O}(p_1 p_2 n)$. Hence it is important to keep the values of p_1 and p_2 as low as possible. Assuming we have a complexity of $\mathcal{O}(p_1 p_2 n)$, in general, for $\{n-1, n-1\}$-Levinson conform matrices this leads to a complexity of $\mathcal{O}(n^3)$.

From now on we will only focus on Levinson conform matrices for which E_k and F_k are equal to the identity matrix of size k. This excludes in some sense important classes of matrices, such as Toeplitz, Hankel and Vandermonde matrices. In the simple formulation these matrices are $\{n-1, n-1\}$-Levinson conform, whereas omitting the assumption of being simple leads to a $\{1, 1\}$-Levinson conform matrix. In this section, however, we will restrict ourselves to the class of simple Levinson conform matrices. This class is already wide enough to admit different types of structures as will be shown in the next section, which includes examples. More information on efficient solvers for Toeplitz, Hankel and Vandermonde matrices, based on their displacement representation, can be found, for example, in [149].

6.3.2   Simple $\{p_1, p_2\}$-Levinson conform matrices

In this section we will construct a Levinson-like solver for solving strongly nonsingular linear systems of equations for which the coefficient matrix is simple $\{p_1, p_2\}$-Levinson conform. The limitation of being strongly nonsingular can be relaxed (see Subsection 6.3.4 on the look-ahead procedure). In this section we will first solve the corresponding Yule-Walker-like systems. The solution of these equations will be used for solving the general system of equations, with an arbitrary right-hand side, based on the Levinson method. We now present consecutively a possible algorithm, followed by complexity-reducing remarks and finally an $\mathcal{O}(p_1 p_2 n)$ method, with a detailed complexity count.

The Yule-Walker-like system

Suppose a simple $\{p_1\}$-Levinson conform matrix A is given. The aim of the Yule-Walker step is to solve the following system of equations $AY_n = -R_n$. The system will be solved by induction. Let us assume we know the solution of the kth order Yule-Walker-like problem (with $1 \leq k \leq n-1$):

$$A_k Y_k = -R_k, \tag{6.13}$$

and we would like to compute the solution of the $(k+1)$th Yule-Walker-like problem. (Note that, in general, Y_k represents a matrix of dimension $k \times p_1$.) The $(k+1)$th system of equations is of the form:

$$A_{k+1} Y_{k+1} = -R_{k+1}.$$

Using the initial conditions put on the matrix A, we can rewrite the equation above as

$$\left[\begin{array}{c|c} A_k & R_k \mathbf{c}_{k+1} \\ \hline \mathbf{d}_{k+1}^T S_k^T & a_{k+1,k+1} \end{array}\right] \left[\begin{array}{c} Z_{k+1} \\ \hline \boldsymbol{\alpha}_{k+1}^T \end{array}\right] = -\left[\begin{array}{c} R_k P_k \\ \hline \boldsymbol{\xi}_{k+1}^T \end{array}\right],$$

with $Z_{k+1} \in \mathbb{R}^{k \times p_1}$ and $\boldsymbol{\alpha}_{k+1} \in \mathbb{R}^{p_1}$. Expanding this equation towards its block-rows, we observe that

$$A_k \, Z_{k+1} + R_k \, \mathbf{c}_{k+1} \, \boldsymbol{\alpha}_{k+1}^T = -R_k P_k, \tag{6.14}$$

and

$$\mathbf{d}_{k+1}^T \, S_k^T Z_{k+1} + a_{k+1,k+1} \boldsymbol{\alpha}_{k+1}^T = -\boldsymbol{\xi}_{k+1}^T. \tag{6.15}$$

Rewriting Equation (6.14) towards Z_{k+1} and using the solution of Equation (6.13) gives

$$\begin{aligned} Z_{k+1} &= -A_k^{-1} R_k \left(P_k + \mathbf{c}_{k+1} \, \boldsymbol{\alpha}_{k+1}^T\right) \\ &= Y_k \left(P_k + \mathbf{c}_{k+1} \, \boldsymbol{\alpha}_{k+1}^T\right). \end{aligned} \tag{6.16}$$

Substituting the latter equation in Equation (6.15) leads to:

$$\mathbf{d}_{k+1}^T \, S_k^T \, Y_k(P_k + \mathbf{c}_{k+1}\boldsymbol{\alpha}_{k+1}^T) + a_{k+1,k+1}\boldsymbol{\alpha}_{k+1}^T = -\boldsymbol{\xi}_{k+1}^T,$$

from which we can extract $\boldsymbol{\alpha}_{k+1}$ as:

$$\boldsymbol{\alpha}_{k+1}^T = -\frac{\boldsymbol{\xi}_{k+1}^T + \mathbf{d}_{k+1}^T S_k^T Y_k P_k}{\mathbf{d}_{k+1}^T S_k^T Y_k \mathbf{c}_{k+1} + a_{k+1,k+1}}.$$

Using now the vector $\boldsymbol{\alpha}_{k+1}$ in Equation (6.16), we can compute the matrix Z_{k+1}. Based on the formulas for $\boldsymbol{\alpha}_{k+1}$ and Z_{k+1}, we can immediately derive a recursive algorithm for solving the Yule-Walker-like problems.

To conclude, we prove that the denominator in the formula for $\boldsymbol{\alpha}_{k+1}$ is always nonzero, i.e., that the computation of $\boldsymbol{\alpha}_{k+1}$ is well defined. Because our matrix A is assumed to be strongly nonsingular, we know that all the leading principal matrices are nonsingular. This means that for every nonzero vector $\mathbf{w}$: $A_{k+1}\mathbf{w} \neq 0$. Taking now $\mathbf{w}^T = [\mathbf{c}_{k+1}^T Y_k^T, 1]$, we have that:

$$\begin{bmatrix} A_k & R_k\mathbf{c}_{k+1} \\ \hline \mathbf{d}_{k+1}^T S_k^T & a_{k+1,k+1} \end{bmatrix} \begin{bmatrix} Y_k\mathbf{c}_{k+1} \\ 1 \end{bmatrix}$$
$$= \begin{bmatrix} A_k Y_k\mathbf{c}_{k+1} + R_k\mathbf{c}_{k+1} \\ \mathbf{d}_{k+1}^T S_k^T Y_k\mathbf{c}_{k+1} + a_{k+1,k+1} \end{bmatrix}$$
$$\neq 0.$$

Using the fact that $A_k Y_k = -R_k$, we obtain that the first k entries of the vector above are zero. As the total vector needs to be different from zero we have that:

$$\mathbf{d}_{k+1}^T S_k^T Y_k\mathbf{c}_{k+1} + a_{k+1,k+1} \neq 0,$$

which states that the calculation of $\boldsymbol{\alpha}_{k+1}$ is well defined.

Based on these results we will now derive the Levinson-like algorithm for solving simple $\{p_1\}$-Levinson conform systems with an arbitrary right-hand side.

The Levinson-like solver

Here a Levinson-like method is proposed for solving systems of equations for which the coefficient matrix A is simple $\{p_1\}$-Levinson conform. The presented solver uses the solution of all the kth order Yule-Walker-like problems and it is based on an inductive procedure.

Suppose a matrix A that is simple $\{p_1\}$-Levinson conform is given; use the notation from Definition 6.3 and the one from the part on the Yule-Walker solver. We would like to compute a solution $\mathbf{x}$ for the following system of equations $A\mathbf{x} = \mathbf{b}$, where $\mathbf{b}^T = [b_1, \ldots, b_n]$ is a general right-hand side. We also assume the matrix A to be strongly nonsingular. As already mentioned, further in the text we will omit this strongly nonsingularity condition.

Assume we know the solution of the kth order Yule-Walker system:

$$A_k Y_k = -R_k, \tag{6.17}$$

and the solution of:

$$A_k\mathbf{x}_k = \mathbf{b}_k, \tag{6.18}$$

where $\mathbf{b}_k^T = [b_1, \ldots, b_k]$. We will now solve $A_{k+1}\ \mathbf{x}_{k+1} = \mathbf{b}_{k+1}$, based on the Equations (6.17) and (6.18).

The system we would like to solve can be rewritten in block form as:

$$\left[\begin{array}{c|c} A_k & R_k\mathbf{c}_{k+1} \\ \hline \mathbf{d}_{k+1}^T S_k^T & a_{k+1,k+1} \end{array}\right] \left[\begin{array}{c} \mathbf{w}_{k+1} \\ \hline \mu_{k+1} \end{array}\right] = \left[\begin{array}{c} \mathbf{b}_k \\ \hline b_{k+1} \end{array}\right], \qquad (6.19)$$

with $\mathbf{w}_{k+1} \in \mathbb{R}^{k \times 1}$ and μ_{k+1} a scalar.

Expanding Formula (6.19) leads to the following two equations

$$A_k\ \mathbf{w}_{k+1} + \mu_{k+1}\ R_k\mathbf{c}_{k+1} = \mathbf{b}_k \qquad (6.20)$$

and

$$\mathbf{d}_{k+1}^T\ S_k^T\ \mathbf{w}_{k+1} + \mu_{k+1}\ a_{k+1,k+1} = b_{k+1}. \qquad (6.21)$$

Equation (6.20) can be rewritten towards $\mathbf{w}_{k+1}$, thereby using $A_k^{-1}\mathbf{b}_k = \mathbf{x}_k$ and $A_k^{-1}(-R_k) = Y_k$ we get:

$$\begin{aligned} \mathbf{w}_{k+1} &= A_k^{-1}\left(\mathbf{b}_k - \mu_{k+1}R_k\mathbf{c}_{k+1}\right) \\ &= \mathbf{x}_k + \mu_{k+1}\ Y_k\mathbf{c}_{k+1}. \end{aligned}$$

Substituting the solution for $\mathbf{w}_{k+1}$ into Equation (6.21) and rewriting this leads to the following expression for μ_{k+1}:

$$\mu_{k+1} = \frac{b_{k+1} - \mathbf{d}_{k+1}^T S_k^T \mathbf{x}_k}{\left(\mathbf{d}_{k+1}^T S_k^T Y_k \mathbf{c}_{k+1} + a_{k+1,k+1}\right)}.$$

The formula for μ_{k+1} is well defined as the denominator is always different from zero. Using the relations for computing μ_{k+1} and $\mathbf{w}_{k+1}$ one can immediately derive a recursive formula for computing the solution. We remark that for solving the system of equations with this Levinson-like algorithm, the solution of the nth Yule-Walker-like equation is not needed; hence, we do not necessarily need to define the matrix R_n. In the next part, this algorithm and the operation count is presented.

A first algorithm of complexity $\mathcal{O}(p_1 p_2 n^2)$

Based on the previous results we can present a first version of a Levinson-like solver for simple $\{p_1\}$-Levinson conform matrices.

The algorithm is given in a simple mathematical formulation. First the problem is initialized and then the main loop is performed. After each computation in the algorithm, the number of flops involved is shown. We remark that the presented algorithm is not the most efficient implementation. It is written in this way to clearly see the computationally most expensive steps. For the operation count, we assume that the multiplication of a row vector with any of the matrices P_k has a flop count bounded by $\kappa_1 p_1 + \kappa_2$, with κ_1 and κ_2 two constants.

Algorithm 6.6. *Initialize*

$$Y_1 = \boldsymbol{\alpha}_1^T = \frac{-R_1}{a_{1,1}}$$

$$\mathbf{x}_1 = \mu_1 = \frac{b_1}{a_{1,1}}$$

For $k = 1, \ldots, n-1$ do

1. *Compute and store the following variables:*

   (a) $S_k^T Y_k$ *Flops: $p_1 p_2 (2k-1)$*

   (b) $S_k^T \mathbf{x}_k$ *Flops: $p_2 (2k-1)$*

   (c) $\mathbf{d}_{k+1}^T (S_k^T Y_k)$ *Flops: $p_1 (2p_2 - 1)$*

   (d) $(\mathbf{d}_{k+1}^T S_k^T Y_k \mathbf{c}_{k+1} + a_{k+1,k+1})$ *Flops: $2p_1$*

2. $\boldsymbol{\alpha}_{k+1}^T = -\dfrac{\boldsymbol{\xi}_{k+1}^T + \mathbf{d}_{k+1}^T S_k^T Y_k P}{\mathbf{d}_{k+1}^T S_k^T Y_k \mathbf{c}_{k+1} + a_{k+1,k+1}}$ *Flops: $2p_1 + \kappa_1 p_1 + \kappa_2$*

3. $Z_{k+1} = Y_k \left(P_k + \mathbf{c}_{k+1} \, \boldsymbol{\alpha}_{k+1}^T \right)$ *Flops: $k(\kappa_1 p_1 + \kappa_2) + k(2p_1 - 1) + 2p_1 k$*

4. $\mu_{k+1} = \dfrac{b_{k+1} - \mathbf{d}_{k+1}^T S_k^T \mathbf{x}_k}{\mathbf{d}_{k+1}^T S_k^T Y_k \mathbf{c}_{k+1} + a_{k+1,k+1}}$ *Flops: $1 + 2p_2$*

5. $\mathbf{w}_{k+1} = \mathbf{x}_k + \mu_{k+1} Y_k \mathbf{c}_{k+1}$ *Flops: $2k + k(2p_1 - 1)$*

6. $Y_{k+1}^T = [Z_{k+1}^T, \boldsymbol{\alpha}_{k+1}]$ *Flops: 0*

7. $\mathbf{x}_{k+1}^T = [\mathbf{w}_{k+1}^T, \mu_{k+1}^T]$ *Flops: 0*

endfor;

     Performing an overall complexity count leads us to an algorithm of complexity $\mathcal{O}(p_1 p_2 n^2)$. This means that as long as the factor $p_1 p_2 < n$, the method will perform better than Gaussian elimination if the matrices in question are large enough. However, taking a closer look at the involved computations, we can see that the bottlenecks, causing the factor n^2 in the operation count, are the computations of the matrices $S_k^T Y_k$, and $S_k^T \mathbf{x}_k$ and the explicit formation of the matrices Z_{k+1} and vectors $\mathbf{w}_{k+1}$. Assume now that one could remove the computation of Z_{k+1} and $\mathbf{w}_{k+1}$ from the most inner loop and compute the final solution, using only the stored values of $\boldsymbol{\alpha}_k$ and μ_k in $\mathcal{O}(n)$ operations (dependent on p_1 and p_2, however). Assume also that one could compute the products $S_k^T \mathbf{x}_k$ and $S_k^T Y_k$ in constant time, independent of k. This would lead to the following algorithm.

Algorithm 6.7. *Initialize*

$$Y_1 = \boldsymbol{\alpha}_1^T = \frac{-R_1}{a_{1,1}}$$

$$\mathbf{x}_1 = \mu_1 = \frac{b_1}{a_{1,1}}$$

For $k = 1, \ldots, n-1$ do the same initializations:

    1. *Compute and store the following variables:*

        (a) $S_k^T Y_k$ *Flops: Independent of k*

        (b) $S_k^T \mathbf{x}_k$ *Flops: Independent of k*

        (c) $(\mathbf{d}_{k+1}^T S_k^T Y_k \mathbf{c}_{k+1} + a_{k+1,k+1})$ *Flops: $p_1(2p_2 - 1) + 2p_1$*

    2. $\boldsymbol{\alpha}_{k+1}^T = -\dfrac{\boldsymbol{\xi}_{k+1}^T + \mathbf{d}_{k+1}^T S_k^T Y_k P}{\mathbf{d}_{k+1}^T S_k^T Y_k \mathbf{c}_{k+1} + a_{k+1,k+1}}$ *Flops: $1 + p_1 + \kappa_1 p_1 + \kappa_2$*

    3. $\mu_{k+1} = \dfrac{b_{k+1} - \mathbf{d}_{k+1}^T S_k^T \mathbf{x}_k}{\mathbf{d}_{k+1}^T S_k^T Y_k \mathbf{c}_{k+1} + a_{k+1,k+1}}$ *Flops: $1 + 2p_2$*

endfor;

    Under these assumptions, the solver has a complexity: $\mathcal{O}(p_1 p_2 n)$.

    Next, we illustrate how to achieve the complexity above by computing the solution in another loop and by computing the products $S_k^T Y_k$ and $S_k^T \mathbf{x}_k$ in an inductive way, thereby making the computational complexity independent of k.

Reduction of the complexity

We know that the computations of the matrices $Y_k^T = [Z_k^T, \boldsymbol{\alpha}_k]$, the vectors $\mathbf{x}_k^T = [\mathbf{w}_k^T, \mu_k]$ and the computations of the matrices $S_k^T \mathbf{x}_k$ and $S_k^T Y_k$ in every step of the algorithm are responsible for the n^2 factor in the complexity count. If we could reduce the complexity of these operations, the overall operation count would decrease by a factor n. Let us start with the computation of the solution vector $\mathbf{x}$. Instead of computing for every k the vector $\mathbf{x}_k$, which incorporates the computation of $\mathbf{w}_k, Z_k$ and Y_k at every step, we will postpone this computation up to the very end, and simply store the factors μ_k and $\boldsymbol{\alpha}_k$ for every k. Extracting the computation of $\mathbf{x}$ out of the loop, the final solution vector can be written in the following form. (Denote with x_i the ith component of the vector $\mathbf{x} = [x_1, x_2, \ldots, x_n]^T$.)

$$
\begin{bmatrix}
\vdots \\
\mu_{n-3} + \boldsymbol{\alpha}_{n-3}^T \left(\mu_{n-2}\mathbf{c}_{n-2} + (P_{n-3} + \mathbf{c}_{n-2}\boldsymbol{\alpha}_{n-2}^T) \left(\mu_{n-1}\mathbf{c}_{n-1} + (P_{n-2} + \mathbf{c}_{n-1}\boldsymbol{\alpha}_{n-1}^T)\mu_n \mathbf{c}_n \right) \right) \\
\mu_{n-2} + \boldsymbol{\alpha}_{n-2}^T \left(\mu_{n-1}\mathbf{c}_{n-1} + (P_{n-2} + \mathbf{c}_{n-1}\boldsymbol{\alpha}_{n-1}^T)\mu_n \mathbf{c}_n \right) \\
\mu_{n-1} + \boldsymbol{\alpha}_{n-1}^T \mu_n \mathbf{c}_n \\
\mu_n
\end{bmatrix}
$$

$$
=
\begin{bmatrix}
\vdots \\
\mu_{n-3} + \boldsymbol{\alpha}_{n-3}^T \left(x_{n-2}\mathbf{c}_{n-2} + P_{n-3} \left(x_{n-1}\mathbf{c}_{n-1} + P_{n-2} x_n \mathbf{c}_n \right) \right) \\
\mu_{n-1} + \boldsymbol{\alpha}_{n-2}^T \left(x_{n-1}\mathbf{c}_{n-1} + P_{n-2} x_n \mathbf{c}_n \right) \\
\mu_{n-1} + \boldsymbol{\alpha}_{n-1}^T \left(x_n \mathbf{c_n} \right) \\
\mu_n
\end{bmatrix}
$$

The computation of the vector above can easily be rewritten in a recursive manner, as the term following the vector $\boldsymbol{\alpha}_{n-1}$ in the computation of x_{n-1} can be used in the computation of x_{n-2}. Consecutively, the term following the vector $\boldsymbol{\alpha}_{n-2}$ in the computation of x_{n-2} can be used in the computation of x_{n-3}, and so on. Implementing this in a recursive way, from bottom to top, requires $\mathcal{O}(p_1 n)$ operations for computing the solution, instead of the $\mathcal{O}(p_1 n^2)$ operations needed if it was incorporated in the main loop. The implementation of this recursion is given in Algorithm 6.9.

Next one needs to reduce the complexity of the computation of the matrices $S_k^T Y_k$ and $S_k^T \mathbf{x}_k$ in the loop. This reduction in complexity is related to the structure in the lower triangular part, as the following example shows:

Example 6.8 As a simple case consider the class of upper triangular matrices. This means that the matrix $S_k = 0$. Hence all the computations involving the matrix S_k, such as $S_k^T \mathbf{x}_k$ and $S_k^T Y^T$, are removed, thereby creating an $\mathcal{O}(p_1 n)$ solver for the upper triangular, simple $\{p_1\}$-Levinson conform matrices. ∎

In this discussion we will derive a Levinson-type solver for systems of equations. The class of matrices admitting this solver will be named simple $\{p_1, p_2\}$-Levinson conform and they admit an order $\mathcal{O}(p_1 p_2 n)$ solver. The derived solver is based on the Levinson algorithm, which is used for solving strongly nonsingular Toeplitz systems.

The solver is constructed in a similar way as the solver for Toeplitz systems: first a Yule-Walker-like equation needs to be solved, and second this solution is used for solving a linear equation with an arbitrary right-hand side.

Various examples will be presented, including different types of matrices. For example, semiseparable, quasiseparable, higher order semiseparable, band matrices, arrowhead matrices, companion matrices, summations of any of the previous matrices, etc.

We would like to place this in a more general framework, however. The class of matrices admitting this reduction in complexity is the class of simple $\{p_1, p_2\}$-Levinson conform matrices. Assume a simple $\{p_1, p_2\}$-Levinson conform matrix A is given. This means that the lower triangular part is structured in a way similar to the upper triangular part. We have that our matrices S_k satisfy the following relations (for $1 \leq k \leq n-1$):

$$S_{k+1} = \left[\begin{array}{c} S_k Q_k \\ \boldsymbol{\eta}_{k+1}^T \end{array} \right],$$

where Q_k is a matrix of dimension $p_2 \times p_2$ and $\boldsymbol{\eta}_{k+1}$ is a vector of length p_2.

Using this relation, the computation of $S_{k+1}^T Y_{k+1}$ can be rewritten in terms of the matrix $S_k^T Y_k$ (use Formula (6.16)), thereby admitting a recursive computation of these products:

$$\begin{aligned} S_{k+1}^T Y_{k+1} &= [Q_k^T S_k^T, \boldsymbol{\eta}_{k+1}] \left[\begin{array}{c} Z_{k+1} \\ \boldsymbol{\alpha}_{k+1}^T \end{array} \right] \\ &= Q_k^T S_k^T Z_{k+1} + \boldsymbol{\eta}_{k+1} \boldsymbol{\alpha}_{k+1}^T \\ &= Q_k^T S_k^T Y_k (P_k + \mathbf{c}_{k+1} \boldsymbol{\alpha}_{k+1}^T) + \boldsymbol{\eta}_{k+1} \boldsymbol{\alpha}_{k+1}^T \\ &= Q_k^T S_k^T Y_k P_k + (Q_k^T S_k^T Y_k \mathbf{c}_{k+1} + \boldsymbol{\eta}_{k+1}) \boldsymbol{\alpha}_{k+1}^T. \end{aligned}$$
$$(6.22)$$

This leads to a recursive calculation for which the computational complexity is independent of k.

In a similar way, a recursive formula for computing the products $S_k^T \mathbf{x}_k$ can be derived:

$$
\begin{aligned}
S_{k+1}^T \mathbf{x}_{k+1} &= [Q_k^T S_k^T, \boldsymbol{\eta}_{k+1}] \begin{bmatrix} \mathbf{w}_{k+1} \\ \mu_{k+1} \end{bmatrix} \\
&= Q_k^T S_k^T \mathbf{x}_k + \mu_{k+1} \left(Q_k^T S_k^T Y_k \mathbf{c}_{k+1} + \boldsymbol{\eta}_{k+1} \right).
\end{aligned}
\tag{6.23}
$$

This recursive formula for computing $S_k^T \mathbf{x}_k$ in each step of the loop is computationally independent of k.

In the first section we proved that every matrix is simple $\{n-1, n-1)\}$-Levinson conform. Solving a system of equations with a strongly nonsingular simple $\{n-1, n-1\}$-Levinson conform matrix therefore costs $\mathcal{O}(n^3)$.

We present now the algorithm for solving strongly nonsingular simple $\{p_1, p_2\}$-Levinson conform systems of equations in $\mathcal{O}(p_1 p_2 n)$ operations.

A second algorithm of complexity $\mathcal{O}(p_1 p_2 n)$

Before giving examples of matrices solvable via these Levinson-like methods, we present here the general algorithm for simple $\{p_1, p_2\}$-Levinson conform matrices. For the computation of $S_k^T Y_k$ and $S_k^T \mathbf{x}_k$ we use the recursive schemes presented in Equations (6.22) and (6.23). We know that for every k the multiplication of a vector with the matrices Q_k and/or P_k can be performed in linear time. Assume that $\kappa_1 p_1 + \kappa_2$ is an upper bound for the number of operations needed to multiply a vector with a matrix P_k, and assume $\gamma_1 p_2 + \gamma_2$ is an upper bound for the number of operations needed to multiply a vector with a matrix Q_k. The algorithm is presented below. After each computation, the number of involved operations is shown (flop count). The operation count presented here is the worst-case scenario, as we do not take into consideration other possible advantages such as sparsity in multiplications and so on. One can see this complexity count as an upper bound.

Algorithm 6.9. *Initialize*

$$
\boldsymbol{\alpha}_1^T = \frac{-R_1}{a_{1,1}} \qquad \textit{Flops: } p_1
$$

$$
\mu_1 = \frac{b_1}{a_{1,1}} \qquad \textit{Flops: } 1
$$

$$
S_1^T Y_1 = S_1^T \boldsymbol{\alpha}_1^T \qquad \textit{Flops: } p_2 p_1
$$

$$
S_1^T x_1 = S_1^T \mu_1 \qquad \textit{Flops: } p_1
$$

For $k = 1, \ldots, n-1$ do:

1. *Compute and store the following variable:*

   (a) $(\mathbf{d}_{k+1}^T S_k^T Y_k)$ *Flops: $(2p_2 - 1)p_1$*

(b) $(\mathbf{d}_{k+1}^T S_k^T Y_k \mathbf{c}_{k+1} + a_{k+1,k+1})$ $Flops:\ 2p_1$

2. $\alpha_{k+1}^T = -\dfrac{\boldsymbol{\xi}_{k+1} + \mathbf{d}_{k+1}^T S_k^T Y_k P_k}{\mathbf{d}_{k+1}^T S_k^T Y_k \mathbf{c}_{k+1} + a_{k+1,k+1}}$ $Flops:\ 2p_1 + \kappa_1 p_1 + \kappa_2$

3. $\mu_{k+1} = \dfrac{b_{k+1} - \mathbf{d}_{k+1}^T S_k^T x_k}{\mathbf{d}_{k+1}^T S_k^T Y_k \mathbf{c}_{k+1} + a_{k+1,k+1}}$ $Flops:\ 2p_2 + 1$

4. *Compute and store the following variables (if $k < n-1$):*

   (a) $Q_k^T S_k^T Y_k c_{k+1}^T + \boldsymbol{\eta}_{k+1}$ $Flops:\ 2p_2 p_1 + \gamma_1 p_2 + \gamma_2$

   (b) $S_{k+1}^T Y_{k+1} = Q_k^T S_k^T Y_k P_k + \left(Q^T S_k^T Y_k \mathbf{c}_{k+1} + \boldsymbol{\eta}_{k+1} \right) \boldsymbol{\alpha}_{k+1}^T$

   $Flops:\ 2p_2 p_1 + p_1(\gamma_1 p_2 + \gamma_2) + p_2(\kappa_1 p_1 + \kappa_2)$

   (c) $S_{k+1}^T \mathbf{x}_{k+1} = Q_k^T S_k^T \mathbf{x}_k + \left(Q_k^T S_k^T Y_k \mathbf{c}_{k+1} + \boldsymbol{\eta}_{k+1} \right) \mu_{k+1}$

   $Flops:\ 2p_2 + \gamma_1 p_2 + \gamma_2$

endfor;

   Computation of the solution vector (h is a dummy variable)

$$x_n = \mu_n$$
$$h = \mu_n \mathbf{c}   \quad Flops:\ p_1$$

For $k = n-1, -1, 1$ do

1. $x_k = \mu_k + \boldsymbol{\alpha}_k^T h$ $Flops:\ 2p_1$

2. $h = x_i \mathbf{c}_i + P_{k-1} h$ $Flops:\ 2p_1 + \kappa_1 p_1 + \kappa_2$

endfor.

It is clear that an overall complexity count leads to method with complexity $\mathcal{O}(p_1 p_2 n)$. A more detailed summation of the involved operations gives us an overall complexity of

$$(n-1)\Big[(6 + \gamma_1 + \kappa_1)p_1 p_2 + (2\kappa_1 + \gamma_2 + 7)p_1 + (2\gamma_1 + \kappa_2 + 4)p_2 + 2\gamma_2 + 2\kappa_2 + 1\Big]$$
$$+(-3 - \gamma_1 - \kappa_1)p_1 p_2 + (3 - \gamma_2)p_1 + (-2 - 2\gamma_1 - \kappa_2)p_2 + (1 - 2\gamma_2)$$

$\{p_1, p_2\}$-**Levinson conform matrices with** $E_k = F_k = I_k$

The considered class of $\{p_1, p_2\}$-Levinson conform matrices was called simple, because the corresponding solver had a complexity of $\mathcal{O}(p_1 p_2 n)$. Suppose now that we omit this simplicity assumption. This would immediately increase the complexity. For example, in the Toeplitz case, we cannot extract the computation of the solution vector $\mathbf{x}$ out of the most inner loop and hence we immediately have a complexity of $\mathcal{O}(n^2)$. If we assume, however, that the matrices $E_k = F_k = I_k$, the algorithm as presented before remains valid, but it will increase in complexity, because we cannot perform multiplications with the matrices P_k or Q_k in linear time anymore. Here, we will assume that the matrices E_k and F_k equal the identity matrix (because then the algorithm presented earlier remains valid), and we will see what the impact of the multiplications with the matrices P_k and Q_k is on the complexity count.

- Assume we cannot multiply vectors with the matrices P_k in linear time, but we can multiply vectors with the matrices Q_k in linear time. This means that in Algorithm 6.9 the following steps increase in complexity:

  - in the main loop, steps

    2. $\boldsymbol{\alpha}_{k+1}^T = -\dfrac{\boldsymbol{\xi}_{k+1}^T + \mathbf{d}_{k+1}^T S_k^T Y_k P_k}{\mathbf{d}_{k+1}^T S_k^T Y_k \mathbf{c}_{k+1} + a_{k+1,k+1}}$ \hspace{1cm} Flops: $(2p_1 + 1)p_1$

    4. (b) $S_{k+1}^T Y_{k+1} = Q_k^T S_k^T Y_k P_k + \left(Q^T S_k^T Y_k \mathbf{c}_{k+1} + \boldsymbol{\eta}_{k+1} \right) \boldsymbol{\alpha}_{k+1}^T$
       \hspace{1cm} Flops: $2p_2 p_1 + p_1(\gamma_1 p_2 + \gamma_2) + p_1 p_2(2p_1 - 1)$

  - in the computation of the solution, step

    2. $h = x_i \mathbf{c}_i + P_{k-1} h$ \hspace{1cm} Flops: $(2p_1 + 1)p_1$.

  This means that we will get an algorithm of complexity $\mathcal{O}(p_1^2 p_2 n)$. More precisely we get the following operation count:

  $$(n-1)\Big[(2p_2 + 4)\, p_1^2 + (\gamma_1 + 5)\, p_1 p_2 + (\gamma_2 + 5)\, p_1 + (2\gamma_1 + 4)\, p_2 + 2\gamma_2 + 1 \Big]$$
  $$+ \mathcal{O}(1)$$

- Assume the multiplication of a vector with the matrix P_k (for all k) can be done in linear time, but not the multiplication with the matrix Q_k (for all k). In Algorithm 6.9, the following steps increase in complexity:

  - in the main loop, steps

    4. (a) $Q_k^T S_k^T Y_k \mathbf{c}_{k+1} + \boldsymbol{\eta}_{k+1}$ \hspace{1cm} Flops: $2p_2 p_1 + p_2(2p_2 - 1)$

    4. (b) $S_{k+1}^T Y_{k+1} = Q_k^T S_k^T Y_k P_k + \left(Q^T S_k^T Y_k \mathbf{c}_{k+1} + \boldsymbol{\eta}_{k+1} \right) \boldsymbol{\alpha}_{k+1}^T$
       \hspace{1cm} Flops: $2p_2 p_1 + p_2(\kappa_1 p_1 + \kappa_2) + p_1 p_2(2p_2 - 1)$

    4. (c) $S_{k+1}^T \mathbf{x}_{k+1} = Q_k^T S_k^T \mathbf{x}_k + \left(Q_k^T S_k^T Y_k \mathbf{c}_{k+1} + \boldsymbol{\eta}_{k+1} \right) \mu_{k+1}$
       \hspace{1cm} Flops: $p_2(2p_2 + 1)$

  - in the computation of the solution nothing changes.

  Overall, we get the following operation count:

  $$(n-1)\Big[2p_1 p_2^2 + 3p_2^2 + (\kappa_1 + 5)\, p_1 p_2 + (2\kappa_1 + 7)\, p_1 + (\kappa_2 + 2)\, p_2 + 2\kappa_2 + 1 \Big]$$
  $$+ \mathcal{O}(1)$$

- Suppose none of the matrices P_k or Q_k admits a multiplication in linear time. Hence the algorithm increases in complexity, and we get:

  - in the main loop, steps

    2. $\boldsymbol{\alpha}_{k+1}^T = -\dfrac{\boldsymbol{\xi}_{k+1}^T + \mathbf{d}_{k+1}^T S_k^T Y_k P_k}{\mathbf{d}_{k+1}^T S_k^T Y_k \mathbf{c}_{k+1} + a_{k+1,k+1}}$ \hspace{1cm} Flops: $(2p_1 + 1)p_1$

    4. (a) $Q_k^T S_k^T Y_k \mathbf{c}_{k+1} + \boldsymbol{\eta}_{k+1}$ \hspace{1cm} Flops: $2p_2 p_1 + p_2(2p_2 - 1)$

    4. (b) $S_{k+1}^T Y_{k+1} = Q_k^T S_k^T Y_k P_k + \left(Q^T S_k^T Y_k \mathbf{c}_{k+1} + \boldsymbol{\eta}_{k+1} \right) \boldsymbol{\alpha}_{k+1}^T$
       \hspace{1cm} Flops: $2p_2 p_1 + p_1 p_2(2p_2 - 1) + p_1 p_2(2p_1 - 1)$

4. (c) $S_{k+1}^T \mathbf{x}_{k+1} = Q_k^T S_k^T \mathbf{x}_k + \left(Q_k^T S_k^T Y_k \mathbf{c}_{k+1} + \boldsymbol{\eta}_{k+1}\right) \mu_{k+1}$

Flops: $p_2(2p_2 + 1)$

– in the computation of the solution, steps

2. $h = x_i \mathbf{c}_i + P_{k-1} h$ Flops: $(2p_1 + 1)p_1$.

If no linear multiplications with vectors are possible, we get the following complexity:

$$(n-1)\left[2p_1^2 p_2 + 2p_1 p_2^2 + 4p_1^2 + 4p_1 p_2 + 3p_2^2 + 5p_1 + 2p_2 + 1\right]$$
$$+\mathcal{O}(1)$$

In Section 6.4 a class of matrices is included (quasiseparable matrices), which are $\{p_1, p_2\}$-Levinson conform, with $E_k = F_k = I_k$, and the matrices P_k and Q_k do not admit a linear multiplication with a vector.

6.3.3 An upper triangular factorization

The Levinson-like algorithm, as presented in this section applied to a matrix A, is closely related to an upper triangular factorization of the involved matrix. This is also the case for the Levinson algorithm related to Toeplitz matrices (see, e.g., [191, 152]). Even though a similar construction is possible for general Levinson conform matrices, we will focus here on the case for which $E_k = F_k = I_k$. Suppose we have a Levinson conform matrix A. Writing down the solution of the kth order Yule-Walker-like problem gives us:

$$A_k Y_k = -R_k.$$

Bringing the matrix R_k to the left-hand side and multiplying this equation on the right with the vector $\mathbf{c}_{k+1}$ gives us:

$$A_k Y_k \mathbf{c}_{k+1} + R_k \mathbf{c}_{k+1} = 0.$$

This equation can be rewritten in terms of the matrix A_{k+1}:

$$\left[\begin{array}{c|c} A_k & R_k \mathbf{c}_{k+1} \\ \hline \mathbf{d}_{k+1}^T S_k^T & a_{k+1,k+1} \end{array}\right] \left[\begin{array}{c} Y_k \mathbf{c}_{k+1} \\ 1 \end{array}\right] = \left[\begin{array}{c} 0 \\ \sigma_{k+1} \end{array}\right],$$

with σ_{k+1} different from zero. If we put the successive equations for the different values of k, in an upper triangular matrix we get (we remark that the products $Y_k \mathbf{c}_{k+1}$ are column vectors of dimension k):

$$A \begin{bmatrix} 1 & Y_1 \mathbf{c}_2 & Y_2 \mathbf{c}_3 & \dots & Y_{n-1} \mathbf{c}_n \\ 0 & 1 & & & \\ 0 & 0 & 1 & & \\ \vdots & \vdots & \ddots & \ddots & \\ 0 & 0 & \dots & 0 & 1 \end{bmatrix} = \begin{bmatrix} \sigma_1 & 0 & 0 & \dots & 0 \\ \times & \sigma_2 & 0 & \dots & 0 \\ \vdots & \times & \sigma_3 & & 0 \\ \vdots & \vdots & \ddots & \ddots & \vdots \\ \times & \times & \dots & \times & \sigma_n \end{bmatrix}.$$

The $\times$ denote arbitrary elements in the matrix. Rewriting the equations above, defining the upper triangular matrix to the right of A as U and rewriting the matrix on the right-hand side as $L\Lambda$, with $\Lambda = \mathrm{diag}(\sigma_1, \sigma_2, \ldots, \sigma_n)$ a diagonal matrix and L lower triangular. We get the following two relations:

$$A = L\Lambda U^{-1},$$
$$A^{-1} = U\Lambda^{-1}L^{-1},$$

where U and L are, respectively, upper and lower triangular matrices, with ones on the diagonal, and Λ a diagonal matrix. Moreover, we do not get the matrices U and L, but in fact we get their generators, in terms of the vectors $\boldsymbol{\alpha}_k$ and $\boldsymbol{\mu}_k$. This factorization can therefore be used to compute the inverse of the matrix A.

6.3.4   The look-ahead procedure

A limitation of the solver presented is the strongly nonsingular assumption. In this subsection we will briefly illustrate how we can overcome this problem by applying a look-ahead technique. Once the main decomposition of the involved matrices is known, it is an easy exercise to derive the complete algorithm. Hence we do not include all the details. We only illustrate it for the Yule-Walker-like equation.

Given a simple $\{p_1\}$-Levinson conform matrix A, suppose that, after having solved the kth order Yule-Walker-like equation, we encounter some singular or almost singular principal leading matrices A_{k+1} up to A_{k+l-1}. This means that the first nonsingular principal leading matrix is A_{k+l}. We will now solve this $(k+l)$th Yule-Walker-like system, based on the solution of the kth Yule-Walker-like problem:

$$A_k Y_k = -R_k.$$

The coefficient matrix A_{k+l} can be decomposed in block form as

$$A_{k+l} = \left[\begin{array}{c|c} A_k & R_k C_{k,l} \\ \hline D_{k,l}^T S_k^T & B_{k,l} \end{array} \right],$$

where $B_{k,l} = A(k+1 : k+l, k+1 : k+l)$ is an $l \times l$ matrix and $C_{k,l}$ is an $p_1 \times l$ matrix of the following form:

$$C_{k,l} = [\mathbf{c}_{k+1}, P_k \mathbf{c}_{k+2}, \ldots, P_k P_{k+1} \cdots P_{k+l-2} \mathbf{c}_{k+l}].$$

Before we can solve the system $A_{k+l} Y_{k+l} = -R_{k+l}$, we also need to block-decompose the matrix R_{k+l}:

$$R_{k+l} = \left[\begin{array}{c} R_k P_k \cdots P_{k+l-1} \\ \hline \Xi_{k,l}^T \end{array} \right]$$

with $\Xi_{k,l} \in \mathbb{R}^{p_1 \times l}$ of the form:

$$\Xi_{k,l}^T = \left[\begin{array}{c} \boldsymbol{\xi}_{k+1}^T P_{k+1} P_{k+2} P_{k+3} \cdots P_{k+l-1} \\ \boldsymbol{\xi}_{k+2}^T P_{k+2} P_{k+3} \cdots P_{k+l-1} \\ \boldsymbol{\xi}_{k+3}^T P_{k+3} \cdots P_{k+l-1} \\ \vdots \\ \boldsymbol{\xi}_{k+l}^T \end{array} \right].$$

Using these block-decompositions of the matrices A_{k+l} and R_{k+l}, the $(k+l)$th Yule-Walker-like equation can be written as

$$\left[\begin{array}{c|c} A_k & R_k C_{k,l} \\ \hline D_{k,l}^T S_k^T & B_{k,l} \end{array}\right] \left[\begin{array}{c} Z_{k,l} \\ \hline \Delta_{k,l}^T \end{array}\right] = \left[\begin{array}{c} R_k P_k \cdots P_{k+l-1} \\ \hline \Xi_{k,l}^T \end{array}\right],$$

with $Z_{k,l} \in \mathbb{R}^{k \times p_1}$ and $\Delta_{k,l} \in \mathbb{R}^{p_1 \times l}$.

Expanding the equations above and solving them towards $Z_{k,l}$ and $\Delta_{k,l}$ leads to the following formulas:

$$Z_{k,l} = Y_k \left(P_k P_{k+1} \cdots P_{k+l-1} + C_{k,l} \Delta_{k,l}^T \right),$$
$$\Delta_{k,l}^T = -\left(D_{k,l}^T S_k^T Y_k C_{k,l} + B_{k,l} \right)^{-1} \left(\Xi_{k,l}^T + D_{k,l}^T S_k^T Y_k P_k \cdots P_{k+l-1} \right).$$

This means that, for computing $\Delta_{k,l}$, we need to solve p_1 systems of size $l \times l$; this can be done by several techniques at a cost of $\mathcal{O}(l^3)$. As long as the value of l, with regard to the problem size is small, there is no significant increase in complexity, but this is of course closely related to the problem.

Having computed the solution of the Yule-Walker-like equation, solving the corresponding Levinson problem is done similarly. Also the complexity-reducing remarks as presented in the Subsection 6.3.2 can be translated towards this block-version.

An important issue, which we will not address here, is to decide when a principal leading matrix is numerically singular, such that it might have a large influence on the accuracy of the final result. An easy way to measure the ill-conditioning is to check the value of the denominator in the computation of α_k; as long as this value is not too close to zero (with regard to the nominator), the computations are numerically sound. A more detailed analysis should however be done case by case for obtaining a fast and accurate practical implementation.

Using the techniques presented in this section we will now present several examples in the next section.

Notes and references

The general framework as presented in this section is based on the following publication.

☞ R. Vandebril, N. Mastronardi, and M. Van Barel. Solving linear systems with a levinson-like solver. *Electronic Transactions on Numerical Analysis*, 26:243–269, 2007.

In the publications [94, 89], the authors derive closely related techniques for solving quasiseparable systems of equations. The authors restricted themselves to the class of block quasiseparable matrices, which also includes the class of band matrices and semiseparable matrices. For more information see Chapter 12. The class of simple $\{p_1, p_2\}$-Levinson conform matrices coincides to some extent with the class of block quasiseparable matrices. The difference lies in the fact that the multiplication of a vector with the matrices P_k and Q_k can be performed in linear time, whereas this is not necessarily the case with quasiseparable matrices. As we have not yet discussed block quasiseparable matrices we will not go into the details.

The algorithm presented is based on an efficient computation of the generators of a triangular factorization of the (inverse) matrix in a recursive fashion and then uses this factorization for computing the solution.

6.4   Examples

In this section we will provide several classes of matrices, which are simple $\{p_1, p_2\}$-Levinson conform. Hence we can apply our previously designed algorithm and come to an $\mathcal{O}(p_1 p_2 n)$ solver for these systems of equations. We do not always derive the complete algorithm, but we define all the necessary matrices to illustrate that the presented matrices are indeed simple $\{p_1, p_2\}$-Levinson conform matrices. As this part of the book is concerned with the simplest classes of structured rank matrices, we postpone some examples, related to higher order rank structures, until Part III.

This section covers the following examples consecutively: Givens-vector-represented semiseparable matrices; quasiseparable matrices; tridiagonal matrices, arrowhead matrices, upper triangular matrices; dense matrices; summations of Levinson conform matrices; companion matrices; fellow matrices; and comrade matrices.

In this section attention is also paid to the representation of the semiseparable matrices. Because the Levinson solver as presented in this chapter is capable of dealing with different types of representations without loss of generality.

6.4.1   Givens-vector-represented semiseparable matrices

Because the structure of the simple $\{p_1, p_2\}$-Levinson conform matrices, as presented in this paper, does not extend towards the diagonal, there is no loss of generality when solving a simple $\{p_1, p_2\}$-Levinson conform matrix plus a diagonal. In fact it does not even increase the complexity. This means that the solver we derive below is also applicable for semiseparable plus diagonal matrices.

Let us consider here the unsymmetric case, namely an unsymmetric semiseparable matrix represented using the decoupled Givens vector representation. Let us denote the first sequence of Givens rotations, and the vector for representing the lower triangular part as:

$$G = \begin{bmatrix} c_1 & c_2 & \dots & c_{n-1} \\ s_1 & s_2 & \dots & s_{n-1} \end{bmatrix},$$

$$\mathbf{v} = \begin{bmatrix} v_1 & v_2 & \dots & v_n \end{bmatrix}^T,$$

and the second sequence of rotations and the vector, representing the strictly upper triangular part as:

$$H = \begin{bmatrix} r_1 & r_2 & \dots & r_{n-2} \\ t_1 & t_2 & \dots & t_{n-2} \end{bmatrix},$$

$$\mathbf{e} = \begin{bmatrix} e_1 & e_2 & \dots & e_{n-1} \end{bmatrix}^T.$$

The matrices G and H contain in the first row, the cosines of Givens transformations and in the second row the sines of the Givens transformation; every column corresponds to one Givens rotation. The resulting semiseparable matrix S is of the following form:

$$
\begin{bmatrix}
c_1 v_1 & & r_1 e_1 & r_2 t_1 e_1 & \cdots & r_{n-2} t_{n-3} \cdots t_1 e_1 & t_{n-2} t_{n-3} \cdots t_1 e_1 \\
c_2 s_1 v_1 & & c_2 v_2 & r_2 e_2 & & r_{n-2} t_{n-3} \cdots t_2 e_2 & t_{n-2} t_{n-3} \cdots t_2 e_2 \\
c_3 s_2 s_1 v_1 & & c_3 s_2 v_2 & c_3 v_3 & \ddots & \vdots & \vdots \\
\vdots & & & \ddots & \ddots & r_{n-2} e_{n-2} & t_{n-2} e_{n-2} \\
c_{n-1} s_{n-2} \cdots s_1 v_1 & & \cdots & & & c_{n-1} v_{n-1} & e_{n-1} \\
s_{n-1} s_{n-2} \cdots s_1 v_1 & & \cdots & & & s_{n-1} v_{n-1} & v_n
\end{bmatrix}.
$$

We will construct here, in this case, only the vectors R_k and the elements $\mathbf{c}_k$ and P_k corresponding to the upper triangular part. The matrices corresponding to the lower triangular part, can be constructed in a similar way.

Put $R_1 = e_1$ and define $\mathbf{c}_{k+1} = r_k$ for $k = 1, \ldots, n-2$ and $\mathbf{c}_n = 1$. If $P_k = t_k$ and $\boldsymbol{\xi}_k = e_k$, we get for $k = 2, \ldots, n-1$:

$$
R_k = \left[\begin{array}{c} R_{k-1} t_{k-1} \\ \hline e_k \end{array} \right] = \begin{bmatrix} e_1 t_1 t_2 \cdots t_{k-1} \\ e_2 t_2 \cdots t_{k-1} \\ \vdots \\ e_{k-1} t_{k-1} \\ e_k \end{bmatrix}
$$

and hence (only the upper triangular part is shown)

$$
A_{k+1} = \left[\begin{array}{c|c} A_k & R_k r_k \\ \hline & c_{k+1} v_{k+1} \end{array} \right] = \left[\begin{array}{c|c} A_k & \begin{array}{c} r_k t_{k-1} \cdots t_1 e_1 \\ r_k t_{k-1} \cdots t_2 e_2 \\ \vdots \\ r_k e_k \end{array} \\ \hline & c_{k+1} v_{k+1} \end{array} \right].
$$

This states that our defined matrices generate the matrix A. Moreover we remark that the multiplication of a vector with any of the matrices $P_k = t_k$ can clearly be performed in linear time. This defines all the matrices, and therefore, this matrix is simple $\{1, 1\}$-Levinson conform and admits an $\mathcal{O}(n)$ Levinson-like method. Using the operation count as presented in Section 6.3.2, we see that the cost of solving a system of equations is bounded by $24n - 29$ operations (with $\kappa_1 = \gamma_1 = 1$ and $\kappa_2 = \gamma_2 = 0$).

6.4.2 Quasiseparable matrices

In this section, the class of quasiseparable matrices is considered represented with the quasiseparable representation. General quasiseparable matrices (see [93]) are investigated in Part III, as the general class is slightly different and we have not yet defined them.

Let us illustrate that quasiseparable matrices can be considered as simple $\{1,1\}$-Levinson conform matrices, and hence admit an $\mathcal{O}(n)$ solver of the Levinson type.

As the structure of the upper triangular part is exactly the same as the structure from the lower triangular part, we will search for the matrices corresponding to the upper triangular part. The upper triangular part of such a $\{1,1\}$-quasiseparable matrix A has the following structure:

$$
A = \begin{bmatrix}
d_1 & q_1\,p_1 & q_2\,r_1\,p_1 & q_3\,r_{2:1}\,p_1 & \cdots & q_{n-1}\,r_{n-2:1}\,p_1 \\
 & d_2 & q_2\,p_2 & q_3 r_2 p_2 & \cdots & q_{n-1}\,r_{n-2:2}\,p_2 \\
 & & d_3 & q_3 p_3 & \cdots & q_{n-1}\,r_{n-2:3}\,p_3 \\
 & & & \ddots & & \vdots \\
 & & & & d_{n-1} & q_{n-1}\,p_{n-1} \\
 & & & & & d_n
\end{bmatrix}.
$$

Initializing $R_1 = p_1$ and defining $\mathbf{c}_{k+1} = q_k$, $P_k = r_k$ and $\boldsymbol{\xi}_k = p_k$ gives us for $k = 1, \ldots, n-1$:

$$
R_k = \left[\begin{array}{c} R_{k-1}P_{k-1} \\ \hline \boldsymbol{\xi}_k \end{array}\right] = \left[\begin{array}{c} R_{k-1}r_{k-1} \\ \hline p_k \end{array}\right] = \begin{bmatrix} p_1 r_1 r_2 \cdots r_{k-1} \\ p_2 r_2 \cdots r_{k-1} \\ \vdots \\ p_{k-1} r_{k-1} \\ p_k \end{bmatrix}
$$

and hence

$$
Q_{k+1} = \left[\begin{array}{c|c} Q_k & R_k\mathbf{c}_{k+1} \\ \hline & d_{k+1} \end{array}\right] = \left[\begin{array}{c|c} A_k & R_k q_k \\ \hline & d_{k+1} \end{array}\right] = \left[\begin{array}{c|c} & p_1 r_1 r_2 \cdots r_{k-1} q_k \\ & p_2 r_2 \cdots r_{k-1} q_k \\ A_k & \vdots \\ & p_{k-1} r_{k-1} q_k \\ & p_k q_k \\ \hline & d_{k+1} \end{array}\right],
$$

which gives us the desired matrices. All the conditions are satisfied (including the demands on the multiplication with P_k) to have a simple $\{1,1\}$-Levinson conform matrix. Using the operation count as presented in Subsection 6.3.2, we see that the cost is bounded by $24n - 29$ operations (with $\kappa_1 = \gamma_1 = 1$ and $\kappa_2 = \gamma_2 = 0$), which is exactly the same number of operations for a semiseparable plus diagonal matrix represented with a sequence of Givens transformations and a vector.

A careful computation of the number of operations involved in Algorithm 5.3 proposed in [94] for solving a $\{1,1\}$-quasiseparable matrix gives us, for the number of flops, the following complexity: $49n + \mathcal{O}(1)$.

6.4.3   Tridiagonal matrices

Because tridiagonal matrices are part of the larger class of quasiseparable matrices, it is natural that they admit the solver as presented in the previous subsection.

However tridiagonal matrices admit also another type of block decomposition, which we will briefly present in this subsection. This type of decomposition will lead to a much faster solver than just applying the quasiseparable method above.

Assume we have a tridiagonal matrix T of the following form:

$$T = \begin{bmatrix} d_1 & d_1^{(u)} & 0 & & & \\ d_1^{(l)} & d_2 & d_2^{(u)} & \ddots & & \\ 0 & d_2^{(l)} & \ddots & \ddots & 0 & \\ & \ddots & \ddots & \ddots & d_{n-1}^{(u)} & \\ & & 0 & d_{n-1}^{(l)} & d_n \end{bmatrix}.$$

Let us define $R_k = \left[0, \ldots, 0, d_k^{(u)} \right]^T$ a vector of length k, with only the last element equal to $d_k^{(u)}$. The vector $\mathbf{c}_k = 1$, $P_k = 0$ and $\boldsymbol{\xi}_k = d_k^{(u)}$ for all k. This gives us for $k = 2, \ldots, n$ clearly the following relation:

$$R_k = \left[\frac{R_{k-1} P_{k-1}}{\boldsymbol{\xi}_k} \right] = \left[\frac{0}{d_k^{(u)}} \right].$$

Solving the Levinson-like system as presented here leads to an overall algorithm of complexity $9n$ plus lower order terms. This is because multiplication with $\mathbf{c}_k$ does not need to be performed, and all summations of products involving P_k and Q_k can be neglected.

6.4.4 Arrowhead matrices

Arrowhead matrices are often an essential tool for the computation of the eigenvalues, via divide and conquer approaches [37, 64]. They also arise, in block form, in domain decomposition methods, when discretizing partial differential equations, where they are used as preconditioners. In this section we will show how arrowhead matrices can be solved efficiently using the presented framework, and we will present the algorithm based on Section 6.3.2. Let us consider the nonsymmetric arrowhead matrix of the following form:

$$A = \begin{bmatrix} a_1 & \bar{a}_2 & \bar{a}_3 & \ldots & \bar{a}_n \\ \underline{a}_2 & a_2 & & & \\ \underline{a}_3 & & a_3 & & \\ \vdots & & & \ddots & \\ \underline{a}_n & & & & a_n \end{bmatrix},$$

where the elements of the form $\bar{a}_i$ denote the elements of the arrow in the upper triangular part and the elements $\underline{a}_i$ denote the elements of the arrow in the lower triangular part. The elements a_i denote the diagonal elements, with $\underline{a}_1 = a_1 = \bar{a}_1$, and the elements not shown are assumed to be zero.

Let us define the matrices R_k and S_k as $S_k^T = R_k^T = [1, 0, \dots, 0]$ which are vectors of length k. Let us define the elements $\mathbf{c}_k = \bar{a}_k$ and $\mathbf{d}_k = \underline{a}_k$, the matrices $P_k = Q_k$ are chosen equal to the identity matrix and the vectors $\xi_k = \eta_k = 0$. One can easily check that this matrix is a simple $\{1, 1\}$-Levinson conform matrix.

Based on the Levinson solver, we derive here the solver for the arrowhead matrix.

Algorithm 6.10. *Initialize*

$$\boldsymbol{\alpha}_1 = \frac{-1}{a_1} \qquad \qquad \textit{Flops: 1}$$

$$\mu_1 = \frac{b_1}{a_1} \qquad \qquad \textit{Flops: 1}$$

$$S_1^T Y_1 = S_1^T \boldsymbol{\alpha}_1 = \boldsymbol{\alpha}_1 \; \textit{Flops: 0}$$

$$S_1^T x_1 = S_1^T \mu_1 = \mu_1 \; \textit{Flops: 0}$$

For $k = 1, \dots, n-1$ do:

1. *Compute and store the following variable:*

   (a) $\underline{a}_{k+1} S_k^T Y_k$ *Flops: 1*

   (b) $(\underline{a}_{k+1} S_k^T Y_k \bar{a}_{k+1} + a_{k+1,k+1})$ *Flops: 2*

2. $\boldsymbol{\alpha}_{k+1} = -\dfrac{\underline{a}_{k+1} S_k^T Y_k}{\underline{a}_{k+1} S_k^T Y_k \bar{a}_{k+1} + a_{k+1,k+1}}$ *Flops: 1*

3. $\mu_{k+1} = \dfrac{b_{k+1} - \underline{a}_{k+1} S_k^T x_k}{\underline{a}_{k+1} S_k^T Y_k \bar{a}_{k+1} + a_{k+1,k+1}}$ *Flops: 3*

4. *Compute and store the following variables (if $k < n-1$):*

   (a) $S_k^T Y_k \bar{a}_{k+1}$ *Flops: 1*

   (b) $S_{k+1}^T Y_{k+1} = S_k^T Y_k + \left(S_k^T Y_k \bar{a}_{k+1} \right) \boldsymbol{\alpha}_{k+1}$ *Flops: 2*

   (c) $S_{k+1}^T \mathbf{x}_{k+1} = S_k^T \mathbf{x}_k + \left(S_k^T Y_k \bar{a}_{k+1} \right) \mu_{k+1}$ *Flops: 2*

endfor;

    Computation of the solution vector (h is a dummy variable)

$$x_n = \mu_n$$

$$h = \mu_n \bar{a} \qquad \qquad \textit{Flops: 1}$$

For $k = n-1, -1, 1$ do

1. $x_k = \mu_k + \boldsymbol{\alpha}_k h$ *Flops: 2*

2. $h = x_i \bar{a}_i + h$ *Flops: 2*

endfor.

The largest term in this algorithm's complexity is $19n$. If one however uses standard Gaussian elimination to solve this system (after having flipped the arrow, so that it points downwards), the complexity is $6n$.

6.4.5 Unsymmetric structures

The examples presented above were matrices having a symmetric structure. This means that if the upper part was semiseparable the lower triangular part was also semiseparable, if the upper triangular part was from a tridiagonal matrix, the lower triangular part was also from a tridiagonal matrix. But in fact, taking a closer look at the conditions, the upper and/or the lower triangular parts of the matrix need not to be related in any way.

Dealing with the upper and the lower triangular part of the matrix separately we can create matrices having a nonsymmetric structure:

- An upper triangular semiseparable matrix is simple $\{1,0\}$-Levinson conform.

- An upper bidiagonal matrix is simple $\{1,0\}$-Levinson conform.

- A matrix for which the upper triangular part is semiseparable, and the lower triangular part is bidiagonal, for example, a unitary Hessenberg matrix, this matrix has the upper triangular part of semiseparable form and only one subdiagonal different from zero. Hence a unitary Hessenberg matrix is simple $\{1,1\}$-Levinson conform.

- A matrix which has a tridiagonal structure in the upper triangular part and, for which the lower triangular part comes from an arrowhead is simple $\{1,1\}$-Levinson conform.

- Moreover, one can also combine matrices, for which the upper or the lower triangular part is quasiseparable.

In the following sections and later on in this book, some interesting classes of matrices having an unsymmetric structure are investigated in more detail.

6.4.6 Upper triangular matrices

Let us apply our Levinson solver to an upper triangular system of equations. We have the matrix $A = (a_{ij})$ where $a_{ij} = 0$ if $i > j$. Let us denote with R_k the matrix of dimension $k \times (n-1)$ where $R_k = [I_k, 0]$ and $\mathbf{c}_k^T = [a_{1,k}, \ldots, a_{k-1,k}, 0, \ldots, 0]$ is a row vector of length $1 \times (n-1)$. Moreover we assume the matrix $S_k = 0$ and $P = I$. The matrix is simple $\{n-1, 0\}$-Levinson conform as:

$$A_{k+1} = \left[\begin{array}{c|c} A_k & R_k \mathbf{c}_{k+1} \\ \hline 0 & a_{k+1,k+1} \end{array} \right].$$

We know that solving the system of equations in this manner will lead to an $\mathcal{O}(n^2)$ method. Moreover, this is a well-known method, as we will show.

Let us take a closer look at the solution generated by this Levinson-like approach. We have all the necessary information to easily calculate the values of $\boldsymbol{\alpha}_k$

and μ_k for every k, as $S_k = 0$. We have that for every k

$$\boldsymbol{\alpha}_k = \frac{\mathbf{e}_k}{a_{k,k}}$$

$$\mu_k = \frac{b_k}{a_{k,k}},$$

where $\mathbf{e}_k$ is the kth vector of the canonical basis of $\mathbb{R}^n$. Using these variables, we can construct the solution vector $\mathbf{x}$ for the system $A\mathbf{x} = \mathbf{b}$. We will consider the computation of the last three components of $\mathbf{x}$, based on Equation (6.22). The last component x_n has the following form:

$$x_n = \mu_n = \frac{b_n}{a_{n,n}}.$$

The component x_{n-1} is of the following form:

$$x_{n-1} = \frac{b_{n-1}}{a_{n-1,n-1}} + \frac{(-1)}{a_{n-1,n-1}} \frac{b_n}{a_{n,n}} \mathbf{e}_{n-1} \mathbf{c}_n$$

$$= \frac{-1}{a_{n-1,n-1}} \left(b_{n-1} - a_{n-1,n} x_n \right).$$

Let us conclude with element x_{n-2}, which gives us the following equations:

$$x_{n-2} = \frac{b_{n-2}}{a_{n-2,n-2}} + \frac{\mathbf{e}_{n-2}}{a_{n-2,n-2}} \left(\frac{b_{n-1}}{a_{n-1,n-1}} \mathbf{c}_{n-1} + (I + \mathbf{c}_{n-1} \frac{-1}{a_{n-1,n-1}} \mathbf{e}_{n-1}) \frac{b_n}{a_{n,n}} \mathbf{c}_n \right)$$

$$= \frac{1}{a_{n-2,n-2}} \left(b_{n-2} - a_{n-2,n-1} \frac{b_{n-1}}{a_{n-1,n-1}} - a_{n-2,n} x_n + \frac{a_{n-2,n-1} a_{n-1,n}}{a_{n-1,n-1}} x_n \right)$$

$$= \frac{1}{a_{n-2,n-2}} \left(b_{n-2} - a_{n-2,n-1} x_{n-1} - a_{n-2,n} x_n \right)$$

This means that rewriting the general formulas for the Levinson-like solver for upper triangular systems of equations gives us the well-known backward substitution algorithm [152]. In a similar way we can derive the solution method for a lower triangular system of equations. This will give us the forward substitution algorithm.

6.4.7 Dense matrices

Using Lemma 6.5 we know that strongly nonsingular systems of equations without structure in the coefficient matrix can also be solved in this way. This gives us an algorithm, requiring $\mathcal{O}(n^3)$ operations, more precisely $6n^3$, which is of course not efficient enough.

6.4.8 Summations of Levinson-conform matrices

If we add up different Levinson-conform matrices, we get again a Levinson-conform matrix. This is proved in the next theorem.

Theorem 6.11. *Suppose we have two matrices $\hat{A}$ and $\tilde{A}$, which are, respectively, $\{\hat{p}_1, \hat{p}_2\}$ and $\{\tilde{p}_1, \tilde{p}_2\}$-Levinson conform. Then the matrix $A = \hat{A} + \tilde{A}$ will be $\{\hat{p}_1 + \tilde{p}_1, \hat{p}_2 + \tilde{p}_2\}$-Levinson conform.*

Proof. Let us denote all the matrices related to the matrix $\hat{A}$ with a hat and the ones related to the matrix $\tilde{A}$ with a tilde. Let us define the matrices $R_k, c_k, d_k, S_k, \xi_k$ and η_k as follows:

$$R_k = \left[\hat{R}_k, \tilde{R}_k\right], \ S_k = \left[\hat{S}_k, \tilde{S}_k\right]$$
$$\mathbf{c}_k^T = \left[\hat{\mathbf{c}}_k^T, \tilde{\mathbf{c}}_k^T\right], \ \mathbf{d}_k^T = \left[\hat{\mathbf{d}}_k^T, \tilde{\mathbf{d}}_k^T\right]$$
$$\boldsymbol{\xi}_k^T = \left[\hat{\boldsymbol{\xi}}_k^T, \tilde{\boldsymbol{\xi}}_k^T\right], \ \boldsymbol{\eta}_k^T = \left[\hat{\boldsymbol{\eta}}_k^T, \tilde{\boldsymbol{\eta}}_k^T\right].$$

Define the operators P_k and Q_k as

$$P_k = \left[\begin{array}{cc} \hat{P}_k & \\ & \tilde{P}_k \end{array}\right] \text{ and } Q_k = \left[\begin{array}{cc} \hat{Q}_k & \\ & \tilde{Q}_k \end{array}\right].$$

Then it is straightforward to prove that these newly defined matrices and vectors satisfy the desired conditions, such that the matrix A is $\{\hat{p}_1 + \tilde{p}_1, \hat{p}_2 + \tilde{p}_2\}$-Levinson conform. □

We remark that if we add up two simple Levinson conform matrices the resulting matrix will also be simple Levinson conform.

Let us illustrate this with some possible structures which now become solvable:

- One can now solve summations of all previously defined Levinson-conform matrices. For example, the sum of a semiseparable matrix plus a tridiagonal matrix.

- Moreover it is not necessary that both matrices be strongly nonsingular. As long as the sum of these matrices is strongly nonsingular, the problem can be solved by the standard Levinson-like solver (for the look-ahead method see Section 6.3.4). In this way, we can also add simple Levinson conform matrices which are singular. For example adding a rank 1 matrix to a Levinson conform matrix is feasible.

- For example an arrowhead matrix with a tridiagonal band is $\{2, 2\}$-Levinson conform, as it can be written as the sum of an arrowhead matrix and tridiagonal matrix.

In the next sections, we will give some more examples of matrices, which can easily be seen as the sum of simple Levinson conform matrices.

6.4.9 Matrices with errors in structures

Quite often one will deal with matrices which do not have a perfect structure; for example, a matrix, which is semiseparable, except for some elements in one row or column. Or a matrix which is almost of tridiagonal form except for some elements which are nonzero. The number of elements desintegrating the structure is often low. If we are now able to write this matrix as the sum of the pure structure (e.g., semiseparable plus tridiagonal) and the elements destroying the structure we can decrease the complexity count of the Levinson method related to this matrix. Let us call these matrices destroying the structure error matrices. If these error matrices are simple $\{e_1, e_2\}$-Levinson conform, the complete matrix will be simple $\{p_1 + e_1, p_2 + e_2\}$-Levinson conform. In case of small e_1 and e_2 this does not lead to a large increase in complexity. Let us illustrate this with some examples of errors in the upper triangular part. (The lower triangular part can be dealt with in a similar way.)

- The error matrix E has only one column different from zero. Suppose column i: $E_i = [e_1, e_2, \ldots, e_n]^T$ contains the only nonzero elements in the matrix E. If we define $R_1 = e_1$, $R_{k+1}^T = [R_k^T, e_{k+1}]$ and all the $\mathbf{c}_k = 0$, except $\mathbf{c}_{i+1} = 1$, this gives us the structure for the upper triangular part. Defining the lower triangular part similarly gives us a simple $\{1, 1\}$-Levinson conform matrix. Similarly one can consider error matrices with more columns different from zero, or error matrices with one or more rows different from zero. For example a matrix which is of unitary Hessenberg form, except for the elements in the last column. These elements do not belong to the semiseparable structure of the upper triangular part. This matrix can be written as the sum of a unitary Hessenberg matrix plus an error matrix, which has only one column different from zero.

- The error matrix has a super diagonal different from zero. Similarly to the band matrix approach, we can prove that this matrix is simple $\{0, 1\}$-Levinson conform. Multiple super diagonals and/or subdiagonals, can also be considered.

- If the error matrix is unstructured, but contains few elements, one might be able to represent it as a simple $\{e_1, e_2\}$-Levinson conform matrix with small e_1 and e_2, but this is of course case dependent.

In the next sections some examples will be given of specific matrices which can be written as the sum of a pure structure plus an error matrix.

6.4.10   Companion matrices

Companion matrices are often used for computing the zeros of polynomials (see [34]). The companion matrix itself is not suitable for applying the Levinson algorithm, as all the leading principal matrices, except possibly the matrix itself, are singular. We can add this matrix however easily to other Levinson-conform matrices as it is simple $\{1, 1\}$-Levinson conform. Let us consider the companion matrix C corresponding to the polynomial $p(x) = x^n + a_{n-1}x^{n-1} + \cdots + a_1 x + a_0$:

$$C = \begin{bmatrix} 0 & 0 & 0 & \cdots & & -a_0 \\ 1 & 0 & 0 & \cdots & & -a_1 \\ 0 & 1 & 0 & \cdots & & -a_2 \\ \vdots & & \ddots & & & \vdots \\ & & & & 1 & -a_{n-1} \end{bmatrix}$$

This matrix is clearly simple $\{1, 1\}$-Levinson conform.

6.4.11 Comrade matrix

Lots of polynomial bases satisfy a three terms recurrence relation. If we would like to compute the roots of a polynomial expressed in such a basis, we can use the comrade matrix [12].

When we have a set of polynomials defined in the following sense:

$$p_i(x) = \sum_{j=0}^{i} p_{ij} x^j, \; i = 0, 1, 2, 3, \ldots$$

which satisfy the following relationships (in fact a three terms recurrence):

$$p_0(x) = 1,$$
$$p_1(x) = \alpha_1 x + \beta_1,$$
$$p_i(x) = (\alpha_i x + \beta_i) p_{i-1}(x) - \gamma_i p_{i-2}(x) \text{ for } i \geq 2,$$

and suppose we have the following polynomial:

$$a(x) = p_n(x) + a_1 p_{n-1}(x) + \ldots + a_n p_0(x),$$

then the comrade matrix is defined as the matrix C:

$$C = \begin{bmatrix} \frac{-\beta_1}{\alpha_1} & \frac{1}{\alpha_1} & 0 & \cdots & & & 0 \\ \frac{\gamma_2}{\alpha_2} & \frac{-\beta_2}{\alpha_2} & \frac{1}{\alpha_2} & 0 & \cdots & & 0 \\ 0 & \frac{\gamma_3}{\alpha_3} & \frac{-\beta_3}{\alpha_3} & \frac{1}{\alpha_3} & 0 & & \vdots \\ \vdots & \ddots & \ddots & \ddots & \ddots & & \\ & & & \frac{\gamma_{n-1}}{\alpha_{n-1}} & \frac{-\beta_{n-1}}{\alpha_{n-1}} & \frac{1}{\alpha_{n-1}} \\ \frac{-a_n}{\alpha_n} & \frac{-a_{n-1}}{\alpha_n} & \cdots & \frac{-a_3}{\alpha_n} & \frac{-a_2+\gamma_n}{\alpha_n} & \frac{-a_1-\beta_n}{\alpha_n} \end{bmatrix}.$$

It is clear that this comrade matrix can be written as the sum of a tridiagonal matrix plus an error matrix, for which one row is different from zero. Hence, the matrix is simple $\{1, 2\}$-Levinson conform.

6.4.12   Fellow matrices

Fellow matrices are rank 1 perturbations of unitary Hessenberg matrices (see [47]). Finding the roots of a polynomial expressed as a linear combination of Szegö polynomials is related to the eigenvalues of a fellow matrix (see [5, 4]). These matrices are naturally written as the sum of two simple Levinson conform matrices. Suppose F is a fellow matrix, then we can write $F = H + \mathbf{u}\mathbf{v}^T$, where H is a unitary Hessenberg matrix, which is simple $\{1,1\}$-Levinson conform, and the rank 1 matrix $\mathbf{u}\mathbf{v}^T$ is also simple $\{1,1\}$-Levinson conform. A fellow matrix is therefore simple $\{2,2\}$-Levinson conform.

Notes and references

The examples of this section and the previous section are mostly based on the manuscripts [212, 289] and the following manuscript.

> ☞ R. Vandebril, N. Mastronardi, and M. Van Barel. A Levinson-like algorithm for symmetric strongly nonsingular higher order semiseparable plus band matrices. *Journal of Computational and Applied Mathematics*, 198:75–97, 2007.

In this manuscript, the authors discuss the Levinson framework applied to higher order generator representable semiseparable plus diagonal matrices. The manuscript [212] discusses generator representable semiseparable plus diagonal matrices (see Section 6.2 in Chapter 6). The manuscript [289] discusses the general framework, with most of the examples presented above (See previous section).

The following references indicate interesting links to the examples included above, which were not already covered in previous sections.

The class of band matrices is examined in several basic books, for example [152] offers an extensive list of references related to band systems, involving LU, QR, inverses and so on for band matrices.

Arrowhead matrices are often used for computing eigenvalue problems via divide-and-conquer algorithms of tridiagonal matrices.

> ☞ C. F. Borges and W. B. Gragg. A parallel divide and conquer algorithm for the generalized real symmetric definite tridiagonal eigenproblem. In L. Reichel, A. Ruttan, and R. S. Varga, editors, *Numerical Linear Algebra and Scientific Computing*, pages 11–29, Berlin, 1993. de Gruyter.

> ☞ J. J. M. Cuppen. A divide and conquer method for the symmetric tridiagonal eigenproblem. *Numerische Mathematik*, 36:177–195, 1981.

An extension of these divide-and-conquer methods for tridiagonal matrices towards the semiseparable case, was proposed by Van Camp, Mastronardi and Van Barel in [211].

If one places the coefficients of polynomials in the standard basis in special positions in a matrix, one gets a companion matrix. Companion matrices have as eigenvalues the roots of the original polynomial. Comrade matrices satisfy the same idea, but, with regard to the Chebyshev basis, more information on these matrices can, for example, be found in the following sources:

> ☞ S. Barnett. *Polynomials and Linear Control Systems*. Marcel Dekker Inc, 1983.

☞ V. Y. Pan. *Structured matrices and polynomials. Unified superfast algorithms*. Birkhäuser Springer, 2001.

Fellow matrices are rank 1 perturbations of unitary Hessenberg matrices.

☞ G. S. Ammar, D. Calvetti, W. B. Gragg, and L. Reichel. Polynomial zerofinders based on Szegö polynomials. *Journal of Computational and Applied Mathematics*, 127:1–16, 2001.

☞ G. S. Ammar, D. Calvetti, and L. Reichel. Continuation methods for the computation of zeros of Szegö polynomials. *Linear Algebra and its Applications*, 249:125–155, 1996.

☞ D. Calvetti, S. Kim, and L. Reichel. The restarted QR-algorithm for eigenvalue computation of structured matrices. *Journal of Computational and Applied Mathematics*, 149:415–422, 2002.

6.5 The Schur algorithm

One of the fundamental concepts in numerical linear algebra is the so-called Schur complement. The Schur complement has been intensively studied in the framework of preconditioning for linear systems (see [246] and the references therein) and in the framework of Toeplitz structured matrices. Indeed, one of the important properties of the Schur complement is the inheritance of the displacement rank. This concept has allowed the development of fast algorithms for structured matrices.

In this section we will show that the Schur complement inherits also the structured rank of a matrix. Exploiting this property, we will construct fast algorithms for computing the LDL^T factorization of structured rank matrices, with L lower triangular and D diagonal. Due to the semiseparable structure of L, linear systems involving such structured matrices can be solved in a fast way. Only a small section is dedicated to Schur-like algorithms. We will only discuss the Schur algorithm for generator representable quasiseparable matrices. Based on these results the reader should be able to derive similar algorithms for more general classes of structured rank matrices.

Without loss of generality the structured rank matrices considered in this chapter are symmetric. The extension of the arguments handled in this chapter to the unsymmetric case is straightforward. Moreover, we also suppose that the matrices are strongly regular (also called strongly nonsingular), i.e., all the principal minors are different from zero. To handle nonstrongly regular matrices, look-ahead techniques [122] can be used.

6.5.1 Basic concepts

The Schur complement is defined as follows.

Definition 6.12. *Let $A \in \mathbb{R}^{n \times n}$ and $1 \leq k < n$. Let us consider the following partitioning of A,*

$$A = \left[\begin{array}{cc} A_{11} & A_{12} \\ A_{21} & A_{22} \end{array} \right],$$

with $A_{11} \in \mathbb{R}^{k \times k}$ *invertible. The Schur complement of A with respect to A_{11} is defined as*

$$\Delta_{A,k} = A_{22} - A_{21} A_{11}^{-1} A_{12}.$$

The Schur complement has been extensively studied in the framework of Toeplitz structured matrices. It has been shown that the Schur complement of a Toeplitz structured matrix inherits its displacement rank. We introduce now the basic ideas of the displacement rank. An extensive treatment of this topic can be found in [191].

Let A be a symmetric matrix and F be a lower triangular matrix. Both matrices are of size $n \times n$. The displacement of A, with respect to F, is defined as

$$\nabla_F A = A - F A F^T. \tag{6.24}$$

The matrix F in the original definition (see [190]) is the lower shift matrix Z of order n,

$$Z = \begin{bmatrix} 0 & & & & \\ 1 & 0 & & & \\ & \ddots & \ddots & & \\ & & 1 & 0 & \\ & & & 1 & 0 \end{bmatrix}.$$

Indeed, the product ZAZ^T displaces A downwards along the main diagonal by one position, explaining the name displacement for $\nabla_Z A$:

$$\nabla_Z A = \begin{bmatrix} a_{1,1} & a_{1,2} & \cdots & a_{1,n} \\ a_{2,1} & a_{2,2} & \cdots & a_{1,n} \\ \vdots & \cdots & \vdots & \vdots \\ a_{n,1} & a_{n,1} & \cdots & a_{n,n} \end{bmatrix} - \left[\begin{array}{c|ccc} 0 & 0 & \cdots & 0 \\ \hline 0 & a_{1,1} & \cdots & a_{1,n-1} \\ \vdots & \vdots & \cdots & \vdots \\ 0 & a_{n-1,1} & \cdots & a_{n-1,n-1} \end{array} \right].$$

The rank r of $\nabla_F A$ is called displacement rank. If $\nabla_F A$ has low displacement rank, independent of n, then A is said to be structured with respect to the displacement ∇_F defined by (6.24), and r is called the displacement rank of A.

The Schur complement of structured matrices inherits the displacement rank. In fact, the following lemma holds.

Lemma 6.13 (from [191]). *Consider $n \times n$ matrices A and F. Assume that F is block-lower triangular (F_{11} and F_{22} need not be triangular),*

$$F = \begin{bmatrix} F_{11} & 0 \\ F_{21} & F_{22} \end{bmatrix},$$

partition A accordingly with F,

$$A = \begin{bmatrix} A_{11} & A_{12} \\ A_{21} & A_{22} \end{bmatrix},$$

and assume that A_{11} is invertible. Then

$$\text{rank}\left(A_{11} - F_{11}A_{11}F_{11}^T\right) \leq \text{rank}\left(A - FAF^T\right),$$

$$\text{rank}\left(\Delta - F_{22}\Delta F_{22}^T\right) \leq \text{rank}\left(A - FAF^T\right),$$

where $\Delta = A_{22} - A_{21}A_{11}^{-1}A_{12}$.

Therefore, the displacement rank of the Schur complement of a matrix can not exceed the displacement rank of the matrix itself.

The displacement rank of a symmetric Toeplitz matrix T,

$$T = \begin{bmatrix} t_0 & t_1 & \cdots & t_{n-2} & t_{n-1} \\ t_1 & t_0 & t_1 & & t_{n-2} \\ \vdots & \ddots & \ddots & \ddots & \\ t_{n-2} & & \ddots & \ddots & t_2 \\ t_{n-1} & t_{n-2} & & t_1 & t_0 \end{bmatrix}$$

with respect to the shift matrix Z is 2, since

$$T - ZTZ^T = \begin{bmatrix} t_0 & t_1 & \cdots & t_{n-2} & t_{n-1} \\ t_1 & 0 & \cdots & & 0 \\ \vdots & \vdots & \ddots & & \\ t_{n-2} & & & & \\ t_{n-1} & 0 & & & 0 \end{bmatrix}.$$

Therefore the Schur complement of symmetric Toeplitz matrices has displacement rank at most 2. Based on this property, fast and superfast Schur-like algorithms have been proposed in the literature (see [191] and the references therein) to solve linear systems involving such matrices.

6.5.2 The Schur reduction

The Schur complement plays an important role in factorizing symmetric matrices in the form LDL^T, with L lower triangular and D diagonal. In this subsection we first show how the latter factorization, also called Schur reduction, can be obtained if strongly nonsingular matrices are involved.

Let $A \in \mathbb{R}^{n \times n}$ and $1 \leq k < n$ and suppose $\Delta_{A,k}$ exists. Then it is easy to check that the following decomposition holds:

$$A = \begin{bmatrix} A_{11} & A_{12} \\ A_{21} & A_{22} \end{bmatrix}$$

$$= \begin{bmatrix} I_k & \\ A_{21}A_{11}^{-1} & I_{n-k} \end{bmatrix} \begin{bmatrix} A_{11} & \\ & \Delta_{A,k} \end{bmatrix} \begin{bmatrix} I_k & A_{11}^{-1}A_{12} \\ & I_{n-k} \end{bmatrix}.$$

Therefore, if all the principal minors of the matrix A are strongly nonsingular, the latter decomposition can be used recursively to compute the LDL^T factorization.

In fact, considering $k = 1$, we have

$$A = \begin{bmatrix} I_1 & \\ A_{21}A_{11}^{-1} & I_{n-1} \end{bmatrix} \begin{bmatrix} A_{11} & \\ & \Delta_{A,1} \end{bmatrix} \begin{bmatrix} I_1 & A_{11}^{-1}A_{12} \\ & I_{n-1} \end{bmatrix}. \qquad (6.25)$$

If all the principal minors of A are different from zero, we can easily prove that all the Schur complements of A, for any k, $1 \leq k < n$, are different from zero. Define $A^{(1)} = \Delta_{A,1}$. Then

$$A^{(1)} = \begin{bmatrix} I_1 & \\ A_{21}^{(1)}A^{(1)}{}_{11}^{-1} & I_{n-2} \end{bmatrix} \begin{bmatrix} A_{11}^{(1)} & \\ & \Delta_{A^{(1)},1} \end{bmatrix} \begin{bmatrix} I_1 & A^{(1)}{}_{11}^{-1}A_{12}^{(1)} \\ & I_{n-2} \end{bmatrix}, \qquad (6.26)$$

with $A_{11}^{(1)} \in \mathbb{R}^{1 \times 1}$. Replacing $\Delta_{A,1}$ in (6.25) with the right-hand side of (6.26), we have

$$A = \left[\begin{array}{c|cc} I_1 & & \\ \hline & I_1 & \\ A_{21}A_{11}^{-1} & A_{21}^{(1)}A^{(1)}{}_{11}^{-1} & I_{n-2} \end{array} \right] \left[\begin{array}{c|cc} A_{11} & & \\ \hline & A_{11}^{(1)} & \\ & & \Delta_{A^{(1)},1} \end{array} \right]$$

$$\left[\begin{array}{c|cc} I_1 & A_{11}^{-1}A_{12} & \\ \hline & I_1 & A^{(1)}{}_{11}^{-1}A_{12}^{(1)} \\ & & I_{n-2} \end{array} \right].$$

Recursively repeating the procedure on the Schur complement, the LDL^T factorization of A can be computed.

The Schur reduction is closely related to the Gaussian elimination for solving linear systems. In fact, considering for instance, the first step of the the Schur reduction (6.25), we observe that the factor L is nothing else than the inverse of the Gauss transform considered in Chapter 4. Therefore,

$$\begin{bmatrix} I_1 & \\ A_{21}A_{11}^{-1} & I_{n-1} \end{bmatrix} = M_1^{-1}.$$

$$M_1 A = \begin{bmatrix} A_{11} & \\ & \Delta_{A,1} \end{bmatrix} \begin{bmatrix} I_1 & A_{11}^{-1}A_{12} \\ & I_{n-1} \end{bmatrix}$$

$$= \begin{bmatrix} A_{11} & A_{12} \\ & \Delta_{A,1} \end{bmatrix}.$$

Hence, instead of computing the LDL^T factorization in the Gaussian elimination the Gaussian transforms are directly applied to the matrix to obtain the final upper triangular matrix.

6.5.3   The Schur complement of quasiseparable matrices

In this subsection we show that the rank structure of a matrix is preserved under the Schur complementation. Exploiting this property of the Schur algorithm, we will construct fast algorithms for the computation of the LDL^T factorization of a generator representable quasiseparable matrix.

Let us introduce the following lemma.

Lemma 6.14 (from [77]). *Let $A, B \in \mathbb{R}^{n \times n}$ be matrices for which the Schur complements $\Delta_{A,k}$ and $\Delta_{B,k}$ exist. If*

$$\operatorname{rank}(A - B) \leq r,$$

then

$$\operatorname{rank}(\Delta_{A,k} - \Delta_{B,k}) \leq r.$$

The latter lemma is now used to show that the rank structure is inherited by the Schur complement of generator representable quasiseparable matrices.

Let us consider a symmetric quasiseparable matrix in generator representable form,

$$A = \begin{bmatrix} d_1 & u_2 v_1 & \cdots & u_n v_1 \\ u_2 v_1 & d_2 & & u_n v_2 \\ \vdots & & \ddots & \vdots \\ u_n v_1 & u_n v_2 & \cdots & d_n \end{bmatrix}.$$

Without loss of generality, we suppose that the principal minors of order k, $1 \leq k < n$, of A are different from zero. The matrix A can be written as

$$
\begin{aligned}
A &= \begin{bmatrix} d_1 & u_2 v_1 & \cdots & u_n v_1 \\ u_2 v_1 & d_2 & & u_n v_2 \\ \vdots & & \ddots & \vdots \\ u_n v_1 & u_n v_2 & \cdots & d_n \end{bmatrix} \\
&= \begin{bmatrix} u_1 v_1 & u_1 v_2 & \cdots & u_1 v_n \\ u_2 v_1 & u_2 v_2 & \cdots & u_2 v_n \\ \vdots & \vdots & \ddots & \vdots \\ u_n v_1 & u_n v_2 & \cdots & u_n v_n \end{bmatrix} \\
&\quad + \begin{bmatrix} d_1 - u_1 v_1 & u_2 v_1 - u_1 v_2 & \cdots & u_n v_1 - u_1 v_n \\ & d_2 - u_2 v_2 & \cdots & u_n v_2 - u_2 v_n \\ & & \ddots & \vdots \\ & & & d_n - u_n v_n \end{bmatrix} \\
&= \mathbf{u}\mathbf{v}^T + \begin{bmatrix} d_1 - u_1 v_1 & u_2 v_1 - u_1 v_2 & \cdots & u_n v_1 - u_1 v_n \\ & d_2 - u_2 v_2 & \cdots & u_n v_2 - u_2 v_n \\ & & \ddots & \\ & & & d_n - u_n v_n \end{bmatrix} \\
&= \mathbf{u}\mathbf{v}^T + R.
\end{aligned}
$$

Hence,

$$A - R = \mathbf{u}\mathbf{v}^T$$

and

$$\text{rank}\,(A - R) = \text{rank}\,\left(\mathbf{u}\mathbf{v}^T\right) = 1.$$

Denote by $\Delta_{A,k}$ and $\Delta_{R,k}$ the Schur complements of the matrices A and R of order k. By Lemma 6.14, also

$$\text{rank}\,(\Delta_{A,k} - \Delta_{R,k}) = 1.$$

Now we observe that the subblock $R(k+1:n, 1:k) = 0$ because R is an upper triangular matrix. As a consequence, the Schur complement of order k of R is equal to the upper triangular block $R(k+1:n, k+1:n)$. Therefore, the lower triangular part of $\Delta_{A,k}$ is the lower triangular part of a rank 1 matrix. We have just proved the following theorem.

Theorem 6.15. *Let A be a strongly nonsingular generator representable quasiseparable matrix. Then, also the Schur complements $\Delta_{A,k}$, $k = 1, \ldots, n-1$ have the same structure.*

Similar theorems hold for more general structured rank matrices (see [77]), some results are presented in Section 6.5.5.

6.5.4   A Schur-like algorithm for quasiseparable matrices

In this subsection we derive a fast LDL^T decomposition of generator representable quasiseparable matrices based on the Schur complementation.

As an example, we describe the Schur reduction applied to a generator representable quasiseparable matrix of order 5.

$$A = \begin{bmatrix} d_1 & u_2 v_1 & u_3 v_1 & u_4 v_1 & u_5 v_1 \\ u_2 v_1 & d_2 & u_3 v_2 & u_4 v_2 & u_5 v_2 \\ u_3 v_1 & u_3 v_2 & d_3 & u_4 v_3 & u_5 v_3 \\ u_4 v_1 & u_4 v_2 & u_4 v_3 & d_4 & u_5 v_4 \\ u_5 v_1 & u_5 v_2 & u_5 v_3 & u_5 v_4 & d_5 \end{bmatrix}. \tag{6.27}$$

Define $D_0 = A$.
Let

$$t_1 = \frac{v_1^2}{d_1} \quad \text{and} \quad \tilde{v}_1 = \frac{v_1}{d_1}.$$

Applying the Schur complementation to D_0 with $k = 1$, we have

$$D_0 = \tilde{L}_1 D_1 \tilde{L}_1^T,$$

with

$$\tilde{L}_1 = \begin{bmatrix} 1 & & & & \\ u_2 \tilde{v}_1 & 1 & & & \\ u_3 \tilde{v}_1 & & 1 & & \\ u_4 \tilde{v}_1 & & & 1 & \\ u_5 \tilde{v}_1 & & & & 1 \end{bmatrix}, \quad D_1 = \begin{bmatrix} d_1 & \\ & \Delta_{D_0,1} \end{bmatrix}$$

and

$$
\Delta_{D_0,1} =
\begin{bmatrix}
d_2 & u_3 v_2 & u_4 v_2 & u_5 v_2 \\
u_3 v_2 & d_3 & u_4 v_3 & u_5 v_3 \\
u_4 v_2 & u_4 v_3 & d_4 & u_5 v_4 \\
u_5 v_2 & u_5 v_3 & u_5 v_4 & d_5
\end{bmatrix}
- t_1
\begin{bmatrix}
u_2 \\
u_3 \\
u_4 \\
u_5
\end{bmatrix}
\begin{bmatrix}
u_2 & u_3 & u_4 & u_5
\end{bmatrix}.
$$

Let $\tilde{D}_1 = \Delta_{D_0,1}$. The generator representable quasiseparable matrix $\tilde{D}_1$ can be written as follows.

$$
\tilde{D}_1 =
\begin{bmatrix}
d_2 - t_1 u_2^2 & u_3(v_2 - t_1 u_2) & u_4(v_2 - t_1 u_2) & u_5(v_2 - t_1 u_2) \\
u_3(v_2 - t_1 u_2) & d_3 - t_1 u_3^2 & u_4(v_3 - t_1 u_3) & u_5(v_3 - t_1 u_3) \\
u_4(v_2 - t_1 u_2) & u_4(v_3 - t_1 u_3) & d_4 - t_1 u_4^2 & u_5 v_4 - t_1 u_4) \\
u_5(v_2 - t_1 u_2) & u_5(v_3 - t_1 u_3) & u_5(v_4 - t_1 u_4) & d_5 - t_1 u_5^2
\end{bmatrix}.
$$

Define

$$
\hat{d}_2 = d_2 - t_1 u_2^2, \quad \tilde{v}_2 = \frac{v_2 - t_1 u_2}{\hat{d}_2} \quad \text{and} \quad t_2 = \frac{\tilde{v}_2^2}{\hat{d}_2}.
$$

Again, applying the Schur complementation to $\tilde{D}_1$ with $k = 1$, we have

$$
\tilde{D}_1 = \tilde{L}_2 D_2 \tilde{L}_2^T,
$$

with

$$
\tilde{L}_2 =
\begin{bmatrix}
1 & & & \\
u_3 \tilde{v}_2 & 1 & & \\
u_4 \tilde{v}_2 & & 1 & \\
u_5 \tilde{v}_2 & & & 1
\end{bmatrix},
\quad
D_2 =
\begin{bmatrix}
\hat{d}_2 & \\
& \Delta_{\tilde{D}_1,1}
\end{bmatrix}
$$

and

$$
\Delta_{\tilde{D}_1,1} =
\begin{bmatrix}
d_3 - t_1 u_3^2 & u_4(v_3 - t_1 u_3) & u_5(v_3 - t_1 u_3) \\
u_4(v_3 - t_1 u_3) & d_4 - t_1 u_4^2 & u_5(v_4 - t_1 u_4) \\
u_5(v_3 - t_1 u_3) & u_5(v_4 - t_1 u_4) & d_5 - t_1 u_5^2
\end{bmatrix}
$$

$$
- t_2
\begin{bmatrix}
u_3 \\
u_4 \\
u_5
\end{bmatrix}
\begin{bmatrix}
u_3 & u_4 & u_5
\end{bmatrix}
$$

$$
=
\begin{bmatrix}
d_3 - (t_1 + t_2) u_3^2 & u_4(v_3 - (t_1 + t_2) u_3) & u_5(v_3 - (t_1 + t_2) u_3) \\
u_4(v_3 - (t_1 + t_2) u_3) & d_4 - (t_1 + t_2) u_4^2 & u_5(v_4 - (t_1 + t_2) u_4) \\
u_5(v_3 - (t_1 + t_2) u_3) & u_5(v_4 - (t_1 + t_2) u_4) & d_5 - (t_1 + t_2) u_5^2
\end{bmatrix}.
$$

Define

$$
\hat{d}_3 = d_3 - (t_1 + t_2) u_3^2, \quad \tilde{v}_3 = \frac{v_3 - (t_1 + t_2) u_3}{\hat{d}_3}, \quad t_3 = \frac{\tilde{v}_3^2}{\hat{d}_3},
$$

and

$$
\tilde{D}_2 = \Delta_{\tilde{D}_1,1}.
$$

Considering the Schur complementation of $\tilde{D}_2$ with $k = 1$, we have

$$
\tilde{D}_2 = \tilde{L}_3 D_3 \tilde{L}_3^T,
$$

with

$$\tilde{L}_3 = \begin{bmatrix} 1 & & \\ u_4\tilde{v}_3 & 1 & \\ u_5\tilde{v}_3 & & 1 \end{bmatrix}, \quad D_3 = \begin{bmatrix} \hat{d}_3 & \\ & \Delta_{\tilde{D}_2,1} \end{bmatrix}$$

and

$$\Delta_{\tilde{D}_2,1} = \begin{bmatrix} d_4 - (t_1 + t_2 + t_3)u_4^2 & u_5(v_4 - (t_1 + t_2 + t_3)u_4) \\ u_5(v_4 - (t_1 + t_2 + t_3)u_4) & d_5 - (t_1 + t_2 + t_3)u_5^2 \end{bmatrix}$$

Again, define

$$\hat{d}_4 = d_4 - (t_1 + t_2 + t_3)u_4^2, \quad \tilde{v}_4 = \frac{v_4 - (t_1 + t_2 + t_3)u_4}{\hat{d}_4}, \quad t_4 = \frac{\tilde{v}_4^2}{\hat{d}_4},$$

and

$$\tilde{D}_3 = \Delta_{\tilde{D}_2,1}.$$

The last step of the reduction is given by considering the Schur complement of $\tilde{D}_3$ with $k = 1$,

$$\tilde{D}_3 = \tilde{L}_4 = \begin{bmatrix} 1 & \\ u_5\tilde{v}_4 & 1 \end{bmatrix}, \quad \begin{bmatrix} \hat{d}_4 & \\ & \hat{d}_5 \end{bmatrix},$$

with

$$\hat{d}_5 = \Delta_{\tilde{D}_2,1} = d_5 - (t_1 + t_2 + t_3 + t_4)v_5^2.$$

Let

$$L_i = \begin{bmatrix} I_{i-1} & \\ & \tilde{L}_i \end{bmatrix}, \quad i = 1, \ldots, 4.$$

Hence, it turns out

$$A = LDL^T,$$

where

$$L = L_1 L_2 \cdots L_4 = \begin{bmatrix} 1 & & & & \\ u_2\tilde{v}_1 & 1 & & & \\ u_3\tilde{v}_1 & u_3\tilde{v}_2 & 1 & & \\ u_4\tilde{v}_1 & u_4\tilde{v}_2 & u_4\tilde{v}_3 & 1 & \\ u_5\tilde{v}_1 & u_5\tilde{v}v_2 & u_5\tilde{v}_3 & u_5\tilde{v}_4 & 1 \end{bmatrix}$$

and

$$D = \begin{bmatrix} \hat{d}_1 & & & & \\ & \hat{d}_2 & & & \\ & & \hat{d}_3 & & \\ & & & \hat{d}_4 & \\ & & & & \hat{d}_5 \end{bmatrix}.$$

The matrix L of the LDL^T factorization is a lower generator representable quasiseparable matrix with generators $\mathbf{u}$ and $\tilde{\mathbf{v}}$.

Suppose we want to solve a linear system $A\mathbf{x} = \mathbf{b}$, where A is a generator representable quasiseparable matrix, with all principal minors different from zero.

Once the LDL^T factorization of the matrix is computed, we need to solve three simple linear systems,

$$L\mathbf{y} = \mathbf{b} \tag{6.28}$$

$$D\mathbf{z} = \mathbf{y} \tag{6.29}$$

$$L^T\mathbf{x} = \mathbf{z}. \tag{6.30}$$

The systems (6.28) and (6.30) are lower and upper triangular having a quasiseparable structure. Therefore, they can be solved with the backward and forward substitution algorithms described in Chapter 4. The matrix of the linear system (6.29) is diagonal. Therefore, this system can be solved in $\mathcal{O}(n)$ flops.

6.5.5 A more general framework for the Schur reduction

In Subsection 6.5.3 it is proved that the Schur complement of a generator representable semiseparable plus diagonal matrix preserves the structure.

In this section we prove more general theorems (see [77]).

Theorem 6.16. *Let $A \in \mathbb{R}^{n \times n}$ and $A_k = A(1:k, 1:k)$ with $\det(A_k) \neq 0$, $1 \leq k < n$. Suppose*

$$\text{rank}\,(A(k+l:n, 1:k+m)) \leq p, \qquad \begin{matrix} 1 \leq l \leq n-k \\ 1 \leq m \leq n-k \end{matrix}.$$

Then

$$\text{rank}\,(\Delta_{A,k}(l:n-k, 1:m)) \leq p.$$

Proof. Define

$$\tilde{C} = A(k+l:n, 1:k), \quad \tilde{D} = A(k+l:n, k+1:k+m) \text{ and } \tilde{B} = A(1:k, k+1:k+m).$$

It turns out that

$$\Delta_{A,k}(l:n-k, 1:m) = \tilde{D} - \tilde{C}A_k^{-1}\tilde{B}.$$

Hence

$$\text{rank}\,(\Delta_{A,k}(l:n-k, 1:m)) \leq p,$$

since the columns of $\Delta_{A,k}(l:n-k, 1:m)$ are linear combinations of the columns of $A(k+l:n, 1:k+m)$. □

A similar theorem holds for low rank blocks in the upper triangular part of the matrix.

Theorem 6.17. *Let $A \in \mathbb{R}^{n \times n}$ and $A_k = A(1:k, 1:k)$ with $\det(A_k) \neq 0$, $1 \leq k < n$. Suppose*

$$\text{rank}\,(A(1:k+l, k+m:n)) \leq p, \qquad \begin{matrix} 1 \leq l \leq n-k \\ 1 \leq m \leq n-k \end{matrix}.$$

Then

$$\text{rank}\left(\Delta_{A,k}(1:l,m:n-k)\right) \le p.$$

We cannot use the previous two theorems for semiseparable plus diagonal matrices as their low rank structure expands below the diagonal. However, Theorems 6.16 and 6.17 can be generalized to include the diagonal. We will give the generalization of Theorem 6.16 without proof. The proof is very similar to the proof of Theorem 6.16.

Theorem 6.18. *Let* $A \in \mathbb{R}^{n \times n}$ *and* $A_k = A(1:k, 1:k)$ *with* $\det(A_k) \ne 0$, $1 \le k < n$. *Let* D *be a diagonal matrix and consider the matrix* $A_D = A - D$. *Suppose*

$$\text{rank}\left(A_D(k+l:n, 1:k+m)\right) \le p, \qquad \begin{matrix} 1 \le l \le n-k \\ 1 \le m \le n-k \end{matrix}.$$

Consider the matrix $B = \Delta_{A,k} - D(1:n-k, 1:n-k)$. *Then*

$$\text{rank}\left(B(l:n-k, 1:m)\right) \le p.$$

Example 6.19   • The entries of a tridiagonal matrix $T \in \mathbb{R}^{n \times n}$ below the subdiagonal and above the superdiagonal are zero, i.e., for any $l \in \{1, 2, \ldots, n-2\}$

$$T(l+2:n, 1:l) = 0$$
$$\Updownarrow$$
$$\text{rank}\left(T(l+2:n, 1:l)\right) = 0.$$

Therefore, applying Theorem 6.16,

$$\text{rank}\left(\Delta_{T,k}(l+2:n-k, 1:l)\right) = 0, \qquad l = 1, 2, \ldots, n-k-2.$$

i.e., the Schur complement, if it exists, inherits the tridiagonal structure.

• For a semiseparable matrix $S \in \mathbb{R}^{n \times n}$, we have

$$\text{rank}\left(S(l:n, 1:l)\right) \le 1, \quad l = 1, 2, \ldots, n.$$

Therefore, by Theorem 6.16, if the Schur complement $\Delta_{S,k}$ exists,

$$\text{rank}\left(\Delta_{S,k}(l:n-k, 1:l)\right) \le 1, \quad l = 1, 2, \ldots, n-k,$$

i.e., the Schur complement inherits the semiseparable structure.

• For a quasiseparable matrix $A \in \mathbb{R}^{n \times n}$, we have

$$\text{rank}\left(A(k+1:n, 1:k)\right) \le 1, \quad k = 1, 2, \ldots, n-1.$$

Therefore, by Theorem 6.16, if the Schur complement $\Delta_{A,k}$ exists,

$$\text{rank}\left(\Delta_{A,k}(l+1:n-k, 1:l)\right) \le 1, \quad l = 1, 2, \ldots, n-k-1,$$

i.e., the Schur complement inherits the quasiseparable structure.

- For a semiseparable plus diagonal matrix $S + D \in \mathbb{R}^{n \times n}$, we have

$$\text{rank}\,(S(k+1:n,1:k)) \leq 1, \quad k = 1, 2, \ldots, n-1.$$

Therefore, by Theorem 6.18, if the Schur complement $\Delta_{S+D,k}$ exists, the structured rank of the matrix $\tilde{S} = \Delta_{S+D,k} - D(1:n-k,1:n-k)$ satisfies

$$\text{rank}\,\left(\tilde{S}(l:n-k,1:l)\right) \leq 1, \quad l = 1, 2, \ldots, n-k-1,$$

i.e., the Schur complement inherits the semiseparable plus diagonal structure.

∎

Since all these matrices have their structures preserved under Schur complementation one can also design Schur-like algorithms for these matrices.

Notes and references

The algorithm for quasiseparable matrices described in this section is based on the following papers.

- ☞ Y. Eidelman and I. C. Gohberg. A look-ahead block Schur algorithm for diagonal plus semiseparable matrices. *Computers & Mathematics with Applications*, 35(10):25–34, 1997.

- ☞ I. C. Gohberg, T. Kailath, and I. Koltracht. Linear complexity algorithms for semiseparable matrices. *Integral Equations and Operator Theory*, 8(6):780–804, 1985.

- ☞ I. C. Gohberg, T. Kailath, and I. Koltracht. A note on diagonal innovation matrices. *Acoustics Speech and Signal Processing*, 7:1068–1069, 1987.

In the 1985 and 1987 papers, an algorithm for computing the LDU-factorization of a higher order generator representable semiseparable plus diagonal matrix and of its inverse with linear complexity in the order of the matrix has been described. The involved higher order semiseparable plus diagonal matrix is strongly regular (strongly nonsingular).

In the 1997 paper the authors extend the latter algorithm to nonsingular higher order semiseparable plus diagonal matrices, relaxing the hypothesis of strong regularity of the involved matrices, considering a look-ahead technique introduced in [122] for solving nonsingular Toeplitz linear systems by a look-ahead Levinson algorithm.

- ☞ S. Delvaux and M. Van Barel. Structures preserved by Schur complementation. *SIAM Journal on Matrix Analysis and its Applications*, 28(1):229–252, 2006.

In this manuscript Delvaux and Van Barel discuss which structures are preserved under Schur complementation. They discuss displacement rank structures and also different types of rank structures. The theorems presented in the last section are based on the ones in this manuscript.

6.6 Conclusions

This chapter was concerned with the development of a general Levinson-like frame-work for structured rank matrices. First, we briefly recapitulated the Levinson-idea for Toeplitz matrices; second, we illustrated the possibility of developing such a solver by applying it to semiseparable plus diagonal matrices. The third section is the most interesting section. It contains a more general formulation of the Levinson-like algorithm. The method was derived for a special class of matrices named $\{p_1, p_2\}$-Levinson conform matrices, proving thereby the existence of an $\mathcal{O}(p_1 p_2 n)$ order solver. The power of the method was illustrated by applying it to different examples. Finally an upper triangular factorization of the inverse of the $\{p_1, p_2\}$-Levinson conform matrices was also presented, as well as hints for applying look-ahead methods, to make the algorithm more robust when numerical problems occur. In the last sections of this chapter we briefly discussed a Schur-like algorithm for quasiseparable matrices. A theorem was also provided for maintaining the rank structure under Schur complementation.

Chapter 7

Inverting semiseparable and related matrices

The inverses of semiseparable and semiseparable plus diagonal matrices have often been used in statistical applications; hence, the extensive list of references related to this subject.

In this chapter, we will not discuss all inversion methods, but we will present some techniques/ideas for inverting semiseparable and related matrices. We will not go too deep into detail, as quite often the matrices one wants to invert in a specific application have a very specific rank structure. It is impossible to discuss all possible inversion methods, so we choose to present some general methods based on factorization, and some, in our opinion, interesting direct inversion methods.

We start by discussing the use of factorization methods for inverting structured rank matrices. We mention the use of the QR-factorization, the LU-decomposition and the Levinson idea. All three methods will provide a factored form for the inverse of the considered structured rank matrix. In Section 7.2 we discuss some direct inversion methods. We start with inverting symmetric tridiagonal and semiseparable matrices, and we finish with unsymmetric tridiagonal and semiseparable matrices. For inverting tridiagonal matrices we discuss different representations, in which we want to obtain our final semiseparable matrix. Section 7.3 provides some standard formulas, which might come in handy when inverting slightly perturbed structured rank matrices. In Section 7.4 we discuss the optimal scaling of a semiseparable matrix. To scale a semiseparable matrix optimally, we need to compute the diagonal elements of its inverse, that is why this section has been placed in this chapter. Section 7.5 discusses the decay rates of the elements of semiseparable matrices coming from the inverse of diagonally dominant tridiagonal matrices. It will be shown that these elements decay exponentially when going away from the diagonal.

✎ *This chapter is concerned with the inversion of structured rank matrices and different aspects related to these structured rank matrices. Inversion of structured rank matrices via factorizations is discussed in Section 7.1. Of course this results in a factored form of the inverse. If one is interested in a direct inversion*

257

method, several algorithms are presented in Section 7.3. One method for symmetric generator representable matrices was already presented in Chapter 3, Section 3.1. These methods discuss the inversion of specific structured rank matrices, depending on their specific representation. Algorithms are presented for inverting symmetric tridiagonal matrices (for which the inverse is represented with the Givens-vector and the generator representation), the inverse of a Givens-vector represented symmetric semiseparable matrix, two methods for inverting a general tridiagonal matrix and the inverse of a specific generator represented semiseparable matrix.

If one wants to invert a matrix, perturbed by a low rank matrix, one needs to read Section 7.3, which presents some standard equations for inverting (low rank) perturbed matrices.

Throughout Section 7.5, the most important theorems are the final ones, from Theorem 7.13 to Theorem 7.15, discussing the decay of the elements in the semiseparable inverse of diagonally dominant tridiagonal matrices.

7.1 Known factorizations

Based on factorizations, one can often easily invert matrices. In this section we briefly discuss some factorizations developed for the class of structured rank matrices.

7.1.1 Inversion via the QR-factorization

Suppose a structured rank matrix A is given, and its QR-factorization is computed. It is straightforward that the following relation also holds:

$$A = QR,$$
$$A^{-1} = R^{-1}Q^{-1}.$$

Hence, if one has computed the QR-factorization, inversion of Q and R leads to a factored form of the inverse of the matrix A. Let us shortly investigate now in more detail the inverses of the factors Q and R.

The inversion of the factor Q is simple. The matrix Q is an orthogonal matrix given in factored form, namely as a product of successive Givens transformations. In total maximum $2n - 1$ Givens transformations are involved for computing the QR-factorization of either a semiseparable, quasiseparable or semiseparable plus diagonal matrix. Instead of computing the full matrix Q, it is more efficient to store these Givens transformations and invert them. The inversion of a Givens transformation is trivial.

Computing the inverse of the factor R can be done via several methods. In case no structure is involved in the matrix R, e.g., when computing the QR-factorization of a Hessenberg-like matrix, one can invert the upper triangular matrix R directly via a standard inversion technique. Formulas for inverting the matrix R are easily computed by using the fact that R^{-1} is also upper triangular and has as diagonal elements the inverse diagonal elements of R. Using $RR^{-1} = R^{-1}R = I$, one can easily compute the remaining elements.

In case the upper triangular matrix is of structured rank form, one can often reduce the complexity of the inversion methods. We know, based on the nullity theorem (see Section 1.5.1), the explicit rank structure of the inverse of the matrix R. Based on this explicit theoretical structure, we can choose an efficient representation of this matrix R and then compute its inverse by simplifying the standard technique; e.g., the inverse of an upper triangular semiseparable matrix will be an upper bidiagonal matrix. (More on inverting bidiagonal matrices was discussed in Chapter 4, Section 4.3.)

7.1.2 Inversion via the LU-factorization

Inversion via the LU-factorization proceeds in a similar manner as the inversion via the QR-factorization. Let us discuss it briefly.

Suppose a structured rank matrix A is given, and its LU-factorization is computed as presented in Chapter 4. This means that we have a factorization of the following form:

$$PA = LU,$$
$$A = PLU,$$
$$A^{-1} = U^{-1}L^{-1}P.$$

Therefore, we can easily compute the inverse of the matrix A once its LU-decomposition is known. Inversion of the permutation[10] matrix P is not necessary as $P^{-1} = P$. The inverse of the matrix L can be computed rather easily as it is given in factored form, where every factor is a Gauss transform. Moreover, when constructing the LU-decomposition, one computes in fact:

$$L^{-1}PA = U,$$

hence, the factor L^{-1} is known, by construction.

Similarly, as in the QR-factorization, one also needs to invert the upper triangular matrix U. One can do this via direct inversion techniques. In case the matrix U is of structured rank form, one can exploit this structure and reduce the complexity of the inversion.

Of course one can also use this technique for inverting tridiagonal and/or Hessenberg matrices. The LU-decomposition of these matrices can be computed rather easily, as only one subdiagonal needs to be annihilated.

7.1.3 Inversion via the Levinson algorithm

The Levinson method as presented in this book yields a specific factorization of the inverse of the matrix A, on which we applied the Levinson method (see Section 6.3.3). The factorization is of the following form

$$A = L\Lambda U^{-1},$$
$$A^{-1} = U\Lambda^{-1}L^{-1},$$

---

[10]The relation $P^{-1} = P$ holds for interchange permutations; these permutations swap two rows. In general a permutation matrix satisfies $P^{-1} = P^{T}$.

where U and L are upper and lower triangular matrices, respectively, with ones on the diagonal, and Λ a diagonal matrix.

More information concerning this factorization can be found in Section 6.3.3 of Chapter 6.

Notes and references

The LU-decomposition for inverting tridiagonal matrices has already been discussed previously in 1953 by Berger and Saibel.

> ☞ W. J. Berger and E. Saibel. On the inversion of continuant matrices. *Journal of the Franklin Institute*, 256:249–253, 1953.

Berger and Saibel provide in this paper an explicit formula for calculating the inverse of a continuant matrix (this is a tridiagonal matrix, not necessarily symmetric). The interest in computing the inverse is based on physical applications in which one desires the smallest instead of the largest eigenvalue. Applying iterative methods on the inverse of this matrix will often converge to the smallest eigenvalues, which is the ultimate goal of this method, according to the authors.

The method of the authors is based on computing a UL-decomposition of the tridiagonal matrix. Indeed, they apply transformations from bottom to top on the tridiagonal matrix to obtain a lower triangular bidiagonal matrix[11]:

$$UA = B,$$

with B a lower bidiagonal matrix. Then a sequence of transformations from top to bottom is used, thereby annihilating the subdiagonal elements of the matrix B (and scaling the remaining diagonal elements):

$$LUA = I.$$

We see that we calculated a UL-decomposition:

$$A = U^{-1}L^{-1}.$$

The authors provide, in fact, a factored form for the inverse of this tridiagonal matrix. Unfortunately, even though explicit formulas for the elements are given, nothing is mentioned about a way to effectively represent the resulting semiseparable matrices. The semiseparable structure is clearly present in the formulas but the authors do not mention the relation between the elements in this matrix. They do however investigate under which assumptions one obtains an inverse matrix having gnomonic symmetry (which are special types of semiseparable matrices). For example, the following matrices are gnomonic symmetric:

$$
\begin{bmatrix} 3 & 3 & 3 & 3 \\ 3 & 7 & 7 & 7 \\ 3 & 7 & 13 & 13 \\ 3 & 7 & 13 & 25 \end{bmatrix}
\text{ and }
\begin{bmatrix} 25 & 13 & 7 & 3 \\ 13 & 13 & 7 & 3 \\ 7 & 7 & 7 & 3 \\ 3 & 3 & 3 & 3 \end{bmatrix}.
$$

Two variants of gnomonic matrices exist; both are depicted here. In the first variant every element below (above) the diagonal is exactly the same as the diagonal element in which column (row) it is standing. The second variant has every element below (above) the diagonal exactly the same as the diagonal element in which row (column) it is standing. It is clear that this is a very specific semiseparable matrix.

---

[11]The transformations are upper triangular Gauss transforms used for annihilating elements in the upper triangular part.

☞ E. Neuman. The inversion of certain band matrices. *Roczniki Polskiego Towarzystwa Matematycznego*, 3(9):15–24, 1977. (In Russian).

Neuman provides in this manuscript algorithms for inverting tri- and pentadiagonal matrices (not necessarily symmetric) via the UL-decomposition. Under some additional conditions the numerical stability of the algorithm is proved. Explicit formulas are provided for computing the inverse, thereby exploiting the semiseparable structure of the inverse.

7.2   Direct inversion methods

There are numerous methods for inverting semiseparable matrices. In the previous section we discussed inversion methods based on the LU- or QR-factorization. Also the Levinson algorithm provided us a factorization of the inverse.

Historically, however, most of the inversion methods were direct methods based on simple computational techniques for deriving the inverse. In this section we present a few of these algorithms, as there is a huge variety of methods.

This section has a rather large footnotes and references section, divided in different parts to easily recover inversion results related to the different classes of structured rank matrices.

We have already explored one direct method for inverting generator representable symmetric matrices in Chapter 3 in Section 3.1. This method by Gantmacher and Kreĭn used determinantal formulas to obtain a closed expression for the inverse.

We will first provide a method for inverting a symmetric tridiagonal matrix to obtain either the Givens-vector representation or the generator representation. Second, we will invert a symmetric Givens-vector represented semiseparable matrix to compute the inverse tridiagonal matrix. In the following two subsections, we will skip the condition of symmetry. In the third subsection we will provide algorithms for inverting a nonsymmetric tridiagonal matrix in two ways. We use the nonsymmetric generator representation from Section 2.7, Equation (2.13) and the standard representation with 4 vectors. Finally, we conclude this section presenting an algorithm for inverting a special semiseparable matrix admitting a representation with three vectors. This section is concluded with an extensive list of references.

7.2.1   The inverse of a symmetric tridiagonal matrix

In this section we construct an algorithm for computing the inverse of symmetric irreducible tridiagonal matrices. The constraint of being irreducible is nonrestrictive as every tridiagonal matrix can be written as a block diagonal matrix for which the blocks are irreducible. Two algorithms are provided. One algorithm gives the semiseparable matrix with the Givens-vector representation, and the second method leads to a semiseparable matrix with the generator representation.

The Givens-vector representation

We want to retrieve at once the Givens-vector representation of the semiseparable matrix. We will demonstrate again this case on a 4×4 example, which will give us enough information to construct the general case. Let us denote the tridiagonal matrix T and its inverse, the semiseparable matrix S, with the Givens-vector representation, in the following way:

$$
T = \begin{bmatrix} a_1 & b_1 & & \\ b_1 & a_2 & b_2 & \\ & b_2 & a_3 & b_3 \\ & & b_3 & a_4 \end{bmatrix} \text{ and } S = \begin{bmatrix} c_1 v_1 & c_2 s_1 v_1 & c_3 s_2 s_1 v_1 & s_3 s_2 s_1 v_1 \\ c_2 s_1 v_1 & c_2 v_2 & c_3 s_2 v_2 & s_3 s_2 v_2 \\ c_3 s_2 s_1 v_1 & c_3 s_2 v_2 & c_3 v_3 & s_3 v_3 \\ s_3 s_2 s_1 v_1 & s_3 s_2 v_2 & s_3 v_3 & v_4 \end{bmatrix}.
$$

We will start by computing the cosines and sines, namely c_i and s_i for the Givens transformations in the representation. Writing down the equations corresponding with $\mathbf{e}_4^T T S$ collapses into two single equations:

$$
\begin{cases} s_3 v_3 b_3 + a_4 v_4 & = & 1 \\ c_3 b_3 + s_3 a_4 & = & 0. \end{cases}
$$

Using the second of these equations, one can easily determine the parameters c_3 and s_3. Let us assume they are the cosine and the sine of the angle θ_3, i.e., $\cos(\theta_3) = c_3$ and $\sin(\theta_3) = s_3$. This leads to $\tan(\theta_3) = -b_3/a_4$. This equation is well-defined, if $a_4 = 0$ we take $c_3 = 0$. As c_3 and s_3 are known now, we can use them to calculate the cosine c_2 and sine s_2 associated with the angle θ_2 by using the equations coming from $\mathbf{e}_3^T T S$:

$$
\tan(\theta_2) = \frac{-b_2}{c_3 a_3 + s_3 b_3}.
$$

Finally we can also calculate the cosine c_1 and sine s_1 associated with the angle θ_1, based on the equations $\mathbf{e}_2^T T S$:

$$
\tan(\theta_1) = \frac{-b_1}{c_2 a_2 + c_3 s_2 b_2}.
$$

In fact we have already determined all the dependencies between the rows of the semiseparable matrix S. Only the calculation of the elements in the vector $\mathbf{v}$, the vector of the Givens-vector representation, remains.

Writing down the equations corresponding to the first column of $TS = I$, namely $T S e_1 = \mathbf{e}_1$, gives us the following four equations:

$$c_1 v_1 a_1 + c_2 s_1 v_1 b_1 = 1 \tag{7.1}$$

$$c_1 v_1 b_1 + c_2 s_1 v_1 a_2 + c_3 s_2 s_1 v_1 b_2 = 0 \tag{7.2}$$

$$c_2 s_1 v_1 b_2 + c_3 s_2 s_1 v_1 a_3 + s_3 s_2 s_1 v_1 b_3 = 0 \tag{7.3}$$

$$c_3 s_2 s_1 v_1 b_3 + s_3 s_2 s_1 v_1 a_4 = 0. \tag{7.4}$$

The first equation gives us enough information to determine v_1. Using Equation (7.1) we can already calculate v_1 as

$$
v_1 = \frac{1}{c_1 a_1 + c_2 s_1 b_1}.
$$

This equation is always well-defined, because v_1 needs to be different from zero.

Next we use the equation $\mathbf{e}_2^T TS\mathbf{e}_2 = 1$ to calculate element v_2. This gives us, using thereby the equality for $\tan(\theta_1)$:

$$c_2 s_1 v_1 b_1 + c_2 v_2 a_2 + c_3 s_2 v_2 b_2 = 1$$
$$v_2 \left(c_2 a_2 + c_3 s_2 b_2 \right) = 1 - c_2 s_1 v_1 b_1$$
$$v_2 \frac{-b_1 c_1}{s_1} = 1 - c_2 s_1 v_1 b_1. \tag{7.5}$$

Remark that $s_1 \neq 0$ as this would imply a block in the semiseparable matrix and therefore also a block in the tridiagonal matrix, which is impossible as all subdiagonal elements are different from zero.

If $c_1 = 0$, we know by Equation (7.1) that the right-hand side of Equation (7.5) will also be zero. Therefore, we define $v_2 = 0$, if $c_1 = 0$, and else we compute v_2 using the expression above. A similar construction can be made for v_3:

$$v_3 \frac{-b_2 c_2}{s_2} = 1 - c_3 s_2 v_2 b_2.$$

The equation for the last vector element v_4 is slightly different:

$$v_4 a_4 = 1 - s_3 v_3 b_3$$
$$v_4 \frac{-b_3 c_3}{s_3} = 1 - s_3 v_3 b_3.$$

This 4×4 example illustrates the general procedure for inverting a tridiagonal matrix and obtaining the Givens-vector representation of the inverse semiseparable matrix.

The generator representation

Another inversion method for symmetric tridiagonal matrices was discussed earlier in Subsection 3.1.3. Let us briefly present an algorithm for computing the generators of the inverse of an irreducible symmetric tridiagonal matrix. This method was presented in [61]. The proof can also be found in this manuscript.

Suppose we have an irreducible tridiagonal matrix T with diagonal elements a_i and sub(super)diagonal elements equal to $-b_i$. The inverse of T is denoted as $S(\mathbf{u}, \mathbf{v})$, where the generators $\mathbf{u}$ and $\mathbf{v}$ can be computed as follows.[12]

---

[12]The choice of $-b_i$ instead of b_i might seem arbitrary. These formulas originate, however, from the section on the decay rates in which the elements b_i are assumed to be positive.

$$v_1 = 1, v_2 = \frac{a_1}{b_1},$$

$$v_i = \frac{a_{i-1}v_{i-1} - b_{i-2}v_{i-2}}{b_{i-1}}, \quad 3 \le i \le n,$$

$$u_n = \frac{1}{-b_{n-1}v_{n-1} + a_n v_n},$$

$$u_i = \frac{1 + b_i v_i u_{i+1}}{a_i v_i - b_{i-1}v_{i-1}}, \quad 2 \le i \le n-1,$$

$$u_1 = \frac{1 + b_1 v_1 u_2}{a_1 v_1}.$$

We have now discussed two variants for computing the inverse of a symmetric tridiagonal matrix. Let us now compute the inverse of a symmetric semiseparable matrix.

7.2.2 The inverse of a symmetric semiseparable matrix

Traditional inversion methods of semiseparable matrices are based on the generator representation. This means that the lower triangular part of a semiseparable matrix as well as the upper triangular part are coming from rank 1 matrices. An inversion method for symmetric generator representable semiseparable matrices was discussed earlier in Subsection 3.1.3. Based on the Givens-vector representation, we will derive here an inversion formula for the class of semiseparable matrices representable in this way. The algorithm will be designed for a 4×4 semiseparable matrix, as this illustrates the general case. Our semiseparable matrix S is of the following form:

$$S = \begin{bmatrix} c_1 v_1 & c_2 s_1 v_1 & c_3 s_2 s_1 v_1 & s_3 s_2 s_1 v_1 \\ c_2 s_1 v_1 & c_2 v_2 & c_3 s_2 v_2 & s_3 s_2 v_2 \\ c_3 s_2 s_1 v_1 & c_3 s_2 v_2 & c_3 v_3 & s_3 v_3 \\ s_3 s_2 s_1 v_1 & s_3 s_2 v_2 & s_3 v_3 & v_4 \end{bmatrix}.$$

This matrix is stored by using $3n - 2$ parameters, namely, the c_i, s_i and the v_i ($v_1 \neq 0$ as the matrix is invertible). Note, however, that the c_i and s_i represent the cosine and sine of the same angle, this means that essentially $2n - 1$ parameters are needed to represent this matrix. The inverse of the semiseparable matrix S is denoted as T:

$$T = \begin{bmatrix} a_1 & b_1 & & \\ b_1 & a_2 & b_2 & \\ & b_2 & a_3 & b_3 \\ & & b_3 & a_4 \end{bmatrix}.$$

Let us start by calculating a_1 and b_1. The 4 equations corresponding to the first column of the product $ST = I$, namely, $ST\mathbf{e}_1 = \mathbf{e}_1$ (the elements $\mathbf{e}_i$ represent the standard basis vectors), are:

$$a_1(c_1 v_1) + b_1(c_2 s_1 v_1) = 1, \tag{7.6}$$

whereas the remaining equations collapse into one single equation. (This happens because the two first columns of S are dependent below the first row.) One can easily verify that even if, e.g., $c_2 = 0$, the equation below still needs to hold. Because $c_2 = 0$ implies $s_2 = 1$ and hence there are two more equations involving c_3 and s_3 from which this equality can be deduced.

$$a_1(s_1 v_1) + b_1 v_2 = 0. \tag{7.7}$$

Rewriting Equation (7.7) towards $a_1 v_1$ we get

$$a_1 v_1 = \frac{-b_1 v_2}{s_1}.$$

If $s_1 = 0$ we can easily see that $b_1 = 0$ and $a_1 = 1/v_1$, so let us assume that s_1 is different from zero. (We note that by definition of the representation the cosines are always positive. If $s_1 = 0$ this implies $c_1 = 1$.) Substituting $a_1 v_1$ in Equation (7.6) and rewriting the equation towards b_1 gives us

$$b_1 = \frac{1}{c_2 s_1 v_1 - v_2 \frac{c_1}{s_1}}.$$

Note that this equation is well defined as s_1 is different from zero and an easy calculation reveals that the denominator in the equation also has to be different from zero, otherwise the semiseparable matrix would be singular. Once we have b_1 the calculation of a_1 is straightforward using Equation (7.7). We have calculated now a_1 and b_1. To continue, we calculate a_2 and b_2. We write down the equations corresponding to the second column of $ST = I$, i.e., $ST\mathbf{e}_2 = \mathbf{e}_2$, let us consider the equations on and below the diagonal. This gives us:

$$\mathbf{e}_2^T ST\mathbf{e}_2 = b_1(c_2 s_1 v_1) + a_2(c_2 v_2) + b_2(c_3 s_2 v_2) = 1, \tag{7.8}$$

while the equations $\mathbf{e}_3^T ST\mathbf{e}_2 = 0$ and $\mathbf{e}_4^T ST\mathbf{e}_2 = 0$ collapse again into one single equation:

$$b_1(s_2 s_1 v_1) + a_2(s_2 v_2) + b_2 v_3 = 0. \tag{7.9}$$

We will distinguish between two cases now: $s_2 = 0$ and $s_2 \neq 0$.

- Let us assume $s_2 = 0$. This implies directly that $b_2 = 0$. Again we have to distinguish between two cases to calculate a_2. Assuming $s_1 = 0$ we get $a_2 = 1/v_2$.

  Assuming now $s_1 \neq 0$, we can write down equation $\mathbf{e}_1^T ST\mathbf{e}_2 = 0$:

  $$b_1(c_1 v_1) + a_2(c_2 s_1 v_1) = 0.$$

  Rewriting this equation towards a_2 we get $a_2 = -b_1 c_1 / (c_2 s_1)$. This means that for $s_2 = 0$ we can calculate a_2 and b_2.

- Let us assume now $s_2 \neq 0$ and $f_2 = b_1 s_1 v_1$, then we get for Equations (7.8) and (7.9):

  $$\begin{cases} c_2 f_2 + a_2(c_2 v_2) + b_2(c_3 s_2 v_2) &= 1 \\ s_2 f_2 + a_2(s_2 v_2) + b_2 v_3 &= 0. \end{cases} \tag{7.10}$$

Extracting $a_2 v_2$ from the second equation of (7.10) gives us:

$$a_2 v_2 = \frac{-b_2 v_3 - s_2 f_2}{s_2},$$

(which is well defined as $s_2 \neq 0$) filling it in in the first equation of (7.10), we get:

$$b_2 = \frac{1}{c_3 s_2 v_2 - v_3 \frac{c_2}{s_2}},$$

which is similar to Equation (7.2.2). The calculation of a_2 is again a little more complicated. Two cases can occur:

- If $v_2 = 0$ we use the equation $\mathbf{e}_1^T S T \mathbf{e}_2 = 0$ and extract a_2 from this equation; this gives us:

$$a_2 = \frac{-b_2 c_3 s_2 s_1 - b_1 c_1}{c_2 s_1}.$$

  We know that $c_2 s_1$ has to be different from zero, otherwise the semiseparable matrix would have been singular.

- If $v_2 \neq 0$ we use Equation (7.9). Giving us

$$a_2 = \frac{-b_2 v_3 - s_2 f_2}{s_2 v_2},$$

  which is well defined because both s_2 and v_2 are different from zero.

This last procedure can similarly be repeated for the third column, leading to the same formulas for calculating a_3 and b_3. Only the last column needs some extra attention. For the last column we consider the equations $\mathbf{e}_4^T S T \mathbf{e}_4 = 1$ and $\mathbf{e}_3^T S T \mathbf{e}_4 = 0$:

$$\begin{cases} b_3(s_3 v_3) + a_4 v_4 & = & 1 \\ b_3(c_3 v_3) + a_4(s_3 v_3) & = & 0 \end{cases}$$

Using these formulas we get

$$a_4 = \frac{1 - b_3 s_3 v_3}{v_4},$$

if $v_4 \neq 0$ or else if $v_4 = 0$

$$a_4 = -\frac{b_3 c_3}{s_3},$$

because if $v_4 = 0$, s_3 has to be different from zero, otherwise the matrix would have been singular.

Even though we calculated only the inverse of a 4×4 matrix, the procedure clearly demonstrates how to calculate inverses of larger semiseparable matrices. Moreover, inverting semiseparable matrices in this way leads to an $\mathcal{O}(n)$ procedure.

We have already discussed the computation of the inverse of a symmetric tridiagonal and a symmetric semiseparable matrix. Let us now consider the nonsymmetric cases.

7.2.3 The inverse of a tridiagonal matrix

In this section we will search for a direct inversion method for not necessarily symmetric tridiagonal matrices. In this subsection we will derive two methods. The methods are different in the sense that they use a different representation for representing the semiseparable matrix. The first representation uses only three vectors and was described in Section 2.7, Equation (2.13). The second representation uses the standard representation for nonsymmetric generator representable semiseparable matrices, based on 4 vectors.

A representation based on three vectors

The algorithm presented in this subsection can be found in more elaborate form in [206].

Suppose we have a tridiagonal matrix T, of the form:

$$T = \begin{bmatrix} a_1 & c_1 & & & \\ b_1 & a_2 & c_2 & & \\ & b_2 & a_3 & \ddots & \\ & & \ddots & \ddots & \end{bmatrix}. \tag{7.11}$$

Let us denote the inverse of the matrix T as S. For the moment, we denote the elements of the matrix S as $s_{i,j}$.

It is clear, that $ST = I = TS$ leads to the following equations (the first equation equals $TSe_j = e_j$ and the second one equals $STe_j = e_j$):

$$\delta_{i,j} = \begin{cases} b_{i-1}s_{i-1,j} + a_i s_{i,j} + c_i s_{i+1,j} \\ b_j s_{i,j+1} + a_j s_{i,j} + c_{j-1}s_{i,j-1}, \end{cases} \tag{7.12}$$

where $\delta_{i,j}$ is the Kronecker delta. This means $\delta_{i,j} = 1$ if $i = j$ and zero otherwise. And let us assume the elements $b_0 = c_0 = c_n = b_n = 0$.

First we will prove a lemma, which links the entries in the lower triangular part of the matrix S, to the entries in the upper triangular part.

Lemma 7.1. *Suppose the inverse S of the tridiagonal matrix T as given in (7.11) exists and the subdiagonal elements of the matrix T are different from zero ($b_i \neq 0$, for $i = 1, \ldots, n-1$). Then we have*

$$s_{i,i+k} = \left(\prod_{l=i}^{i+k-1} \frac{c_l}{b_l} \right) s_{i+k,i}.$$

This lemma naturally collapses into $s_{i,i+k} = s_{i+k,i}$, in case the matrix is symmetric.

Proof. The proof is by induction on both k and i.

- Assume $k = 1$. This means that we have to prove that

$$s_{i,i+1} = \frac{c_i}{b_i} s_{i+1,i}.$$

Let us do this by induction on i.

  - Assume $i = 1$. This follows from Equation (7.12) for $i = 1, j = i$, using thereby both equations.

  - By induction on l, we assume the statement is valid for all $i \leq l - 1$. Let us prove now that the statement also holds for $i = l$. The top equality in Equation (7.12) gives us for $i = j = l$:

$$1 = b_{l-1} s_{l-1,l} + a_l s_{l,l} + c_l s_{l+1,l}$$
$$= b_{l-1} \frac{c_{l-1}}{b_{l-1}} s_{l,l-1} + a_l s_{l,l} + c_l s_{l+1,l}$$
$$= c_{l-1} s_{l,l-1} + a_l s_{l,l} + c_l s_{l+1,l},$$

and we also have, using the bottom equality in Equation (7.12):

$$1 = c_{l-1} s_{l,l-1} + a_l s_{l,l} + b_l s_{l,l+1}.$$

Combining these equations leads to $s_{l,l+1} = (c_l s_{l+1,l})/b_l$.

- By induction on k, we suppose the theorem is true for all $k \leq l$. The bottom equality in Equation (7.12) for $j = i + l$ becomes:

$$0 = c_{i+l-1} s_{i,i+l-1} + a_{i+l} s_{i,i+l} + b_{i+l} s_{i,i+l+1}$$
$$= \left(\prod_{t=i}^{i+l-1} \frac{c_t}{b_t} \right) \left(b_{i+l-1} s_{i+l-1,i} + a_{i+l} s_{i+l,i} \right) + b_{i+l} s_{i,i+l+1}.$$

We can also multiply the top equality in Equation (7.12) (where i is substituted by $i + l$ and j is substituted by i), leading to

$$0 = \left(\prod_{t=i}^{i+l-1} \frac{c_t}{b_t} \right) \left(b_{i+l-1} s_{i+l-1,i} + a_{i+l} s_{i+l,i} + c_{i+l} s_{i+l+1,i} \right).$$

Combining the last two equations provides us with the proof.

□

This lemma can come in handy when inverting tridiagonal matrices; if the subdiagonal (superdiagonal) entries are different from zero, we only need to calculate the lower (upper) triangular part. Further on we will prove that we can always overcome this constraint if a sub(super)diagonal element is zero.

Lemma 7.2. *Suppose a tridiagonal matrix T is given, whose elements are defined by Equation (7.11) with subdiagonal elements different from zero. Define $\mu_n = 1$*

and $\mu_{n-1} = -f_n$. We define

$$\mu_i = -f_{i+1}\mu_{i+1} - g_{i+1}\mu_{i+2}, \quad i = 1, 2, \ldots, n-1,$$

$$f_i = \frac{a_i}{b_{i-1}}, \quad i = 2, 3, \ldots, n,$$

$$g_i = \frac{c_i}{b_{i-1}}, \quad i = 2, 3, \ldots, n.$$

Suppose T to be invertible (this is true if, and only if, $a_1\mu_1 + c_1\mu_2 \neq 0$.) The elements of $T^{-1} = S = (s_{i,j})_{i,j}$ are defined as:

$$s_{i,j} = \mu_i s_{n,j}, \text{ for } i \geq j$$

$$s_{n,1} = (a_1\mu_1 + c_1\mu_2)^{-1}$$

and

$$s_{n,j} = \begin{cases} \dfrac{c_{j-1}\mu_j s_{n,j-1} - 1}{b_{j-1}\mu_{j-1}} & \text{if} \quad \mu_{j-1} \neq 0 \\ -\dfrac{c_{j-2}s_{n,j-2} + a_{j-1}s_{n,j-1}}{b_{j-1}} & \text{if} \quad \mu_{j-1} = 0 \end{cases} \quad j = 2, 3, \ldots, n.$$

Proof. The proof is a direct calculation involving Equation (7.12) and the results of the previous lemma. The proof can be found in [206]. □

In fact, one can decompose every tridiagonal matrix as a block upper triangular matrix, for which the blocks on the diagonals are of tridiagonal form, with subdiagonal elements different from zero. Using the following relation, we can easily invert any tridiagonal matrix:

$$\begin{bmatrix} T_1 & R \\ 0 & T_2 \end{bmatrix}^{-1} = \begin{bmatrix} T_1^{-1} & -T_1^{-1}RT_2^{-1} \\ 0 & T_2^{-1} \end{bmatrix}.$$

The matrices T_1^{-1} and T_2 are tridiagonal matrices; hence, their inverses are of semiseparable form. The matrix R has at most one element in the lower left corner different from zero, namely, c_i for a certain i. Hence we obtain

$$T_1^{-1}RT_2^{-1} = c_i\boldsymbol{\alpha}\boldsymbol{\beta}^T,$$

where $\boldsymbol{\alpha}$ is the last column of the matrix T_1 and $\boldsymbol{\beta}^T$ is the first row of the matrix T_2^{-1}. Hence the complete upper triangular part of the matrix will be of semiseparability rank 1.

Similarly as the previous lemma we have, in case the superdiagonal elements are different from zero.

Lemma 7.3. *Suppose a tridiagonal matrix T is given, with superdiagonal elements different from zero. The matrix T is invertible if and only if $b_{n-1}\nu_{n-1} + a_n\nu_n \neq 0$.*

Define $\nu_1 = 1$ and $\nu_2 = -h_1$. We define

$$\nu_j = -h_{j-1}\nu_{j-1} - \frac{\nu_{j-2}}{g_{j-1}}, \quad j = 3, \ldots, n$$

$$h_j = \frac{a_j}{c_j} \text{ and } g_j = \frac{c_j}{b_{j-1}}, \quad j = 2, 3, \ldots, n-1.$$

Suppose T to be invertible , the elements of $T^{-1} = S = (s_{i,j})_{i,j}$ are defined as:

$$s_{i,j} = \nu_i s_{1,j}, \text{ for } i \leq j$$
$$s_{1,n} = (b_{n-1}\nu_{n-1} + a_n\nu_n)^{-1}$$

and

$$s_{1,j} = \begin{cases} \dfrac{b_j \nu_j s_{1,j+1} - 1}{c_j \nu_{j+1}} & \text{if } \nu_{j+1} \neq 0 \\ -\dfrac{b_{j+1} s_{1,j+2} + a_{j+1} s_{1,j+1}}{c_j} & \text{if } \nu_{j+1} = 0 \end{cases} \quad j = n-1, n-2, \ldots, 1.$$

Combining the last two lemmas and assuming (which is in fact no constraint) that all diagonal elements are different from zero, we get the following theorem for the inverse of the tridiagonal matrix.

Theorem 7.4. *Suppose T is an invertible irreducible tridiagonal matrix. Then we have that $S = T^{-1}$ is of the following form. (The variables $\mu_i, \nu_j, s_{n,j}$ are defined as in the previous lemmas):*

$$s_{i,j} = \begin{cases} \left(\prod_{l=1}^{j-1} \frac{c_l}{b_l}\right) \mu_i \nu_j s_{n,1}, & i \geq j \\ \left(\prod_{l=1}^{j-1} \frac{c_l}{b_l}\right) s_{j,i} = \left(\prod_{l=1}^{j-1} \frac{c_l}{b_l}\right) \mu_j \nu_i s_{n,1} & i < j \end{cases}.$$

Proof. The proof is a combination of the previous lemmas.   □

It is clear that the representation of the semiseparable matrix is of the special form proposed in Section 2.7, Equation (2.13). Another method, representing the lower and upper part with two vectors is discussed next.

A representation based on four vectors

We present here an inversion method for irreducible, not necessarily symmetric tridiagonal matrices T. The inverse of the matrix T will be represented in the standard form $S(\mathbf{u}, \mathbf{v}, \mathbf{p}, \mathbf{q})$. This inversion method was presented in [221] and expands in some sense the inversion method presented in [61].

Assume we have an irreducible tridiagonal matrix T, with diagonal elements a_i, subdiagonal elements $-b_i$ and superdiagonal elements $-c_i$. The inverse of T is represented as $S(\mathbf{u}, \mathbf{v}, \mathbf{p}, \mathbf{q})$. The generators are computed as follows:

$$p_1 = 1, p_2 = \frac{a_1}{c_1},$$

$$p_i = \frac{a_{i-1}p_{i-1} - b_{i-2}p_{i-2}}{c_{i-1}}, \quad 3 \le i \le n,$$

$$v_1 = 1, v_2 = \frac{a_1}{b_1},$$

$$v_i = \frac{a_{i-1}v_{i-1} - c_{i-2}v_{i-2}}{b_{i-1}}, \quad 3 \le i \le n,$$

$$q_n = \frac{1}{-b_{n-1}p_{-1} + a_n p_n},$$

$$q_i = \frac{1 + c_i v_i u_{i+1}}{a_i p_i - b_{i-1}p_{i-1}}, \quad 2 \le i \le n-1,$$

$$q_1 = \frac{1 + c_1 v_1 u_2}{a_1 p_1},$$

$$u_i = \frac{p_i q_i}{v_i}.$$

We have provided some algorithms for inverting tridiagonal matrices, and obtaining semiseparable matrices, represented using a specific representation. In the next subsection a method is presented for inverting a very specific, not necessarily symmetric semiseparable matrix represented with three vectors.

7.2.4 The inverse of a specific semiseparable matrix

We pursue in this subsection our analysis of the previous subsection, based on this specific representation for the generator representable semiseparable matrix. We only formulate the theorem as the proof involves calculations which do not lead to more insight (the proof can be found in [206].)

Theorem 7.5. *Suppose we have a semiseparable matrix S of the following form:*

$$s_{i,j} = \begin{cases} x_i y_j z_j, & i \ge j \\ y_i x_j z_j, & i \le j. \end{cases}$$

The elements of the tridiagonal matrix $T = S^{-1}$ are of the following form (with a_i the diagonal, b_i the subdiagonal and c_i the superdiagonal elements):

$$a_i = \begin{cases} -c_1 \nu_2, & i = 1 \\ -c_{i-1} b_i \left(s_{i+1,i-1} - \nu_{i+1} s_{1,i-1} \right), & i = 2, \dots, n-1 \\ -b_{n-1} \mu_{n-1}, & \end{cases}$$

$$c_i = \left((x_{i+1}y_i - y_{i+1}x_i) z_i \right)^{-1} = \left(s_{i+1,i} - \nu_{i+1} s_{1,i} \right)^{-1}$$

$$b_i = \left((x_{i+1}y_i - y_{i+1}x_i) z_{i+1} \right)^{-1} = \left(s_{i,i+1} - \mu_i s_{n,i+1} \right)^{-1},$$

where

$$\mu_i = \frac{x_i}{x_n} = \frac{s_{i,1}}{s_{n,1}}$$

$$\nu_j = \frac{y_j}{y_1} = \frac{s_{j,n}}{s_{1,n}}.$$

Notes and references

The literature related to the inversion of semiseparable, generator representable semiseparable and quasiseparable matrices is extensive. In this section we will present an overview of interesting publications divided into a few categories. Most of the references were already discussed elsewhere; hence, we present pointers to these places.

General references on the inversion of structured rank matrices include: Vandebril et al. [291], Barrett et al. [15], Gustafson [167], Fiedler [114] in Section 1.5 on relations under inversion; Berger et al. [19] in Section 1.1 on the definition of symmetric semiseparable matrices.

Generator representable semiseparable matrices

Manuscripts discussed earlier on the inverse of irreducible tridiagonal matrices include: Gantmacher and Kreĭn [126, 131] in Section 1.1 on the definition of semiseparable matrices; Asplund S.O. [7], Bukhberger et al. [42] in Section 2.2 on the symmetric generator representation; Torii [263, 264], Capovani et al. [49, 50, 24], Baranger et al. [11], Mallik [207] in Section 2.7 on the nonsymmetric generator representation; Moskovitz [193] in Section 3.5; Golub, Concus and Meurant, provide an explicit formula for computing the inverse of a tridiagonal matrix [61] in Section 7.5 on the decay rate of the elements in the inverse; da Fonseca and Petronilho provide formulas for inverting tridiagonal k-Toeplitz matrices based on results for inverting tridiagonal matrices in Section 14.1 on inverting Toeplitz matrices.

In the next article, the author Kershaw provides explicit formulas for inverting a special tridiagonal matrix.

☞ D. Kershaw. The explicit inverses of two commonly occuring matrices.
Mathematics of Computation, 23(105):189–191, January 1969.

The matrix is of the following form:

$$T = \begin{bmatrix} -\alpha & 1 & & & & \\ 1 & -2\beta & 1 & & & \\ & 1 & -2\beta & 1 & & \\ & & \ddots & \ddots & \ddots & \\ & & & 1 & -2\beta & 1 \\ & & & & 1 & -\alpha \end{bmatrix},$$

with α and β arbitrary elements.

The following manuscript of Schlegel generalizes the article of Kershaw discussed above by removing the constraints on the diagonal elements.

☞ P. Schlegel. The explicit inverse of a tridiagonal matrix. *Mathematics of Computation*, 24(111):665–665, July 1970.

In this manuscript the author presents explicit formulas for inverting the following tridiagonal matrix:

$$T = \begin{bmatrix} a_1 & 1 & & & \\ 1 & a_2 & 1 & & \\ & 1 & a_3 & 1 & \\ & & \ddots & \ddots & \ddots \end{bmatrix}.$$

The author Usmani provides an algorithm for inverting tridiagonal matrices based on the principal minors. In [275] also a method for inverting a tridiagonal matrix based on the associated difference equation is presented (see Section 3.5 in Chapter 3).

☞ R. A. Usmani. Inversion of a tridiagonal Jacobi matrix. *Linear Algebra and its Applications*, 212/213:413–414, 1994.

☞ R. A. Usmani. Inversion of Jacobi's tridiagonal matrix. *Computers & Mathematics with Applications*, 27(8):59–66, 1994.

The inversion method by Lewis, using the representation with three vectors, was discussed thoroughly in Section 7.2 of this Chapter.

☞ J. W. Lewis. Inversion of tridiagonal matrices. *Numerische Mathematik*, 38:333–345, 1982.

The article [201] focuses on the inversion of block tridiagonal matrices and includes the irreducible tridiagonal case as an example; it will be discussed in Section 14.4 in Chapter 14.

Semiseparable matrices

The following references were discussed earlier: Barrett et al. [14] in Section 3.2 on covariance matrices; Barrett et al. in Section 1.5 on inversion; Meurant [215] in Section 7.5 in this Chapter; Golub et al. in Section 7.4 (this manuscript contains the inversion of the tridiagonal matrix, with regard to Givens-vector representation as discussed in this section).

Semiseparable plus diagonal and quasiseparable matrices

Most of the results for these classes of matrices are related to higher order generator representable semiseparable plus diagonal matrices and also higher order quasiseparable matrices. These references will be discussed in forthcoming chapters. More information can be found in Chapter 12, which is completely dedicated to higher order (block) quasiseparable matrices and Section 14.5 in Chapter 14, which presents different methods for inverting higher order semiseparable plus band matrices. In the manuscript [115], Fiedler proves, based on the nullity Theorem, that the off-diagonal rank of a matrix is maintained under inversion.

Covariance matrices

Covariance matrices are matrices arising in statistical applications. Quite often statisticians are interested in computing the inverse of these matrices.

The following references are discussed elsewhere: Roy et al. [240] in Section 2.2 on the symmetric generator representation; Mustafi [220], Valvi [278], Graybill [158], Barrett et al. [14], Uppuluri et al. [273], Greenberg et al. [160] in Section 3.2 on covariance matrices.

7.3   General formulas for inversion

In this section we will propose some general formulas, which might come in handy when inverting specific structured rank matrices, or slightly perturbed structured rank matrices. Most of the proofs of the presented formulas can be verified by direct computations.

The first formula shows how the inverse of a matrix changes if the matrix changes. Suppose we have two invertible matrices A and B:

$$B^{-1} = A^{-1} - B^{-1}(B - A)A^{-1}.$$

The following formula is the well-known Sherman-Morrison-Woodbury formula. This formula gives an expression for the inverse of a matrix when it is modified by a low rank perturbation:

$$\left(A + UV^T\right)^{-1} = A^{-1} - A^{-1}U(I + V^T A^{-1}U)^{-1}V^T A^{-1}.$$

The reader can verify the correctness of these equations by directly checking that the computed inverses do the job. We note that in the equation above the invertibility of both A and $I + V^T A^{-1}U$ are assumed.

The Sherman-Morrison-Woodbury formula leads also to some interesting variants and generalizations. The following formulas are based on an overview article [174].

Let us first show the rank 1 update of an invertible matrix (assume the equation is well-defined, i.e., that all inverses exist):

$$\left(A + \alpha \mathbf{u}\mathbf{v}^T\right)^{-1} = A^{-1} - \alpha \frac{A^{-1}\mathbf{u}\mathbf{v}^T A^{-1}}{1 + \alpha \mathbf{v}^T A^{-1}\mathbf{u}}.$$

This formula is quite useful for inverting specific patterned matrices as discussed in Chapter 3 and the matrices from, e.g., [158, 240].

A formula similar to the Sherman-Morrison-Woodbury formula is the following one, with the assumption that the matrix B is not necessarily invertible:

$$(A + UBV)^{-1} = A^{-1} - A^{-1}U\left(I + BVA^{-1}U\right)^{-1}BVA^{-1}.$$

In case the matrix B is invertible, we also have the following formulation:

$$(A + UBV)^{-1} = A^{-1} - A^{-1}U\left(B^{-1} + VA^{-1}U\right)^{-1}VA^{-1}.$$

Using the following simple formulas, for any P assuming $I + P$ and $I + PQ$ are nonsingular,

$$(I + P)^{-1} = I - P(I + P)^{-1} = I - (I + P)^{-1},$$
$$(I + PQ)^{-1}P = P(I + QP)^{-1},$$

leads us to the following 5 alternative formulations for the Sherman-Morrison-

Woodbury formula:

$$
\begin{aligned}
(A + UBV)^{-1} &= A^{-1} - A^{-1} \left(I + UBVA^{-1} \right)^{-1} UBVA^{-1} \\
&= A^{-1} - A^{-1}U \left(I + BVA^{-1}U \right)^{-1} BVA^{-1} \\
&= A^{-1} - A^{-1}UB \left(I + VA^{-1}UB \right)^{-1} VA^{-1} \\
&= A^{-1} - A^{-1}UBV \left(I + A^{-1}UBV \right)^{-1} A^{-1} \\
&= A^{-1} - A^{-1}UBVA^{-1} \left(I + UBVA^{-1} \right)^{-1}.
\end{aligned}
$$

Let us propose now some formulas for inverting partitioned matrices. In this equation, B is possibly singular:

$$
\begin{bmatrix} A & U \\ V & B \end{bmatrix}^{-1} \tag{7.13}
$$
$$
= \begin{bmatrix} A^{-1} + A^{-1}U \left(B - VA^{-1}U \right)^{-1} VA^{-1} & -A^{-1}U \left(B - VA^{-1}U \right)^{-1} \\ - \left(B - VA^{-1}U \right)^{-1} VA^{-1} & \left(B - VA^{-1}U \right)^{-1} \end{bmatrix}.
$$

If B is invertible we get:

$$
\begin{aligned}
\begin{bmatrix} A & U \\ V & B \end{bmatrix}^{-1} &= \begin{bmatrix} \left(A - UB^{-1}V \right)^{-1} & - \left(A - UB^{-1}V \right)^{-1} UB^{-1} \\ - \left(B - VA^{-1}U \right)^{-1} VA^{-1} & \left(B - VA^{-1}U \right)^{-1} \end{bmatrix} \\
&= \begin{bmatrix} \left(A - UB^{-1}V \right)^{-1} & -A^{-1}U \left(B - VA^{-1}U \right)^{-1} \\ -B^{-1}V \left(A - UB^{-1}V \right)^{-1} & \left(B - VA^{-1}U \right)^{-1} \end{bmatrix}.
\end{aligned}
$$

When all matrices, U, V, B and A are invertible we even get:

$$
\begin{bmatrix} A & U \\ V & B \end{bmatrix}^{-1} = \begin{bmatrix} \left(A - UB^{-1}V \right)^{-1} & \left(V - BU^{-1}A \right)^{-1} \\ \left(U - AV^{-1}B \right)^{-1} & \left(B - VA^{-1}U \right)^{-1} \end{bmatrix}.
$$

Notes and references

A more elaborate study of the equalities and references discussed in this section can be found in the manuscript by Henderson and Searle.

☞ H. V. Henderson and S. R. Searle. On deriving the inverse of a sum of matrices. *SIAM Review*, 23(1):53–59, January 1981.

The Sherman-Morrison-Woodbury formula is based on the following articles.

☞ J. Sherman and K. E. Morrison. Adjustment of an inverse matrix corresponding to changes in the elements of a given column or a given row of the original matrix. *Annals of Mathematical Statistics*, 20:621–621, 1949.

☞ J. Sherman and W. J. Morrison. Adjustment of an inverse matrix corresponding to a change in one element of a given matrix. *Annals of Mathematical Statistics*, 21:124–127, 1950.

☞ M. A. Woodbury. Inverting modified matrices. Memorandum Report 42, Statistical Research Group, Princeton, N.J., 1950.

Several methods for inverting matrices were discussed by Householder in the following manuscript.

☞ A. S. Householder. A survey of closed methods for inverting matrices. *Journal of the Society for Industrial and Applied Mathematics*, 5:155–169, 1957.

Equation (7.13) is often contributed to Schur because he formulated quite close results. Others, Bodewig and Aitken, also derived very close results.

☞ A. C. Aitken. *Determinants and matrices.* Oliver and Boyd, Edinburgh, fourth edition, 1939.

☞ E. Bodewig. *Matrix Calculus.* North-Holland, Amsterdam, second edition, 1959.

☞ E. Bodewig. Comparison of some direct methods for computing determinants and inverse matrices. In *Koninklijke Nederlandse Akademie van Wetenschappen, Proceedings of the Section of Sciences*, volume 50, pages 49–57, 2003.

☞ I. Schur. Über potenzreihen, die im innern des einheitskreises beschränkt sind.I. *Journal für die Reine und Angewandte Mathematik*, 147:205–232, 1917. (In German).

It seems that the first presentation of the formula (Equation (7.13)) is due to Banachiewicz. Similar results were obtained by Frazer, Waugh and Jossa. Oulette called this the "Schur-Banachiewicz" inversion formula.

☞ T. Banachiewicz. Sur l'inverse d'un cracovien et une solution générale d'un système d'équations linéaires. *Comptes Rendus Mensuels des Séances de la Classe des Sciences Mathématiques et Naturelles de l'Académie Polonaise des Sciences et des Lettres*, 4:3–4, April 1937. (In French).

☞ R. A. Frazer, W. J. Duncan, and A. R. Collar. *Elementary matrices and some applications to dynamics and differential equations.* Cambridge University Press, Cambridge, 1938.

☞ F. Jossa. Risoluzione progressiva di un sistema di equazioni lineari, analogia con un problema meccanico. *Reale Accadademia di Scienze Fisiche e Matematiche, Società Reale di Napoli*, 4(10):346–352, 1940. (In Italian).

☞ D. V. Oulette. Schur complements and statistics. Master's thesis, McGill University, Montreal, 1978.

☞ F. Waugh. A note concerning Hotelling's method of inverting a partitioned matrix. *Annals of Mathematics and Statistics*, 16:216–217, 1945.

The alternative formulations of Equation (7.13), in the case that matrix B is invertible,

☞ W. J. Duncan. Some devices for the solution of large sets of simultaneous linear equations (with an appendix on the reciprocation of partitioned matrices). *The London, Edinburgh and Dublin Philosophical Magazine and Journal of Science*, 7(35):660–670, 1944.

☞ H. Hotelling. Further points on matrix calculation and simultaneous equations. *Annals of Mathematical Statistics*, 14:440–441, 1943.

☞ H. Hotelling. Some new methods in matrix calculation. *Annals of Mathematical Statistics*, 14:1–34, 1943.

7.4   Scaling of symmetric positive definite semiseparable matrices

In this section we will briefly present some results related to the optimal diagonal scaling of a semiseparable matrix. By optimal scaling we mean applying a two-sided diagonal multiplication onto the matrix in question to get its condition number as close as possible to 1. The semiseparable matrices considered in this section are symmetric and positive definite.

First we will present some introductionary theory. A symmetric matrix A is said to have **property A** when a permutation P exists, such that PAP^T is of the following form:

$$\hat{A} = PAP^T = \begin{bmatrix} D_1 & F \\ F^T & D_2 \end{bmatrix}, \tag{7.14}$$

where F is an arbitrary matrix and D_1 and D_2 are diagonal matrices. These matrices are known to be best scaled if the diagonal matrices D_1 and D_2 are equal to the identity matrix (see [119, 16, 153]). This can be achieved rather easily by applying a diagonal scaling of the matrix. Defining the scaling matrix $\hat{D}$ as

$$\hat{D} = \begin{bmatrix} \frac{1}{\sqrt{(D_1)_{i,i}}} & \\ & \frac{1}{\sqrt{(D_2)_{i,i}}} \end{bmatrix}$$

gives us the desired solution.

It is easy to see that a tridiagonal matrix satisfies property A. For the permutation P^T one uses a reshuffling of the columns, thereby first placing the even columns ($[2, 4, \ldots]$) and after the even, the odd columns ($[1, 3, 5, \ldots]$). Applying the same reshuffling technique P for the rows, one gets a matrix PTP^T of the desired form.

Using the knowledge related to matrices having property A, the optimal diagonal scaling of a tridiagonal matrix T can be obtained by choosing the diagonal $\hat{D}$ in such a way that the diagonal elements of the tridiagonal matrix $\hat{D}T\hat{D}$ are equal to one. This means that for a tridiagonal matrix T with diagonal elements a_i, our optimal scaling diagonal matrix $\hat{D}$ has as diagonal elements $1/\sqrt{a_i}$.

Using this result for tridiagonal matrices we can derive the best diagonal scaling for semiseparable matrices. This optimal scaling matrix $\hat{D}$ for semiseparable matrices is calculated by using an inversion formula for semiseparable matrices.

Defining the condition number κ of a matrix A as $\kappa(A) = \|A\|\|A^{-1}\|$ for any consistent matrix norm, we have that $\kappa(A) = \kappa(A^{-1})$. This means that for an invertible symmetric, positive definite tridiagonal matrix T, with inverse S and optimal diagonal scaling matrix $\hat{D}$ for T, we have:

$$\kappa(\hat{D}^{-1}S\hat{D}^{-1}) = \kappa(\hat{D}T\hat{D}) \leq \kappa(DTD) = \kappa(D^{-1}SD^{-1}),$$

for any matrix D. Using any of the inversion formulas for semiseparable matrices as presented in this chapter, we can easily calculate in $O(n)$ flops the optimal diagonal scaling of a semiseparable matrix.

Notes and references

Scaling matrices is an interesting subject, which can improve the accuracy of the resulting computations, without significantly changing the complexity of the involved method.

The results of this section, with numerical experiments and an inversion formula for semiseparable matrices, were presented by Vandebril, Golub and Van Barel in the manuscript of 2006. In the manuscript of 1963 Golub investigates the optimal scaling of some variance matrices from statistical applications, which are of semiseparable form, based on the scaling of their inverse.

☞ G. H. Golub. Comparison of the variance of minimum variance and weighted least squares regression coefficients. *Annals of Mathematical Statistics*, 34(3):984–991, September 1963.

☞ R. Vandebril, G. H. Golub, and M. Van Barel. A small note on the scaling of positive definite semiseparable matrices. *Numerical Algorithms*, 41:319–326, 2006.

Interesting manuscripts related to the optimal scaling of matrices are provided by Bauer, Forsythe and Strauss and Golub and Varah, in the following manuscripts.

☞ F. L. Bauer. Optimally scaled matrices. *Numerische Mathematik*, 5:73–87, 1963.

☞ G. E. Forsythe and E. G. Straus. On best conditioned matrices. *Proceedings of the American Mathematical Society*, 6:340–345, 1955.

☞ G. H. Golub and J. M. Varah. On the characterization of the best L_2-scaling of a matrix. *SIAM Journal on Numerical Analysis*, 11:472–479, 1974.

7.5 Decay rates for the inverses of tridiagonal matrices

It is well-known by now that inverting tridiagonal matrices leads to semiseparable matrices. Under some additional conditions on the tridiagonal matrix, the elements of their inverse matrices decay, starting from the diagonal towards the corners. Let us present here some of these results. The information presented here can be found in a more elaborate form in the manuscripts [221, 222] by Nabben. We divided this section in three parts. First, we define M-matrices. Second, we will provide some bounds for the inverses of tridiagonal M-matrices, satisfying some extra conditions. Finally, we will provide some general bounds for the inverses of tridiagonal matrices satisfying some constraints.

7.5.1 M-matrices

M-matrices are special Z-matrices. Let us define these matrices in general form.

Definition 7.6. *A matrix $A = (a_{i,j})_{i,j}$ is called an Z-matrix if all its off-diagonal entries are less than or equal to zero:*

$$a_{i,j} \leq 0 \ if \ i \neq j.$$

An M-matrix is a special Z-matrix. An M-matrix can be defined in several different ways (e.g., in [20] 50 equivalent definitions are presented). Different definitions are discussed in [20, 110, 175]. We do not go into the details and present only one definition.

Definition 7.7. *A Z-matrix A is called an M-matrix if it is invertible and its inverse is a nonnegative[13] matrix.*

Let us focus now on tridiagonal M-matrices.

Definition 7.8. *Suppose T is a symmetric irreducible tridiagonal matrix. The following conditions are equivalent:*

1. *The matrix T is an M-matrix.*

2. *T^{-1} is of the form $S(\mathbf{u}, \mathbf{v})$, where all u_i and v_i have the same sign and[14]*

$$\frac{v_1}{u_1} < \frac{v_2}{u_2} < \ldots < \frac{v_n}{u_n}.$$

For nonsymmetric tridiagonal matrices we have a similar definition in terms of the generators of the inverse.

Definition 7.9. *Suppose T is an irreducible tridiagonal matrix. The following definitions are equivalent:*

1. *The matrix T is an M-matrix.*

2. *The inverse T^{-1} is of the form $S(\mathbf{u}, \mathbf{v}, \mathbf{p}, \mathbf{q})$, with all u_i, v_i, p_i and q_i of the same sign and*

    $$\alpha_i \beta_i < 1 \text{ for all } i,$$

    where

    $$\alpha_i = \frac{p_i q_{i+1}}{p_{i+1} q_i} \text{ and } \beta_i = \frac{v_i u_{i+1}}{v_{i+1} u_i}.$$

3. *There exists a diagonal matrix D, such that the matrix DT is symmetric.*

Let us provide now some bounds for tridiagonal M-matrices, which are diagonally dominant.

---

[13] A matrix is said to be nonnegative if all its elements are larger than or equal to zero.

[14] This is almost the same equation as encountered in the section on oscillation matrices, Section 3.1, Equation 3.1.

7.5.2   Decay rates for the inverse of diagonally dominant tridiagonal M-matrices

Let us provide now some of the decay results for diagonally dominant matrices.

Definition 7.10. *A matrix A is said to be diagonally dominant by columns if*

$$a_{j,j} \geq \sum_{k \neq j} |a_{k,j}| \text{ for all } j.$$

A matrix A is said to be diagonally dominant by rows if

$$a_{i,i} \geq \sum_{k \neq i} |a_{i,k}| \text{ for all } i.$$

The first theorem on the decay of the elements, was proved by Concus, Golub and Meurant [61] and formulated for symmetric tridiagonal M-matrices. In the following theorem we denote with $\{\alpha_i\}_{i=1}^n$ the sequence of numbers $\alpha_1, \alpha_2, \ldots, \alpha_n$.

Theorem 7.11. *Suppose T is a symmetric irreducible tridiagonal M-matrix, with diagonal elements a_i and subdiagonal elements $-b_i$ (note the minus sign), with $b_i \geq 0$. The matrix is assumed to be diagonally dominant by columns and $a_1 > b_1$ and $a_n > b_{n-1}$. When denoting the inverse of T as $S(\mathbf{u}, \mathbf{v})$ (with the normalization $v_1 = 1$; see, e.g., Subsection 7.2.1, the part on the generator representation), we have that the sequence $\{u_i\}_{i=1}^n$ is strictly decreasing, whereas the sequence $\{v_i\}_{i=1}^n$ is strictly increasing.*

Moreover if

$$\rho^{-1} = \min_{k \geq 2} \frac{a_k - b_{k-1}}{b_k},$$

which is strictly larger than 1, as the matrix is diagonally dominant by columns, we have that

$$\begin{cases} u_i v_j \leq u_i v_i \rho^{|i-j|} & \text{if} \quad i \geq j \\ v_i u_j \leq u_i v_i \rho^{|i-j|} & \text{if} \quad i \leq j. \end{cases}$$

It is clear that there is a decay in the size of the elements when going away from the diagonal.

Proof. The proof can be found in [61, Lemma 3].   □

This theorem is generalized in Lemma 3.1 in [221] for nonsymmetric matrices as follows (the proof is not included).

Theorem 7.12. *Suppose T is an irreducible tridiagonal M-matrix, with diagonal elements a_i, subdiagonal elements $-b_i$ and superdiagonal elements $-c_i$, with $b_i, c_i \geq 0$. Denote the inverse of T as $S(\mathbf{u}, \mathbf{v}, \mathbf{p}, \mathbf{q})$, with the proper normalization $p_1 = 1$ and $v_1 = 1$.*

We can formulate two statements.

- *If the matrix is diagonally dominant by rows and $a_1 > c_1$ and $a_n > b_{n-1}$, then we have that $\{p_i\}$ is strictly increasing, while $\{u_i\}$ is strictly decreasing.*

  Moreover if

  $$\rho_1^{-1} = \min_{k \geq 2} \frac{a_k - b_{k-1}}{c_k} \ \ and \ \ \rho_2^{-1} = \min_{k \geq 2} \frac{a_k - c_k}{b_{k-1}}.$$

  We have that

  $$\begin{cases} p_i q_j \leq p_i q_i \rho_1^{|i-j|} & if \quad i \leq j \\ u_i v_j \leq u_i v_i \rho_2^{|i-j|} & if \quad i \geq j. \end{cases}$$

- *If the matrix is diagonally dominant by columns and $a_1 > b_1$ and $a_n > c_{n-1}$, then we have that $\{v_i\}$ is strictly increasing, while $\{q_i\}$ is strictly decreasing.*

  Moreover if

  $$\rho_3^{-1} = \min_{k \geq 2} \frac{a_k - b_k}{c_{k-1}} \ \ and \ \ \rho_4^{-1} = \min_{k \geq 2} \frac{a_k - c_{k-1}}{b_k}.$$

  We have that

  $$\begin{cases} p_i q_j \leq p_i q_i \rho_3^{|i-j|} & if \quad i \leq j \\ u_i v_j \leq u_i v_i \rho_4^{|i-j|} & if \quad i \geq j. \end{cases}$$

It is clear that also in the nonsymmetric case, under some constraints, we can provide statements concerning the decay of the elements in the matrix T^{-1}.

It is not always necessary to have an M-matrix. Let us provide more general results.

7.5.3 Decay rates for the inverse of diagonally dominant tridiagonal matrices

In the manuscript [222] of Nabben, results of Shivakumar et al. [254] were generalized. Let us present some of the main results.

Theorem 7.13 (Theorem 3.1 in [222]). *Suppose we have a nonsingular tridiagonal matrix T with diagonal entries a_i, subdiagonal entries b_i and superdiagonal entries c_i.[15] If the matrix T is row diagonally dominant[16], we obtain for the inverse $S(\mathbf{u}, \mathbf{v}, \mathbf{p}, \mathbf{q})$ the following bounds:*

$$\begin{cases} |p_i q_j| \leq |p_i q_i| \prod_{k=i}^{j-1} \tau_k^- & if \quad i \leq j \\ |u_i v_j| \leq |u_i v_i| \prod_{k=j+1}^{i} \omega_k^- & if \quad i \geq j, \end{cases}$$

---

[15]The elements are not denoted by $-c_i$ or $-b_i$ anymore, as the matrix is not necessarily an M-matrix anymore.

[16]In the following theorems of [222], only row diagonally dominant results are considered, they can easily be generalized to the column dominant form by considering the transpose.

where τ_k^- and ω_k^- are defined as:

$$\tau_k^- = \frac{|c_k|}{|a_k| - |b_{k-1}|} \quad \text{for } k = 1, \ldots, n-1,$$

$$\omega_k^- = \frac{|b_{k-1}|}{|a_k| - |c_k|} \quad \text{for } k = 1, \ldots, n-1.$$

We can bound also the diagonal elements.

Theorem 7.14 (Theorem 3.2 in [222]). *Using the same assumptions and notations as in Theorem 7.13, we obtain that for $S = (s_{i,j})_{i,j} = T^{-1}$:*

$$\frac{1}{|a_i| + \tau_{i-1}^-|b_{i-1}| + \omega_{i+1}^-|c_i|} \leq |s_{i,i}| \leq \frac{1}{|a_i| - \tau_{i-1}^-|b_{i-1}| - \omega_{i+1}^-|c_i|}.$$

Moreover lower bounds on the elements are also provided.

Theorem 7.15 (Theorem 3.3 in [222]). *Using the same assumptions and notations as in Theorem 7.13, we obtain that*

$$\begin{cases} |p_i q_i| \prod_{k=i}^{j-1} \tau_k^+ \leq |p_i q_j| & \text{if} \quad i \leq j \\ |u_i v_i| \prod_{k=j+1}^{i} \omega_k^+ \leq |u_i v_j| & \text{if} \quad i \geq j, \end{cases}$$

where τ_k^+ and ω_k^+ are defined as:

$$\omega_k^+ = \frac{|c_i|}{|a_i| + |b_{i-1}|} \quad \text{for } k = 1, \ldots, n-1,$$

$$\tau_k^+ = \frac{|b_{i-1}|}{|a_i| + |c_i|} \quad \text{for } k = 1, \ldots, n-1.$$

Notes and references

The main results of this section are based on the manuscripts by Nabben. Let us mention some other manuscripts closely related to the results proved in this section.

☞ D. Kershaw. Inequalities on the elements of the inverse of a certain tridiagonal matrix. *Mathematics of computation*, 24:155–158, 1970.

Kershaw provides bounds between which the elements of the inverse of a specific tridiagonal matrix with positive off-diagonal elements will lie. The elements of the super and the subdiagonal of the tridiagonal are related in the following way. Denote with T the tridiagonal matrix, then we have: $T_{i,i+1} = 1 - T_{i-1,i}$ for $1 < i < n$. The author proves that there is a uniform exponential decay if $|i - j| \to \infty$, the rate of decay depending on the strength of the diagonal dominance.

☞ S. Demko. Inverses of band matrices and local convergence of spline projections. *SIAM Journal on Numerical Analysis*, 14(4):616–619, 1977.

Demko proves theorems bounding the size of the elements of the inverse of strictly diagonally dominant band matrices in terms of the norm of the original matrix, the distance towards the diagonal and the bandwidth. In particular it is shown that the size of the elements decays exponentially to zero if one goes further and further away from the diagonal. More precisely if $S = (s_{i,j})_{i,j}$ is the inverse of such a strictly dominant band matrix, then its elements are bounded as follows (with K a constant and r depending on the norm of the matrix):

$$s_{i,j} \leq K r^{|i-j|} \text{ for all } i,j.$$

☞ C. De Boor. Odd degree spline interpolation at a biinfinite knot sequence. In R. Schaback and K. Scherer, editors, *Approximation Theory*, volume 556 of *Lecture Notes in Mathematics*, pages 30–53. Springer, Heidelberg, 1976.

☞ C. De Boor. Dichotomies for band matrices. *SIAM Journal on Numerical Analysis*, 17:894–907, 1980.

☞ S. Demko, W. F. Moss, and P. W. Smith. Decay rates for inverses of band matrices. *Mathematics of Computation*, 43:491–499, 1984.

The manuscripts of de Boor are based on the manuscript of Demko described above and provide explicit bounds for the decay of the inverse depending on the bandwidth and the condition number of the matrix. The authors Demko, Moss and Smith provide bounds, based on spectral theory and a result on the best approximation of $(x - a)^{-1}$, a result of Chebyshev. The bound presented in this paper appears to be sharper than the bound presented in the other two manuscripts. The results as presented in these manuscripts do not take into consideration that the upper and lower part of the matrix might have different decay factors, a global decay rate for the upper as well as the lower part is presented.

☞ V. Eijkhout and B. Polman. Decay rates of inverses of banded M-matrices that are near to Toeplitz matrices. *Linear Algebra and its Applications*, 109:247–277, 1988.

Eijkhout and Polman provide decay rates for the inverse of band matrices which are close to Toeplitz matrices. This class contains, for example, the diagonally dominant Toeplitz M-matrices.

☞ G. Meurant. A review of the inverse of symmetric tridiagonal and block tridiagonal matrices. *SIAM Journal on Matrix Analysis and its Applications*, 13:707–728, 1992.

This paper by Meurant revisits and expands some results concerning the inverses of irreducible symmetric tridiagonal and block tridiagonal matrices. Based on the UL- and LU-decomposition of the tridiagonal matrix, he presents a method for computing the generators of the inverse. These results are extended towards the higher order block form, in which he exploits also the block UL, LU and a twisted decomposition. The twisted decomposition is some kind of hybrid form between the UL- and LU-decomposition. This technique gives the block generators of the inverse.

The author also pays attention to the computation of inverses of specific matrices arising from differential equations. The considered matrices are specific (block) tridiagonal Toeplitz matrices with minus (identity matrices) ones on the off-diagonals. Decay rates on the elements of these matrices are presented.

Finally some results concerning the decay of symmetric block tridiagonal matrices are presented.

☞ P. Concus, G. H. Golub, and G. Meurant. Block preconditioning for the
conjugate gradient method. *SIAM Journal on Scientific and Statistical
Computation*, 6:220–252, 1985.

In this manuscript, the authors Concus, Golub and Meurant investigate block preconditioners for positive definite block tridiagonal systems of equations. Based on the decay
rates on the elements of the inverse of the tridiagonal matrix they are able to construct
a preconditioner. The manuscript contains a method for computing the generators of the
inverse of an irreducible tridiagonal matrix, as well as Theorem 7.11.

☞ J. J. McDonald, R. Nabben, M. Neumann, H. Schneider, and M. J. Tsatsomeros. Inverse tridiagonal Z-matrices. *Linear and Multilinear Algebra*,
45(1):75–97, 1998.

McDonald, Nabben, Neumann, Schneider and Tsatsomeros pose properties on the class of
generator semiseparable matrices such that their tridiagonal inverses belong to the class
of Z-matrices.

☞ R. Nabben. Two sided bounds on the inverse of diagonally dominant
tridiagonal matrices. *Linear Algebra and its Applications*, 287:289–305,
July 1999.

Nabben provides upper and lower bounds for the entries of the inverse of diagonally dominant tridiagonal matrices, where the inverse is represented using the generator representation. Moreover, as the upper and lower bounds are highly related to each other, it is
possible to update them one after another. This leads to an iterative scheme updating in
every step the upper and lower bound. This iterative procedure converges rapidly to the
inverse of the tridiagonal matrix.

☞ R. Nabben. Decay rates of the inverse of nonsymmetric tridiagonal and
band matrices. *SIAM Journal on Matrix Analysis and its Applications*,
20(3):820–837, 1999.

In this manuscript by Nabben, the results as provided in this section concerning irreducible
tridiagonal M-matrices are presented. The manuscript expands Theorem 7.12 even more.
More precisely it is investigated how the elements of the inverse will behave in case the
matrix is not necessarily diagonally dominant. Finally some corollaries are formulated
concerning the decay of the inverse of banded M-matrices.

Applications of the bounds described in this section have been used for proving
local rates of convergence of spline approximations and bounding the norm of orthogonal
projections on spline spaces. These results were established in the manuscripts by Demko
(1977), Kershaw (1970) and the following publications.

☞ C. De Boor. A bound on the L_∞-norm of the L_2-approximation by splines
in term of a global mesh ratio. *Mathematics of Computation*, 30:687–694,
1976.

☞ W. J. Kammerer and G. W. Reddien, Jr. Local convergence of smooth
cubic spline interpolants. *SIAM Journal on Numerical Analysis*, 9:687–
694, 1972.

☞ B. Mityagin. Quadratic pencils and least-squares piecewise-polynomial
approximation. *Mathematics of Computation*, 40:283–300, 1983.

7.6 Conclusions

In this chapter some inversion methods related to the most simple classes of structured rank matrices were discussed. We discussed inversion of matrices via factorizations, but also presented some direct inversion methods. Briefly we showed some important results related to the scaling of semiseparable matrices, the decay of the elements in the inverse of a diagonally dominant tridiagonal matrix.

Part III

Structured rank matrices

The third part of this book is dedicated to the generalizations of semiseparable, tridiagonal and quasiseparable matrices, namely the classes of $\{p_1, p_2\}$-semiseparable, $\{p_1, p_2\}$-quasiseparable and band matrices. This includes in some sense also the classes of band, semiseparable plus band and $\{p_1, p_2\}$-generator representable semiseparable matrices. This part is not divided in two parts, where one part would focus on the definition and representation, and another part would focus on the solving of the systems. The last chapters of this part discuss also more general structured rank matrices such as block quasiseparable, block band, block semiseparable $\mathcal{H}$-matrices and hierarchically semiseparable. These last chapters discuss references related to these classes.

We choose to put everything together in one part as we will not pay so much attention anymore to the specific representation of the classes of matrices. Representations of these classes are highly subjected to the application and the origin of the matrices. Hence it would be unwise in some sense to devote all attention to a specific representation. Moreover, it would also be impossible trying to cover all of the many possible representations. Therefore we focus on the one thing all these matrices have in common, and that is the rank structure. All these different matrices obey certain rank restrictions in their structure. From this viewpoint we would like to work with these matrices. We would like to see what, e.g., the effect of a sequence of Givens transformations on a specific rank structure is. As we know the effect on a specific rank structure, this automatically gives us the effect on a semiseparable, quasiseparable, semiseparable plus band matrix and so forth. Therefore, the main focus of this part will be to develop tools, essential theorems, for working with structured rank matrices. When using these matrices for any kind of application, the first question is: "How can we exploit the rank structure of these matrices?" Second, one should try to search for a suitable representation. Using the knowledge for manipulating representations gained in the first two parts of the book, the reader should be capable of implementing the different algorithms.

This part of the book is divided in different chapters, covering, respectively, the definition and some properties of these matrices, the QR-factorization and LU-factorization of a structured rank matrix and more examples for the Levinson-like solver are included. The last three chapters of this part present an overview, with lots of references related to block quasiseparable and $\mathcal{H}$-matrices. The last chapter focuses on the inversion of matrices such as band, semiseparable, block band, band Toeplitz and so forth.

The first chapter of this part defines structured rank matrices. These matrices are defined in a general context. This means that a specific part of the matrix, a so-called structure, will be subjected to some rank structural constraints. One or more of these structured rank constraints define in some sense a structured rank matrix. Based on this general definition, we briefly show how matrices such as semiseparable, quasiseparable and generator representable semiseparable matrices fit into this framework. The nullity theorem is revisited and is applied to different classes of matrices to predict the rank structure of its inverse. An interesting part in this chapter is the part on completion problems. We investigate the relation between the generator definition and the semiseparable definition, just as we did in the rank 1 case. Now, however, as the rank of the matrices can of course be larger than 1, it becomes more complicated to address questions like:

> What is the minimum number of rank 1 matrices needed to represent a rank structure?

or

> What is the minimum number of semiseparable matrices needed to represent a rank structure?

It will be proved that, in general, a structure of rank r, needs r structures of rank 1 to represent this matrix. This decomposition will come in handy when searching for factorizations of these matrices. Finally some generalizations of the representations proposed in the first part of the book are presented, such as the quasiseparable, the higher order generator, the sum of Givens-vector represented matrices.

The second chapter in this part computes the QR-factorization of a general structured rank matrix. The factorization will be computed only by using Givens transformations. We will prove the existence of two types of Givens transformations. A first kind will be the well-known type of rank-decreasing Givens transformations. These Givens transformations are chosen to create zeros if performed on a matrix. The second type of Givens transformations will be the rank-expanding Givens transformations. They will create rank 1 blocks, when performed on 3×2 matrices. These Givens transformations will be combined to obtain sequences of rank-decreasing Givens transformations and sequences of rank-expanding Givens transformations. The global number of sequences to make the matrix upper triangular will be computed. The interaction between these sequences of Givens transformations is also very interesting. It will be shown that sequences of transformations can be interchanged. This means that, for example, a sequence of ascending Givens transformations, followed by a sequence of descending Givens transformations can be changed such that first a descending sequence and next an ascending sequence is performed. This possibility for interchanging the order leads to interesting results and different patterns for computing the upper triangular factor R. This chapter will be concluded by some remarks, resulting from the theoretical results deduced in this chapter. An alternative interpretation to the standard QR-algorithm for unstructured matrices will be given. Also a parallel QR-factorization will be discussed. Finally, some results will be presented predicting the structure of the multiplication of different structured rank matrices.

In the chapter on the LU-decomposition we try to derive results similar to those in the previous chapter on the QR-decomposition. Instead of using Givens transformations we try to do it now with Gauss transforms. Unfortunately we will see that Gauss transformation matrices alone are not enough to guarantee the existence of rank-decreasing and rank-expanding sequences of transformations. In general pivoting needs to be incorporated to guarantee the existence of the above-mentioned sequences. Pivoting has with regard to Gauss transforms a deep impact on the structured rank of the upper triangular structure. Gauss transforms leave the rank of this structure untouched, whereas these permutation matrices tend to increase it. The interactions with Gauss transforms and permutation matrices will be investigated and a graphical scheme for representing sequences of Gauss transforms will also be presented.

In Chapter 11 some new examples of applications of the Levinson-like solver are discussed. The general framework as discussed earlier is now used to solve systems of equations involving quasiseparable, higher order generator representable, band matrices and other matrices of higher order structured rank form.

Chapter 12 describes the class of block quasiseparable matrices. Historically, these matrices were defined by their representation. We will give such a definition. However, as we will show these matrices satisfy very nice structured rank properties, they could even be defined using these properties. This would give more freedom in the choice of a suitable representation depending on the specific situation in which such a matrix is used.

In Chapter 13 the class of $\mathcal{H}^2$-matrices is defined as a subclass of the class of $\mathcal{H}$-matrices, also called hierarchical matrices. Hierarchical matrices are a very useful tool, e.g., when solving integral equations. A similarly structured class of matrices is the class of hierarchically semiseparable matrices.

The final chapter of this part presents an overview of applications related to structured rank matrices and several pointers to manuscripts on the inversion of structured rank matrices. The inversion of structured rank matrices is not covered in detail, as most methods in the literature focus onto one specific subclass. Therefore we present an overview of the available literature and briefly describe the contents of these manuscripts. We cover for example semiseparable, quasiseparable, band, band Toeplitz, block band and other types of matrices belonging to the class of higher order structured rank matrices.

✎ *The main aim of this part is to extend the results discussed in the first and second parts for semiseparable and related matrices towards their higher order generalizations: higher order semiseparable, higher order quasiseparable and so forth. The more general classes such as the block quasiseparable and the hierarchically semiseparable matrices are also discussed, but in the last chapters of this part.*

Chapter 8 defines the new classes of matrices we will consider. Section 8.1 contains the most important setting of this chapter, it defines the concept "structured rank" as a rank constraint posed on a structure in a matrix. The algorithms proposed in the remainder of this part are designed for such structured rank matrix parts. The definition of some particular structured rank classes such as higher order generator representable semiseparable, higher order quasiseparable, higher order semiseparable and so forth are presented (Section 8.2). A second important issue addressed in this chapter is the decomposition of structured rank parts; Theorem 8.71 summarizes the complete derivation, namely that a semiseparable rank part of rank r can be decomposed as the sum of r terms of semiseparable rank 1. This idea of decomposing the rank part of a matrix is being used extensively in the development of the proposed structured rank algorithms. Section 8.5 briefly discusses possible representations.

Chapter 9 is a large chapter in which most of the attention is paid to the effect of sequences of Givens transformations on structured rank parts in the matrices. The effect of a sequence of transformations from bottom to top is presented in Theorem 9.26, where a rank-decreasing sequence of Givens transformations is defined in Definition 9.12 and a rank-expanding Givens transformation is defined in Definition 9.25. Combining this with different types of sequences as discussed in Subsection 9.1.8 leads to different types of factorizations involving orthogonal matrices. Combining the results in Subsections 9.2.1 and Section 9.2.2 provides information on the number of Givens transformations needed for computing the QR-factorization. The remaining part of the chapter discusses possibilities with Givens transformations and a more general existence of rank-expanding Givens transformations (Section 9.4). Exploiting this can lead to different patterns for computing orthogonal factorizations. The most general theorem for ascending sequences of Givens transformations is therefore presented in Theorem 9.54 in Subsection 9.4.4. Based on the results presented in this chapter, Theorem 9.56 is derived in Section 9.7, predicting the rank structure of a matrix matrix product of two structured rank matrices.

Chapter 10 discusses how to compute the LU-factorization of higher order structured rank matrices. Corollary 10.15 and Corollary 10.9 discuss the effect of a sequence of Gauss transforms both on the lower and on the upper triangular structure of the structured rank matrix. Remark that in this case no pivoting was involved. The effect of sequences with pivoting is discussed in Corollary 10.19. Combining results for the pivoting and nonpivoting sequences, one can use this method to solve systems of equations (Subsection 10.4.1). Unfortunately it is not guaranteed that one computes an LU-factorization (Subsection 10.4.2). To conclude the interested reader can figure out the possibilities of designing patterns with Gauss transforms (Section 10.5).

Chapter 11 discusses some extra examples for which the Levinson-like solver discussed in Chapter 6 is suitable.

Chapter 12 summarizes some results on block quasiseparable matrices. The definition of this class of matrices is given in terms of their representation as well as of their structured rank properties.

Chapter 13 introduces several classes of hierarchically structured rank matrices.

Chapter 14 gives an overview of the main results concerning the inversion of higher order structured rank matrices.

Chapter 8

Definitions of higher order semiseparable matrices

In this chapter we will discuss some new classes of structured rank matrices. In fact we will extend in a natural way the classes as defined in Chapter 1 of this book towards their higher order generalizations. The definitions, however, will be placed in a very general context, namely the context of structured rank matrices; hence, we do not restrict ourselves in the beginning to symmetric matrices. First, we will define what exactly is meant by a rank structured matrix. This will give us the building blocks for defining the classes of higher order semiseparable, higher order quasiseparable, band matrices and so forth. We will show that they all fit in a natural way into this general framework. We will not investigate in detail the inner relations between all these classes anymore, however. A profound study of the relations for the easy classes was presented in Chapter 1; a repetition of similar results for the newly defined matrices is not included, only partial interesting results, as, e.g., the relation between generator representable semiseparable and semiseparable matrices.

The nullity theorem will also be reconsidered. Using this theorem we will investigate the relations under inversion of these new higher order structured rank matrices.

The chapter is organized as follows. First we will present definitions of structures and rank structures. This leads to a general theory for defining all kinds of structured rank matrices. In a second section we will use these definitions for defining higher order semiseparable, quasiseparable, generator representable, and other classes of interesting structured rank matrices. In Section 8.3 we investigate, based on the nullity theorem, the inverses of general structured rank matrices. Based on these observations it is an easy exercise to predict the structure of the inverses of previously defined classes of matrices. Section 8.4 presents interesting results concerning the decomposition of a structured rank block in a matrix into the sum of lower order structured rank blocks. To prove these results some completion problems are addressed. In these completion problems we investigate the minimum number of rank 1 matrices that is needed to write a structured rank part of a matrix as a sum of these rank 1 matrices. In this section we also generalize the results of

the beginning of the book, stating whether a matrix is generator representable or
not. The final section shows some possible representations for these higher order
structured rank matrices. The presented representations are generalizations of previously discussed representations for structured rank matrices of semiseparability
rank 1.

✎ *The concept "structured rank" is the most important one in this chapter.*
Definition 8.1 defines these concepts and examples are presented in Definition 8.3
and Example 8.4. Even though the remainder of the book focuses on general struc-
tured rank matrices (and no knowledge of specific classes is required), some examples
of specific 'important' classes of structured rank matrices are given. Higher order
semiseparable matrices are considered in Definition 8.8 and Example 8.10. Higher
order quasiseparable matrices are considered in Definition 8.12 and the higher order
generator representable semiseparable matrices in Definition 8.14. Also Hessenberg-
like and sparse matrices fit naturally into this context (Definition 8.22 and Subsec-
tion 8.2.7). Particular inverses of some structured rank matrices are considered
in Subsection 8.3.2. Section 8.4 contains very important results for a good un-
derstanding of the remainder of the book. After a lot of theoretical investigations
a powerful theorem on the decomposition of structured rank matrices is proved,
namely, Theorem 8.71. The theorem states that one can decompose a structured
rank part into a sum of other structured rank parts of lower rank. Examples of this
decomposition can be found in Subsection 8.4.4. This theorem and the following
examples result from the investigations concerning the relation between semisepara-
ble and generator representable semiseparable matrices. For the interested reader,
Theorem 8.60 provides conditions, whether a structured rank part is generator rep-
resentable or not. This is an extension of results presented in the first chapter. The
proof of this theorem uses results related to several completion problems discussed
in Subsection 8.4.2.

8.1   Structured rank matrices

Previously defined matrices, such as semiseparable matrices, quasiseparable matrices and tridiagonal matrices, have often been referred to as structured rank matrices.
As already mentioned, this means that specific parts of the matrix, the so-called
structure, satisfy certain rank properties. Before continuing the use of structured
rank matrices, we will define in a rigid way what is meant by a structure and a
structured rank.

Definition 8.1 (Structures; Structured rank). *Let A be an $m \times n$ matrix.*
Denote with M the set of numbers $\{1, 2, \ldots, m\}$ and with N the set of numbers
$\{1, 2, \ldots, n\}$. Let α and β be nonempty subsets of M and N, respectively. Then, we
denote with the matrix $A(\alpha; \beta)$ the submatrix of A with row indices in α and column
indices in β. A structure Σ is defined as a nonempty subset of $M \times N$. Based on
a structure, the structured rank $\mathrm{r}(\Sigma; A)$ is defined as (where $\alpha \times \beta$ denotes the set

$\{(i,j)|i \in \alpha, j \in \beta\}$):

$$r(\Sigma; A) = \max\{\text{rank}\,(A(\alpha; \beta))\,|\alpha \times \beta \subseteq \Sigma\}.$$

Definition 8.2. *A matrix A is called a structured rank matrix if there are upper bounds posed on one or more structured ranks of the matrix. (We assume of course that one or more structured ranks of the matrix have a rank smaller than the rank one would generically expect.)*

Even though the definition is very general it clearly illustrates what is meant with a structured rank matrix. Often we refer to the semiseparability rank or the quasiseparability rank of a structure in the matrix. This just refers to the structured rank related to that specific structure.

Let us present, as an example, some structures. In Section 8.3 some more structures will be constructed and investigated.

Definition 8.3. *For $M = \{1, \ldots, m\}$ and $N = \{1, \ldots, n\}$ we define the following structures:*

- *The subset*
$$\Sigma_l = \{(i,j)|i \geq j, i \in M, j \in N\}$$
is called the lower triangular structure; in fact, the elements of the structure correspond to the indices from the lower triangular part of the matrix.

- *The subset*
$$\Sigma_{wl} = \{(i,j)|i > j, i \in M, j \in N\}$$
is called the weakly lower triangular structure.

- *The subset*
$$\Sigma_l^{(p)} = \{(i,j)|i > j - p, i \in M, j \in N\}$$
is called the p-lower triangular structure and corresponds with all the indices of the matrix A, below the pth diagonal. The 0th diagonal corresponds to the main diagonal, while the pth diagonal refers to the pth superdiagonal (for $p > 0$) and the $-p$th diagonal refers to the pth subdiagonal (for $p > 0$).

Note that $\Sigma_l^{(1)} = \Sigma_l$, $\Sigma_l^{(0)} = \Sigma_{wl}$ and $\Sigma_{wl} \subsetneq \Sigma_l$. Note that the structure $\Sigma_l^{(p)}$ for $p > 1$ contains all the indices from the lower triangular part, but it also contains some superdiagonals of the strictly upper triangular part of the matrix. The weakly lower triangular structure is sometimes also called the strictly lower triangular structure or the subdiagonal structure. For the upper triangular part of the matrix, the structures Σ_u, Σ_{wu} and $\Sigma_u^{(p)}$ are defined similarly, and are called the upper triangular structure, the weakly upper triangular structure and the p-upper triangular structure, respectively. The structured rank connected to the lower triangular structure is called the lower triangular rank. Similar definitions are assumed for the other structures.

Let us present some examples of classes of matrices already considered in the book.

Example 8.4 For the following examples, we rewrite their definitions in terms of the constraints posed on certain structures corresponding to the matrix.

- A matrix S is called a semiseparable matrix if and only if

$$\text{r}(\Sigma_l; A) \leq 1 \text{ and } \text{r}(\Sigma_u; A) \leq 1.$$

- A matrix A is called quasiseparable if and only if

$$\text{r}(\Sigma_{wl}; A) \leq 1 \text{ and } \text{r}(\Sigma_{wu}; A) \leq 1.$$

- A matrix D is a diagonal matrix, if and only if

$$\text{r}(\Sigma_{wl}; A) \leq 0 \text{ and } \text{r}(\Sigma_{wu}; A) \leq 0.$$

- A matrix T is a tridiagonal matrix, if and only if

$$\text{r}\left(\Sigma_l^{(-1)}; A\right) \leq 0 \text{ and } \text{r}\left(\Sigma_u^{(+1)}; A\right) \leq 0.$$

- A matrix S is a semiseparable plus diagonal matrix, if and only if (for a certain diagonal matrix D)

$$\text{r}(\Sigma_l; S - D) \leq 0 \text{ and } \text{r}(\Sigma_u; S - D) \leq 0.$$

∎

The reader can see that these relations can easily be expanded to obtain, for example, a band matrix. In the next section we will define what is meant with higher order semiseparable, quasiseparable and related matrices.

Notes and references

The concept of structured rank is due to Fiedler. In the following manuscripts, he defines different types of rank structures.

☞ M. Fiedler. Structure ranks of matrices. *Linear Algebra and its Applications*, 179:119–127, 1993.

The manuscript of 1993 defines the concept of structure and structured rank as used throughout this book. Fiedler investigates different types of rank structures (such as off-diagonal and block off-diagonal rank), with regard to inversion and the LU-decomposition.

The manuscripts [112, 118], investigate special properties of different types of structured rank matrices. They cover basic matrices, which are matrices having both sub- and superdiagonal rank 1; these are, in fact, quasiseparable matrices. The class of generalized

Hessenberg matrices is investigated in the manuscript [118] and covers the class of matrices having subdiagonal rank 1. This corresponds to the class of Hessenberg-like matrices as defined in this book. We did not choose to adopt the name generalized Hessenberg matrices because this is often used in different contexts (e.g., the generalized eigenvalue problem). More information on these matrices and the contents of these manuscripts can be found in Section 4.6 in Chapter 4.

Another type of more general structured rank matrix is considered in the manuscripts by Delvaux and Van Barel. Let us recapitulate their definition of a structured rank matrix. In their context (see, e.g., [78, 75]) we have the following definition.

Definition 8.5 (From [78]). *A rank structure on $\mathbb{C}^{n \times n}$ is defined as a collection of so-called structure blocks $\mathcal{R} = \{\mathcal{B}_k\}_k$. Each structure block $\mathcal{B}_k$ is characterized by a 4-tuple as follows:*

$$\mathcal{B}_k = (i_k, j_k, r_k, \lambda_k), \tag{8.1}$$

with i_k and j_k denoting, respectively, a row and column index, r_k a rank upper bound and $\lambda_k \in \mathbb{C}$ is the shift element of the structure block. A matrix $A \in \mathbb{C}^{n \times n}$ is said to satisfy this rank structure if for each k the following relation is satisfied:

$$\text{rank}\,(A_k(i_k : n, 1 : j_k)) \leq r_k \text{ with } A_k = A - \lambda_k I. \tag{8.2}$$

In this context a rank structure is called pure, if it has the shift element equal to zero. The definition presented here is the 'basic' definition. In some of the manuscripts the authors extend the shift element to a shift block ([76]) which has to be subtracted, and also towards blocks starting in the upper right part.

Let us include two examples: for a semiseparable and a semiseparable plus diagonal matrix. (We only consider the structure of the lower triangular part. For the upper triangular part one needs to define the structure blocks also starting from the upper right corner.) Hence we have for the structure of a lower semiseparable matrix A of dimension $n \times n$ (named Hessenberg-like matrix in our setting), the following structure blocks:

$$\mathcal{B}_k = (k, k, 1, 0), \text{ with } k = 1, \ldots, n.$$

For a lower semiseparable plus diagonal matrix (a Hessenberg-like plus diagonal matrix in our setting), the structure blocks are defined as (where $D = \text{diag}(d_1, \ldots, d_n)$ denotes the diagonal matrix):

$$\mathcal{B}_k = (k, k, 1, d_k), \text{ with } k = 1, \ldots, n.$$

The definition as presented here naturally fits into the definition proposed in the beginning of this section (see Definition 8.1 and Definition 8.2). Considering the following index sets

$$\Sigma_k = (1 : i_k, j_k : n),$$

the structured rank constraint

$$\text{r}(\Sigma_k; A - \lambda_k I) \leq r_k,$$

coincides with the constraint of block k: $\mathcal{B}_k = (i_k, j_k, r_k, \lambda_k)$ proposed in Definition 8.5. Hence all constraints posed by the different blocks can be rewritten in terms of demands posed on the structured ranks related to different Σ_k's.

8.2 Definition of higher order semiseparable and related matrices

In this section we will expand the definition of several classes of matrices, including semiseparable and quasiseparable matrices. Moreover we will also define some new classes of matrices such as extended semiseparable matrices. Also, a new class for which only the lower (or upper) triangular part satisfies a specific rank condition will be considered. These matrices are the so-called Hessenberg-like matrices.

In the first part of this book, we examined in great detail the inner relations between all different classes of matrices. We will not continue such analysis here. Provided with the tools in the first part, readers know the essential differences and can pursue this analysis if they desire to do so.

8.2.1 Some inner structured rank relations

Before defining the different types of higher order structured rank matrices, we will present some theorems related to the structured rank. These theorems will provide us with interesting information to see whether, e.g., a semiseparable matrix of semiseparability rank 1 will also be of semiseparability rank 2.

We will investigate here the inner relation between different rank structures. The question we will address is the following: What happens, in general, with the rank, if a diagonal is added to, or removed from the structure.

In this section we will only investigate these properties for lower triangular structured ranks; upper triangular ranks can be dealt with similarly.

Theorem 8.6. *Suppose we have a matrix A satisfying the following constraint:*

$$\mathrm{r}\left(\Sigma_l^{(p)}; A\right) = r_1.$$

Then, also the following constraints are satisfied:

$$\mathrm{r}\left(\Sigma_l^{(p+1)}; A\right) \leq r_1 + 1 \quad and \quad \mathrm{r}\left(\Sigma_l^{(p-1)}; A\right) \leq r_1.$$

Proof. Let us start with the first statement:

$$\mathrm{r}\left(\Sigma_l^{(p+1)}; A\right) \leq r_1 + 1. \tag{8.3}$$

Suppose we have the following block B taken out of the matrix A:

$$B = \left[\begin{array}{cc} A_{11} & \times \\ A_{21} & A_{22} \end{array} \right],$$

where the element $\times$ is an element of the pth diagonal, which we want to include in the structure. We know by construction that the matrix $[A_{21}, A_{22}]$ is of rank at most r_1. Adding now the row $[A_{11}, \times]$ on top of the matrix $[A_{21}, A_{22}]$ increases the rank at most by 1. Hence the rank of the complete block B is at most $r_1 + 1$.

This procedure can be repeated for every block containing an element of the pth diagonal. Hence we can conclude that Equation (8.3) needs to be satisfied. Let us prove now the relation:

$$\text{r}\left(\Sigma_l^{(p-1)}; A\right) \leq r_1. \tag{8.4}$$

This statement as formulated here is always valid.   □

Equation (8.4) is always satisfied, but as adding a diagonal increases the rank, one might wonder that removing a diagonal decreases the rank. This, however, is not true in general, as illustrated by the following 4×4 example:

$$A = \begin{bmatrix} 0 & & & \\ 0 & 0 & & \\ 1 & 1 & 0 & \\ 0 & 1 & 0 & 1 \end{bmatrix}.$$

This matrix satisfies the following rank constraint:

$$\text{r}\left(\Sigma_l^{(1)}; A\right) = 2.$$

Excluding the diagonal from the structure gives us the following constraint:

$$\text{r}\left(\Sigma_l^{(0)}; A\right) = 2.$$

Hence it is clear that removing a diagonal does not necessarily reduce the structured rank, leading to the fact that Equation (8.4) is true in general.

In the following examples we illustrate that the new constraints presented in the theorem above are indeed upper bounds.

Example 8.7 Suppose we have a rank r matrix. This matrix satisfies the following constraint:

$$\text{r}\left(\Sigma_l^{(p)}; A\right) \leq r,$$

for every value of $-n \leq p \leq n$. Hence we see that including more diagonals, does not necessarily increase the rank.

Suppose we have a diagonal matrix D, having nonnegative diagonal elements. This matrix is of semiseparable form, satisfying the following constraint:

$$\text{r}\left(\Sigma_l^{(1)}; A\right) = 1.$$

This matrix satisfies also the following relation, when excluding the diagonal from the structure.

$$\text{r}\left(\Sigma_l^{(0)}; A\right) = 0.$$

This shows that excluding a diagonal, can indeed decrease the structured rank.
                                                                                      ■

The previous theorem, will help us in the following subsections to clarify the relation between, for example, a higher order semiseparable and a lower order semiseparable matrix.

8.2.2 Semiseparable matrices

With the previously defined structures we define semiseparable matrices as follows.

Definition 8.8. *An $n \times n$ matrix S is called a $\{p, q\}$-semiseparable matrix[1], with $p \geq 0$ and $q \geq 0$, if the following two properties are satisfied:*

$$\mathrm{r}\left(\Sigma_l^{(p)}; S\right) \leq p \quad and \quad \mathrm{r}\left(\Sigma_u^{(-q)}; S\right) \leq q.$$

This means that the p-lower triangular rank is less than or equal to p and the q-upper triangular rank is less than or equal to q.

The definition above says that the maximum rank of all subblocks which one can take out of the matrix below the pth superdiagonal is less than or equal to p and the maximum rank of all subblocks which one can take above the qth subdiagonal is less than or equal to q. When speaking about a $\{p\}$-semiseparable matrix or a semiseparable matrix of semiseparability rank p, we mean a $\{p, p\}$-semiseparable matrix. When briefly speaking about a semiseparable matrix, we refer to a semiseparable matrix of semiseparability rank 1.

Note 8.9. *We would like to remark that the p in $\{p, q\}$-semiseparable refers to the part below the pth superdiagonal. The p in $\{p, q\}$-Levinson conform, refers to the structure above the diagonal. Hence one should be careful with these definitions (see Chapter 11).*

Let us present some examples of $\{p, q\}$-semiseparable matrices.

Example 8.10 Consider the following matrix

$$S_1 = \begin{bmatrix} 1 & 1 & 1 & 1 \\ 0 & 1 & 1 & 1 \\ 0 & 0 & 1 & 1 \\ 1 & 0 & 0 & 1 \end{bmatrix}.$$

This matrix is $\{3, 1\}$-semiseparable. The lower triangular part is especially intriguing as it seems to be of rank 2, but due to the inclusion of the superdiagonals, this becomes in fact of rank 3 semiseparable form. The matrix S_2 is $\{2, 2\}$-semiseparable and S_3 is of $\{3, 3\}$-semiseparable form:

$$S_2 = \begin{bmatrix} 1 & 0 & 0 & 0 \\ 0 & 1 & 0 & 2 \\ 1 & 0 & 1 & 0 \\ 0 & 0 & 0 & 1 \end{bmatrix} \text{ and } S_3 = \begin{bmatrix} 1 & 0 & 0 & 4 \\ 0 & 1 & 0 & 0 \\ 0 & 0 & 1 & 0 \\ 1 & 0 & 0 & 1 \end{bmatrix}.$$

■

---

[1] A motivation about why we choose this definition can be found in the notes and references section.

Higher order semiseparable matrices as defined here have a so-called maximal invertible rank structure. This means that if a matrix is of $\{p, q\}$-semiseparable form, that no element in the pth superdiagonal and no element in the qth subdiagonal can be included in the rank structure of the lower and upper structure, respectively, without making the matrix singular. For example, a nonsingular semiseparable matrix cannot have an element of the superdiagonal, or an element of the subdiagonal includable in the lower/upper semiseparable part, respectively, without being singular. A proof of this statement is straightforward, as this would create a rank block in the matrix, which is in fact too large. A more rigid approach to these statements will be provided in Chapter 9, Subsection 9.2.3.

Before defining other classes of semiseparable matrices, we want to take a closer look at the relation between semiseparable matrices of different orders. Using Theorem 8.6, we can easily see that a $\{1\}$-semiseparable matrix is also a $\{2\}$-semiseparable matrix, similarly as a tridiagonal matrix is also a pentadiagonal matrix.

In general, we have the following relations between semiseparable matrices.

Corollary 8.11. *Suppose we have matrix A, which is of $\{p, q\}$-semiseparable form, with $p, q \geq 0$. Then this matrix is also of $\{\hat{p}, \hat{q}\}$-semiseparable form, with $p \leq \hat{p}$ and $q \leq \hat{q}$.*

This means that, for example, a $\{0\}$-semiseparable matrix (which is a diagonal matrix), is also a $\{1\}$-semiseparable as well as a $\{2\}$-semiseparable matrix. Further in the text we will show that the inverse of a $\{2\}$-semiseparable matrix will be a pentadiagonal matrix. In this viewpoint it is clear that a $\{1\}$-semiseparable matrix, which is the inverse of a tridiagonal matrix, should also be a $\{2\}$-semiseparable matrix.

8.2.3 Quasiseparable matrices

Whereas in the higher order semiseparable case, the rank structure of the semiseparable matrix expanded also above and below the diagonal, this is not the case for the quasiseparable matrices. The structured part remains below and above the diagonal. In a forthcoming chapter we will also discuss the class of block quasiseparable matrices (see Chapter 12).

Definition 8.12. *An $n \times n$ matrix A is called a $\{p, q\}$-quasiseparable matrix, with $p \geq 0$ and $q \geq 0$, if the following two properties are satisfied:*

$$\mathrm{r}\left(\Sigma_{wl}; A\right) \leq p \quad and \quad \mathrm{r}\left(\Sigma_{wu}; A\right) \leq q.$$

This means that the strictly (or weakly) lower triangular rank is less than or equal to p and the strictly (or weakly) upper triangular rank is less than or equal to q.

Similar assumptions concerning terminology as in the semiseparable case are used here. Hence we have immediately defined $\{p\}$-quasiseparable or quasiseparable matrices of quasiseparability (semiseparability[2]) rank p, whereas speaking about a quasiseparable matrix refers to the simple class of $\{1\}$-quasiseparable matrices.

The following corollary for quasiseparable matrices is straightforward, and shows the relation between matrices within the class of quasiseparable matrices.

Corollary 8.13. *Suppose we have matrix A, which is of $\{p, q\}$-quasiseparable form, with $p, q \geq 0$. Then this matrix is also of $\{\hat{p}, \hat{q}\}$-quasiseparable form, with $p \leq \hat{p}$ and $q \leq \hat{q}$.*

The proof is straightforward and therefore not included.

At the end of this section, in the notes and references part, we discuss the manuscript [219], in which a lot of interesting properties of quasiseparable matrices are derived.

8.2.4   Generator representable semiseparable matrices

Within the class of generator representable semiseparable matrices one can make a distinction between two viewpoints. Let us consider both, and make a clear distinction. First we will define the class which we will name the class of higher order generator representable semiseparable matrices. In the following subsection we will define the second class.

Definition 8.14.   *An $n \times n$ matrix S is called a $\{p, q\}$-generator representable semiseparable matrix, with $p \geq 0$ and $q \geq 0$, if the following two properties are satisfied*

$$\text{tril}(S, p - 1) = \text{tril}(UV^T, p - 1),$$
$$\text{triu}(S, -q + 1) = \text{triu}(PQ^T, -q + 1),$$

with U and V of size $n \times p$ and P and Q of size $n \times q$. This means that the lower triangular part of the matrix A, up to and including the $(p - 1)$th superdiagonal, is coming from a rank p matrix. Similarly the upper triangular part starting from the $(q - 1)$th subdiagonal is coming from a rank q matrix.

Similarly as in the previous section we adopt the notation of a $\{p\}$-generator representable semiseparable matrix and a generator representable matrix of semiseparability rank p. We also sometimes briefly denote such a generator representable matrix as $S(U, V, P, Q)$ and a symmetric one as $S(U, V)$, just as in the rank 1 case.

It is clear that the class of $\{p, q\}$-generator representable semiseparable matrices is a subset of the class of $\{p, q\}$-semiseparable matrices. More details, and theorems stating when a semiseparable matrix is of generator representable form and when not, are presented in Section 8.4 of this chapter.

For the semiseparable and quasiseparable matrices we proved that lower rank matrices in these classes were includable in the higher rank classes, for example, a

---

[2]With semiseparability or quasiseparability rank we simply refer to the rank related to the structure. The term semiseparability or quasiseparability rank, does not at all refer to constraints posed on the structure.

{1}-semiseparable was also a {2}-semiseparable matrix. For the generator representable matrices this statement is false. A trivial example of this is a diagonal matrix (with nonzero diagonal elements), which is {0}-generator representable semiseparable, but which is clearly not of {1}-generator representable form. Let us illustrate this also for another example.

Example 8.15 Suppose we have the {1}-generator representable semiseparable matrix S:

$$S = \begin{bmatrix} 1 & 2 & 3 \\ 2 & 2 & 3 \\ 3 & 3 & 3 \end{bmatrix}.$$

The question is now, whether this matrix is also of {2}-generator representable form. If this matrix is a {2}-generator representable semiseparable matrix, this means that there is a value α, such that the matrix A is of rank 2^3:

$$A = \begin{bmatrix} 1 & 2 & \alpha \\ 2 & 2 & 3 \\ 3 & 3 & 3 \end{bmatrix}.$$

The reader can easily check that no such value of α exists, such that the matrix becomes of rank 2. Hence, this matrix cannot be of {2}-generator representable form. ∎

In general we can prove the following lemma:

Lemma 8.16. *Suppose the matrix A is nonsingular and of $\{p, q\}$-generator representable semiseparable form ($p, q \leq n - 1$). This matrix can never be of $\{\hat{p}, \hat{q}\}$-generator representable form, with $\hat{p} \neq p$ or $\hat{q} \neq q$ and ($\hat{p}, \hat{q} \leq n - 1$).*

Proof. We postpone this proof until Section 8.3 after Theorem 8.47. Using the knowledge gained in that section, the proof will be easy. □

The assumption in Lemma 8.16 is important: for singular matrices we can easily find examples for which the statement in the lemma is not true.

Example 8.17 Let us provide some examples of belonging to several $\{p\}$-generator representable semiseparable classes. These matrices will be singular, or not satisfy the constraints placed on $p, q, \hat{p}$ and $\hat{q}$.

- An almost trivial example is the zero matrix. This matrix clearly belongs to each class of $\{p, q\}$-generator representable semiseparable matrices.

- Suppose we have the following 3×3 matrix A:

$$A = \begin{bmatrix} 1 & 2 & 3 \\ 1 & 2 & 3 \\ 1 & 2 & 3 \end{bmatrix}.$$

---

[3]Adding/changing elements in a matrix, such that the complete matrix will satisfy a certain constraint (e.g., rank constraints), are called completion problems. Some of these problems will be discussed in Section 8.4.

This matrix is clearly of rank 1 and hence singular. The matrix is of $\{1\}$, $\{2\}$ and of $\{3\}$-generator representable form.

- Suppose we have a rank 2 matrix A:

$$A = \begin{bmatrix} 1 & 1 & 0 \\ 1 & 1 & 1 \\ 0 & 1 & 1 \end{bmatrix}.$$

This matrix is of $\{2\}$ and of $\{3\}$-generator representable form, even though the matrix is invertible. This is because $\hat{p} = 3$ does not satisfy $\hat{p} \leq n - 1$, with $n = 3$.

■

8.2.5   Extended semiseparable matrices

Let us define here two other types of structured rank matrices: namely, the class of extended semiseparable matrices and the class of extended generator representable semiseparable matrices. These classes are called the extended classes, because they cover also the regular classes of semiseparable matrices, and the class of generator representable semiseparable matrices. The only difference is that the ranks do not go across the diagonal, but just up to and including the diagonal. Let us first define the class of extended semiseparable matrices.

Definition 8.18. *An $n \times n$ matrix S is called an extended $\{p, q\}$-semiseparable matrix, with $p \geq 0$ and $q \geq 0$, if the following two properties are satisfied:*

$$\mathrm{r}\left(\Sigma_l; S\right) \leq p \quad and \quad \mathrm{r}\left(\Sigma_u; S\right) \leq q.$$

This means that the lower triangular rank is less than or equal to p and the upper triangular rank is less than or equal to q.

Second, we can apply a similar idea to the generator representable structured rank matrices.

Definition 8.19. *An $n \times n$ matrix S is called an extended $\{p, q\}$-generator representable semiseparable matrix, with $p \geq 0$ and $q \geq 0$, if the following two properties are satisfied:*

$$\mathrm{tril}(S) = \mathrm{tril}(UV^T),$$
$$\mathrm{triu}(S) = \mathrm{triu}(PQ^T),$$

with U and V of size $n \times p$ and P and Q of size $n \times q$. This means that the lower triangular part of the matrix A is coming from a rank p matrix. Similarly the upper triangular part is coming from a rank q matrix.

These classes of matrices are quite often considered as the natural generalization of either semiseparable and/or generator representable semiseparable matrices.

In this text, however, we follow the idea that these matrices originate from the inverse of a tridiagonal, and hence in this case from the inverse of a band matrix. It will be proven further in this chapter that semiseparable matrices as defined above are the inverses of band matrices and also that the inverses of generator representable semiseparable matrices are the inverses of strict band matrices, i.e., band matrices not having zeros on the extreme diagonals. More information on this choice of nomenclature can be found in the notes and references section.

Before defining some other classes of matrices, we would like to state briefly, by an example, that the class of extended semiseparable matrices and the class of semiseparable matrices are substantially different from each other.

Example 8.20 Consider the matrix S_1

$$S_1 = \begin{bmatrix} 1 & 1 & 1 & 1 \\ 0 & 1 & 1 & 1 \\ 0 & 0 & 1 & 1 \\ 1 & 0 & 0 & 1 \end{bmatrix},$$

from Example 8.10. This matrix is of $\{3, 1\}$-semiseparable form, but it is of extended $\{2, 1\}$-semiseparable form. Hence it is clear that both classes are essentially different. In fact the class of extended matrices is larger than the class of regular (generator representable) semiseparable matrices. ∎

The class of extended matrices includes also all classes of lower order extended matrices. This is formulated in the following corollary.

Corollary 8.21. *If the matrix A is of $\{p, q\}$-extended (generator representable) semiseparable form, the matrix A will also be of $\{\hat{p}, \hat{q}\}$-extended (generator representable) semiseparable form, with $\hat{p} \geq p$ and $\hat{q} \geq q$.*

Proof. The proof is trivial. □

8.2.6 Hessenberg-like matrices

It is not necessary for a structured rank matrix to take the structure from the upper as well as from the lower triangular part of the matrix.

Definition 8.22. *A matrix Z is called an upper $\{p\}$-Hessenberg-like matrix if the p-lower triangular rank of Z is less than or equal to p:*

$$\mathrm{r}\left(\Sigma_l^{(p)}; Z\right) \leq p.$$

A lower $\{q\}$-Hessenberg-like matrix is defined in a similar way, having the structured rank part in the upper part of the matrix.

Often these matrices are also named lower semiseparable and/or upper semi-separable matrices. Like in the semiseparable case, when speaking about a Hessenberg-like, a {1}-Hessenberg-like matrix is meant. When it is clear from the context, we omit the notation "upper".

Similarly as in the semiseparable case, we can formulate the following corollary:

Corollary 8.23. *Suppose the matrix A to be of $\{p\}$-Hessenberg-like form, then the matrix A will also be of $\{\hat{p}\}$-Hessenberg-like form, with $\hat{p} \geq p$.*

Proof. The proof is based on Theorem 8.6.   □

Note 8.24. *We remark that the name of Hessenberg-like matrices is not wide spread. Often the name lower semiseparable is used, referring to the lower part of the matrix which should be of semiseparable form. One should not confuse lower semiseparable matrices with lower triangular semiseparable matrices, which have the strictly upper triangular part equal to zero.*

8.2.7   Sparse matrices

Even though we did not yet explicitly mention it, some sparse matrices can also be considered in a natural manner as being structured rank. Let us use the following definition of band matrices.

Definition 8.25. *An $n \times n$ matrix B is called a $\{p,q\}$-band matrix, with $p \geq 0$ and $q \geq 0$, if the following two properties are satisfied:*

$$\mathrm{r}\left(\Sigma_l^{(-p)}; B\right) = 0 \quad and \quad \mathrm{r}\left(\Sigma_u^{(q)}; B\right) = 0.$$

This means that below subdiagonal p, the matrix equals zero, and similarly above superdiagonal q there are also zeros.

The relation between different types of band matrices is straightforward. Let us formulate this as a corollary.

Corollary 8.26. *A $\{p,q\}$-band matrix B, is also a $\{\hat{p},\hat{q}\}$-band matrix, with $\hat{p} \geq p$ and $\hat{q} \geq q$.*

When only the lower part satisfies the zero pattern, this leads to the definition of a $\{p\}$-generalized Hessenberg matrix.

Definition 8.27. *An $n \times n$ matrix Z is called a $\{p\}$-generalized Hessenberg matrix, with $p \geq 0$, if the following property is satisfied:*

$$\mathrm{r}\left(\Sigma_l^{(-p)}; Z\right) = 0.$$

This means that below subdiagonal p, the matrix equals zero.

This is a direct generalization of a Hessenberg matrix, which has only one subdiagonal possibly different from zero. It is straightforward that a $\{p\}$-generalized Hessenberg matrix is also a $\{\hat{p}\}$-generalized Hessenberg matrix, with $\hat{p} \geq p$.

Note 8.28. *To conclude this section, we remark that a $\{p, q\}$-band matrix can also be seen as a $\{p, q\}$-quasiseparable matrix. We leave the proof of this statement to the reader.*

8.2.8 What is in a name

In the first part of this book attention was also paid to the class of semiseparable plus diagonal matrices. In this context, we can combine different types of matrices, leading, for example, to:

- $\{p, q\}$-semiseparable plus diagonal or even $\{p, q\}$-semiseparable plus $\{r, s\}$-band matrices;

- generator representable $\{p, q\}$-quasiseparable matrices; generator representable $\{p\}$-Hessenberg-like matrices;

- extended $\{p\}$-Hessenberg-like matrices;

- extended generator representable $\{p, q\}$-semiseparable plus $\{r, s\}$-band matrices;

- ...

In the next sections we will prove that the inverse of an invertible $\{p, q\}$-semiseparable matrix is a $\{p, q\}$-band matrix and the inverse of an invertible $\{p\}$-Hessenberg-like matrix is an invertible $\{p\}$-generalized Hessenberg matrix.

Notes and references

In this section several new types of structured rank matrices were presented. Of course there are even more variants, but they will be discussed in forthcoming chapters. These variants specify even more precisely the structure of each rank block, or they link different blocks within the matrix to each other. For example, the classes of sequentially semiseparable, hierarchically semiseparable and $\mathcal{H}$-matrices are of this special form.

Once more we would like to draw attention to the fact that the classes of matrices considered here are the easiest possible generalizations of the previously defined classes having semiseparability rank 1. The classes presented here have natural extensions, when not considering elements anymore but blocks. This immediately gives us block band matrices, block semiseparable, block quasiseparable matrices and so on. The building blocks are now not elements anymore but matrices, which also can be linked within a matrix to each other. These classes of matrices will be briefly discussed in a forthcoming part. In this part of the book, we will however first examine the easiest extensions as defined in this chapter. Following this part we will provide some interesting solution schemes for some of these classes, and also give pointers to the literature connected with these variants.

In this book we choose to link the name semiseparable to matrices that are the inverses of band matrices. In other manuscripts semiseparable refers quite often to the class of extended generator representable semiseparable matrices. This name finds its origin in the discretization of semiseparable kernels.

We will now describe some references related to higher order semiseparable, quasiseparable and generator representable semiseparable matrices. Some references related to higher order quasiseparable matrices have already been discussed, in Section 2.5 and Section 2.9 on the representation of quasiseparable matrices. More extensive discussions related to the classes of block quasiseparable and related matrices can be found in Chapter 12. Similarly, some remarks concerning higher order generator representable matrices appeared already in Sections 2.2 and 2.7.

> ☞ E. E. Tyrtyshnikov. Mosaic ranks for weakly semiseparable matrices. In
> M. Griebel, S. Margenov, and P. Y. Yalamov, editors, *Large-Scale Scientific
> Computations of Engineering and Environmental Problems II*, volume 73
> of *Notes on numerical fluid mechanics*, pages 36–41. Vieweg, 2000.

Tyrtyshnikov expands the class of generator representable matrices towards a class called weakly semiseparable matrices. He proves that the inverse of a $\{p, q\}$-band matrix is a $\{p, q\}$-weakly semiseparable matrix, but the converse does not necessarily hold. This means that the class of $\{p, q\}$-weakly semiseparable matrices is a little more general than our class of $\{p, q\}$-semiseparable matrices. The class of weakly semiseparable matrices as defined by Tyrtyshnikov corresponds to the class of quasiseparable matrices. Results are also presented on the *mosaic* ranks of matrices. A mosaic rank is connected with a partitioning of the matrix, and a representation of this partitioned blocks with low rank matrices (so-called *skeletons* in the manuscript).

In 1959, higher order generator representable semiseparable matrices occurred, to our knowledge for the first time, in the manuscript of E. Asplund [6]. He proved a theorem stating that the inverse of an invertible lower $\{p\}$-generalized Hessenberg matrix (with no restrictions on the elements and called a $\{p\}$-band matrix in the paper) is an invertible lower $\{p\}$-Hessenberg-like matrix (with the definition of the ranks of subblocks) and vice versa. See also Section 14.2 in Chapter 14.

> ☞ A. P. Mullhaupt and K. S. Riedel. Low grade matrices and matrix fraction
> representations. *Numerical Linear Algebra with Applications*, 342:187–201,
> 2002.

The article by Mullhaupt and Riedel discusses from a theoretical viewpoint the class of quasiseparable matrices. Different results concerning summation, multiplication and factorization of these matrices are provided. We summarize here some of the results.

The authors name the considered matrices low grade matrices. The *lgrade* and *ugrade* denote respectively the lower and upper quasiseparability rank. Several results relating the lower and upper bandwith of matrices with their *lgrade* and *ugrade* are presented. Note that the authors point out that band matrices can be considered as low grade matrices, but they do implicitly always take into consideration the bandwith. Considering the bandwith instead of the *lgrade* rank leads to a more efficient representation.

We restrict ourselves to formulating some theorems related to the *lgrade*. Theorems for *ugrade* can be formulated similarly. The first theorem provides a kind of additive and multiplicative relation for quasiseparable matrices (These properties will be discussed further in the manuscript in a more general context.):

$$\mathrm{lgrade}\,(A_1 + A_2) \leq \mathrm{lgrade}\,(A_1) + \mathrm{lgrade}\,(A_2)$$
$$\mathrm{lgrade}\,(A_1 A_2) \leq \mathrm{lgrade}\,(A_1) + \mathrm{lgrade}\,(A_2)\,.$$

The authors also prove that the *lgrade* is maintained under inversion. Moreover, the authors also prove that the class of *lgrade* matrices of a certain rank is closed, similarly as we proved some relations for the rank 1 case in Chapter 1. Finally, they also prove a theorem, stating that an *LU*-decomposition of an *lgrade* matrix passes its structure onto its factors, L and U. The following theorem provides an approximation of an *lgrade* matrix with a matrix of a certain rank.

Theorem 8.29. *Suppose A to have* $\mathrm{lgrade}\,(A) = r$. *Then for every $\epsilon > 0$, there exist an upper triangular matrix R_ϵ and a low rank matrix U_ϵ, of rank r such that:*

$$\| A - (U_\epsilon + R_\epsilon) \| < \epsilon.$$

It is important to stress that the relation above only holds for the summation. The values of $\|U_\epsilon\|$ and $\|R_\epsilon\|$ can become arbitrarily large.

The following theorem is again some kind of generalization of results we presented earlier in Chapter 1. The proof of this theorem is based on the so-called *Pasting Lemma*.

Lemma 8.30 (Pasting Lemma). *Suppose a matrix A is given, decomposed as follows*

$$A = \left[\begin{array}{cc} \mathbf{x}^T & \cdot \\ A_{21} & \mathbf{y} \end{array} \right],$$

where A_{21} is of dimension $p \times q$ and the matrix A of dimension $(p+1) \times (q+1)$. If the following conditions are satisfied

$$\mathrm{rank} \left[\begin{array}{c} \mathbf{x}^T \\ A_{21} \end{array} \right] = r,$$
$$\mathrm{rank}\,[A_{21}, \mathbf{y}] \leq r,$$

then a value α exists such that

$$\mathrm{rank} \left[\begin{array}{cc} \mathbf{x}^T & \alpha \\ A_{21} & \mathbf{y} \end{array} \right] \leq r.$$

This lemma solves a so-called completion problem. We will discuss these problems in more detail later in this chapter.

Furthermore Mullhaupt and Riedel consider different types of factorizations of the considered *lgrade* matrices. We provide here some of these theorems.

Theorem 8.31. *Suppose A is a matrix having* $\mathrm{lgrade}\,(A) \leq r$ *and upper bandwith q. Then two matrices L and B exist, such that*

$$L^{-1}A = B,$$

where L is lower triangular and has lower bandwith smaller than or equal to r and B is a $\{r, q\}$-band matrix.

The presented factorization is a kind of partial LU-decomposition. This type of factorization is discussed in Chapter 10.

Theorem 8.32. *Suppose A is a matrix having* lgrade $(A) \leq r$. *Then two matrices Q and H exist, such that*

$$Q^T A = H,$$

where Q is a unitary matrix constructed with Householder transformations and the lower bandwith is bounded above by r. Also H has the lower bandwith bounded above by r.

This factorization can be considered as a kind of partial QR-decomposition. We will discuss the QR-decomposition in Chapter 9. This QR-decomposition, will be based on the application of Givens transformations. It is possible, however, to perform similar reduction schemes via Householder transformations.

Finally some theorems are provided for factoring matrices in a very specific way. More precisely, suppose we have the following equation:

$$K A = L,$$

where K and L have a small lower bandwith. Then it is possible to write the matrix as $A = U + H$, for which U is a low rank matrix and H has a low bandwith depending on those of K and L. A similar theorem is proved in the case K is a unitary matrix.

8.3   Inverses of structured rank matrices

In this section we will reuse the nullity theorem as it was defined in Chapter 1. Initially this theorem was used for proving that the inverses of semiseparable matrices were tridiagonal matrices, and so on, but, in fact, the theorem is very general. It does not take into account the complete matrix, but only specific subblocks, which need to satisfy a structured rank condition. Hence this makes the theorem suitable for predicting also the inverses of structured rank matrices as they were defined in the previous section.

This section is divided in two subsections. First, the inverses of structured rank matrices, in general, will be investigated based on the nullity theorem. In a second subsection the nullity theorem will be applied on the previously defined classes of matrices to determine the structure of their inverses.

8.3.1   Inverse of structured rank matrices

In the publications [6, 13, 15, 101, 111, 112, 114, 115, 118, 260] the connection between the structured rank of a matrix and its inverse is investigated for different classes of matrices. In some papers the nullity theorem is used to obtain these connections while in others they are based on determinantal formulas. In this section, we reformulate the most important results and illustrate them by different examples. To refresh the basic results from the beginning of the book, we provide some easy examples.

Using Corollaries 1.36 and 1.37 from Chapter 1, one can prove the results of the following examples by choosing the correct submatrices. Note that in this section only statements considering the lower triangular part are made: similar results hold for the upper triangular part.

Example 8.33

- Suppose the nonsingular matrix $A \in \mathbb{R}^{n \times n}$ is lower triangular, then its inverse A^{-1} will also be lower triangular. This can be seen rather easily, because for every $k \in \{1, \ldots, n\}$, with $\alpha = \{1, \ldots, k\}$, applying Corollary 1.36, we have that:

$$\text{rank}\left(A^{-1}(\alpha; N\backslash\alpha)\right) = \text{rank}\left(A(\alpha; N\backslash\alpha)\right) = 0.$$

  Hence,

$$\text{r}(\Sigma_{wu}; A) = \text{r}(\Sigma_{wu}; A^{-1}) = 0.$$

- Suppose we have a matrix for which the (first) subdiagonal is different from zero, and all the other entries in the strictly lower triangular part are zero. Then we know by Corollary 1.36 that the blocks coming from the strictly lower triangular part of the inverse of this matrix have maximum rank 1. Hence, the weakly lower triangular rank is maintained. ∎

Before we can deduce theorems connected to other structured rank matrices, we will define some more structures. The structure

$$\Sigma_\sigma = (N \times N)\backslash\{(1,1), (2,2), \ldots, (n,n)\}$$

is called the off-diagonal structure. Connected with this off-diagonal structure we have the following theorem. (When the theorems are straightforward extensions of the results presented in Section 1.5.1 of Chapter 1, the proofs are not included.)

Theorem 8.34 (Theorem 2.2 in [115]). *Assume A is a nonsingular matrix. Then the off-diagonal rank of A equals the off-diagonal rank of A^{-1}:*

$$\text{r}\left(\Sigma_\sigma; A\right) = \text{r}\left(\Sigma_\sigma; A^{-1}\right).$$

We remark that the off-diagonal structure admits blocks crossing the diagonal.

Example 8.35 Using Theorem 8.34, we can easily prove that the inverse of an invertible diagonal plus rank k matrix is again a diagonal plus a rank k matrix. ∎

A partition of a set N is a decomposition of $N = N_1 \dot\cup N_2 \dot\cup \cdots \dot\cup N_p$, where $\dot\cup$ denotes the disjunct union: this means that $N_i \cap N_j$ is the empty set, $\forall i, j$ with $i \neq j$. A generalization of Theorem 8.34, from diagonal to block diagonal, is as follows.

Theorem 8.36 (Theorem 2 in [111]). *Let $N = N_1 \dot\cup N_2 \dot\cup \cdots \dot\cup N_p$ a partition of N with $N = \{1, 2, \ldots, n\}$. Let*

$$\Sigma_{\sigma b} = (N \times N)\backslash \bigcup_{i=1}^{p} (N_i \times N_i).$$

Then, for every nonsingular $n \times n$ matrix A we have:

$$r\left(\Sigma_{\sigma b}; A^{-1}\right) = r\left(\Sigma_{\sigma b}; A\right).$$

This structure is referred to as the block off-diagonal structure.

With this theorem one can generalize the previous example.

Example 8.37 The inverse of a rank k matrix plus a block diagonal matrix is again a rank k matrix plus a block diagonal matrix for which the sizes of the blocks of the first and the latter diagonal are the same. ∎

The main disadvantage of both of Theorems 8.34 and 8.36 is the fact that the theorems only deal with matrices from which a diagonal, or a block diagonal matrix is subtracted. Nothing is mentioned about matrices for which the upper triangular part of rank k comes from a matrix other than the lower triangular part of rank k. This is generalized in the following theorem. Note that this theorem is already in the block version, similar to Theorem 8.36.

Theorem 8.38 (Theorem 3 in [111]). *Let $N = N_1 \dot\cup N_2 \dot\cup \cdots \dot\cup N_p$ a partition of N with $N = \{1, 2, \ldots, n\}$. Let*

$$\Sigma = \bigcup_{(i,j),i>j} (N_i \times N_j).$$

Then, for every nonsingular $n \times n$ matrix A we have:

$$r\left(\Sigma; A^{-1}\right) = r\left(\Sigma; A\right).$$

Note first of all that the entries of N_k, for example, do not necessarily need to be smaller than the entries of N_{k+1}. It can be seen, however, that the structure Σ_{wl} fits in this theorem. We can generalize this structure towards a weakly lower triangular block structure:

$$\Sigma_{wlb} = \bigcup_{1 \leq j < i \leq p} (N_i \times N_j)$$

for which all indices in N_k are smaller than all indices in N_{k+1} with $1 \leq k \leq p-1$.

Example 8.39 The inverse of a matrix plus a block diagonal matrix, for which the weakly block upper triangular rank is k and the weakly block lower triangular rank is l, is again such a matrix with the same weakly block upper and weakly block lower triangular rank. This means also that the inverse of a $\{p, q\}$-semiseparable matrix is a $\{p, q\}$-block band matrix. ∎

The theorems above already provide a lot of information connected to the class of semiseparable matrices. However, nothing has yet been mentioned about the connection with the diagonal. In all the theorems the diagonal elements are excluded from the structure. We know, for example, that a quasiseparable matrix again has a quasiseparable matrix as an inverse. However, the definition of the semiseparable matrices includes the diagonal, or even exceeds the diagonal. This is the point of interest in the following theorems.

Admitting the diagonal entries as well in the structure Σ_{wl}, which is the structure Σ_l, gives rise to the following theorem.

Theorem 8.40. *The lower triangular ranks of A and A^{-1} differ at most by one:*

$$\left| r(\Sigma_l; A) - r\left(\Sigma_l; A^{-1}\right) \right| \leq 1.$$

***Proof* (from [111]).** Choosing the correct blocks and applying Corollary 1.36 gives us the result immediately. We have

$$r(\Sigma_l; A) = \max_{k=1,\ldots,n} \operatorname{rank} A[\{k, k+1, \ldots, n\}; \{1, \ldots, k\}].$$

Also the following relation holds (use the nullity theorem):

$$\operatorname{rank} A(k : n; 1 : k) - 1$$
$$= \operatorname{rank} A^{-1}(k+1 : n; 1 : k-1)$$

(which is just 1 on the right-hand side if $k = 1$ or $k = n$). Combining the previous relations gives us

$$\operatorname{rank} A^{-1}(k : n; 1 : k)$$
$$\leq \operatorname{rank} A^{-1}(k+1 : n; 1 : k-1) + 2$$
$$= \operatorname{rank} A(k : n; 1 : k) + 1.$$

Therefore, the maxima also differ at most by 1. ☐

We will now reconsider the class of tridiagonal and semiseparable matrices to investigate in more detail the importance of the diagonal in the semiseparable matrix.

Example 8.41 Suppose we have the following invertible tridiagonal matrix:

$$A = \begin{bmatrix} 1 & 1 & 0 \\ 1 & 1 & 1 \\ 0 & 1 & 1 \end{bmatrix},$$

which has lower triangular rank 2. Its inverse however is a semiseparable matrix

$$A^{-1} = \begin{bmatrix} 0 & 1 & -1 \\ 1 & -1 & 1 \\ -1 & 1 & 0 \end{bmatrix},$$

with lower triangular rank 1. ∎

From now on we will investigate the connection between the class of tridiagonal matrices and the class of semiseparable matrices in more detail with respect to taking the inverse. We know already that the inverse of a tridiagonal matrix is a semiseparable matrix and vice versa, but let us reconsider these theoretical results in the framework of structured rank matrices. This means that we will restrict ourselves to the matrices having weakly upper and lower triangular rank 1. Note that in this class of matrices the tridiagonal matrices as well as the semiseparable matrices are included. Nevertheless, these theorems can be generalized towards higher rank semiseparable and band matrices. For example, the following behavior needs investigation.

Example 8.42 Suppose we have a tridiagonal matrix T, with its inverse S:

$$T = \frac{1}{5} \begin{bmatrix} 2 & 1 & 0 & 0 \\ 1 & 2 & 1 & 0 \\ 0 & 1 & 2 & 1 \\ 0 & 0 & 1 & 2 \end{bmatrix} \quad S = T^{-1} = \begin{bmatrix} 4 & -3 & 2 & -1 \\ -3 & 6 & -4 & 2 \\ 2 & -4 & 6 & -3 \\ -1 & 2 & -3 & 4 \end{bmatrix}.$$

It can be seen clearly that T^{-1} has weakly upper and weakly lower triangular rank equal to 1. However, the inverse of a matrix satisfying these conditions is not necessarily a tridiagonal matrix. Taking the matrix A and its inverse A^{-1}

$$A = \begin{bmatrix} 1 & 2 & 2 & 2 & 2 \\ 1 & 1 & 2 & 2 & 2 \\ 1 & 1 & 1 & 2 & 2 \\ 1 & 1 & 1 & 1 & 2 \\ 1 & 1 & 1 & 1 & 1 \end{bmatrix} \quad A^{-1} = \begin{bmatrix} -1 & 0 & 0 & 0 & 2 \\ 1 & -1 & 0 & 0 & 0 \\ 0 & 1 & -1 & 0 & 0 \\ 0 & 0 & 1 & -1 & 0 \\ 0 & 0 & 0 & 1 & -1 \end{bmatrix}.$$

The matrix A as well as its inverse have weakly upper and weakly lower triangular rank equal to 1. However, A^{-1} is not at all a tridiagonal matrix. ∎

Investigations concerning these questions can be found in a more general framework in [112, 118]. It is interesting that the framework in the latter papers has very strong connections with oscillation matrices and totally nonnegative matrices [103, 116]. We will give a theorem now, explaining the structure of the matrices in the previous example.

Theorem 8.43 (Lemma 2.5 in [118]). *Suppose A is a nonsingular $n \times n$ matrix with weakly lower triangular rank 1. For k with $1 < k < n$, the following two statements are equivalent.*

- *The weakly lower triangular rank of A remains 1 even if we extend the weakly lower triangular structure by the diagonal position (k, k).*

- *In A^{-1} there is a block of zeros $A^{-1}(N\backslash N_k; N_{k-1})$ with $N_k = \{1, \ldots, k\}$.*

Proof. Straightforward using Corollary 1.36.   □

We can immediately construct some very interesting examples connected with this theorem.

Example 8.44

- The inverse of a lower bidiagonal matrix is a lower triangular semiseparable matrix and vice versa.

- We have the following matrix A and its inverse:

$$A = \begin{bmatrix} 1 & 0 & 0 & 0 & 0 \\ 1 & 1 & 0 & 0 & 0 \\ 1 & 1 & 1 & 0 & 0 \\ 0 & 0 & 1 & 1 & 0 \\ 0 & 0 & 1 & 1 & 1 \end{bmatrix}, \quad A^{-1} = \begin{bmatrix} 1 & 0 & 0 & 0 & 0 \\ -1 & 1 & 0 & 0 & 0 \\ 0 & -1 & 1 & 0 & 0 \\ 0 & 1 & -1 & 1 & 0 \\ 0 & 0 & 0 & -1 & 1 \end{bmatrix}.$$

What one can see very easily looking at the two matrices is the fact that, in the first matrix, the diagonal positions $(2, 2)$ and $(4, 4)$ can be added to the weakly lower triangular structure without changing the rank. This means that in its inverse there will be two zero blocks that can be identified easily. It can also be seen that the first matrix has a zero block in the lower left position. Hence, its inverse has now a diagonal element in the position $(3, 3)$ which can be included in the structure without changing the rank.

- Another example showing this relation between zero submatrices and extended rank properties is the following. Take, e.g., the next matrix and its inverse:

$$A = \begin{bmatrix} 1 & 0 & 0 & 0 & 0 \\ 1 & 1 & 0 & 0 & 0 \\ 1 & 0 & 1 & 0 & 0 \\ 1 & 0 & 1 & 1 & 0 \\ 1 & 0 & 1 & 1 & 1 \end{bmatrix}, \quad A^{-1} = \begin{bmatrix} 1 & 0 & 0 & 0 & 0 \\ -1 & 1 & 0 & 0 & 0 \\ -1 & 0 & 1 & 0 & 0 \\ 0 & 0 & -1 & 1 & 0 \\ 0 & 0 & 0 & -1 & 1 \end{bmatrix}.$$

It can be seen that, for the extended diagonal positions $(3, 3)$ and $(4, 4)$, the corresponding zero blocks appear in the matrix.

- Let us look again at matrix A from Example 8.42. Note that the inverse of a matrix whose lower triangular rank is 1, and whose weakly upper triangular rank is 1, is not necessarily a tridiagonal matrix. The inverse matrix will only have the subdiagonal different from zero in the lower triangular part, and the weakly upper triangular rank will be 1, as the weakly upper triangular rank is maintained under inversion. ∎

Using the knowledge gained in this section on the inverse of structured rank matrices, it can often become very easy to predict the structure of the inverses of particular structured rank matrices. In the following subsection we investigate the inverses of some particular structured rank matrices.

8.3.2 Some particular inverses

We will present some results concerning the inverses of the structured rank matrices as defined in the beginning of this chapter. The proofs are straightforward applications of the results presented above and the nullity theorem, and are hence not included.

Band matrices and semiseparable matrices are related via inversion.

Theorem 8.45. *The inverse of an invertible $\{p, q\}$-semiseparable matrix, with $p \geq 0$ and $q \geq 0$ is a $\{p, q\}$-band matrix.*

The quasiseparable structure is maintained under inversion.

Theorem 8.46. *The inverse of an invertible $\{p, q\}$-quasiseparable matrix, with $p \geq 0$ and $q \geq 0$, is again a $\{p, q\}$-quasiseparable matrix.*

In fact there are many subclasses in the class of quasiseparables. Some of these subclasses have more structured rank information than others. All this information should be taken into consideration when computing the inverse. For example, for some of these structures the diagonal can be incorporated in the inverse or not. The inclusion of the diagonal into the matrix structure has a strong impact on the rank structure of the inverse.

Because a $\{p, q\}$-generator representable semiseparable matrix is a special $\{p, q\}$-semiseparable matrix, its inverse will also be a special band matrix.

Theorem 8.47. *The inverse of an invertible $\{p, q\}$-generator representable semiseparable matrix is a strict $\{p, q\}$-band matrix.*[4]

Using this theorem, we can reconsider the proof of a lemma, formulated in the section on the definition of higher order generator representable semiseparable matrices.

Lemma 8.48 (Proof of Lemma 8.16). *Suppose the matrix A is nonsingular and of $\{p, q\}$-generator representable semiseparable form. This matrix can never be of $\{\hat{p}, \hat{q}\}$-generator representable form, with $\hat{p} \neq p$ or $\hat{q} \neq q$.*

Proof. Using Theorem 8.47, the proof is trivial, as a strict $\{p, q\}$-band matrix cannot be a strict $\{\hat{p}, \hat{q}\}$-band matrix with $\hat{p} \neq p$ or $\hat{q} \neq q$. $\square$

The class of $\{p, q\}$-extended semiseparable and $\{p, q\}$-extended generator representable semiseparable matrices are special cases of the class of $\{p, q\}$-quasiseparables. These are quasiseparable matrices for which the diagonal is includable in the structure. In fact, we already know that the inverse of both matrices will

---

[4]A band matrix is called strict, if the elements on the extreme diagonals are all different from zero.

be again of the $\{p, q\}$-quasiseparable form. But using all the information provided leads to the following theorem. The class of $\{p, q\}$-extended generator representable semiseparable matrices is left to the reader as an exercise.

Theorem 8.49. *The inverse of an extended $\{p, q\}$-semiseparable matrix, with $p \geq 0$ and $q \geq 0$, is a $\{p - 1, q - 1\}$-quasiseparable matrix.*

As already indicated, the inverse of a Hessenberg-like matrix will be related to the inverse of a generalized Hessenberg matrix, also for higher rank structures.

Theorem 8.50. *The inverse of an upper $\{p\}$-Hessenberg-like matrix is a $\{p\}$-generalized Hessenberg matrix.*

The ideas as provided in this section can naturally be expanded to matrices plus band or plus block band matrices.

Notes and references

The manuscript [160] by Greenberg and Sarhan presents a sufficient condition such that the inverse of a matrix is of band form. This condition corresponds with being of generator representable semiseparable form. For more information, see Section 3.2 in Chapter 3.

Most of the references related to the nullity theorem have already been discussed in Section 1.5 in Chapter 1, see, e.g., the manuscripts [167, 114, 13, 15, 111, 115, 260, 76, 291]; see also [112, 118] in Section 4.6 in Chapter 4.

☞ P. Rózsa, F. Romani, and R. Bevilacqua. On generalized band matrices and their inverses. In D. J. Brown, M. T. Chu, D. C. Ellison, and R. J. Plemmons, editors, *Proceedings of the Cornelius Lanczos International Centenary Conference*, volume 73 of *Proceedings in Applied Mathematics*, pages 109–121, Philadelphia PA, 1994. SIAM Press.

In this manuscript the authors, Rózsa, Romani and Bevilacqua provide theoretical results on the inverses of so-called generalized band matrices. The results are based on the nullity theorem, which they prove in an alternative way, based on the Schur complement and a slight perturbation of the matrix. Taking the limit proves the nullity Theorem in an alternative fashion. The authors are well aware that the inverse of band matrices has a rank structure expanding above and below the diagonal. Moreover, they even define structures for working with these matrices. Their structures are called shapes and correspond to $\{p\}$-upper(lower) triangular structures and their block version. They define the class of generalized band matrices as a structured rank matrix satisfying an upper and lower triangular rank constraint plus a band matrix. This class is very general and also covers semiseparable, band plus semiseparable, quasiseparable and so forth. The only restriction posed by the authors on the matrices is that the rank parts need to be of generator representable form. Different examples on the rank structure when inverting these matrices are considered. The examples include structured rank blocks crossing the diagonal, block diagonal structured rank matrices and so forth. Also, results related to factorizations based on Schur complementation are presented.

In the next section we will discuss the relation between higher order generator representable semiseparable and semiseparable matrices. Issues between these matrices will be solved based on several completion problems. In completion problems one has a matrix with partially specified elements and one tries to complete this matrix, keeping the rank bounded. A generalization of the nullity theorem, based on such completion problems will be proposed in the footnotes and references following the next section. This theorem is discussed in [305, 308].

8.4   Generator representable semiseparable matrices

In Chapter 1, we proved that every semiseparable matrix could be written as a block diagonal matrix, for which the blocks were generator representable semiseparable. Moreover, we also showed under which assumptions a matrix was not generator representable semiseparable. In this section an extension of these results for higher order structured rank matrices will be provided. Let us first define the higher order generator representation.

8.4.1   Generator representation

The generator representation for higher order structured rank matrices is, in fact, a straightforward extension of the representation for generator representable semiseparable matrices of semiseparability rank 1.

The part of rank r is represented by the multiplication of two matrices. Suppose A is a $\{p, q\}$-generator representable semiseparable matrix. Then we name the matrices U, V, P and Q, from Definition 8.14 the generators of the matrix A. In general, every part of rank r in a structured rank matrix, which is representable by generators, admits a representation in terms of two matrices U, V.

This representation can easily be interpreted as the sum of rank 1 generator representations. We will briefly illustrate this fact for the class of extended generator representable semiseparable matrices. An extended $\{p\}$-generator representable semiseparable matrix is characterized by the fact that the part below (and including) the diagonal is coming from a certain rank p matrix; the same statement holds for the upper triangular part including the diagonal. This means, in fact, that the lower triangular part of the higher order semiseparable matrix can be written as the sum of generator representable semiseparable matrices of rank 1. We remark that in this discussion we decouple the representation of the lower and the upper triangular parts. This means that in the coming analysis we will only consider the representation of either the lower or the upper triangular parts, we will omit the connection. Let us consider in the remaining text only the lower triangular part, unless otherwise stated.

For the $\{p\}$-generator representable semiseparable matrix S, we get

$$S = S_1 + S_2 + S_3 + \ldots + S_p, \tag{8.5}$$

where all the matrices S_j are generator representable semiseparable matrices of semiseparability rank 1, i.e., they are of the following form (only the lower triangular part of the matrices is shown):

$$
S_j = \begin{bmatrix}
u_{1j}v_{1j} & & & \\
u_{2j}v_{1j} & u_{2j}v_{2j} & & \\
\vdots & & \ddots & \\
u_{nj}v_{1j} & u_{nj}v_{2j} & \cdots & u_{nj}v_{nj}
\end{bmatrix}.
$$

Let us define the following two matrices $U = (u_{ij})$ and $V = (v_{ij})$, with $i = 1, \ldots, n$ and $j = 1, \ldots, p$ and let us use the following notation

$$
U = \begin{bmatrix} \mathbf{u}_1^T \\ \vdots \\ \mathbf{u}_n^T \end{bmatrix} \text{ and } V = \begin{bmatrix} \mathbf{v}_1^T \\ \vdots \\ \mathbf{v}_n^T \end{bmatrix}, \tag{8.6}
$$

where for $i = 1, \ldots, n$ $\mathbf{v}_i^T = [v_{i1}, \ldots, v_{ip}]$ and $\mathbf{u}_i^T = [u_{i1}, \ldots, u_{ip}]$ denote the ith row out of V and U, respectively. Using this new notation, one can easily rewrite the matrix S of Equation (8.5) as

$$
S = S(U, V) = \begin{bmatrix}
\mathbf{u}_1^T\mathbf{v}_1 & & & \\
\mathbf{u}_2^T\mathbf{v}_1 & \mathbf{u}_2^T\mathbf{v}_2 & & \\
\vdots & & \ddots & \\
\mathbf{u}_n^T\mathbf{v}_1 & \mathbf{u}_n^T\mathbf{v}_2 & \cdots & \mathbf{u}_n^T\mathbf{v}_n
\end{bmatrix}.
$$

This gives us a structure similar to that for the generator representable semiseparable matrices of semiseparability rank 1, where the elements are now replaced by vectors.

One can conclude this subsection by stating that the lower (upper) part of a higher order generator representable semiseparable matrix can be represented as the sum of lower order generator representable semiseparable matrices. This is a construction which we will use quite often.

We know now that the generator representation of the lower triangular part of a higher order generator representable semiseparable matrices is in fact a summation of generator representable semiseparable matrices of rank 1. This means that, if we split lower and strictly upper triangular parts of a $\{p, q\}$-generator representable semiseparable matrix, we can write the lower triangular part of this matrix as the sum of $\{p\}$ lower triangular generator representable semiseparable matrices of semiseparability rank 1 plus the sum of q strictly upper triangular generator representable semiseparable matrices of semiseparability rank 1.

In the following subsections, we will investigate whether we can always split up higher structured rank blocks into a sum of lower rank structured blocks. This question is highly related to the question whether a matrix is of generator representable form or not. We start the investigation by looking at the decomposition of a block 2×2 matrix in terms of lower rank block 2×2 matrices. This decomposition is only possible under some additional assumptions on the block 2×2

matrix. These assumptions lead to a generalization of Theorem 1.12, stating when it is possible to represent a semiseparable matrix with the generator representation. Then we investigate some completion problems, presenting us the minimum number of generators for representing a semiseparable matrix with generators. More precisely, given a 2×2 lower block triangular matrix, we investigate the minimum rank which can be obtained by changing the upper right block. Finally we combine all these results to formulate some theorems related to the decomposition of a general structured rank matrix in terms of lower order structured rank matrices. In this decomposition we do not restrict ourselves to low rank matrices, but to matrices of low semiseparability rank.

8.4.2   When is a matrix generator representable?

In this subsection we investigate the conditions which must be satisfied for a matrix to be of generator representable form. For ease of understanding, we first focus our attention on blocks of rank structures. More precisely we will investigate how a 2×2 block matrix with a specific rank constraint can be written as the sum of lower order structured rank matrices. Later on, we will generalize this analysis dedicated to blocks and to general structured rank matrices. This is done by taking into consideration the fact that a structured rank matrix is composed of several blocks satisfying a specific rank constraint.

Partition the matrix A in the following form:

$$A = \left[\begin{array}{c|c} A_{11} & 0 \\ \hline A_{21} & A_{22} \end{array} \right],$$

where the rank of $[A_{11}^T, A_{21}^T]^T$, $[A_{21}, A_{22}]$ as well as A_{21} is equal to r.

We will first show that in this case we can decompose the different block matrices A_{ij} as follows:

$$A_{11} = U_1 V_1^T,$$
$$A_{21} = U_2 V_1^T,$$
$$A_{22} = U_2 V_2^T,$$

with U_1, V_1, U_2, V_2 matrices with r columns.

We know that the first block column of the matrix A is of rank r. This means that we can write the first block column in the following form:

$$\left[\begin{array}{c} A_{11} \\ A_{21} \end{array} \right] = \left[\begin{array}{c} U_1 \\ U_2 \end{array} \right] V_1^T,$$

with the matrices U_1, U_2 and V_1 having r columns. As the matrix A_{21} is of rank r, we know that both U_2 and V_1 must have r columns different from zero. Combining this with the fact that the rank of $[A_{21}, A_{22}] = [U_2 V_1^T, A_{22}]^T$ equals r, we know that A_{22} is linearly dependent of U_2. Hence, a matrix V_2 exists such that $A_{22} = U_2 V_2^T$, with V_2 having r columns.[5]

---

[5]The restriction on the number of columns is, in fact, not necessary. We use the term *can decompose* to indicate this. However, using this restriction on the number of columns is very useful for indicating bounds on the rank.

In fact this matrix A can be considered as coming from the following rank r matrix UV^T, with $U^T = [U_1^T, U_2^T]$ and $V^T = [V_1^T, V_2^T]$:

$$A = UV^T - \left[\begin{array}{c|c} 0 & U_2 V_2^T \\ \hline 0 & 0 \end{array} \right].$$

This means that the matrix A is a part of the rank r matrix UV^T. We say that the specified part in the matrix is coming from a rank r matrix, or the specified part in the matrix is of generator representable form.

It is clear that this matrix A can be decomposed as the sum of r matrices, with blocks of rank 1:

$$A = \left[\begin{array}{c|c} A_{11} & 0 \\ \hline A_{21} & A_{22} \end{array} \right] = \left[\begin{array}{c|c} A_{11}^{(1)} & 0 \\ \hline A_{21}^{(1)} & A_{22}^{(1)} \end{array} \right] + \left[\begin{array}{c|c} A_{11}^{(2)} & 0 \\ \hline A_{21}^{(2)} & A_{22}^{(2)} \end{array} \right] + \cdots + \left[\begin{array}{c|c} A_{11}^{(r)} & 0 \\ \hline A_{21}^{(r)} & A_{22}^{(r)} \end{array} \right].$$

The blocks $A_{11}^{(i)}, A_{21}^{(i)}$ and $A_{22}^{(i)}$ have the following generators:

$$A_{11}^{(i)} = U_1^{(i)} V_1^{(i)^T},$$
$$A_{21}^{(i)} = U_2^{(i)} V_1^{(i)^T},$$
$$A_{22}^{(i)} = U_2^{(i)} V_2^{(i)^T},$$

with $U_1^{(i)}, V_1^{(i)}, U_2^{(i)}, V_2^{(i)}$ the ith column of U_1, V_1, U_2, V_2 respectively. This is a similar approach as for the higher order generator representation.

We proved in this subsection that for blocks satisfying certain rank constraints, that we can always decompose the matrix as a sum of low rank matrices, but we do not yet know what happens in case the rank of A_{21} is strictly lower than r. We know from Chapter 1, that a decomposition of this part, into rank 1 matrices is not necessarily possible anymore. This is illustrated in the following example.

Example 8.51 Consider the following 2×2 matrix A. The element marked with $\times$ is not specified:

$$A = \left[\begin{array}{c|c} 1 & \times \\ \hline 0 & 1 \end{array} \right] = \left[\begin{array}{c|c} A_{11} & \times \\ \hline A_{21} & A_{22} \end{array} \right].$$

The matrix A has the left block $[A_{11}^T, A_{21}^T]^T$ of rank 1 and also the lower block $[A_{21}, A_{22}]$ is of rank 1, but the block A_{21} is of rank 0. It is clear that we cannot write this matrix as coming from a rank 1 matrix. This was clearly formulated and proved in Theorem 1.12. ∎

The next theorems will present some completion problems. Let us adopt the following conventions. The elements marked with $\times$ in the matrix are not specified. They represent the elements which should be completed to obtain a rank as low as possible in the complete matrix. These problems are called completion problems. One wants to change some elements in a matrix to make the complete matrix satisfying a certain condition. The completion problems as considered here involve the changing of the elements marked with $\times$ of the matrix A, such that the complete matrix becomes of the lowest possible rank. If we say that the matrix A is coming from a rank r matrix, or that the matrix can be written as a rank r matrix, this means that it is possible to change the elements $\times$ in the matrix such that the complete matrix becomes of rank r.

In fact, the elements $\times$ do not pose specific constraints on our matrix as they are quite often unimportant. Consider, for example, the generator definition, in which the lower triangular part is coming from a rank 1 matrix and the upper triangular part is coming from a rank 1 matrix. In the representation of the lower(upper) triangular part, the strictly upper(lower) triangular part of the rank 1 matrix makes no significant contribution to the final matrix.

Even though the elements $\times$ do not pose specific constraints, we will see that it is not always possible to choose them such that the complete matrix satisfies a certain rank condition.

We will now investigate, in more detail, what happens if the block in the lower left corner has a rank strictly smaller than r.

Proposition 8.52. *Suppose we have a matrix A partitioned as follows:*

$$A = \left[\begin{array}{c|c} A_{11} & \times \\ \hline A_{21} & A_{22} \end{array} \right],$$

with the rank of $[A_{11}^T, A_{21}^T]^T$ and $[A_{21}, A_{22}]$ equal to r, and the rank of A_{21} equal to s. This matrix can be written as coming from a rank r matrix if and only if $s = r$.

Proof. We know already that if $s = r$, the matrix A can be written as coming from a rank r matrix. We prove the other direction; namely, if A comes from a rank r matrix, then A_{21} needs to be of rank r.
Assume the matrix A comes from a rank r matrix:

$$A = UV^T - \left[\begin{array}{c|c} 0 & \times \\ \hline 0 & 0 \end{array} \right].$$

With the matrices U and V having r columns and partitioned as follows:

$$U = \left[\begin{array}{c} U_1 \\ U_2 \end{array} \right] \text{ and } V = \left[\begin{array}{c} V_1 \\ V_2 \end{array} \right],$$

this means that

$$[A_{21}, A_{22}] = U_2 \left[\begin{array}{c} V_1 \\ V_2 \end{array} \right]^T,$$

which naturally implies that the matrix U_2 needs to be of rank r. As the matrix U_2 has r columns and is of rank r, this implies that the columns of the matrix U_2 are linearly independent. Similarly we can deduce that the columns of the matrix V_1 are linearly independent. Hence, the multiplication between U_2 and V_1^T leads to a rank r matrix. $\square$

An extension of the previous theorem, if the first block column and last block row have different ranks, is formulated in the next proposition.

Proposition 8.53. *Suppose we have a matrix A partitioned as follows:*

$$A = \left[\begin{array}{c|c} A_{11} & \times \\ \hline A_{21} & A_{22} \end{array} \right],$$

with $\mathrm{rank}[A_{11}^T, A_{21}^T]^T = r_1$, $\mathrm{rank}[A_{21}, A_{22}] = r_2$ *and* $\mathrm{rank}(A_{21}) = s$. *This matrix can be written as coming from a rank* $r = \max(r_1, r_2)$ *matrix if and only if* $s = \min(r_1, r_2)$.

Proof. Without loss of generality we assume $r_1 > r_2$.

- Assume $s = \min(r_1, r_2) = r_2$. This means that we can write the matrix A as the following sum:

$$A = A^{(1)} + A^{(2)} = \left[\begin{array}{c|c} A_{11}^{(1)} & \times \\ \hline A_{21}^{(1)} & A_{22}^{(1)} \end{array} \right] + \left[\begin{array}{c|c} A_{11}^{(2)} & \times \\ \hline 0 & 0 \end{array} \right],$$

  where $\left[A_{11}^{(1)^T}, A_{21}^{(1)^T} \right]^T, \left[A_{21}^{(1)}, A_{22}^{(1)} \right]$, and $A_{21}^{(1)}$ all are of rank r_2. The rank of $A_{11}^{(2)}$ equals $r_1 - r_2$. The matrix $A^{(1)}$ satisfies Proposition 8.53 and can therefore be decomposed as the sum of r_2 matrices of rank 1. The matrix $A_{11}^{(2)}$ is of rank $r_1 - r_2$, and hence A can be written as the sum of $r_2 + r_1 - r_2 = \max(r_1, r_2)$ terms of rank 1.

- Let us decompose the matrix A. In general, the decomposition will then be of the following form:

$$A = A^{(1)} + A^{(2)} + A^{(3)} = \left[\begin{array}{c|c} A_{11}^{(1)} & \times \\ \hline A_{21}^{(1)} & A_{22}^{(1)} \end{array} \right] + \left[\begin{array}{c|c} A_{11}^{(2)} & \times \\ \hline 0 & A_{22}^{(2)} \end{array} \right] + \left[\begin{array}{c|c} 0 & \times \\ \hline A_{21}^{(3)} & 0 \end{array} \right],$$

  where $\mathrm{rank} \left[A_{11}^{(1)^T}, A_{21}^{(1)^T} \right]^T = \hat{r}$, $\mathrm{rank} \left[A_{21}^{(1)}, A_{22}^{(1)} \right] = \hat{r}$ and $\mathrm{rank} \left[A_{21}^{(1)} \right] = \hat{r}$. For the second matrix we have $\mathrm{rank} \left[A_{11}^{(2)} \right] = r_1 - \hat{r}$ and $\mathrm{rank} \left[A_{22}^{(2)} \right] = r_2 - \hat{r}$. The last matrix will have the lower left block of rank $\left[A_{21}^{(3)} \right] = \hat{r} - s$, with $\min(r_1, r_2) \geq \hat{r} \geq s$.

  We clearly remark that the last block $A_{21}^{(3)}$ cannot be included in the structure of matrix $A^{(2)}$. This is because the corresponding rank 1 blocks of A_{11} and A_{22} are already stored in the first term of the summation.

  Counting all the terms leads to a global number of rank 1 matrices equal to:

$$\hat{r} + (r_1 - \hat{r}) + (r_2 - \hat{r}) + (\hat{r} - s) = r_1 + r_2 - s.$$

Hence it is clear that the number of terms equal to r_1 implies that $s = r_2 = \min(r_1, r_2)$.

$\square$

To address the question whether a matrix is generator representable or not, we provide the following converse statement.

Corollary 8.54. *Suppose we have a matrix A partitioned as follows:*

$$A = \left[\begin{array}{c|c} A_{11} & \times \\ \hline A_{21} & A_{22} \end{array}\right],$$

with $\operatorname{rank}[A_{11}^T, A_{21}^T]^T = r_1$, $\operatorname{rank}[A_{21}, A_{22}] = r_2$, *and the rank of A_{21} equal to s. This matrix cannot be written as coming from a rank $r = \max(r_1, r_2)$ matrix if and only if $s < \min(r_1, r_2)$.*

This means that if the rank of the block A_{21} is strictly smaller than r, a rank r matrix cannot contain enough information anymore to represent the complete block matrix. This was already clear in the proof of the previous proposition.

We have now formulated some theorems for block 2×2 matrices. We already know when a block 2×2 matrix does not admit the generator representation. However, this 2×2 matrix perhaps might admit a generator representation of an order higher than r. Hence, we cannot really say that this 2×2 matrix does not admit a generator representation, but we should say: "It does not admit a generator representation of rank r."

One might correctly wonder now what the lowest possible rank $\hat{r}$ is that can be reached by changing the upper right block in the matrix. This means that this matrix will then admit a generator representation of rank $\hat{r}$.

The following completion problem involves the changing of the upper right block A_{12} of the matrix A, such that the complete matrix becomes of the lowest possible rank.

Theorem 8.55 (Completion problem 1). *Suppose we have a matrix A partitioned as follows:*

$$A = \left[\begin{array}{c|c} A_{11} & \times \\ \hline A_{21} & A_{22} \end{array}\right],$$

with the rank of $[A_{11}^T, A_{21}^T]^T$ and $[A_{21}, A_{22}]$ equal to r, and the rank of A_{21} equal to s. This matrix can be written as coming from a matrix of rank $\hat{r} = s + 2(r - s)$ and not lower.

Proof. The proof, in fact, is similar to the proof of Theorem 8.53. The matrix A can be written as:

$$A = \left[\begin{array}{c|c} A_{11}^{(1)} & \times \\ \hline A_{21}^{(1)} & A_{22}^{(1)} \end{array}\right] + \left[\begin{array}{c|c} A_{11}^{(2)} & \times \\ \hline 0 & 0 \end{array}\right] + \left[\begin{array}{c|c} 0 & \times \\ \hline 0 & A_{22}^{(2)} \end{array}\right], \tag{8.7}$$

where $\left[A_{11}^{(1)T}, A_{21}^{(1)T}\right]^T$, $\left[A_{21}^{(1)}, A_{22}^{(1)}\right]$, and $A_{21}^{(1)}$ all are of rank s. The matrices $A_{11}^{(2)}$ and $A_{22}^{(2)}$ are of rank $r - s$. The first term in the right-hand side of the equation can be completed to a rank s matrix, only by changing the upper left block of the matrix. The last two terms in the right-hand side of the equation are of rank $r - s$ and cannot be cumulated. Hence, the completion problem leads to a matrix of rank $s + 2(r - s)$. □

Note 8.56. *We remark that this theorem remains valid, also in the case of $r = s$.*

In fact, we can easily extend this problem to different ranks for each of the blocks.

Theorem 8.57 (Completion problem 2). *Suppose we have a matrix A partitioned as follows:*

$$A = \left[\begin{array}{c|c} A_{11} & \times \\ \hline A_{21} & A_{22} \end{array}\right],$$

with the rank of $[A_{11}^T, A_{21}^T]^T$ equal to r_1 and the rank of $[A_{21}, A_{22}]$ equal to r_2, and the rank of A_{21} equal to s. This matrix can be written as coming from a matrix of rank $\hat{r} = s + (r_1 - s) + (r_2 - s)$.

The proof is a straightforward extension of the proof of the first completion problem. One more completion problem is needed. This generalizes once more the previous completion problem. We assume now that the ranks of the upper left and the lower right block are not completely specified, but a lower bound is given. This is possible when, for example, some elements of these matrices also need to be completed. (Even though only one more completion problem is necessary for our analysis the results as presented here were presented in much more general form in [304]. See the notes and references for more information on a more general theorem for completion problems.)

Theorem 8.58 (Completion problem 3). *Suppose we have a matrix A partitioned as follows:*

$$A = \left[\begin{array}{c|c} A_{11} & \times \\ \hline A_{21} & A_{22} \end{array}\right],$$

with $\text{rank}[A_{11}^T, A_{21}^T]^T \geq r_1$, $\text{rank}[A_{21}, A_{22}] \geq r_2$ and the rank of A_{21} equal to s. This matrix can be written as coming from a matrix of rank $\hat{r} \geq s + (r_1 - s) + (r_2 - s)$ and not lower.

The proof is a natural extension of the proof of the previous completion problem. Let us illustrate more precisely what is meant in this theorem by an example.

Example 8.59 Suppose the following 3×3 matrix is given, for which the elements

$\times$ are not specified:

$$
A = \left[\begin{array}{cc|c} 1 & \times & \times \\ 0 & 1 & \times \\ \hline 0 & 0 & 1 \end{array}\right].
$$

The division in subblocks is already marked, and it is clear that this matrix can be written as coming from a rank $\hat{r}$ matrix with $\hat{r} \geq 0 + (2 - 0) + (1 - 0)$. We remark that a statement like this one is not always possible by only using the second completion problem. Considering the second completion problem and only using blocks of matrices which are completely determined, we would only be able to use the first two columns and the last two rows:

$$
\left[\begin{array}{c|c} 1 & \times \\ \hline 0 & 1 \\ 0 & 0 \end{array}\right] \quad \text{and} \quad \left[\begin{array}{cc|c} 0 & 1 & \times \\ 0 & 0 & 1 \end{array}\right].
$$

This gives us a rank 2 matrix for the first two columns, and another rank 2 matrix for the last two rows. Combining these two rank 2 matrices leads, in general, to a rank 2, 3 or rank 4 matrix.

Consider the last two 4×4 examples:

$$
A_1 = \left[\begin{array}{cc|cc} 1 & \times & \times & \times \\ 0 & 1 & \times & \times \\ \hline 1 & 1 & 0 & \times \\ 1 & 1 & 0 & 1 \end{array}\right] \quad \text{and } A_1 = \left[\begin{array}{cc|cc} 1 & \times & \times & \times \\ 1 & 1 & \times & \times \\ \hline 1 & 1 & 0 & \times \\ 1 & 1 & 0 & 1 \end{array}\right].
$$

The first matrix is coming from a rank 2 matrix, whereas the second matrix is coming from a rank 1 matrix. The readers can try themselves to write such a decomposition. ∎

We have now all the essential tools to prove under which assumptions a matrix is of generator representable form. We formulate the theorem only for the p-lower triangular structure (p-upper triangular structure is similar), as the matrices discussed in this book all satisfy a certain p-lower triangular rank constraint. The theorem can easily be generalized to other structures, but as matrices of these specific forms are not covered in this book, they are beyond its scope.

Theorem 8.60. *Suppose we have a matrix $A \in \mathbb{R}^{n \times n}$ satisfying the following structured rank condition:*

$$
\mathrm{r}\left(\Sigma_l^{(p)}; A\right) \leq r.
$$

Denote $N = \{1, \ldots, n\}$.

The matrix cannot be written as coming from a matrix of rank $\leq r$ if and only if the following statement holds.

There exist an $\alpha_1, \alpha_2, \beta_1$ and β_2 as subsets of N, such that $\alpha_2 \times \beta_1 \subset \Sigma_l^{(p)}$, *and we have*

$$\text{rank} \begin{bmatrix} A(\alpha_1, \beta_1) \\ A(\alpha_2, \beta_1) \end{bmatrix} \geq r_1,$$

$$\text{rank}\,[A(\alpha_2, \beta_1), A(\alpha_2, \beta_2)] \geq r_2,$$

$$\text{rank}(A(\alpha_2, \beta_1)) = s,$$

with $\hat{r} = s + (r_1 - s) + (r_2 - s) > r$.

It is clear that this theorem is an extension of Theorem 1.12, which was formulated for matrices of semiseparability rank 1.

Proof. We distinguish between the two directions.

- Suppose $\alpha_1, \alpha_2, \beta_1$ and β_2 exist such that the formulated statements are true. This means that there is a block 2×2 part in the structure, which is coming from a rank $\hat{r}$ matrix. As this embedded part in the matrix admits a generator representation of rank $\hat{r}$ and not lower, the complete matrix needs a generator representation of at least $\hat{r} > r$. Hence, the matrix cannot be written as coming from a rank r matrix.

- For the other direction we prove the converse statement: if for all $\alpha_1, \alpha_2, \beta_1$ and β_2 we can find the value of $\hat{r} \leq r$, this implies that the matrix is of generator representable form of rank $\leq r$.

  Choose $\alpha_1, \alpha_2, \beta_1$ and β_2 such that $\alpha_2 \times \beta_1$ contains one element of the $p - 1$th diagonal and

  $$\Sigma_l^{(p)} \subset (\alpha_1 \times \beta_1 \cup \alpha_2 \times \beta_1 \cup \alpha_2 \times \beta_2).$$

  For example, for a 5×5 matrix, the following division satisfies these properties:

  $$\begin{bmatrix} \boxtimes & \boxtimes & \times & \times & \times \\ \boxtimes & \boxtimes & \boxtimes & \times & \times \\ \boxtimes & \boxtimes & \boxtimes & \boxtimes & \times \\ \boxtimes & \boxtimes & \boxtimes & \boxtimes & \boxtimes \\ \boxtimes & \boxtimes & \boxtimes & \boxtimes & \boxtimes \end{bmatrix}.$$

  As all elements of the matrix belonging to the structure $\Sigma_l^{(p)}$ are included in this block division of the matrix, and this block division is coming from a rank $\leq r$ matrix, the complete matrix can also be written as coming from a rank $\leq r$ matrix.

  $\square$

It is clear that the previous theorem is a direct generalization of Theorem 1.12, stating whether a matrix of semiseparability rank was generator representable or not. Following Theorem 1.12 in Chapter 1, we proved that it was always possible (in the symmetric case) to write the semiseparable matrix as a block diagonal matrix, for which all the blocks were of generator representable form. In the following subsection we will generalize this statement. First, some examples are presented.

Example 8.61 The following two matrices have both the lower part of $\{2\}$-semi-separable form. The first matrix A_1 will be of generator representable form, whereas the second matrix A_2 is not:

$$
A_1 = \begin{bmatrix} 1 & 0 & & & \\ 1 & 0 & 1 & & \\ 0 & 1 & 0 & 1 & \\ 1 & 0 & 1 & 0 & 0 \\ 0 & 1 & 0 & 1 & 1 \end{bmatrix} \quad \text{and} \quad A_2 = \begin{bmatrix} 1 & 0 & & & \\ 1 & 0 & 1 & & \\ 0 & 0 & 0 & 1 & \\ 0 & 1 & 0 & 0 & 0 \\ 0 & 0 & 0 & 1 & 1 \end{bmatrix}.
$$

∎

8.4.3 Decomposition of structured rank matrices

In general, structured rank matrices have several blocks of different ranks, intersecting each other. In the simple case of rank 1 semiseparable matrices, we proved Proposition 1.13. In this proposition it was proven that a rank 0 block, under some conditions, could be expanded towards a larger rank 0 block extending from the bottom left corner towards the diagonal. The following theorem is a generalization of this proposition.

Theorem 8.62. *Suppose we have a matrix A partitioned as follows:*

$$
A = \left[\begin{array}{c|c} A_{11} & \times \\ \hline A_{21} & A_{22} \end{array} \right],
$$

with $A_{11} \in \mathbb{R}^{n_1 \times m_1}$, $A_{21} \in \mathbb{R}^{n_2 \times m_1}$, $A_{22} \in \mathbb{R}^{n_2 \times m_2}$, $\mathrm{rank}[A_{11}^T, A_{21}^T]^T = r_1$ and $\mathrm{rank}[A_{21}, A_{22}] = r_2$.

Denote $N_i = \{1, \ldots, n_i\}$ and $M_i = \{1, \ldots, m_i\}$, for $i = 1, 2$. Suppose there exist $\alpha \subset N_2$ and $\beta \subset M_1$, such that for the following block 2×2 matrix

$$
A = \left[\begin{array}{c|c} A_{11}(1 : n_1; \beta) & \times \\ \hline A_{21}(\alpha; \beta) & A_{22}(\alpha; 1 : m_2) \end{array} \right],
$$

the following conditions are satisfied: $\mathrm{rank}[A_{11}(1 : n_1; \beta)^T, A_{21}(\alpha; \beta)^T]^T = r_1$, $\mathrm{rank}[A_{21}(\alpha; \beta), A_{22}(\alpha; 1 : m_2)] = r_2$ and $\mathrm{rank}\,(A_{21}(\alpha; \beta)) = s$.

Then we have that the complete block A_{21} will be of rank s: $\mathrm{rank}(A_{21}) = s$.

We can say that if a block within the structure of the matrix is of a certain rank and some extra conditions are satisfied, that one can extend this block to a larger block from the bottom left to the end of the structure of the same rank. Quite often the end of the structure corresponds to the pth diagonal (e.g., in the structure $\Sigma_l^{(p)}$).

Proof. Due to the rank assumptions in the theorem, we have that the columns of the matrix on the left are linearly dependent on the columns of the matrix on the right:

$$\left[\begin{array}{c} A_{11}(1:n_1;N_1\backslash\beta) \\ A_{21}(\alpha;N_1\backslash\beta) \end{array} \right] \text{ and } \left[\begin{array}{c} A_{11}(1:n_1;\beta) \\ A_{21}(\alpha;\beta) \end{array} \right].$$

Hence we have that the columns of $A_{21}(\alpha;N_1\backslash\beta)$ are linearly dependent on the columns of $A_{21}(\alpha;\beta)$. Therefore, the complete matrix $A_{21}(\alpha;1:m_1)$ needs to be of rank s.

Similar arguments can now be used to extend the rank s block in the vertical direction, such that $A_{21}(1:n_2;1:m_1)$ becomes of rank s.   □

Higher order structured rank matrices, e.g., quasiseparable and semisepara-ble, have much more structured rank blocks than in the previous subsection. In general these structured rank blocks are in some sense nested. Hence, before we can derive theorems for splitting a higher order structured rank matrix into lower order structured rank matrices, we need to see what happens in the case of nested blocks. The following theorem generalizes in some sense the previous theorem on the extension of low rank blocks.

Theorem 8.63. *Suppose we have a matrix A partitioned as follows:*

$$\left[\begin{array}{c|c|c} A_{11} & \times & \times \\ \hline A_{21} & A_{22} & \times \\ \hline A_{31} & A_{32} & A_{33} \end{array} \right],$$

and the following relations hold

$$\text{rank} \left[\begin{array}{cc} A_{21} & A_{22} \\ A_{31} & A_{32} \end{array} \right] = r,$$

$$\text{rank}\,(A_{31}) = s.$$

If $s < r$, then we will have at least one of the following two inequalities satisfied

$$\text{rank}\,[A_{31}, A_{32}] < r,$$

$$\text{rank}\,\left[A_{21}^T, A_{31}^T \right]^T < r.$$

This means that if there is a low rank block in the bottom left corner, that this block has an extension of a block of rank strictly smaller than r, up to the diagonal. This theorem will be essential for decomposing higher order structured rank matrices into lower order structured rank matrices.

Proof. We assume that the block $\left[A_{21}^T, A_{31}^T \right]^T$ is not extendible towards the diagonal, otherwise the theorem is already satisfied. This means that rank $\left[A_{21}^T, A_{31}^T \right]^T = r$.

We also have that the following block 2×2 matrix is of rank r:

$$\text{rank} \left[\begin{array}{cc} A_{21} & A_{22} \\ A_{31} & A_{32} \end{array} \right] = r.$$

Therefore, we have that $\left[A_{22}^T, A_{32}^T\right]^T$ is linearly dependent on $\left[A_{21}^T, A_{31}^T\right]^T$. Hence, A_{32} is linearly dependent on A_{31} which means that:

$$\operatorname{rank}[A_{31}, A_{32}] = s < r.$$

$\square$

The following example illustrates this expansion of the low rank structure.

Example 8.64 Consider the following matrix:

$$\left[\begin{array}{cc|cc|c} 1 & 0 & \times & \times & \times \\ \hline 0 & 1 & 0 & 1 & \times \\ \hline 0 & 0 & 0 & 0 & \times \\ \hline 0 & 0 & 0 & 1 & 0 \\ 0 & 0 & 0 & 0 & 1 \end{array}\right],$$

that satisfies the conditions of the theorem with $r = 2$ and $s = 0$. It is clear that both equations

$$\operatorname{rank}[A_{31}, A_{32}] < r,$$
$$\operatorname{rank}\left[A_{21}^T, A_{31}^T\right]^T < r,$$

are satisfied.   ∎

Before we continue our analysis, we would like to translate the completion problems towards the semiseparable case. More precisely, in the completion problems we discussed the minimum rank needed to represent the specified part in the matrix. Now we will discuss the minimum number of matrices of semiseparability rank equal to 1 needed to represent this matrix.

Suppose, we have a structure Σ given, then we say that this structure is of semiseparable form (of semiseparability rank 1), if $\operatorname{r}(\Sigma, A) \leq 1$.

Corollary 8.65. *Suppose we have a matrix A partitioned as follows:*

$$A = \left[\begin{array}{c|c} A_{11} & \times \\ \hline A_{21} & A_{22} \end{array}\right],$$

with the rank of $[A_{11}^T, A_{21}^T]^T$ equal to r_1 and the rank of $[A_{21}, A_{22}]$ equal to r_2, and the rank of A_{21} equal to s. Suppose Σ is the structure corresponding with the indices of the three blocks A_{11}, A_{21} and A_{22}.

Then the matrix A can be written as the sum of $\max(r_1, r_2)$ matrices of semiseparable form, corresponding to the structure Σ. More precisely, the matrix can be written as the sum of s matrices of rank 1 and $\max(r_1, r_2) - s$ matrices of semiseparable form with a lower left zero block.

Proof. Following the results of the completion theorems, we know that we can decompose the matrix A as the sum of three block matrices (see Equation (8.7)). For our semiseparable structure, corresponding with the structure Σ, we can however recombine the last two terms in this equation. This leads to the sum of $s + \max(r_1 - s, r_2 - s)$ matrices, which clearly gives us the desired result. $\square$

The previous corollary states that for very easy higher order semiseparable structures, we can decompose the matrix as the sum of semiseparable matrices. But, a general semiseparable matrix itself has not only two blocks with intersections, but many more blocks intersecting each other. Hence we need to generalize this to more blocks intersecting each other. Let us first prove another decomposition theorem.

Theorem 8.66. *Suppose we have a matrix A partitioned as follows:*

$$\left[\begin{array}{c|c} A_{11} & \times \\ \hline A_{21} & A_{22} \\ A_{31} & A_{32} \end{array} \right],$$

and the following relations hold

$$\operatorname{rank}[A_{11}^T, A_{21}^T, A_{31}^T]^T = r_1,$$
$$\operatorname{rank} \left[\begin{array}{cc} A_{21} & A_{22} \\ A_{31} & A_{32} \end{array} \right] = r_2,$$
$$\operatorname{rank}[A_{31}, A_{32}] = t.$$

Then this matrix can be written as the sum of t matrices of semiseparable form (rank 1 matrices) and the sum of $\max(r_1, r_2) - t$ matrices of semiseparable form, having the last block row, equal to zero.

We remark that there are at least t matrices of rank 1. In general it is possible that within the remaining $\max(r_1, r_2) - t$ matrices of semiseparable form, there are still matrices of rank 1.

Proof. Assume that the matrix A, split as above, has $\operatorname{rank}(A_{31}) = s$. The matrix A can naturally be written as follows:

$$A = \left[\begin{array}{c|c} A_{11}^{(1)} & \times \\ \hline A_{21}^{(1)} & A_{22}^{(1)} \\ A_{31}^{(1)} & A_{32}^{(1)} \end{array} \right] + \left[\begin{array}{c|c} \tilde{A}_{11}^{(2)} & \times \\ \hline \tilde{A}_{21}^{(2)} & \tilde{A}_{22}^{(2)} \\ 0 & \tilde{A}_{32}^{(2)} \end{array} \right],$$

where the matrix $A^{(1)}$ (an abbreviation for the first block matrix), is of rank s, and hence can be written as the sum of s matrices of semiseparable form and where we have the following rank relations for the matrix $A^{(2)}$.

$$\operatorname{rank}\left[\left(\tilde{A}_{11}^{(2)}\right)^{T},\left(\tilde{A}_{21}^{(2)}\right)^{T},0\right]^{T} = r_1 - s,$$

$$\operatorname{rank}\left[\begin{array}{cc} \tilde{A}_{21}^{(2)} & \tilde{A}_{22}^{(2)} \\ 0 & \tilde{A}_{32}^{(2)} \end{array}\right] = r_2 - s,$$

$$\operatorname{rank}\left[0,\tilde{A}_{32}^{(2)}\right] = t - s.$$

Extracting the rank $t - s$ bottom block into another matrix leads to the following sum for the matrix A:

$$A = A^{(1)} + \left[\begin{array}{c|c} 0 & \times \\ \hline 0 & \hat{A}_{22}^{(2)} \\ 0 & \hat{A}_{32}^{(2)} \end{array}\right] + \left[\begin{array}{c|c} \tilde{A}_{11}^{(3)} & \times \\ \tilde{A}_{21}^{(3)} & \tilde{A}_{22}^{(3)} \\ \hline 0 & 0 \end{array}\right],$$

where

$$\operatorname{rank}\left[\left(\tilde{A}_{11}^{(3)}\right)^{T},\left(\tilde{A}_{21}^{(3)}\right)^{T},0\right]^{T} = r_1 - s,$$

$$\operatorname{rank}\left[\begin{array}{cc} \tilde{A}_{21}^{(3)} & \tilde{A}_{22}^{(3)} \\ 0 & 0 \end{array}\right] = r_2 - t.$$

The matrix $\hat{A}^{(2)}$ can be written as the sum of $t - s$ matrices of rank 1. To prove the final result we distinguish between two different cases.

- First, suppose $r_1 - s \le r_2 - t$, this leads to the following number of terms (remember that $t \ge s$) and use Corollary 8.65:

$$s + (t - s) + \max(r_1 - s, r_2 - t) = r_2 = \max(r_1, r_2).$$

- Second, suppose $r_1 - s \ge r_2 - t$. This means that there exists a matrix of rank u namely $A_{11}^{(2)}$, with $\tilde{A}_{11}^{(3)} = A_{11}^{(2)} + A_{11}^{(3)}$, and $A_{11}^{(3)}$ of rank $r_1 - s - u$, such that $r_1 - s - u = r_2 - t$. Hence we can write the matrix decomposition of A as follows:

$$A = A^{(1)} + \left[\begin{array}{c|c} A_{11}^{(2)} & \times \\ \hline 0 & \hat{A}_{22}^{(2)} \\ 0 & \hat{A}_{32}^{(2)} \end{array}\right] + \left[\begin{array}{c|c} A_{11}^{(3)} & \times \\ \tilde{A}_{21}^{(3)} & \tilde{A}_{22}^{(3)} \\ \hline 0 & 0 \end{array}\right].$$

Counting now the total number of terms gives us:

$$s + \max(t - s, u) + \max(r_1 - s - u, r_2 - t).$$

Distinguishing between cases $t - s \leq u$ and $t - s \geq u$, and using the fact that $r_2 - r_1 = t - s - u$ leads to the equality:

$$s + \max(t - s, u) + \max(r_1 - s - u, r_2 - t) = \max(r_2, r_1).$$

Hence, in either case we obtain the desired number of terms of the specific form.

□

Let us formulate now a theorem for a 3×3 block matrix to determine the number of necessary matrices of semiseparability rank 1.

Theorem 8.67. *Suppose we have a matrix A partitioned as follows:*

$$\left[\begin{array}{c|c|c} A_{11} & \times & \times \\ \hline A_{21} & A_{22} & \times \\ \hline A_{31} & A_{32} & A_{33} \end{array}\right],$$

and the following relations hold

$$\operatorname{rank}[A_{11}^T, A_{21}^T, A_{31}^T]^T = r_1,$$

$$\operatorname{rank}\left[\begin{array}{cc} A_{21} & A_{22} \\ A_{31} & A_{32} \end{array}\right] = r_2,$$

$$\operatorname{rank}[A_{31}, A_{32}, A_{33}] = r_3,$$

$$\operatorname{rank}(A_{31}) = s.$$

Then this matrix can be written as the sum of $\max(r_1, r_2, r_3)$ 3×3 block matrices of semiseparable form.

Proof. Suppose $\operatorname{rank}[A_{31}, A_{32}] = t$. Considering the first two block columns, and applying Theorem 8.66 to them, leads to a decomposition of the matrix as follows.

$$A = \left[\begin{array}{c|c|c} A_{11}^{(1)} & \times & \times \\ \hline A_{21}^{(1)} & A_{22}^{(1)} & \times \\ \hline A_{31}^{(1)} & A_{32}^{(1)} & 0 \end{array}\right] + \left[\begin{array}{c|c|c} A_{11}^{(2)} & \times & \times \\ \hline A_{21}^{(2)} & A_{22}^{(2)} & \times \\ \hline 0 & 0 & 0 \end{array}\right] + \left[\begin{array}{c|c|c} 0 & 0 & 0 \\ \hline 0 & 0 & 0 \\ \hline 0 & 0 & \tilde{A}_{33}^{(3)} \end{array}\right].$$

The first matrix can be written as the sum of t matrices of semiseparable form, the second matrix of $\max(r_1, r_2) - t$ matrices of semiseparable form, and the last matrix contains the block $\tilde{A}_{33}^{(3)} = A_{33}$. We can, however, add a rank t matrix $A_{33}^{(1)}$ into the first matrix such that we get the following equations:

$$A = \left[\begin{array}{c|c|c} A_{11}^{(1)} & \times & \times \\ \hline A_{21}^{(1)} & A_{22}^{(1)} & \times \\ \hline A_{31}^{(1)} & A_{32}^{(1)} & A_{33}^{(1)} \end{array}\right] + \left[\begin{array}{c|c|c} A_{11}^{(2)} & \times & \times \\ \hline A_{21}^{(2)} & A_{22}^{(2)} & \times \\ \hline 0 & 0 & 0 \end{array}\right] + \left[\begin{array}{c|c|c} 0 & 0 & 0 \\ \hline 0 & 0 & 0 \\ \hline 0 & 0 & \hat{A}_{33}^{(3)} \end{array}\right],$$

with the rank of $\hat{A}_{33}^{(3)}$ equal to $r_3 - t$. If now $\max(r_1, r_2) - t \geq r_3 - t$ the problem is already solved, as we can incorporate the matrix $\hat{A}_{33}$ into the sum of the $\max(r_1, r_2) - t$ matrices of semiseparable form. If, however, $\max(r_1, r_2) - t \leq r_3 - t$, we can only add a rank $\max(r_1, r_2) - t$ part of the matrix $\hat{A}_{33}^{(3)}$ into the second term, leaving us a matrix of the form

$$
\left[
\begin{array}{cc|c}
0 & 0 & 0 \\
\hline
0 & 0 & 0 \\
\hline
0 & 0 & A_{33}^{(3)}
\end{array}
\right],
$$

where the rank of $A_{33}^{(3)} = r_3 - \max(r_1, r_2)$. In either case, counting the complete number of matrices of semiseparable form leads to $\max(r_1, r_2, r_3)$ terms. $\qquad\square$

We remark that there are in the decomposition of the previous theorem at least s matrices of rank 1.

To generalize this theorem for more than three blocks we need to generalize first Theorem 8.66 for more blocks. We formulate the theorem here for three nested blocks, but the applied procedure can be generalized to more blocks, as the proof can be converted to an inductive proof.

Theorem 8.68. *Suppose we have a matrix A partitioned as follows:*

$$
\left[
\begin{array}{c|c|c}
A_{11} & \times & \times \\
\hline
A_{21} & A_{22} & \times \\
\hline
A_{31} & A_{32} & A_{33} \\
\hline
A_{41} & A_{42} & A_{43}
\end{array}
\right],
$$

and the following relations hold

$$
\operatorname{rank}[A_{11}^T, A_{21}^T, A_{31}^T, A_{41}^T]^T = r_1,
$$

$$
\operatorname{rank}
\left[
\begin{array}{cc}
A_{21} & A_{22} \\
A_{31} & A_{32} \\
A_{41} & A_{42}
\end{array}
\right] = r_2,
$$

$$
\operatorname{rank}
\left[
\begin{array}{ccc}
A_{31} & A_{32} & A_{33} \\
A_{41} & A_{42} & A_{43}
\end{array}
\right] = r_3,
$$

$$
\operatorname{rank}[A_{41}, A_{42}, A_{43}] = t.
$$

Then this matrix can be written as the sum of t matrices of semiseparable form (more precisely rank 1 matrices) and the sum of $\max(r_1, r_2, r_3) - t$ matrices of semiseparable form, having the last block row equal to zero.

Proof. To start the proof we apply Theorem 8.66 on the following block matrix

$$
\left[
\begin{array}{cc|c}
A_{21} & A_{22} & \times \\
\hline
A_{31} & A_{32} & A_{33} \\
\hline
A_{41} & A_{42} & A_{43}
\end{array}
\right].
$$

Decomposing this matrix gives us a summation of t terms with the last block row different from zero and $\max(r_2, r_3) - t$ terms with the last block row equal to zero. As the matrix A_{11} is linked to the blocks A_{41}, A_{31} and A_{21}, however, we can decompose the matrix A_{11} into low rank blocks and add some of these blocks to the first terms and some of these blocks to the last terms. (This is done similarly as in the proof of Theorem 8.67.) In the case $r_1 \leq \max(r_2, r_3)$, no terms will be left and we have solved the problem. Otherwise, in the case $r_1 > \max(r_2, r_3)$ we will have t terms with the last block row different from zero, and $r_1 - t$ terms with the last block row equal to zero. $\qquad \square$

Let us now generalize Theorem 8.67 for 4 blocks.

Theorem 8.69. *Suppose we have a matrix A partitioned as follows:*

$$\left[\begin{array}{c|c|c|c} A_{11} & \times & \times & \times \\ \hline A_{21} & A_{22} & \times & \times \\ \hline A_{31} & A_{32} & A_{33} & \times \\ \hline A_{41} & A_{42} & A_{43} & A_{44} \end{array} \right],$$

and the following relations hold

$$\operatorname{rank}[A_{11}^T, A_{21}^T, A_{31}^T, A_{41}^T]^T = r_1,$$

$$\operatorname{rank} \left[\begin{array}{cc} A_{21} & A_{22} \\ A_{31} & A_{32} \\ A_{41} & A_{42} \end{array} \right] = r_2,$$

$$\operatorname{rank} \left[\begin{array}{ccc} A_{31} & A_{32} & A_{33} \\ A_{41} & A_{42} & A_{43} \end{array} \right] = r_3,$$

$$\operatorname{rank}[A_{41}, A_{42}, A_{43}, A_{44}] = r_4.$$

Then this matrix can be written as the sum of $\max(r_1, r_2, r_3, r_4)$ 4×4 block matrices of semiseparable form.

Proof. The proof proceeds similar to the proof of Theorem 8.67 by first applying Theorem 8.68. $\qquad \square$

Let us present some examples of decompositions of matrices into terms of semiseparability rank 1.

Example 8.70 Consider the following examples, for which the blocks are separated by the vertical and horizontal lines. The elements marked by $\times$ are the elements to be completed to obtain a matrix with the smallest possible rank.

$$A_1 = \left[\begin{array}{cc|cc} 1 & 0 & \times & \times \\ 1 & 1 & \times & \times \\ \hline 0 & 1 & 1 & 0 \\ 1 & 0 & 1 & 1 \end{array} \right] = \left[\begin{array}{cc|cc} 1 & 0 & 1 & 1 \\ 1 & 0 & 1 & 1 \\ \hline 0 & 0 & 0 & 0 \\ 1 & 0 & 1 & 1 \end{array} \right] + \left[\begin{array}{cc|cc} 0 & 0 & 0 & 0 \\ 0 & 1 & 1 & 0 \\ \hline 0 & 1 & 1 & 0 \\ 0 & 0 & 0 & 0 \end{array} \right]$$

The next example shows that the decomposition is not necessarily unique as it provides another decomposition for the same matrix.

$$
A_1 = \left[\begin{array}{cc|cc} 1 & 0 & \times & \times \\ 1 & 1 & \times & \times \\ \hline 0 & 1 & 1 & 0 \\ 1 & 0 & 1 & 1 \end{array}\right] = \left[\begin{array}{cc|cc} 1 & 0 & 1 & 1 \\ 2 & 0 & 2 & 2 \\ \hline 1 & 0 & 1 & 1 \\ 1 & 0 & 1 & 1 \end{array}\right] + \left[\begin{array}{cc|cc} 0 & 0 & 0 & 0 \\ -1 & 1 & 0 & -1 \\ \hline -1 & 1 & 0 & -1 \\ 0 & 0 & 0 & 0 \end{array}\right]
$$

Let us consider now an example of a matrix having the lower left block of rank strictly smaller than 2.

$$
A_2 = \left[\begin{array}{cc|cc} 1 & 0 & \times & \times \\ 1 & 1 & \times & \times \\ \hline 0 & 0 & 1 & 0 \\ 1 & 0 & 1 & 1 \end{array}\right] = \left[\begin{array}{cc|cc} 1 & 0 & 1 & 1 \\ 1 & 0 & 1 & 1 \\ \hline 0 & 0 & 0 & 0 \\ 1 & 0 & 1 & 1 \end{array}\right] + \left[\begin{array}{cc|cc} 0 & 0 & \times & \times \\ 0 & 1 & \times & \times \\ \hline 0 & 0 & 1 & 0 \\ 0 & 0 & 0 & 0 \end{array}\right]
$$

It is clear that the right matrix in the equation above is of rank 2; hence, this matrix can be decomposed as the sum of two rank 1 matrices. This gives us in total 3 rank 1 matrices. This is equal to the number of terms in the summation, according to Theorem 8.55, namely $1 + 2(2 - 1)$. But writing this as a sum of matrices satisfying the semiseparable structure constraints, this gives us the sum of 2 matrices as the right matrix in the equation above is of semiseparable form. This is according to Corollary 8.65.

In the following examples the left and the bottom block do not necessarily have the same rank conditions:

$$
A_3 = \left[\begin{array}{cc|cc} 1 & 0 & \times & \times \\ 1 & 1 & \times & \times \\ \hline 0 & 0 & 1 & 0 \\ 0 & 0 & 1 & 1 \end{array}\right] \text{ and } A_4 = \left[\begin{array}{cc|cc} 1 & 0 & \times & \times \\ 1 & 1 & \times & \times \\ \hline 0 & 0 & 0 & 0 \\ 1 & 0 & 0 & 1 \end{array}\right].
$$

Considering the rank decompositions, the matrix A_3 can be written as the sum of $0 + 2(2 - 0) = 4$ rank 1 matrices. The matrix A_4 can be written as the sum of $1 + (2 - 1) + (1 - 1) = 2$ rank one matrices. When considering the semiseparable structure (remark that the structure in this case includes also the first and the last superdiagonal element), we have that matrix A_3 as well as A_4 can be written as the sum of 2 matrices of semiseparable form. The reader can verify this by trying to write down some decompositions of these last two matrices. ∎

In general we can conclude now with the following general theorem, which we only formulate for p lower triangular ranks. This includes all important classes covered in this book such as semiseparable, quasiseparable, band matrices and so forth. Of course similar statements are true when posing a rank condition on a p-upper triangular structure.

Theorem 8.71. *Suppose we have a matrix A, satisfying the following rank structure:*

$$
\mathrm{r}\left(\Sigma_l^{(p)}; A\right) \leq r
$$

Then this matrix A can be written as

$$A = A_1 + A_2 + \cdots + A_r,$$

for which all matrices A_i satisfy the following rank constraints.

$$r\left(\Sigma_l^{(p)}; A_i\right) \leq 1$$

Proof. The proof is a generalization of Theorem 8.69. $\qquad\qquad\square$

Note 8.72.

- *The theorem above can easily be expanded to the block p-lower triangular structures.*

- *The theorem is also valid for decomposing matrices satisfying a (block) p-upper triangular structure.*

Let us conclude this part with some examples.

Example 8.73 In the following examples we illustrate the decomposition possibilities of some lower triangular semiseparable matrices of higher order semiseparability rank into a sum of semiseparable matrices of semiseparability rank 1. We also illustrate once more that the decompositions are not necessarily unique. A lower bidiagonal matrix can be decomposed as two {1}-semiseparable matrices. The decomposition is clearly not unique:

$$S_1 = \begin{bmatrix} 1 & & & & \\ 1 & 1 & & & \\ & 1 & 1 & & \\ & & 1 & 1 & \\ & & & 1 & 1 \end{bmatrix} = \begin{bmatrix} 1 & & & & \\ 1 & 0 & & & \\ & 0 & 1 & & \\ & & 1 & 0 & \\ & & & 0 & 1 \end{bmatrix} + \begin{bmatrix} 0 & & & & \\ 0 & 1 & & & \\ & 1 & 0 & & \\ & & 0 & 1 & \\ & & & 1 & 0 \end{bmatrix}$$

$$= \begin{bmatrix} 1 & & & & \\ 0 & 0 & & & \\ & 1 & 1 & & \\ & & 0 & 0 & \\ & & & 1 & 1 \end{bmatrix} + \begin{bmatrix} 0 & & & & \\ 1 & 1 & & & \\ & 0 & 0 & & \\ & & 1 & 1 & \\ & & & 0 & 0 \end{bmatrix}.$$

Consider the following lower triangular {2}-semiseparable matrix. This matrix can be decomposed as the sum of two lower semiseparable matrices.

$$S_2 = \begin{bmatrix} 1 & & & & \\ 1 & 2 & & & \\ 1 & 1 & 1 & & \\ 0 & 0 & 0 & 1 & \\ 1 & 0 & 1 & 1 & 1 \end{bmatrix} = \begin{bmatrix} 1 & & & & \\ 1 & 0 & & & \\ 1 & 0 & 1 & & \\ 0 & 0 & 0 & 0 & \\ 1 & 0 & 1 & 1 & 1 \end{bmatrix} + \begin{bmatrix} 0 & & & & \\ 0 & 2 & & & \\ 0 & 1 & 0 & & \\ 0 & 0 & 0 & 1 & \\ 0 & 0 & 0 & 0 & 0 \end{bmatrix}$$

This subsection focused on the splitting of matrices satisfying one specific structured rank condition. In general, however, structured rank matrices will often satisfy more than 1 constraint. Let us investigate the classes of structured rank matrices as defined in this chapter in the following subsection.

8.4.4   Decomposition of semiseparable and related matrices

In this subsection we will briefly investigate the classes of matrices as defined in this chapter. We start with the most simple classes. With simple classes we mean matrices obeying structures which are not intersecting each other.

We will investigate in this subsection what the sum of two structured matrices will be, as well as the possibility to decompose a structured rank matrix into the sum of lower order structured rank matrices.

Quasiseparable matrices

The class of $\{p, q\}$-quasiseparable matrices is characterized by the fact that the parts of the matrix below and above the diagonal satisfy a specific rank constraint. The structures of the upper and lower part do not interfere with each other.

It is an easy exercise to prove that the sum of a $\{p_1, q_1\}$-quasiseparable matrix and a $\{p_2, q_2\}$-quasiseparable matrix results in a $\{p_1 + p_2, q_1 + q_2\}$-quasiseparable matrix.

The decomposition of a quasiseparable matrix is also straightforward. Suppose we have a $\{p, q\}$-quasiseparable matrix A. This matrix can be written as the sum of $\max(p, q)$ $\{1, 1\}$-quasiseparable matrices. In fact we can even be more precise. Suppose $p \leq q$, then the matrix A can be written as the sum of p $\{1, 1\}$-quasiseparable matrices and $q - p$ $\{0, 1\}$-quasiseparable matrices. A similar statement holds for the case $q \leq p$.

Hessenberg-like matrices

Hessenberg-like matrices only satisfy one structural constraint, hence their analysis should not be too difficult. Suppose a $\{p\}$-Hessenberg-like matrix Z is given. This matrix can always be written as the sum of p $\{1\}$-Hessenberg-like matrices.

The converse statement is false, however. The sum of p $\{1\}$-Hessenberg-like matrices is not necessary a $\{p\}$-Hessenberg-like matrix. This is illustrated by the following example.

Example 8.74 Consider the sum of the following two Hessenberg-like matrices of semiseparability rank 1:

$$
\begin{bmatrix} 0 & 0 & 0 & 0 \\ 0 & 0 & 0 & 0 \\ 0 & 0 & 0 & 0 \\ 1 & 0 & 0 & 0 \end{bmatrix}
+
\begin{bmatrix} 1 & 0 & 0 & 0 \\ 0 & 1 & 0 & 0 \\ 0 & 0 & 1 & 0 \\ 0 & 0 & 0 & 1 \end{bmatrix}
=
\begin{bmatrix} 1 & 0 & 0 & 0 \\ 0 & 1 & 0 & 0 \\ 0 & 0 & 1 & 0 \\ 1 & 0 & 0 & 1 \end{bmatrix}.
$$

The resulting matrix is clearly not Hessenberg-like of semiseparability rank 2. Because the superdiagonal should be included in the structure we obtain that the lower left 3×3 block of the sum is of rank 3.   ∎

Extended semiseparable matrices

The class of extended $\{p, q\}$-semiseparable matrices has, as different from the class of $\{p, q\}$-semiseparable matrices, that the number of sub(super)diagonals includable in the structure is independent of the value of p and q; only the diagonal is includable in both the upper and the lower triangular part.

This makes it possible to prove the following statement. The sum of an extended $\{p_1, q_1\}$-semiseparable matrix with an extended $\{p_2, q_2\}$-semiseparable matrix gives us an extended $\{p_1 + p_2, q_1 + q_2\}$-semiseparable matrix.

Semiseparable matrices

Semiseparable matrices are the most difficult ones to deal with. The structure depends on the values p and q of the $\{p, q\}$-semiseparable matrix and, moreover, the structured rank constraints have structures intersecting each other.

Unfortunately we have already shown that the sum of Hessenberg-like matrices of semiseparability rank 1 does not necessarily result in Hessenberg-like matrices of a higher order semiseparability rank. The same is true for semiseparable matrices. This is illustrated by Example 8.74, which also works for semiseparable matrices.

Notes and references

In [115], Fiedler and Markham investigate some conditions to complete the diagonal elements of a matrix with off-diagonal rank r, such that the rank of the resulting matrix is as small as possible. (See also Section 2.2 in Chapter 2.) Results similar to the one presented in this section, namely that a low rank part of rank k can be decomposed as the sum of k low rank parts of rank 1, was formulated and proved by Mullhaupt and Riedel in [219]. More information on this article can be found in Section 8.2 of Chapter 8.

Reconsidering Theorem 8.57, we see that this is just a corollary of a much more general theorem presented by Woerdeman in the following manuscripts.

☞ H. J. Woerdeman. The lower order of lower triangular operators and minimal rank extensions. *Integral Equations and Operator Theory*, 10:859–879, 1987.

☞ H. J. Woerdeman. Minimal rank completions for block matrices. *Linear Algebra and its Applications*, 121:105–122, 1989.

The 1989 article discusses the matrix version, whereas the 1987 article is focused on operator theory. Let us present the results of interest for our survey. Suppose a matrix A of the following form is given (the elements marked with $\times$ are not specified):

$$A = \begin{bmatrix} A_{11} & \times & \dots & \times \\ A_{21} & A_{22} & \ddots & \vdots \\ \vdots & & \ddots & \times \\ A_{m1} & A_{m2} & \dots & A_{mm} \end{bmatrix}. \tag{8.8}$$

This is an $m \times m$ block matrix of dimension $n \times n$. Remark that in the previous formula the matrix A is divided into subblocks A_{ij} which are also matrices (just like in Theorem 8.57 for a 2×2 block division). The provided theorem states that the minimal rank completion of this matrix has rank l, where l is defined as:

$$l = \sum_{p=1}^{m} \text{rank} \begin{bmatrix} A_{p,1} & \cdots & A_{p,p} \\ \vdots & & \vdots \\ A_{m,1} & \cdots & A_{m,p} \end{bmatrix} - \sum_{p=1}^{m-1} \text{rank} \begin{bmatrix} A_{p+1,1} & \cdots & A_{p+1,p} \\ \vdots & & \vdots \\ A_{m,1} & \cdots & A_{m,p} \end{bmatrix}. \tag{8.9}$$

In a system theoretical context l is called the lower order. For our purposes we will denote l as $\min \text{rank}(A)$, which is more descriptive and is defined as the minimum rank of all finite rank extensions of the matrix A. Reconsidering Theorem 8.57 in this context we see that the value of l computed as above equals $r_1 + r_2 - s$, which is identical to $(r_1 - s) + (r_2 - s) + s$ as derived in this chapter. One can also rewrite Equation (8.9), such that the connection with the results presented in this chapter are even more clear:

$$\min \text{rank}(A) = \sum_{p=1}^{m-1} \left(\text{rank} \begin{bmatrix} A_{p,1} & \cdots & A_{p,p} \\ \vdots & & \vdots \\ A_{m,1} & \cdots & A_{m,p} \end{bmatrix} - \text{rank} \begin{bmatrix} A_{p+1,1} & \cdots & A_{p+1,p} \\ \vdots & & \vdots \\ A_{m,1} & \cdots & A_{m,p} \end{bmatrix} \right)$$
$$+ \text{rank} \begin{bmatrix} A_{m,1} & \cdots & A_{m,m} \end{bmatrix}.$$

In the manuscript a method for constructing a minimal rank extension is also presented.

An interesting reference (covering also the theoretical results presented above and the ones presented below) is the following book.

☞ H. J. Woerdeman. *Matrix and operator extensions*, volume 68 of *CWI Tract*. Centre for Mathematics and Computer Science, Amsterdam, 1989.

Chapter IV of Woerdeman's book is devoted entirely to minimal rank extensions for matrices. The result discussed above is included in matrix terminology. Results related to partial extension, to extensions for lower triangular Toeplitz matrices and to general extension problems are included. Related to the theorem above, we will provide now a theorem (Theorem 2.1 in the chapter) on the uniqueness of the extension.

Theorem 8.75. *Suppose a matrix A, partitioned as in Equation (8.8), is given. Let us denote $A^{(p,q)}$ as follows:*

$$A^{(p,q)} = \begin{bmatrix} A_{p1} & \cdots & A_{pq} \\ \vdots & & \vdots \\ A_{m1} & \cdots & A_{mq} \end{bmatrix}.$$

Then the following conditions are equivalent.

1. *The matrix A has a unique minimal rank extension.*

2. $\text{rank}\left(A^{(p,p)}\right) = \text{rank}\left(A^{(p+1,p)}\right) = \text{rank}\left(A^{(p+1,p+1)}\right)$ *for $p = 1, \ldots, m-1$.*

3. *The values of $\text{rank}\left(A^{(p,q)}\right)$, with $1 \leq p \leq q \leq m$, are all identical.*

4. *The value of $\text{rank}(A_{m1})$ equals $\min \text{rank}(A)$.*

Furthermore in Chapter IV of the same book, a powerful theorem (Theorem 5.1), which is in some sense an extension of the nullity theorem, is proved. Let us formulate this theorem. The following presentation is mainly based on the manuscript by Woerdeman.

☞ H. J. Woerdeman. A matrix and its inverse: revisiting minimal rank completions, 2006. http://arxiv.org/abs/math.NA/0608130.

Let us present the theorem and how one can derive the nullity theorem from it.

Theorem 8.76. *Suppose an invertible $m \times m$ block matrix T of dimension $n \times n$ is given with blocks T_{ij}. Partition the inverse of the matrix T according to the one of T: $T^{-1} = (S_{ij})_{ij}$. Denote with $\hat{T}$ and $\hat{S}$ the partially specified block matrices:*

$$\hat{T} = (T_{ij})_{j \leq i} \quad \text{and} \quad \hat{S} = (S_{ij})_{j < i}.$$

Then we have the following equality:

$$\min \operatorname{rank}(\hat{T}) + \min \operatorname{rank}(\hat{S}) = n.$$

Rewriting the theorem above we get in fact the following, for $T^{-1} = (S_{ij})$, an $m \times m$ block matrix of dimension $n \times n$:

$$\min \operatorname{rank} \begin{bmatrix} T_{1,1} & \times & \cdots & \times \\ T_{2,1} & T_{2,2} & \ddots & \vdots \\ \vdots & & \ddots & \times \\ T_{m,1} & T_{m,2} & \cdots & T_{m,m} \end{bmatrix} + \min \operatorname{rank} \begin{bmatrix} \times & \times & \cdots & \times \\ S_{2,1} & \times & \ddots & \vdots \\ \vdots & \ddots & \ddots & \times \\ S_{m,1} & \cdots & S_{m,m-1} & \times \end{bmatrix} = n.$$

This theorem can be used to prove the nullity theorem (Theorem 1.33) in a straightforward way. We formulate the nullity theorem here again to use the same notation.

Theorem 8.77 (The nullity theorem). *Suppose we have the following invertible matrix $A \in \mathbb{R}^{n \times n}$ partitioned as*

$$A = \begin{bmatrix} A_{11} & A_{12} \\ A_{21} & A_{22} \end{bmatrix}$$

with A_{11} of size $p \times q$. The inverse B of A is partitioned as

$$B = \begin{bmatrix} B_{11} & B_{12} \\ B_{21} & B_{22} \end{bmatrix}$$

with B_{11} of size $q \times p$. Then the nullities $\operatorname{n}(A_{11})$ and $\operatorname{n}(B_{22})$ are equal.

The nullity theorem coincides with the case of a 2×2 block division ($m = 2$) of the matrix T in Theorem 8.76. Consider

$$\hat{T} = \begin{bmatrix} A_{11} & \times \\ A_{21} & A_{22} \end{bmatrix} \quad \text{and} \quad \hat{S} = \begin{bmatrix} \times & \times \\ B_{21} & \times \end{bmatrix},$$

with A and B from the nullity theorem above. We know by Theorem 8.76 that

$$\min \operatorname{rank}(\hat{T}) + \min \operatorname{rank}(\hat{S}) = n.$$

Using the information from Equation (8.9), we get:

$$\left(\operatorname{rank} \begin{bmatrix} A_{11} \\ A_{21} \end{bmatrix} + \operatorname{rank} [A_{21} \ A_{22}] - \operatorname{rank} [A_{21}] \right) + \operatorname{rank} B_{21} = n.$$

Because the matrices A and B are invertible, we get that

$$\mathrm{rank} \left[\begin{array}{c} A_{11} \\ A_{21} \end{array} \right] = q \quad \text{and} \quad \mathrm{rank}\,[A_{21}\ A_{22}] = n - p.$$

This leads to:

$$q + n - p - \mathrm{rank}\,[A_{21}] + \mathrm{rank}\,B_{21} = n.$$

Rewriting this leads to the following equality

$$0 = (q - \mathrm{rank}\,[A_{21}]) - (p - \mathrm{rank}\,B_{21}) = \mathrm{n}(A_{21}) - \mathrm{n}(B_{21}),$$

which is clearly equivalent to the statement from the nullity theorem. Theorem 8.76 is more general than the nullity theorem. It is obvious that many results on the structure of matrices under inversion can also be proved via this approach.

The results presented above are the ones mostly related to matrices and the results presented in the book. The following manuscripts by Woerdeman et al. cover related results, often focused on operators. The results as presented in the book on completion problems related to Toeplitz matrices are presented in the manuscript of 1994. The manuscript together with Rodman discusses the distance to the closest partial triangular matrix with prescribed rank.

☞ H. Dym and I. C. Gohberg. Extensions of band matrices with band inverses. *Linear Algebra and its Applications*, 36:1–24, 1981.

☞ I. C. Gohberg and M. A. Kaashoek. Minimal representations of semiseparable kernels and systems with separable boundary conditions. *Journal of Mathematical Analysis and Applications*, 124(2):436–458, June 1987.

☞ I. C. Gohberg, M. A. Kaashoek, and H. J. Woerdeman. A note on extensions of band matrices with invertible maximal and submaximal blocks. *Linear Algebra and its Applications*, 150:157–166, 1991.

☞ M. A. Kaashoek and H. J. Woerdeman. Unique minimal rank extensions of triangular operators. *Journal of Mathematical Analysis and Applications*, 131(2):501–516, May 1988.

☞ M. A. Kaashoek and H. J. Woerdeman. Minimal lower separable representations: characterization and construction. *Operator Theory: Advances and Applications*, 41:329–344, 1989.

☞ L. Rodman and H. J. Woerdeman. Perturbations, singular values, and ranks of partial triangular matrices. *SIAM Journal on Matrix Analysis and its Applications*, 16(1):278–288, January 1995.

☞ H. J. Woerdeman. Toeplitz minimal rank completions. *Linear Algebra and its Applications*, 202:267–278, 1994.

8.5   Representations

Just as for the easy classes of structured rank matrices, a huge variety of possible representations for the higher order types of structured rank matrices exists. It is impossible to cover all types of representations, as representations are invented to suit different needs. In this section, we would like to present some of the most common representations of the classes defined previously. Again we clearly distinguish between the types of matrices we work with and the representations used.

8.5.1 Givens-vector representation

The Givens-vector representation for semiseparable matrices was in fact based on the QR-factorization by means of Givens transformations of the structured rank part of the original semiseparable matrix. The higher order Givens-vector representation can be defined as the sum of Givens-vector representations of semiseparable matrices. This is possible, as we illustrated in the previous section that, for every higher order semiseparable matrix, its rank structured part can be decomposed as the sum of order 1 structured rank matrices.

Hence, this is a generalization, just as in the case of the generator representation. The sum of these representations was in fact used previously in the construction of the QR-factorization of semiseparable matrices as discussed in Chapter 5.

8.5.2 Quasiseparable representation

The quasiseparable representation has a nice expansion towards the higher order case. In fact, just as in the $\{1\}$-quasiseparable case, quasiseparable matrices are defined by their representation in several manuscripts.

The part satisfying the rank conditions in the matrix can be represented in the following way. Suppose the elements $a_{i,j}$ represent the elements of the matrix satisfying the low rank conditions. The indices i, j belong to a set Σ, and suppose we have quasiseparability (equals semiseparability) rank p,

$$a_{i,j} = \mathbf{u}_i^T t_{i-1:j+1} \mathbf{v}_j.$$

where $t_{i-1:j+1} = t_{i-1} t_{i-2} \ldots t_{j+1}$, $\mathbf{u}_i, \mathbf{v}_i \in \mathbb{R}^p$, and $t_i \in \mathbb{R}^{p \times p}$ for all i.

Taking a closer look at this representation, one can easily see that it is closely related to the generator representation. The extra factors are the matrices a_i, which are put in between the generators.

Once more we can say that the quasiseparable representation is the most general one as it contains both the Givens-vector and the generator representation for higher order structured rank matrices. We do not examine this in further detail. These statements can easily be verified by the reader.

Let us briefly illustrate this representation applied to a quasiseparable matrix.

Example 8.78 A general $\{p, q\}$-quasiseparable matrix A with the quasiseparable representation is of the following form

$$a_{i,j} = \begin{cases} \mathbf{q}_j^T r_{j-1:i+1} \mathbf{p}_i, & 1 \leq i < j \leq n, \\ d_i, & 1 \leq i = j \leq n, \\ \mathbf{u}_i^T t_{i-1:j+1} \mathbf{v}_j, & 1 \leq j < i \leq n, \end{cases}$$

where $r_{j-1:i+1} = r_{j-1} r_{j-2} \ldots r_{i+1}$, $t_{i-1:j+1} = t_{i-1} t_{i-2} \ldots t_{j+1}$, $\mathbf{p}_i, \mathbf{q}_i \in \mathbb{R}^q$, $\mathbf{u}_i, \mathbf{v}_i \in \mathbb{R}^p$, $r_i \in \mathbb{R}^{q \times q}$ and $t_i \in \mathbb{R}^{p \times p}$ for all i.[6] ∎

Even though at first sight this definition might seem different from the one used in Section 2.9 in Chapter 2, some index reshuffling indicates that they are the same.

---

[6] We remark that in most of the articles based on quasiseparable matrices, the authors use a different notation, they denote $t_{i-1:j+1} = t_{i,j}^\times$, and they also use $r_{i,j}^\times$.

8.5.3　Split representations

It was illustrated in the previous section that we can always decompose the structured rank part of the matrix into the sum of lower order structured rank matrices. This means that we can use different types of representations for representing these matrices. In this way we can get combinations of quasiseparable, generator and other types of representations.

In the spirit of mixing representations, we can also use different types of representations for representing the lower and upper triangular part of a matrix.

Notes and references

The different types of representations as discussed in this and the previous section are commonly used in several manuscripts. The higher order generator representation was used for solving systems of equations in the manuscripts [209, 56]. The algorithms are based on a kind of hybrid URV-factorization (see Section 5.6 in Chapter 5). Also methods for reducing higher order generator representable matrices to tridiagonal or bidiagonal form are known [108, 210] (see Section 3.5 in Chapter 3). The articles [145, 148, 150, 91, 90, 92] contain methods for inverting higher order generator representable quasiseparable matrices. They present direct techniques as well as techniques based on an upper triangular factorization. More information can be found in Section 14.5 from Chapter 14 and in Chapter 12. The higher order quasiseparable representation is used in several manuscripts [93, 89, 96] (see Section 14.5 from Chapter 14 and Chapter 12).

Recently an extension of the Givens-vector representation for higher order semiseparable matrices was proposed in the following manuscript by Delvaux and Van Barel.

> ☞　S. Delvaux and M. Van Barel. A Givens-weight representation for rank structured matrices. Technical Report TW453, Department of Computer Science, Katholieke Universiteit Leuven, Celestijnenlaan 200A, 3000 Leuven (Heverlee), Belgium, March 2006. (To appear in SIMAX).

The idea to expand the Givens-vector representation is based on the idea of the QR-factorization. In fact one obtains a unitary-weight representation. The unitary transformations replace the Givens transformation, and the vector is replaced by weights; this can be blocks in the matrix and so forth. Algorithms are presented on how to retrieve this representation for a full matrix, how to reduce the order of the representation when needed. Finally it is also illustrated how the representation can be updated under the action of Givens transformations, and it is shown how a fast multiplication with a vector can be performed. The upcoming section on computing the QR-factorization is in some sense closely related to the Givens-weight representation. Following the chapter on the QR-factorization we will present results on how to compute the LU-factorization of matrices. In [72, Remark 12], it is also mentioned that one can use Gauss transforms and weights for representing structured rank matrices. This leads to a so-called Gauss-weight representation.

This Givens-weight representation can then be used for computing the Hessenberg reduction, the QR-factorization, and the explicit QR-algorithm as described in the manuscripts [74, 73, 71]. For manuscripts [73, 71], see Section 3.5 Chapter 3, and for manuscript [74] see the notes and references at the end of Chapter 9.

8.6   Conclusions

In this chapter we introduced the concept of structures and rank structures to define structured rank matrices in an easy manner. These tools seemed to be very useful for defining commonly known classes of structured rank matrices, such as quasiseparable, semiseparable, band matrices and so on. We reconsidered in this chapter also the nullity theorem to investigate the rank structure of the inverses of the previously defined matrices. Several completion problems were formulated and proven to show that every structured rank matrix can be decomposed into the sum of lower order structured rank matrices. This provided also a theorem stating whether a semiseparable matrix is of generator representable form or not. Finally, we concluded this chapter by showing some representations for higher order structured rank matrices.

Chapter 9

A QR-factorization for structured rank matrices

In the beginning of the book we discussed the QR-factorization for the easiest classes of structured rank matrices, e.g., semiseparable, semiseparable plus diagonal and quasiseparable matrices. For quasiseparable matrices, we investigated the computation of the Givens transformations from bottom to top for reducing the rank 1 structure, and then we needed to perform an extra sequence of Givens transformations from top to bottom to bring the Hessenberg structure to upper triangular form.

In this chapter we will investigate the QR-decomposition for higher order structured rank matrices in more detail. Even though this might seem much more complicated at first glance, we will try to present easy admissible techniques to tackle this problem. We will, however, in contradiction to the easy case, not present ready-to-use algorithms. We chose not to overload this part of the book with cluttered formulas, but we did choose to present a theory giving insight in the structural behavior of these matrices during the reduction to upper triangular form. This viewpoint will hand the reader interesting tools and will help him to program his own QR-solver for higher order structured rank matrices in an efficient way.

The most important idea used throughout this chapter is the decomposition of a structured rank part in a matrix into a sum of lower order structured rank matrices. Even though this decomposition might not be known, in general, the ideas remain valid and are valuable tools in the computation of the QR-factorization. Working on each of the matrices in the decomposition separately gives insight in the complete reduction algorithm. Hence, to investigate in detail the structure of the factors Q and R, of the QR-factorization of structured rank matrices, we decompose the original matrix and investigate in detail the structure of the terms in this decomposition separately.

The transformations used for making the involved structured rank matrix of upper triangular form are Givens transformations. In Section 9.1 we investigate the effect of sequences of Givens transformations on order 1 structured rank matrices. Based on these observations we derive general theorems for the performance of a sequence of Givens transformations on a general structured rank matrix. As in the quasiseparable case, sequences are often also needed from top to bottom and we investigate the effect of such sequences on structured rank matrices. To conclude the first section we investigate the application of Givens transformations on the right of the matrix in sequences from left to right and right to left.

In Section 9.2, we will combine our gained knowledge to effectively reduce a structured rank matrix to upper triangular form by combining sequences from bottom to top and sequences from top to bottom. We analyze the number of ascending as well as descending sequences of transformations that are essential for bringing the matrix to its desired, upper triangular form.

In Section 9.3 we investigate the order of the different sequences of transformations. In the previous section we introduced a pattern of Givens transformations consisting of a number of ascending sequences of transformations, followed by a number of descending sequences of transformations. But it seems that we can change the order of the Givens transformations involved. This gives us, for example, the leaf and pyramid form for removing the rank structure, as well as the leaf and diamond form for removing the subdiagonals. More important is the shift through lemma. This lemma proves that we can rearrange the order of some specifically ordered Givens transformations. This lemma is very interesting as it creates the possibility to perform first the sequences of descending Givens transformations, followed by the sequences of ascending Givens transformations. A lot of flexibility in creating the upper triangular matrix R is created by this lemma.

Section 9.4 investigates in more detail the structure of the Givens transformations, if one performs first the sequence of descending Givens transformations for making the matrix upper triangular. It is proven that these Givens transformations do not decrease the rank of the lower part, but in some sense they expand the structure of the lower structured rank part. This means that after performing a sequence of rank expanding Givens transformations from top to bottom, the rank structure in the lower triangular part has grown. Theorems related to the existence of this transformation and its effect on a general structured rank matrix are formulated.

Section 9.5 investigates in more detail the Givens-vector representation with regard to performance of sequences of Givens transformations. Some tools are provided to make the manipulation of the Givens-vector representation easier. Also a computational method is provided for effectively computing the rank expanding sequence of Givens transformations.

The last section of this chapter is a kind of blend of material related of the QR-factorization. A new interpretation to the QR-factorization for unstructured matrices is provided, also a new type of factorization; namely, a QZ-factorization is proposed. More interesting is the construction of a parallel QR-algorithm for structured rank matrices. The final subsection in this chapter focuses on the multiplication between structured rank matrices. Some theorems are provided for deducing the rank structure of products of structured rank matrices.

✎   *This chapter contains important results related to the performance of sequences of Givens transformations on structured rank matrices. The effect of these sequences of transformations on these matrices is deduced from the very simple rank 1 case towards the general case. The results are interesting as they link structured* rank matrices A with factorizations $A = Q\hat{A}$, where Q is a matrix consisting of specific Givens transformations and $\hat{A}$ is another structured rank matrix. This chapter contains a lot of theoretical derivations to obtain these results. We will summarize the key theorems with some examples.

It is essential to understand the difference between the different types of sequences of Givens transformations. First a rank-annihilating sequence of Givens transformations is defined (Definition 9.12). A direct extension of the annihilating sequence is the rank-decreasing sequence of Givens transformations (Definition 9.12). The example of the transformation of the Hessenberg matrix to triangular form, just before Definition 9.25 and the definition itself, illustrates what is meant by a rank-expanding sequence of Givens transformations. All the intermediate theorems and their proofs are necessary to formulate the global result depicting the effect of the different types of sequences of Givens transformations from bottom to top on a structured rank matrix: Theorem 9.26. The note following the theorem explains the effect of the different sequences. If one wishes to also perform sequences in other directions, Subsection 9.1.8 explains the effect on the rank structure for sequences from bottom to top, from left to right and from right to left. To clearly illustrate the effect of some of these sequences one can take a look at Section 9.1.9, showing the effect for semiseparable, band, quasiseparable and some other classes of structured rank matrices. Section 9.1.10 summarizes once more the concept for all types of sequences, regardless of their direction. Combining the results in Subsection 9.2.1 and Subsection 9.2.2 one can understand the upper triangularization of the matrices in the examples in Section 9.2.4. In fact one knows now all the essential information to compute the QR-factorization of structured rank matrices. The idea of using the QR-factorization for solving systems of equations is in fact straightforward, but discussed briefly in Section 9.2.5. If one is interested in also exploiting different directions of sequences for computing factorizations, Section 9.2.6 provides some possibilities.

The remaining part of the chapter focuses towards the interaction of Givens transformations with each other, leading towards the existence of rank-expanding sequences of Givens transformations for higher order matrices. Briefly one can say that the important results are captured in Section 9.4.4, stating the effect of a sequence of Givens transformations in its most general form (more general than the one discussed above). How to compute the transformation is described throughout the preceding results, which are of a more theoretical nature. Important and interesting results in this preceding part are the graphical representation of the Givens transformations and their mutual interaction. The graphical representation provides a flexible tool for depicting interactions between Givens transformations and the matrices and can also significantly simplify the implementation. This representation is also used in the next chapter, in which we will use it to depict the interaction between Gauss transforms. The graphical representation is depicted in the beginning of Section 9.3, where the example of computing the QR-factorization of a semiseparable plus diagonal matrix is depicted. Following this representation one can jump right ahead to Section 9.3.5 in which the interaction between Givens transformations and their graphical representation are investigated. Especially the "Shift through Lemma" (Lemma 9.38) will be used extensively. Its graphical inter-

pretation is also important. To understand the power of the shift through lemma one should read through Subsection 9.3.6 to Subsection 9.3.8. If one is interested in the relation between the $\vee$-pattern and the rank-expanding sequence of Givens transformations one should check out Subsection 9.3.9.

The next section considers the rank-expanding sequence of transformations discussed above and the remaining sections before Section 9.7 consider some special cases and examples. Section 9.7, however, contains an important theorem, namely Theorem 9.56 predicting the structure of a structured rank matrix when one multiplies different structured rank matrices with each other. Some examples of the effect of multiplying matrices are given in Subsection 9.7.2.

9.1 A sequence of Givens transformations from bottom to top

In Chapter 5 of the book we discussed the QR-decomposition of structured rank matrices of easy form. We discussed the factorization of semiseparable, semiseparable plus diagonal and quasiseparable matrices. For these classes a sequence of Givens transformations going from bottom to top and one sequence going from top to bottom was enough to reduce the matrix to upper triangular form. As we are now working with higher order structured rank matrices, one might correctly wonder whether one sequence of ascending and one sequence of descending Givens transformations will still suffice our needs. The answer is no. In general, the number of ascending sequences will depend on the rank structure of the lower triangular part, and the number of descending sequences will depend on the extent with regard to the diagonal of the rank structure and on the order of the rank structure. Therefore, in this section, we will investigate, the effect of the application of sequences of transformations on the rank structure of structured rank matrices.

Initially, we discuss in detail the sequence from bottom to top. We start by applying these transformations on lower triangular semiseparable matrices, followed by an analysis of this sequence of transformations on upper triangular semiseparable matrices. We first investigate matrices with structured ranks equal to one. Based on these simple cases we can go to more general, higher order structured ranks.

We restrict ourselves to lower/upper triangular semiseparable matrices, as this illustrates the general case. After each example we briefly show the effect on more general structured rank blocks in matrices.

Finally we will show the effect on general structured rank matrices, and moreover, we will also investigate the effect of descending sequences as well as sequences going from left to right or from right to left. To conclude the section some examples of the applications of such sequences on semiseparable, quasiseparable, band and semiseparable plus band matrices are given.

Note that we will focus most of the attention on p-lower/upper triangular structures. This is natural as all the classes of structured rank matrices discussed have constraints posed on p-lower/upper triangular structures. Most of the theorems remain valid, however, if applied on more general structures. Sometimes this is mentioned, but in any case, the reader should be able to derive similar theorems, based on the ideas proposed in this chapter. We will start by deducing results for matrices having the lower triangular part of a certain rank structure. Finally we will expand these results to p-lower triangular structures expanding across the diagonal.

9.1.1 Annihilating Givens transformations on lower rank 1 structures

Let us investigate the QR-factorization of a lower triangular semiseparable matrix S. One sequence of Givens transformations from bottom to top is performed for making the matrix S of upper triangular form. To create the upper triangular matrix R we use Givens transformations. Sequences of Givens transformations acting on structured rank matrices can be divided in several classes; let us define the most commonly used sequence.

Definition 9.1. *A sequence of Givens transformations $G_k^T \ldots G_l^T$ (with $2 \leq k \leq l \leq n$) applied on the matrix A is called an annihilating sequence of Givens transformations, if for all i, the application of the Givens transformation G_i^T (acting on row $i-1$ and i) onto $G_{i+1}^T \ldots G_n^T A$ creates at least one zero in row i in the matrix $G_{i+1}^T \ldots G_n^T A$.[7]*

A sequence of Givens transformations is called a rank-annihilating sequence of Givens transformations, if it annihilates elements in a rank 1 structure.

The performance of a rank-annihilating sequence of Givens transformations results in the creation of many more zeros than just one. A standard sequence has $k = 2$ and $l = n$, i.e., we perform the following sequence of Givens transformations: $G_2^T \ldots G_n^T$.

Theorem 9.2. *Suppose S is an $n \times n$ lower triangular semiseparable matrix, on which a sequence of $n-1$ rank-annihilating Givens transformations G_i^T, acting on row $i-1$ and i, is performed:*

$$G_2^T \ldots G_n^T S = R.$$

Then the matrix R will be an upper triangular semiseparable matrix.

Proof. Even though this is straightforward, we would like to illustrate this easy case by a proof based on figures, as the following theorems have similar proofs.
Let us illustrate this for a 5×5 lower triangular semiseparable matrix $S = S_0$, as this illustrates the general case. Let us denote the lower semiseparable part of the matrix with elements $\boxtimes$, the newly created elements in the upper triangular part are denoted with $\boxplus$, the elements that will be annihilated with the Givens transformation are marked with $\otimes$. The invisible elements are assumed to be zero. Let us start with the annihilation of the last row by performing the Givens transformation G_5^T:

---

[7]We are flexible in this definition, i.e., if the row i is zero, we choose the Givens transformation equal to the identity matrix.

$$
\begin{bmatrix}
\boxtimes & & & & \\
\boxtimes & \boxtimes & & & \\
\boxtimes & \boxtimes & \boxtimes & & \\
\boxtimes & \boxtimes & \boxtimes & \boxtimes & \\
\otimes & \otimes & \otimes & \otimes & \boxtimes
\end{bmatrix}
\xrightarrow{G_5^T S_0}
\begin{bmatrix}
\boxtimes & & & & & \\
\boxtimes & \boxtimes & & & & \\
\boxtimes & \boxtimes & \boxtimes & & & \\
\boxtimes & \boxtimes & \boxtimes & \boxtimes & \boxplus \\
& & & & \boxplus
\end{bmatrix}
$$

$$\Updownarrow$$

$$S_0 \xrightarrow{G_5^T S_0} S_1.$$

Continuing this process, by annihilating row 4, leads to:

$$
\begin{bmatrix}
\boxtimes & & & & \\
\boxtimes & \boxtimes & & & \\
\boxtimes & \boxtimes & \boxtimes & & \\
\otimes & \otimes & \otimes & \boxtimes & \boxplus \\
& & & & \boxplus
\end{bmatrix}
\xrightarrow{G_4^T S_1}
\begin{bmatrix}
\boxtimes & & & & & \\
\boxtimes & \boxtimes & & & & \\
\boxtimes & \boxtimes & \boxtimes & \boxplus & \boxplus \\
& & & \boxplus & \boxplus \\
& & & & \boxplus
\end{bmatrix}
$$

$$\Updownarrow$$

$$S_1 \xrightarrow{G_4^T S_1} S_2.$$

It is clear that this annihilating process in the lower triangular part gradually fills up the upper triangular part of this matrix. Moreover all the newly created rows in the upper triangular part are dependent and hence the upper triangular part will be of semiseparable form. Finally, after performing the last transformation G_2^T we get:

$$
\begin{bmatrix}
\boxtimes & & & & \\
\otimes & \boxtimes & \boxplus & \boxplus & \boxplus \\
& & \boxplus & \boxplus & \boxplus \\
& & & \boxplus & \boxplus \\
& & & & \boxplus
\end{bmatrix}
\xrightarrow{G_2^T S_3}
\begin{bmatrix}
\boxplus & \boxplus & \boxplus & \boxplus & \boxplus \\
& \boxplus & \boxplus & \boxplus & \boxplus \\
& & \boxplus & \boxplus & \boxplus \\
& & & \boxplus & \boxplus \\
& & & & \boxplus
\end{bmatrix}
$$

$$\Updownarrow$$

$$S_3 \xrightarrow{G_2^T S_3} S_4.$$

This clearly shows that after performing $n-1$ annihilating Givens transformations from bottom to top, we gradually filled up the upper triangular part with a semiseparable structure of rank 1. $\qquad\square$

As already mentioned, the Givens transformations as performed here for creating zeros in the lower triangular part are called annihilating Givens transformations. In this case they create many more zeros than just one. This sequence of Givens transformations is often also called an eliminating Givens pattern.

Let us formulate now the structure after performing such a Givens transformation in the context of structured rank matrices.

Corollary 9.3. *Suppose we have a structured rank matrix A satisfying the following rank structural constraints (with $p \leq 1$):*

$$\mathrm{r}\left(\Sigma_l^{(p)}; A\right) \leq 1 \quad and \quad \mathrm{r}\left(\Sigma_u^{(p-1)}; A\right) = 0.$$

Performing $\min(n - 1, n + p - 2)$ rank-annihilating Givens transformations from bottom to top leads to a structured rank matrix $\hat{A}$ of the following form:

$$\mathrm{r}\left(\Sigma_l^{(p-1)}; \hat{A}\right) = 0 \quad and \quad \mathrm{r}\left(\Sigma_u^{(p-2)}; \hat{A}\right) \leq 1.$$

We remark that the rank structure in the upper triangular part has shifted down one diagonal position and has increased in structured rank with value 1.

Note 9.4. *Some remarks concerning the corollary above:*

- *We illustrated here the case for structures below a certain diagonal in the matrix. This theorem can easily be extended to other types of structures such as block structures, etc. For our purposes, however, this theorem is sufficient.*

- *In the corollary it was mentioned that only $\min(n - 1, n + p - 2)$ rank-annihilating Givens transformations needed to be performed. This is because p can also be negative or equal to zero. In case p is negative the structure does not expand above the diagonal and hence less Givens transformations need to be performed.*

- *Even though we mentioned the performance of $\min(n - 1, n + p - 2)$ Givens transformations, the corollary still remains valid in case more Givens transformations than only $n + p - 2$ ones (in case $p \leq 0$) are performed. In this case, only $n + p - 2$ rank-annihilating Givens transformations are performed, and the remaining $1 - p$ Givens transformations are arbitrary. Therefore, in the following corollaries, when speaking about a rank-annihilating (or later on a rank-decreasing) sequence of Givens transformations, we mean that $\min(n - 1, n + p - 2)$ rank-annihilating Givens transformations were performed and the remaining ones are arbitrary.*

- *An important remark has to do with the fact that we can also admit p to be larger than 1. In this case the remark that the structure shifts down remains completely valid. But one should of course be careful now that the structure is not necessarily a lower triangular structure anymore! For example, consider the following 5×5 matrix A having the part including the first superdiagonal of structured rank 1 (this means $p = 2$ in Corollary 9.3). Performing a sequence of Givens transformations annihilating thereby the rank structure results in* *the matrix $\hat{A}$:*

$$
A = \begin{bmatrix} \boxtimes & \boxtimes & & & \\ \boxtimes & \boxtimes & \boxtimes & & \\ \boxtimes & \boxtimes & \boxtimes & \boxtimes & \\ \boxtimes & \boxtimes & \boxtimes & \boxtimes & \boxtimes \\ \boxtimes & \boxtimes & \boxtimes & \boxtimes & \boxtimes \end{bmatrix} \quad and \quad \hat{A} = \begin{bmatrix} \boxplus & \boxplus & \boxplus & \boxplus & \boxplus \\ & & \boxplus & \boxplus & \boxplus \\ & & & \boxplus & \boxplus \\ & & & & \boxplus \\ & & & & \end{bmatrix}.
$$

It is clear that the upper right structure in the right matrix has shifted down one position and has decreased in rank. But the structure is not a p-lower triangular structure anymore. As mentioned earlier, we will focus on transformations which give us again a p-lower/upper triangular structure as the classes of matrices we are working with belong to these classes.

We illustrated here what happened when a rank-annihilating sequence of Givens transformations was performed on a lower triangular semiseparable matrix. Let us investigate now what the effect of an arbitrary sequence of Givens transformations on the rank structure of a lower triangular semiseparable matrix will be. This will be essential when later considering higher order structured rank matrices.

9.1.2   Arbitrary Givens transformations on lower rank 1 structures

In the previous section we illustrated the change in structured rank of a lower semiseparable matrix if a sequence of rank-annihilating Givens transformations was performed. Suppose however, that we have, for example, a lower triangular {2}-semiseparable matrix. We know that this matrix can be written as the sum of 2 lower triangular {1}-semiseparable matrices by the results of Section 8.4 in Chapter 8. Performing a rank-annihilating sequence of Givens transformations on one of these lower triangular semiseparable matrices creates zeros in this matrix, but changes also the second matrix without necessarily creating zeros. We will now show in this section, how the structure of this second matrix will change, as these Givens transformations are not necessarily of rank-annihilating form for this matrix.

Theorem 9.5. *Suppose S is an $n \times n$ lower triangular semiseparable matrix. A sequence of arbitrary $n - 1$ Givens transformations G_i^T, acting on row $i - 1$ and i, is performed on the matrix S:*

$$
G_2^T \dots G_n^T S = \hat{S}.
$$

The matrix $\hat{S}$ will have the upper triangular part of semiseparable form and also the strictly lower triangular part will be of semiseparable form.

Proof. The main difference from Theorem 9.2 is that zeros will not necessarily be created here. We illustrate this again for a 5×5 lower triangular semiseparable matrix, as this illustrates the general case.

We use a similar notation as in Theorem 9.2, namely, that the elements in boxes, $\boxtimes$ and $\boxplus$, belong to different semiseparable structures depending on the sign (+ and ×) in the box. The invisible elements are assumed to be equal to zero.

We perform a first Givens transformation G_5^T acting on the last two rows of the 5×5 matrix:

$$
\begin{bmatrix}
\boxtimes & & & & \\
\boxtimes & \boxtimes & & & \\
\boxtimes & \boxtimes & \boxtimes & & \\
\boxtimes & \boxtimes & \boxtimes & \boxtimes & \\
\boxtimes & \boxtimes & \boxtimes & \boxtimes & \boxtimes
\end{bmatrix}
\xrightarrow{G_5^T S_0}
\begin{bmatrix}
\boxtimes & & & & \\
\boxtimes & \boxtimes & & & \\
\boxtimes & \boxtimes & \boxtimes & & \\
\boxtimes & \boxtimes & \boxtimes & \boxtimes & \boxplus \\
\boxtimes & \boxtimes & \boxtimes & \boxtimes & \boxplus
\end{bmatrix}
$$

$$\Updownarrow$$

$$S_0 \xrightarrow{G_5^T S_0} S_1.$$

Continuing this process, leads to:

$$
\begin{bmatrix}
\boxtimes & & & & \\
\boxtimes & \boxtimes & & & \\
\boxtimes & \boxtimes & \boxtimes & & \\
\boxtimes & \boxtimes & \boxtimes & \boxtimes & \boxplus \\
\boxtimes & \boxtimes & \boxtimes & \boxtimes & \boxplus
\end{bmatrix}
\xrightarrow{G_4^T S_1}
\begin{bmatrix}
\boxtimes & & & & \\
\boxtimes & \boxtimes & & & \\
\boxtimes & \boxtimes & \boxtimes & \boxplus & \boxplus \\
\boxtimes & \boxtimes & \boxtimes & \boxplus & \boxplus \\
\boxtimes & \boxtimes & \boxtimes & \boxtimes & \boxplus
\end{bmatrix}
$$

$$\Updownarrow$$

$$S_1 \xrightarrow{G_4^T S_1} S_2.$$

The only difference from the previous theorem is that, in this case, no zeros are created. The upper triangular part is similarly filled up with a rank 1 structure. Performing the final transformation G_2^T gives us:

$$
\begin{bmatrix}
\boxtimes & & & & \\
\boxtimes & \boxtimes & \boxplus & \boxplus & \boxplus \\
\boxtimes & \boxtimes & \boxplus & \boxplus & \boxplus \\
\boxtimes & \boxtimes & \boxtimes & \boxplus & \boxplus \\
\boxtimes & \boxtimes & \boxtimes & \boxtimes & \boxplus
\end{bmatrix}
\xrightarrow{G_2^T S_3}
\begin{bmatrix}
\boxplus & \boxplus & \boxplus & \boxplus & \boxplus \\
\boxtimes & \boxplus & \boxplus & \boxplus & \boxplus \\
\boxtimes & \boxtimes & \boxplus & \boxplus & \boxplus \\
\boxtimes & \boxtimes & \boxtimes & \boxplus & \boxplus \\
\boxtimes & \boxtimes & \boxtimes & \boxtimes & \boxplus
\end{bmatrix}
$$

$$\Updownarrow$$

$$S_3 \xrightarrow{G_2^T S_3} S_4.$$

The upper triangular as well as the strictly lower triangular part of this matrix are now of semiseparable form.   □

Similarly as in the previous section, we can easily generalize this towards different structured ranks.

Corollary 9.6. *Suppose we have a structured rank matrix A satisfying the following rank structural constraints[8]:*

$$\mathrm{r}\left(\Sigma_l^{(p)}; A\right) \leq 1 \quad and \quad \mathrm{r}\left(\Sigma_u^{(p-1)}; A\right) = 0.$$

Performing a sequence of arbitrary Givens transformations from bottom to top leads to a structured rank matrix $\hat{A}$ of the following form:

$$\mathrm{r}\left(\Sigma_l^{(p-1)}; \hat{A}\right) \leq 1 \quad and \quad \mathrm{r}\left(\Sigma_u^{(p-2)}; \hat{A}\right) \leq 1.$$

We remark, similarly as in the previous case, that the rank structure in the upper triangular part has shifted down one position and has increased in structured rank with value 1.

Note 9.7.

- *Similarly as in the previous subsection, we remark that this theorem can easily be expanded to different rank structures, such as a block lower triangular structure, for example.*

- *It is clear that Theorem 9.2 can be interpreted as a special case of the previous Theorem. In this case the strictly lower triangular rank is bounded by 1, whereas in the previous theorem, it was bounded by 0.*

9.1.3   Givens transformations on lower rank structures

In the previous two sections we examined the effect of a sequence of Givens transformations from bottom to top on lower triangular semiseparable matrices of semiseparability rank 1. Let us free the rank 1 condition and investigate what happens if, for example, we have a lower triangular $\{p\}$-semiseparable matrix. We formulate a theorem for the $p = 2$ case, and extend this afterwards to higher order ranks. We formulate the theorem for an arbitrary sequence of Givens transformations, and we formulate a similar theorem afterwards for rank-annihilating patterns of Givens transformations.

Theorem 9.8. *Suppose S is an $n \times n$ lower triangular $\{2\}$-semiseparable matrix. A sequence of arbitrary $n - 1$ Givens transformations G_i^T acting on row $i - 1$ and i is performed on the matrix S:*

$$G_2^T \dots G_n^T S = \hat{S}.$$

The matrix $\hat{S}$ will have the upper triangular part of semiseparable form and the strictly lower triangular part will be of $\{2\}$-semiseparable form.

---

[8]Remark that the constraint $p \leq 1$ is not necessary in this case.

This means that the {2}-semiseparable structure has shifted down one position, and the rank of the upper triangular part has increased by 1.

Proof. We know that we can write the lower triangular part as the sum of two separate lower triangular {1}-semiseparable matrices. Performing the Givens transformations on both lower triangular semiseparable matrices separately gives us (by Theorem 9.5) two matrices with the upper triangular part of semiseparable form and the strictly lower triangular part of semiseparable form. Recombining these two matrices gives us a matrix with a strict lower triangular part of {2}-semiseparable form and, in general, also the upper triangular part of {2}-semiseparable form. But because the Givens transformations working on both of the matrices are the same ones, the upper triangular part will be of {1}-semiseparable form. We will illustrate this with a 5×5 example.

Assume our matrix S is decomposed as the sum of two lower triangular semiseparable matrices. We will write the matrix S as the sum of three separate matrices. The first two matrices will afterwards be combined and give the strictly lower triangular part. The last matrix will contain the upper triangular part of $G_2^T \ldots G_n^T S$. The elements of the matrices are denoted with $\times$ and $+$. When they are placed in a box, they satisfy a semiseparable structure.

$$
S = \left[\begin{array}{ccccc} \boxtimes & & & & \\ \boxtimes & \boxtimes & & & \\ \boxtimes & \boxtimes & \boxtimes & & \\ \boxtimes & \boxtimes & \boxtimes & \boxtimes & \\ \boxtimes & \boxtimes & \boxtimes & \boxtimes & 0 \end{array}\right] + \left[\begin{array}{ccccc} \boxtimes & & & & \\ \boxtimes & \boxtimes & & & \\ \boxtimes & \boxtimes & \boxtimes & & \\ \boxtimes & \boxtimes & \boxtimes & \boxtimes & \\ \boxtimes & \boxtimes & \boxtimes & \boxtimes & 0 \end{array}\right] + \left[\begin{array}{ccccc} 0 & & & & \\ & 0 & & & \\ & & 0 & & \\ & & & 0 & \\ & & & & \times \end{array}\right]
$$

The first Givens transformation gives us

$$
G_5^T S = \left[\begin{array}{ccccc} \boxtimes & & & & \\ \boxtimes & \boxtimes & & & \\ \boxtimes & \boxtimes & \boxtimes & & \\ \boxtimes & \boxtimes & \boxtimes & \boxtimes & \\ \boxtimes & \boxtimes & \boxtimes & \boxtimes & 0 \end{array}\right] + \left[\begin{array}{ccccc} \boxtimes & & & & \\ \boxtimes & \boxtimes & & & \\ \boxtimes & \boxtimes & \boxtimes & & \\ \boxtimes & \boxtimes & \boxtimes & \boxtimes & \\ \boxtimes & \boxtimes & \boxtimes & \boxtimes & 0 \end{array}\right] + \left[\begin{array}{ccccc} 0 & & & & \\ & 0 & & & \\ & & 0 & & \\ & & & 0 & \boxplus \\ & & & & \boxplus \end{array}\right].
$$

We rewrite this equation by now placing the sum of the elements in position $(4, 4)$ of the first two matrices in the last matrix. This gives us the following equation (we can immediately incorporate this diagonal element in the semiseparable structure of the last matrix):

$$
G_5^T S = \left[\begin{array}{ccccc} \boxtimes & & & & \\ \boxtimes & \boxtimes & & & \\ \boxtimes & \boxtimes & \boxtimes & & \\ \boxtimes & \boxtimes & \boxtimes & 0 & \\ \boxtimes & \boxtimes & \boxtimes & \boxtimes & 0 \end{array}\right] + \left[\begin{array}{ccccc} \boxtimes & & & & \\ \boxtimes & \boxtimes & & & \\ \boxtimes & \boxtimes & \boxtimes & & \\ \boxtimes & \boxtimes & \boxtimes & 0 & \\ \boxtimes & \boxtimes & \boxtimes & \boxtimes & 0 \end{array}\right] + \left[\begin{array}{ccccc} 0 & & & & \\ & 0 & & & \\ & & 0 & & \\ & & & \boxplus & \boxplus \\ & & & & \boxplus \end{array}\right].
$$

Performing Givens transformation 4 on rows 3 and 4, gives us:

$$G_4^T G_5^T S =$$

$$\begin{bmatrix} \boxtimes & & & & \\ \boxtimes & \boxtimes & & & \\ \boxtimes & \boxtimes & \boxtimes & & \\ \boxtimes & \boxtimes & \boxtimes & 0 & \\ \boxtimes & \boxtimes & \boxtimes & \boxtimes & 0 \end{bmatrix} + \begin{bmatrix} \boxtimes & & & & \\ \boxtimes & \boxtimes & & & \\ \boxtimes & \boxtimes & \boxtimes & & \\ \boxtimes & \boxtimes & \boxtimes & 0 & \\ \boxtimes & \boxtimes & \boxtimes & \boxtimes & 0 \end{bmatrix} + \begin{bmatrix} 0 & & & & \\ & 0 & & & \\ & & 0 & \boxplus & \boxplus \\ & & & \boxplus & \boxplus \\ & & & & \boxplus \end{bmatrix}.$$

We can clearly see that the right matrix fills up to become an upper triangular semi-separable matrix of rank 1, whereas the first two matrices maintain their structure, but the structure shifts down one position.

This procedure can easily be continued to give us the final rank structure, as proposed in the theorem, namely an upper triangular part of semiseparable form and the strictly lower triangular part of {2}-semiseparable form.   □

The previous theorem concentrated on matrices having the lower triangular form of {2}-semiseparable form. But, in fact the proof of the theorem can easily be adapted to matrices having the lower triangular form of a higher order. This is formulated in the following corollary in the general context of structured rank matrices.

Corollary 9.9. *Suppose we have a structured rank matrix A satisfying the following rank structural constraints:*

$$\mathrm{r}\left(\Sigma_l^{(p)};A\right) \le r \quad and \quad \mathrm{r}\left(\Sigma_u^{(p-1)};A\right) = 0.$$

Performing a sequence of arbitrary Givens transformations from bottom to top leads to a structured rank matrix $\hat{A}$ of the following form:

$$\mathrm{r}\left(\Sigma_l^{(p-1)};\hat{A}\right) \le r \quad and \quad \mathrm{r}\left(\Sigma_u^{(p-2)};\hat{A}\right) \le 1.$$

This means that the rank structure of the lower triangular part stays of the same rank but shifts down one position, whereas the rank structure of the upper triangular part increases by 1.

Note 9.10.

- *Again we can formulate this theorem for other types of rank structures, for example, if we have a lower triangular {2}-semiseparable matrix which can be written as the sum of a lower triangular generator representable semiseparable matrix plus a lower triangular block-diagonal semiseparable matrix. Even though we formulated the corollary for a matrix obeying a certain rank structure, we can also formulate the corollary for specific blocks in the matrix of a specific rank. This can also be seen in the next example, where there are specific blocks of rank 0.*

- *As mentioned before, in the case of an arbitrary sequence of Givens transformations, it is not necessary to put a constraint on the value of p for the corollary to be valid. We do remark, however, that if $p \geq 1$ the previous corollary is in some sense too strict. We have a larger structured rank block satisfying structured rank condition.*

Example 9.11 The theorems above can be extended to more general forms of rank structures, for example, making them applicable to more hybrid structured rank matrices. For example, the reader can try to investigate the structure after performing a sequence of Givens transformations on one of the following matrices:

$$
S_1 = \begin{bmatrix} \boxtimes & & & & \\ \boxtimes & \boxtimes & & & \\ \boxtimes & \boxtimes & \boxtimes & & \\ \boxtimes & \boxtimes & \boxtimes & \boxtimes & \\ \boxtimes & \boxtimes & \boxtimes & \boxtimes & \boxtimes \end{bmatrix} + \begin{bmatrix} \boxtimes & & & & \\ \boxtimes & \boxtimes & & & \\ 0 & 0 & \boxtimes & & \\ 0 & 0 & \boxtimes & \boxtimes & \\ 0 & 0 & \boxtimes & \boxtimes & \boxtimes \end{bmatrix},
$$

$$
S_2 = \begin{bmatrix} \boxtimes & & & & \\ \boxtimes & \boxtimes & & & \\ \boxtimes & \boxtimes & \boxtimes & & \\ 0 & 0 & 0 & \boxtimes & \\ 0 & 0 & 0 & \boxtimes & \boxtimes \end{bmatrix} + \begin{bmatrix} \boxtimes & & & & \\ \boxtimes & \boxtimes & & & \\ 0 & 0 & \boxtimes & & \\ 0 & 0 & \boxtimes & \boxtimes & \\ 0 & 0 & \boxtimes & \boxtimes & \boxtimes \end{bmatrix},
$$

$$
S_3 = \begin{bmatrix} \boxtimes & & & & \\ \boxtimes & \boxtimes & & & \\ \boxtimes & \boxtimes & \boxtimes & & \\ \boxtimes & \boxtimes & \boxtimes & \boxtimes & \\ \boxtimes & \boxtimes & \boxtimes & \boxtimes & \boxtimes \end{bmatrix} + \begin{bmatrix} \boxtimes & & & & \\ \boxtimes & 0 & & & \\ \boxtimes & 0 & 0 & & \\ \boxtimes & 0 & 0 & \boxtimes & \\ \boxtimes & 0 & 0 & \boxtimes & \boxtimes \end{bmatrix}.
$$

A similar conclusion as in the previous cases holds. The rank structure shifts down one position in the lower triangular part, and it increases with 1 in the upper triangular part. ∎

In Theorem 9.8 and in the following corollary, we investigated only the effect of an arbitrary sequence of Givens transformations. If we want to compute a *QR*-factorization, however, we want to create zeros in the original matrix to bring the matrix to upper triangular form. Reconsider now the splitting of the lower triangular {2}-semiseparable matrix into the sum of two lower triangular {1}-semiseparable matrices. Performing a sequence of rank-annihilating Givens transformations on one of the two lower triangular {1}-semiseparable matrices corresponds, in fact, to reducing the rank in the lower triangular part. Let us define a rank-decreasing sequence of Givens transformations as follows.

Definition 9.12. *A sequence of Givens transformations from bottom to top is called a rank-decreasing sequence of Givens transformations, if the performance of these Givens transformations on the matrix causes the rank in the lower part of the matrix to decrease by at least one.*

The term rank-decreasing is, in fact, an extension of the concept rank annihilating. The term rank-annihilating worked on a matrix of semiseparability rank 1, and in fact removed this rank. In some sense this is also a rank-decreasing operation as the rank goes from 1 to 0.

Performing a rank-annihiliting sequence on the first term of a sum of two semiseparable matrices causes the global rank to decrease. Hence, we performed a rank-decreasing transformation on the sum of these two matrices as the rank went from 2 to 1.

In general, when applying such a rank-decreasing sequence of operations, we have the following corollary.

Corollary 9.13. *Suppose we have a structured rank matrix A satisfying the following rank structural constraints (with $p \leq 1$):*

$$\mathrm{r}\left(\Sigma_l^{(p)}; A\right) \leq r \quad and \quad \mathrm{r}\left(\Sigma_u^{(p-1)}; A\right) = 0.$$

Performing a sequence of Givens transformations from bottom to top, thereby annihilating the lower triangular rank structure of one of the terms in the decomposition of the matrix A, means performing a sequence of rank-decreasing Givens transformations leads to a structured rank matrix $\hat{A}$ of the following form:

$$\mathrm{r}\left(\Sigma_l^{(p-1)}; \hat{A}\right) \leq r - 1 \quad and \quad \mathrm{r}\left(\Sigma_u^{(p-2)}; \hat{A}\right) \leq 1.$$

We can say that the upper triangular part has increased in rank, whereas the lower triangular part has decreased in rank. The only constraint we posed on all the structured rank matrices having a p-lower triangular structure was $p \leq 1$. In the following subsection we will somewhat loosen this constraint.

9.1.4   Can the value of p in $\Sigma_l^{(p)}$ be larger than 1?

The main question we will address here is the following. What is the maximum value of p, such that Corollary 9.13 remains valid. The following theorem neatly addresses this question.

Theorem 9.14. *Suppose we have a structured rank matrix A, satisfying the following constraint:*

$$\mathrm{r}\left(\Sigma_l^{(p)}; A\right) \leq r.$$

In case $p \leq r$, performing a rank-decreasing sequence of Givens transformations from bottom to top on the matrix A leads to a transformed matrix $\hat{A}$ satisfying the following constraint:

$$\mathrm{r}\left(\Sigma_l^{(p-1)}; \hat{A}\right) \leq r - 1.$$

Proof. We will prove the statement here for $r = p = 3$. The other cases are completely similar. Suppose we have decomposed our initial matrix A as the sum of three matrices having the lower triangular part including two superdiagonals of rank 1. After having performed a rank-decreasing sequence of Givens transformations on the first term of this summation, we obtain the following situation (The elements marked with ⊠ are includable in the semiseparable structure; the elements × are just arbitrary elements.):

$$
\begin{bmatrix}
⊠ & ⊠ & ⊠ & × & × \\
0 & 0 & 0 & × & × \\
0 & 0 & 0 & 0 & × \\
0 & 0 & 0 & 0 & 0 \\
0 & 0 & 0 & 0 & 0
\end{bmatrix}
+
\begin{bmatrix}
⊠ & ⊠ & ⊠ & × & × \\
⊠ & ⊠ & ⊠ & × & × \\
⊠ & ⊠ & ⊠ & ⊠ & × \\
⊠ & ⊠ & ⊠ & ⊠ & ⊠ \\
⊠ & ⊠ & ⊠ & ⊠ & ⊠
\end{bmatrix}
+
\begin{bmatrix}
⊠ & ⊠ & ⊠ & × & × \\
⊠ & ⊠ & ⊠ & × & × \\
⊠ & ⊠ & ⊠ & ⊠ & × \\
⊠ & ⊠ & ⊠ & ⊠ & ⊠ \\
⊠ & ⊠ & ⊠ & ⊠ & ⊠
\end{bmatrix}
$$

It is not directly clear that the following constraint is satisfied:

$$
\mathrm{r}\left(\Sigma_l^{(2)}; \hat{A}\right) \le 2.
$$

The problem elements are located in the first term of the summation: namely, in row 1, the first two elements. If we could get rid of them, we would satisfy the constraint. Let us rewrite the summation above as follows:

$$
\begin{bmatrix}
⊠ & ⊠ & × & × & × \\
0 & 0 & 0 & × & × \\
0 & 0 & 0 & 0 & × \\
0 & 0 & 0 & 0 & 0 \\
0 & 0 & 0 & 0 & 0
\end{bmatrix}
+
\begin{bmatrix}
⊠ & ⊠ & × & × & × \\
⊠ & ⊠ & ⊠ & × & × \\
⊠ & ⊠ & ⊠ & ⊠ & × \\
⊠ & ⊠ & ⊠ & ⊠ & ⊠ \\
⊠ & ⊠ & ⊠ & ⊠ & ⊠
\end{bmatrix}
+
\begin{bmatrix}
⊠ & ⊠ & × & × & × \\
⊠ & ⊠ & ⊠ & × & × \\
⊠ & ⊠ & ⊠ & ⊠ & × \\
⊠ & ⊠ & ⊠ & ⊠ & ⊠ \\
⊠ & ⊠ & ⊠ & ⊠ & ⊠
\end{bmatrix}.
$$

It is clear that the constraint

$$
\mathrm{r}\left(\Sigma_l^{(2)}; \hat{A}\right) \le 2,
$$

is always satisfied, as the rank of the first two columns, can never exceed 2. □

If one really wants to remove these two elements marked with ⊠ in the proof of the previous theorem, one has to distinguish between two cases. Denote with $[a, b]$ the two elements which can generate the first two columns of the second matrix and denote with $[c, d]$ the two elements which generate the first two columns of the third term in the summation.

- If $[a, b]$ and $[c, d]$ are linearly independent, then one can simply write the remaining elements ⊠ in the first term as a linear combination of the elements $[a, b]$ and $[c, d]$, and hence remove them from the first term by adding the split up parts to terms 2 and 3..

- If the elements $[a, b]$ are linearly dependent, there are two possibilities.

    – At least one of the terms 2 or 3 has the first two columns equal to zero. In this case, one can easily add the elements $\boxtimes$ to this matrix, without changing the semiseparability rank.

    – None of these terms has the first two columns equal to zero. Then one can easily add the first two columns of the second matrix to the first two columns of the third matrix, without destroying the semiseparability rank 1 of this third matrix. One has now created two zero columns in term 2. This can then be solved similarly to the case just discussed.

We have already investigated the effect of sequences of Givens transformations from bottom to top on matrices satisfying a particular p-lower triangular structure. In the next section we investigate what the effect is of such a sequence on an upper triangular semiseparable matrix.

To conclude the part on transformations on the lower triangular part we formulate the following theorem. For simplicity we always assumed in the previous corollaries and theorems that a part above a certain diagonal was equal to zero, but, in fact, we can just omit this constraint and focus to the lower triangular part. This means that the upper triangular part does not need to be equal to zero, it can be dense as well.

Theorem 9.15. *Suppose we have a structured rank matrix A satisfying the following rank structural constraint (with $p \leq r$):*

$$\mathrm{r}\left(\Sigma_l^{(p)}; A\right) \leq r,$$

Performing a sequence of Givens transformations from bottom to top leads to a structured rank matrix $\hat{A}$ of the following form:

$$\mathrm{r}\left(\Sigma_l^{(p-1)}; \hat{A}\right) \leq r - 1,$$

if a rank-decreasing sequence of Givens transformations was performed, and to

$$\mathrm{r}\left(\Sigma_l^{(p-1)}; \hat{A}\right) \leq r,$$

if an arbitrary sequence of Givens transformations was done.

Proof. The proof is straightforward by decomposing again the structured rank part of the matrix. The upper part, which can be dense, can be stored in another term. After performing the transformations, one can just recombine all the terms as we did previously. $\quad\square$

In this subsection we mainly focused on the lower triangular part. In fact, we know exactly how the lower triangular structure behaves under the performance of an ascending sequence of Givens transformations, but the upper part also can obey a structured rank constraint. Let us therefore investigate what the effect of an ascending sequence is on the upper part.

9.1.5   Givens transformations on upper rank 1 structures

Suppose we have a semiseparable matrix. This matrix satisfies also some structural constraints related to the upper triangular part. Performing now a sequence of annihilating Givens transformations on the lower triangular structure, one might wonder what the effect is of these transformations on the structure in the upper part. Let us first see what the effect is of an arbitrary sequence of Givens transformations on an upper triangular semiseparable matrix of semiseparability rank 1. To make the proofs more simple, we distinguish between two cases. First, we deal with the case in which the upper triangular part is generator representable; secondly, we deal with the case in which the upper part is of general semiseparable form.

The generator representable case

We start the simple case by working on an upper triangular generator representable semiseparable matrix S.

Theorem 9.16. *Suppose S is an upper triangular semiseparable matrix of generator representable form. A sequence of arbitrary $n-1$ Givens transformations G_i^T, acting on row $i-1$ and i, is performed on the matrix S:*

$$G_2^T \ldots G_n^T S = \hat{S}.$$

The matrix $\hat{S}$ will have the upper triangular part, including the first subdiagonal of $\{2\}$-semiseparable form.

This means that the rank structure has shifted down one position, and the rank of the upper part has increased by 1.

Proof. We assume the upper triangular part to be of generator representable form. This gives us an intuitive proof of the fact that the rank structure shifts down one position but increases its rank by one. As the upper triangular part is of generator representable semiseparable form of semiseparability rank 1, we can easily modify the subdiagonal elements to incorporate them in the rank structure of the upper triangular part. This is the procedure we will follow to illustrate that the upper triangular part will increase in rank, but will incorporate the subdiagonal.
Assume our matrix S, written as the sum of a matrix for which the last subdiagonal element can be incorporated in the structure and a matrix having only this subdiagonal element different from zero:

$$
S = \begin{bmatrix}
\boxtimes & \boxtimes & \boxtimes & \boxtimes & \boxtimes \\
 & \boxtimes & \boxtimes & \boxtimes & \boxtimes \\
 & & \boxtimes & \boxtimes & \boxtimes \\
 & & & \boxtimes & \boxtimes \\
 & & & \boxtimes & \boxtimes
\end{bmatrix}
+
\begin{bmatrix}
0 & & & & \\
 & 0 & & & \\
 & & 0 & & \\
 & & & 0 & \\
 & & & + & 0
\end{bmatrix}.
$$

Performing the first Givens transformation on the sum of two matrices of the following form, using thereby the same notation as in the previous theorems:

$$
G_5^T S = \begin{bmatrix} \boxtimes & \boxtimes & \boxtimes & \boxtimes & \boxtimes \\ & \boxtimes & \boxtimes & \boxtimes & \boxtimes \\ & & \boxtimes & \boxtimes & \boxtimes \\ & & & \boxtimes & \boxtimes \\ & & & \boxtimes & \boxtimes \end{bmatrix} + \begin{bmatrix} 0 & & & & \\ & 0 & & & \\ & & 0 & & \\ & & & \boxplus & 0 \\ & & & \boxplus & 0 \end{bmatrix}.
$$

Before performing the next Givens transformation, we rewrite the matrix $G_5^T S$ as follows (we can immediately incorporate the changed subdiagonal element of the second matrix into its semiseparable structure):

$$
G_5^T S = \begin{bmatrix} \boxtimes & \boxtimes & \boxtimes & \boxtimes & \boxtimes \\ & \boxtimes & \boxtimes & \boxtimes & \boxtimes \\ & & \boxtimes & \boxtimes & \boxtimes \\ & & \boxtimes & \boxtimes & \boxtimes \\ & & & \boxtimes & \boxtimes \end{bmatrix} + \begin{bmatrix} 0 & & & & \\ & 0 & & & \\ & & 0 & & \\ & & \boxplus & \boxplus & 0 \\ & & & \boxplus & 0 \end{bmatrix}.
$$

Performing transformation G_4^T leads to the following equations.

$$
G_4^T G_5^T S = \begin{bmatrix} \boxtimes & \boxtimes & \boxtimes & \boxtimes & \boxtimes \\ & \boxtimes & \boxtimes & \boxtimes & \boxtimes \\ & & \boxtimes & \boxtimes & \boxtimes \\ & & \boxtimes & \boxtimes & \boxtimes \\ & & & \boxtimes & \boxtimes \end{bmatrix} + \begin{bmatrix} 0 & & & & \\ & 0 & & & \\ & & \boxplus & \boxplus & 0 \\ & & \boxplus & \boxplus & 0 \\ & & & \boxplus & 0 \end{bmatrix}.
$$

This process can be continued and completion of this sequence of transformations gives us the following structure of the matrices.

$$
G_2^T G_3^T G_4^T G_5^T S = \begin{bmatrix} \boxtimes & \boxtimes & \boxtimes & \boxtimes & \boxtimes \\ \boxtimes & \boxtimes & \boxtimes & \boxtimes & \boxtimes \\ & \boxtimes & \boxtimes & \boxtimes & \boxtimes \\ & & & \boxtimes & \boxtimes \\ & & & \boxtimes & \boxtimes \end{bmatrix} + \begin{bmatrix} \boxplus & \boxplus & \boxplus & \boxplus & 0 \\ \boxplus & \boxplus & \boxplus & \boxplus & 0 \\ & \boxplus & \boxplus & \boxplus & 0 \\ & & & \boxplus & \boxplus & 0 \\ & & & & \boxplus & 0 \end{bmatrix}.
$$

A summation of both these matrices shows that the rank of the upper triangular part is increased by 1, but in fact the structure of this new rank part has also grown. This structure has increased by incorporating the subdiagonal.   □

The theorem, as formulated here, is of course not valid for general semisepara-ble matrices, because not every semiseparable matrix is of generator representable semiseparable form. The general semiseparable case needs a slightly different ap-proach.

The semiseparable case

We will investigate in the next theorem what happens if the semiseparable matrix is not of generator representable semiseparable form. We see that the theorem above remains valid, but we need a different proof, which will also provide us with more information on the structure if the matrix is not of generator representable form.

Theorem 9.17. *Suppose S is an upper triangular semiseparable matrix. A sequence of arbitrary $n - 1$ Givens transformations G_i^T, acting on row $i - 1$ and i, is performed on the matrix S:*

$$G_2^T \ldots G_n^T S = \hat{S}.$$

The matrix $\hat{S}$ will have the upper triangular part, including the first subdiagonal of $\{2\}$-semiseparable form.

Proof. We know from Chapter 1 that, if an upper triangular semiseparable matrix is not of generator representable semiseparable form, it can always be written as a block diagonal matrix for which all the blocks are of generator representable form. For simplicity we assume in our example that we have a 6×6 matrix, consisting of two blocks on the diagonal. This illustrates the general case. Our matrix S is of the following form:

$$S = \begin{bmatrix} \boxtimes & \boxtimes & \boxtimes & & & \\ & \boxtimes & \boxtimes & & & \\ & & \boxtimes & & & \\ & & & \boxtimes & \boxtimes & \boxtimes \\ & & & & \boxtimes & \boxtimes \\ & & & & & \boxtimes \end{bmatrix}$$

The upper left 3×3 block and the lower right 3×3 block are both assumed to be of generator representable semiseparable form.

Hence, after applying the Givens transformations G_6^T and G_5^T, we know that we have the following structure (similarly as in the proof of Theorem 9.16:

$$G_5^T G_6^T S = \begin{bmatrix} \boxtimes & \boxtimes & \boxtimes & & & \\ & \boxtimes & \boxtimes & & & \\ & & \boxtimes & & & \\ & & & \boxtimes & \boxtimes & \boxtimes \\ & & & \boxtimes & \boxtimes & \boxtimes \\ & & & & \boxtimes & \boxtimes \end{bmatrix} + \begin{bmatrix} 0 & & & & & \\ & 0 & & & & \\ & & 0 & & & \\ & & & \boxplus & \boxplus & 0 \\ & & & \boxplus & \boxplus & 0 \\ & & & & \boxplus & 0 \end{bmatrix}.$$

Unfortunately, in general, it is not possible anymore to change the subdiagonal element $(4, 3)$, to include it in the rank structure of the first term, without changing its rank. In general (if all surrounding elements $\boxtimes$ differ from zero), including this element in the structure, the rank increases. To overcome this problem, we extract row 4 out of both terms and place this row in a new matrix. This gives us the following equation for the matrix $G_5^T G_6^T S$.

$$\begin{bmatrix} \boxtimes & \boxtimes & \boxtimes & & & \\ & \boxtimes & \boxtimes & & & \\ & & \boxtimes & & & \\ & & & 0 & 0 & 0 \\ & & & \boxtimes & \boxtimes & \boxtimes \\ & & & & \boxtimes & \boxtimes \end{bmatrix} + \begin{bmatrix} 0 & & & & & \\ & 0 & & & & \\ & & 0 & & & \\ & & & 0 & 0 & 0 \\ & & & \boxplus & \boxplus & \boxplus \\ & & & & \boxplus & \boxplus \end{bmatrix} + \begin{bmatrix} 0 & & & & & \\ & 0 & & & & \\ & & 0 & & & \\ & & & 0 & \boxdot & \boxdot \\ & & & & 0 & 0 \\ & & & & & 0 \end{bmatrix}$$

The elements $\boxdot$ denote the summation of the elements $\boxplus$ and $\boxtimes$. They are put in a box, as they will make up a part of the matrix of semiseparable form (more precisely, in this case, of rank 1). Performing transformation G_4^T leads us to the following equation.

$$\begin{bmatrix} \boxtimes & \boxtimes & \boxtimes & & & \\ & \boxtimes & \boxtimes & & & \\ & & \boxtimes & & & \\ & & \boxtimes & 0 & 0 & 0 \\ & & & \boxtimes & \boxtimes & \boxtimes \\ & & & & \boxtimes & \boxtimes \end{bmatrix} + \begin{bmatrix} 0 & & & & & \\ & 0 & & & & \\ & & 0 & 0 & 0 & 0 \\ & & 0 & 0 & 0 & 0 \\ & & & \boxplus & \boxplus & \boxplus \\ & & & & \boxplus & \boxplus \end{bmatrix} + \begin{bmatrix} 0 & & & & & \\ & 0 & & & & \\ & & 0 & \boxdot & \boxdot & \boxdot \\ & & & \boxdot & \boxdot & \boxdot \\ & & & & 0 & 0 \\ & & & & & 0 \end{bmatrix}$$

We can now continue working on the first two terms of the summation, changing again the subdiagonal elements of the first term and adding this to the second term. In this way we get, before performing transformation G_3^T:

$$\begin{bmatrix} \boxtimes & \boxtimes & \boxtimes & & & \\ & \boxtimes & \boxtimes & & & \\ & \boxtimes & \boxtimes & & & \\ & & \boxtimes & 0 & 0 & 0 \\ & & & \boxtimes & \boxtimes & \boxtimes \\ & & & & \boxtimes & \boxtimes \end{bmatrix} + \begin{bmatrix} 0 & & & & & \\ & 0 & & & & \\ & \boxplus & 0 & & & \\ & & & 0 & 0 & 0 \\ & & & \boxplus & \boxplus & \boxplus \\ & & & & \boxplus & \boxplus \end{bmatrix} + \begin{bmatrix} 0 & & & & & \\ & 0 & & & & \\ & & 0 & \boxdot & \boxdot & \boxdot \\ & & & \boxdot & \boxdot & \boxdot \\ & & & & 0 & 0 \\ & & & & & 0 \end{bmatrix}$$

Performing transformation G_3^T hence gives us:

$$\begin{bmatrix} \boxtimes & \boxtimes & \boxtimes & & & \\ & \boxtimes & \boxtimes & & & \\ & \boxtimes & \boxtimes & & & \\ & & \boxtimes & 0 & 0 & 0 \\ & & & \boxtimes & \boxtimes & \boxtimes \\ & & & & \boxtimes & \boxtimes \end{bmatrix} + \begin{bmatrix} 0 & & & & & \\ & \boxplus & & & & \\ & \boxplus & 0 & & & \\ & & & 0 & 0 & 0 \\ & & & \boxplus & \boxplus & \boxplus \\ & & & & \boxplus & \boxplus \end{bmatrix} + \begin{bmatrix} 0 & & & & & \\ & 0 & & \boxdot & \boxdot & \boxdot \\ & & 0 & \boxdot & \boxdot & \boxdot \\ & & & \boxdot & \boxdot & \boxdot \\ & & & & 0 & 0 \\ & & & & & 0 \end{bmatrix}$$

Continuing now, by changing the first subdiagonal element in the first two terms, and then performing transformation G_2^T gives us:

$$\begin{bmatrix} \boxtimes & \boxtimes & \boxtimes & & & \\ \boxtimes & \boxtimes & \boxtimes & & & \\ & \boxtimes & \boxtimes & & & \\ & & \boxtimes & 0 & 0 & 0 \\ & & & \boxtimes & \boxtimes & \boxtimes \\ & & & & \boxtimes & \boxtimes \end{bmatrix} + \begin{bmatrix} \boxplus & \boxplus & & & & \\ \boxplus & \boxplus & & & & \\ & \boxplus & 0 & & & \\ & & & 0 & 0 & 0 \\ & & & \boxplus & \boxplus & \boxplus \\ & & & & \boxplus & \boxplus \end{bmatrix} + \begin{bmatrix} 0 & & & \boxdot & \boxdot & \boxdot \\ & 0 & & \boxdot & \boxdot & \boxdot \\ & & 0 & \boxdot & \boxdot & \boxdot \\ & & & \boxdot & \boxdot & \boxdot \\ & & & & 0 & 0 \\ & & & & & 0 \end{bmatrix}$$

Recombining these matrices we can clearly see that the upper triangular part, including the first subdiagonal, is of {2}-semiseparable form.

Moreover, we would like to remark that the original zero block in the matrix S has increased in rank by 1 and has also included one row extra in its structure (it shifted down one position). The initial matrix had the upper right 3×3 block of rank 0 and the resulting matrix has the upper left 4×3 block of rank 1.

More general structures with more diagonal blocks can be considered similarly; every time the end of a diagonal block is encountered, the top row of the completed block can be extracted out of the first two matrices and can be stored in the third matrix. The reader can verify that adding these rows to the last matrix does not destroy the rank structure of this last matrix. □

Note 9.18.

- *Similarly as in the previous cases we can say that the upper triangular structure has shifted down one position and has increased its rank by 1.*

- *Note that the previous remark also seems to be true for lower rank blocks in the upper triangular part. In the proof the block of rank 0 added the subdiagonal to its structure (shifted down one position) but increased its rank by 1.*

- *Again this theorem can be generalized to more hybrid forms of rank structures, e.g., block upper triangular structures.*

We can write this in a more general structured rank context as follows.

Corollary 9.19. *Suppose we have a structured rank matrix A satisfying the following rank structural constraints:*

$$\mathrm{r}\left(\Sigma_l^{(q+1)}; A\right) = 0 \quad and \quad \mathrm{r}\left(\Sigma_u^{(q)}; A\right) \leq 1.$$

Performing a sequence of arbitrary Givens transformations from bottom to top leads to a structured rank matrix $\hat{A}$ of the following form:

$$\mathrm{r}\left(\Sigma_l^{(q)}; \hat{A}\right) = 0 \quad and \quad \mathrm{r}\left(\Sigma_u^{(q-1)}; \hat{A}\right) \leq 2.$$

We proved now a theorem for an upper triangular rank structure equal to 1. This theorem can easily be generalized to higher order rank structures.

9.1.6 Givens transformations on upper rank structures

Generalizing the order 1 case to higher order upper triangular rank structures leads to the following theorem.

Theorem 9.20. *Suppose S is an upper triangular $\{p\}$-semiseparable matrix. A sequence of arbitrary $n-1$ Givens transformations G_i^T, acting on row $i-1$ and i, is performed on the matrix S:*

$$G_2^T \ldots G_n^T S = \hat{S}.$$

The matrix $\hat{S}$ will have the upper triangular part, including the first subdiagonal of $\{p+1\}$-semiseparable form.

Proof. The proof proceeds as follows, and the details are left to the reader. First one decomposes the higher order structured rank matrix into a sum of matrices having semiseparability rank 1. Then one should apply the sequence of Givens transformations from bottom to top on all these matrices separately. The changed subdiagonal elements should be combined into one matrix and the extracted rows in the case of blocks should all be put in another separate matrix. Then combining all the matrices leads to the desired result. □

This leads to the following corollary in terms of structured rank matrices.

Corollary 9.21. *Suppose we have a structured rank matrix A satisfying the following rank structural constraints:*

$$\mathrm{r}\left(\Sigma_l^{(q+1)}; A\right) = 0 \quad and \quad \mathrm{r}\left(\Sigma_u^{(q)}; A\right) \le r.$$

Performing a sequence of arbitrary Givens transformations from bottom to top leads to a structured rank matrix $\hat{A}$ of the following form:

$$\mathrm{r}\left(\Sigma_l^{(q+1)}; \hat{A}\right) = 0 \quad and \quad \mathrm{r}\left(\Sigma_u^{(q-1)}; \hat{A}\right) \le r+1.$$

Similarly, as at the end of the lower triangular cases, we remark that the lower triangular part does not influence the upper triangular part. We formulate this again as a theorem. This means that the lower triangular part can also be dense.

Theorem 9.22. *Suppose we have a structured rank matrix A satisfying the following rank structural constraint:*

$$\mathrm{r}\left(\Sigma_u^{(q)}; A\right) \le r.$$

Performing a sequence of arbitrary Givens transformations from bottom to top leads to a structured rank matrix $\hat{A}$ of the following form:

$$\mathrm{r}\left(\Sigma_u^{(q-1)}; \hat{A}\right) \le r+1.$$

To conclude this subsection we would like to remark briefly that in this subsection we did not speak about rank-annihilating or rank-decreasing sequences. This is because in general, a sequence from bottom to top to annihilate the upper rank structure does not exist. Later on it will become clear that one can annihilate the upper rank structure, but then one needs to perform a sequence from top to bottom.

In the next subsection, we combine all these results for upper and lower structures and observe the following structural theorem.

9.1.7 Givens transformations on rank structures

The following theorem combines the results of the previous sections and holds the most important issues related to the rank structure of a matrix after performing a sequence of Givens transformations from bottom to top.

Theorem 9.23. *Suppose we have a structured rank matrix A satisfying the following rank structural constraints:*

$$\mathrm{r}\left(\Sigma_l^{(p)}; A\right) \leq r_1 \quad and \quad \mathrm{r}\left(\Sigma_u^{(q)}; A\right) \leq r_2.$$

Performing a sequence of arbitrary Givens transformations from bottom to top leads to a structured rank matrix $\hat{A}$ of the following form:

$$\mathrm{r}\left(\Sigma_l^{(p-1)}; \hat{A}\right) \leq r_1 \quad and \quad \mathrm{r}\left(\Sigma_u^{(q-1)}; \hat{A}\right) \leq r_2 + 1.$$

Performing a sequence of rank-decreasing Givens transformations (in case $p \leq r_1$) from bottom to top leads to a structured rank matrix $\hat{A}$ of the following form:

$$\mathrm{r}\left(\Sigma_l^{(q-1)}; \hat{A}\right) \leq r_1 - 1 \quad and \quad \mathrm{r}\left(\Sigma_u^{(q-1)}; \hat{A}\right) \leq r_2 + 1.$$

This theorem is important in the sense that it also provides information concerning intertwined rank structures. For example, a $\{2, 5\}$-semiseparable matrix has an interaction between the lower structured rank and the upper structured rank; this theorem neatly shows the rank structure of the matrix after a sequence of Givens transformations from bottom to top has been performed.

Moreover these effects of a sequence of Givens transformations from top to bottom are also valid for different types of structured blocks inside the matrix! This can be seen when studying the following example.

Example 9.24 The reader can investigate what the rank structure will be of the following upper triangular $\{2\}$-semiseparable matrix:

$$
\begin{bmatrix}
\boxtimes & \boxtimes & 0 & 0 & \boxtimes & \boxtimes \\
 & \boxtimes & 0 & 0 & \boxtimes & \boxtimes \\
 & & 0 & 0 & \boxtimes & \boxtimes \\
 & & & 0 & \boxtimes & \boxtimes \\
 & & & & \boxtimes & \boxtimes \\
 & & & & & \boxtimes
\end{bmatrix}
+
\begin{bmatrix}
\boxplus & \boxplus & \boxplus & \boxplus & \boxplus & \boxplus \\
 & 0 & 0 & 0 & 0 & 0 \\
 & & 0 & 0 & 0 & 0 \\
 & & & \boxplus & \boxplus & \boxplus \\
 & & & & \boxplus & \boxplus \\
 & & & & & \boxplus
\end{bmatrix}.
$$

Before continuing our analysis, we need to pay attention to a special type of Givens transformations acting on the upper triangular part: for example, a lower Hessenberg matrix, for which the ascending sequence of Givens transformations creates zeros in the superdiagonal. It is clear that this sequence of transformations cannot always be performed. For example, for a lower triangular matrix (with nonzero diagonal elements), it is not possible to annihilate the diagonal by performing an ascending sequence of Givens transformations.

Suppose we have a lower Hessenberg matrix H, and $\hat{H}$ is the matrix after having annihilated the diagonal elements:

$$
H = \begin{bmatrix}
\boxtimes & \boxtimes & & & \\
\boxtimes & \boxtimes & \boxtimes & & \\
\boxtimes & \boxtimes & \boxtimes & \boxtimes & \\
\boxtimes & \boxtimes & \boxtimes & \boxtimes & \boxtimes \\
\boxtimes & \boxtimes & \boxtimes & \boxtimes & \boxtimes
\end{bmatrix}
\quad \text{and} \quad
\hat{H} = \begin{bmatrix}
\boxtimes & & & & \\
\boxtimes & \boxtimes & & & \\
\boxtimes & \boxtimes & \boxtimes & & \\
\boxtimes & \boxtimes & \boxtimes & \boxtimes & \\
\boxtimes & \boxtimes & \boxtimes & \boxtimes & \boxtimes
\end{bmatrix}.
$$

In fact we see that the upper zero structure of the matrix H has expanded in $\hat{H}$. Extra zeros are introduced in the superdiagonal position, thereby expanding the zero structured rank block. Hence, the following definition:

Definition 9.25. *A sequence of Givens transformations from bottom to top is called a rank expanding sequence of Givens transformations, if performing these Givens transformations on the matrix causes the zero structured rank block in the upper triangular part of the matrix to expand.* [9]

Hence we can conclude this section with the following general theorem covering all, currently known situations. This theorem is slightly more precise than the preceding theorem.

Theorem 9.26 (Ascending sequences of Givens transformations). *Suppose we have a structured rank matrix A satisfying the following rank structural constraints:*

$$
\mathrm{r}\left(\Sigma_l^{(p)}; A\right) \leq r_1 \quad \text{and} \quad \mathrm{r}\left(\Sigma_u^{(q)}; A\right) \leq r_2.
$$

Performing a sequence of arbitrary Givens transformations from bottom to top leads to a structured rank matrix $\hat{A}$ of the following form:

$$
\mathrm{r}\left(\Sigma_l^{(p-1)}; \hat{A}\right) \leq r_1 \quad \text{and} \quad \mathrm{r}\left(\Sigma_u^{(q-1)}; \hat{A}\right) \leq r_2 + 1.
$$

Performing a sequence of rank-decreasing Givens transformations (in case $p \leq r_1$) from bottom to top leads to a structured rank matrix $\hat{A}$ of the following form:

$$
\mathrm{r}\left(\Sigma_l^{(p-1)}; \hat{A}\right) \leq r_1 - 1 \quad \text{and} \quad \mathrm{r}\left(\Sigma_u^{(q-1)}; \hat{A}\right) \leq r_2 + 1.
$$

---

[9]We restrict ourselves to blocks having rank zero. Later on we will see that in more general cases of higher order ranks, this kind of transformation also exists to expand the higher order rank blocks.

In the case $r_2 = 0$ and it is possible to perform a rank-expanding sequence of Givens transformations[10], we obtain a structured rank matrix $\hat{A}$ of the following form:

$$\mathrm{r}\left(\Sigma_l^{(p-1)}; \hat{A}\right) \leq r_1 \quad and \quad \mathrm{r}\left(\Sigma_u^{(q-1)}; \hat{A}\right) = 0.$$

Note 9.27. *One might say that performing a sequence of Givens transformations from bottom to top on a structured rank matrix has the following effect on this matrix:*

- *The structure of the lower part shifts down one position.*

  - *The rank of the lower part stays the same in the case of an arbitrary sequence of Givens transformations.*

  - *The rank decreases by one in the case of a rank-decreasing pattern of Givens transformations.*

- *The structure of the upper part incorporates an extra subdiagonal (the diagonal which was lost by the lower part).*

  - *The rank of this structure increases by 1 in the case of an arbitrary sequence of Givens transformations.*

  - *The rank structure stays equal to zero in the case the lower parts structured rank was equal to zero and a sequence of rank-expanding transformations was performed.*

We did not mention it explicitly, but a rank-decreasing and a rank-expanding are not exclusive. This means that a sequence of Givens transformations can be rank-expanding as well as rank-decreasing. This is shown in a forthcoming example (Example 9.34).

In this section we investigated what the effect was of a sequence of Givens transformations from bottom to top on a structured rank matrix. Moreover we showed that we can also annihilate terms in the structured rank decomposition of these matrices. The annihilation of such a term results in a rank reduction of the lower structured rank part. This is very similar to computing the QR-decomposition of a semiseparable matrix as discussed in Part II of this book, where we reduced the rank from 1 to 0.

In the following section we will show that, in general, more sequences of Givens transformations from bottom to top are needed for transforming the lower triangular part of the matrix to rank 0. Unfortunately, these sequences can create nonzero subdiagonals; these subdiagonals will be removed by performing sequences of Givens transformations from top to bottom.

Before discussing the computation of the QR-factorization we will briefly show the effect of sequences of Givens transformations on matrices if we perform them

---

[10]More information on the existence of such a sequence of transformations will be given in Section 9.4.

from top to bottom and also from left to right and right to left (if working on the right-hand side of the matrix). We present the results related to rank-decreasing and rank-expanding sequences of transformations similarly as we did in this section with regard to the sequences from bottom to top.

9.1.8   Other directions of sequences of Givens transformations

In the previous part of this section, attention was paid to the performance of a sequence of Givens transformations on a structured rank matrix, where the direction of the transformations was from bottom to top. Similarly, we can deduce theorems, for transformations acting on the left of the matrix but going from top to bottom. Moreover, if we are interested in performing actions on the matrix from the right, we can construct sequences of Givens transformations from right to left and from left to right.

We will not go into detail as done in the previous section, but we will present similar results, in a general structured rank context. The proofs proceed completely similarly. Again we formulate these theorems for p-upper(lower) triangular ranks; similar results hold for more general rank structured blocks.

A sequence of Givens transformations from top to bottom

Corollary 9.28. *Suppose we have a structured rank matrix A satisfying the following rank structural constraints:*

$$\mathrm{r}\left(\Sigma_l^{(p)}; A\right) \leq r_1 \quad and \quad \mathrm{r}\left(\Sigma_u^{(q)}; A\right) \leq r_2.$$

Performing a sequence of arbitrary Givens transformations from top to bottom leads to a structured rank matrix $\hat{A}$ of the following form:

$$\mathrm{r}\left(\Sigma_l^{(p+1)}; \hat{A}\right) \leq r_1 + 1 \quad and \quad \mathrm{r}\left(\Sigma_u^{(q+1)}; \hat{A}\right) \leq r_2.$$

Performing a sequence of rank-decreasing Givens transformations (in case $-q \leq r_2$) (acting on the upper part) from top to bottom leads to a structured rank matrix $\hat{A}$ of the following form:

$$\mathrm{r}\left(\Sigma_l^{(p+1)}; \hat{A}\right) \leq r_1 + 1 \quad and \quad \mathrm{r}\left(\Sigma_u^{(q+1)}; \hat{A}\right) \leq r_2 - 1.$$

If $r_1 = 0$ and it is possible to perform a rank-expanding sequence of Givens transformations (acting on the lower part), we obtain a structured rank matrix $\hat{A}$ of the following form:

$$\mathrm{r}\left(\Sigma_l^{(p+1)}; \hat{A}\right) = 0 \quad and \quad \mathrm{r}\left(\Sigma_u^{(q+1)}; \hat{A}\right) \leq r_2.$$

To better understand why $-q \leq r_2$ (and not $q \leq r_2$), one can take the following relation into consideration. If $\mathrm{r}\left(\Sigma_u^{(q)}; A\right) \leq r$, then we have that $\mathrm{r}\left(\Sigma_l^{(-q)}; A^T\right) \leq r$.

Note 9.29. *One can say that performing a sequence of Givens transformations from top to bottom on a structured rank matrix has the following effect on this matrix:*

- *The structure of the lower part shifts up one position.*

    - *The rank of this structure increases by 1 in the case of an arbitrary sequence of Givens transformations.*
    - *The rank structure stays equal to zero if the lower parts structured rank was equal to zero and a sequence of rank-expanding transformations was performed.*

- *The structure of the upper part shifts up one position.*

    - *The rank of the upper part stays the same in the case of an arbitrary sequence of Givens transformations.*
    - *The rank decreases by one in the case of a rank-decreasing pattern of Givens transformations.*

For example, a rank-expanding sequence of Givens transformations from top to bottom can be applied for bringing a Hessenberg matrix to upper triangular form.

A sequence of Givens transformations from right to left

Remark that this sequence of Givens transformations is performed on the right side of the structured rank matrix.

Corollary 9.30. *Suppose we have a structured rank matrix A satisfying the following rank structural constraints:*

$$\mathrm{r}\left(\Sigma_l^{(p)}; A\right) \leq r_1 \quad and \quad \mathrm{r}\left(\Sigma_u^{(q)}; A\right) \leq r_2.$$

Performing a sequence of arbitrary Givens transformations from right to left leads to a structured rank matrix $\hat{A}$ of the following form:

$$\mathrm{r}\left(\Sigma_l^{(p+1)}; \hat{A}\right) \leq r_1 + 1 \quad and \quad \mathrm{r}\left(\Sigma_u^{(q+1)}; \hat{A}\right) \leq r_2.$$

Performing a sequence of rank-decreasing Givens transformations (in case $-q \leq r_2$) (acting on the right part) from right to left leads to a structured rank matrix $\hat{A}$ of the following form:

$$\mathrm{r}\left(\Sigma_l^{(p+1)}; \hat{A}\right) \leq r_1 + 1 \quad and \quad \mathrm{r}\left(\Sigma_u^{(q+1)}; \hat{A}\right) \leq r_2 - 1.$$

If $r_1 = 0$ and it is possible to perform a rank-expanding sequence of Givens transformations (acting on the left part), we obtain a structured rank matrix $\hat{A}$ of the following form:

$$\mathrm{r}\left(\Sigma_l^{(p+1)}; \hat{A}\right) = 0 \quad and \quad \mathrm{r}\left(\Sigma_u^{(q+1)}; \hat{A}\right) \leq r_2.$$

Corollary 9.31. *One can say that performing a sequence of Givens transformations from right to left on a structured rank matrix has the following effect on this matrix:*

- *The structure of the left part shifts to the right by one position.*

  - *The rank of this structure increases by 1 in the case of an arbitrary sequence of Givens transformations.*
  - *The rank structure stays equal to zero if the left part's structured rank was equal to zero and a sequence of rank-expanding transformations was performed.*

- *The structure of the right part shifts to the right by one position.*

  - *The rank of the right part stays the same in the case of an arbitrary sequence of Givens transformations.*
  - *The rank decreases by one in the case of a rank-decreasing pattern of Givens transformations.*

A sequence of Givens transformations from left to right

Remark that this sequence of Givens transformations is performed on the right side of the structured rank matrix.

Corollary 9.32. *Suppose we have a structured rank matrix A satisfying the following rank structural constraints:*

$$\mathrm{r}\left(\Sigma_l^{(p)}; A\right) \leq r_1 \quad and \quad \mathrm{r}\left(\Sigma_u^{(q)}; A\right) \leq r_2.$$

Performing a sequence of arbitrary Givens transformations from left to right leads to a structured rank matrix $\hat{A}$ of the following form:

$$\mathrm{r}\left(\Sigma_l^{(p-1)}; \hat{A}\right) \leq r_1 \quad and \quad \mathrm{r}\left(\Sigma_u^{(q-1)}; \hat{A}\right) \leq r_2 + 1.$$

Performing a sequence of rank-decreasing Givens transformations (in case $p \leq r_1$) (acting on the left part) from left to right leads to a structured rank matrix $\hat{A}$ of the following form:

$$\mathrm{r}\left(\Sigma_l^{(p-1)}; \hat{A}\right) \leq r_1 - 1 \quad and \quad \mathrm{r}\left(\Sigma_u^{(q-1)}; \hat{A}\right) \leq r_2 + 1.$$

If $r_2 = 0$ and it is possible to perform a rank-expanding sequence of Givens transformations (acting on the right part), we obtain a structured rank matrix $\hat{A}$ of the following form:

$$\mathrm{r}\left(\Sigma_l^{(p-1)}; \hat{A}\right) \leq r_1 - 1 \quad and \quad \mathrm{r}\left(\Sigma_u^{(q-1)}; \hat{A}\right) = 0.$$

Note 9.33. *One can say that performing a sequence of Givens transformations from left to right on a structured rank matrix has the following effect on this matrix:*

- *The structure of the left part shifts to the left by one position.*

  - *The rank of this part stays the same in the case of an arbitrary sequence of Givens transformations.*

  - *The rank decreases by one in the case of a rank-decreasing pattern of Givens transformations.*

- *The structure of the right part shifts to the left by one position.*

  - *The rank of this structure increases by 1 in the case of an arbitrary sequence of Givens transformations.*

  - *The rank structure stays equal to zero if the right part's structured rank was equal to zero and a sequence of rank-expanding transformations was performed.*

9.1.9 Examples

Before continuing our search for the QR-decomposition of these structured rank matrices, we present some well-known classes, and the effect of a sequence of Givens transformations on them.

Semiseparable matrices

Suppose S is a $\{p, q\}$-semiseparable matrix. This means that S is of the following form.

$$\mathrm{r}\left(\Sigma_l^{(p)}; S\right) \leq p \quad \text{and} \quad \mathrm{r}\left(\Sigma_u^{(-q)}; S\right) \leq q.$$

After performing a rank-decreasing sequence of Givens transformations from bottom to top on the matrix S we obtain that the resulting matrix $\hat{S}$ will satisfy the following rank constraints:

$$\mathrm{r}\left(\Sigma_l^{(p-1)}; \hat{S}\right) \leq p - 1 \quad \text{and} \quad \mathrm{r}\left(\Sigma_u^{(-q-1)}; \hat{S}\right) \leq q + 1.$$

This means that this matrix will again be a semiseparable matrix, but of a different order. The resulting matrix $\hat{S}$ will be a $\{p-1, q+1\}$-semiseparable matrix. For example, applying a sequence of annihilating Givens transformations on a $\{1\}$-semiseparable matrix clearly results in a $\{0, 2\}$-semiseparable matrix. The resulting matrix is of $\{0, 2\}$-semiseparable form, meaning that it is upper triangular; moreover, we also know that the upper triangular part will be of $\{2\}$-semiseparable form, just as we observed in Chapter 5 on the QR-factorization of the easy structured rank matrices.

Applying an arbitrary sequence of Givens transformations does not lead to a semiseparable matrix again. The resulting matrix $\tilde{S}$ will then be of the following structured rank form:

$$\mathrm{r}\left(\Sigma_l^{(p-1)}; \tilde{S}\right) \leq p \quad \text{and} \quad \mathrm{r}\left(\Sigma_u^{(-q-1)}; \tilde{S}\right) \leq q + 1.$$

Band matrices

We mentioned earlier that band matrices can also be considered as structured rank matrices. Consider a band matrix B of the following form:

$$\mathrm{r}\left(\Sigma_l^{(-p)}; B\right) = 0 \quad \text{and} \quad \mathrm{r}\left(\Sigma_u^{(q)}; B\right) = 0.$$

After performing a sequence of arbitrary Givens transformations from bottom to top we obtain:

$$\mathrm{r}\left(\Sigma_l^{(-p-1)}; \hat{B}\right) = 0 \quad \text{and} \quad \mathrm{r}\left(\Sigma_u^{(q-1)}; \hat{B}\right) = 1.$$

Hence we clearly see that the band has shifted down one position, and moreover the upper triangular part, which was zero, has filled up and has now become of rank 1.

Suppose now that we perform on the same band matrix (with $p < 0$) a sequence of descending rank-expanding Givens transformations. This will remove one subdiagonal in the lower part and will fill up one extra superdiagonal with nonzero elements. The resulting band matrix $\tilde{B}$ will be of the following form:

$$\mathrm{r}\left(\Sigma_l^{(-p+1)}; \tilde{B}\right) = 0 \quad \text{and} \quad \mathrm{r}\left(\Sigma_u^{(q+1)}; \tilde{B}\right) = 0.$$

Hence it is clear that the effect of this sequence of Givens transformations has created a new band matrix, namely a $\{-p+1, q+1\}$ one. In fact, the complete band of the matrix is lifted up one position.

Quasiseparable matrices

Suppose we have a quasiseparable matrix A satisfying the following constraints:

$$\mathrm{r}\left(\Sigma_l^{(0)}; A\right) \le p \quad \text{and} \quad \mathrm{r}\left(\Sigma_u^{(0)}; A\right) \le q.$$

This means that the strictly lower triangular part and the strictly upper triangular part satisfy certain rank conditions. Performing a sequence of Givens transformations from bottom to top leads to the following result:

$$\mathrm{r}\left(\Sigma_l^{(-1)}; \hat{A}\right) \le p \quad \text{and} \quad \mathrm{r}\left(\Sigma_u^{(-1)}; \hat{A}\right) \le q + 1.$$

The reader can try to verify this. Note that we do not have to take into account the diagonal; the structured rank conditions as posed on the quasiseparable matrix regulate this themselves.

Extended/decoupled semiseparable matrices

Similarly as we could apply the theorems for semiseparable and band matrices, we can apply them for either extended or decoupled semiseparable matrices. We illustrate the technique for extended semiseparable matrices. Suppose S to be an extended $\{p, q\}$-semiseparable matrix:

$$\mathrm{r}\left(\Sigma_l^{(1)}; S\right) \le p \quad \text{and} \quad \mathrm{r}\left(\Sigma_u^{(-1)}; S\right) \le q.$$

The application of a sequence of Givens transformations from bottom to top leads to the following result:

$$\mathrm{r}\left(\Sigma_l^{(0)}; S\right) \le p \quad \text{and} \quad \mathrm{r}\left(\Sigma_u^{(-2)}; S\right) \le q + 1.$$

Semiseparable plus band matrices

Suppose, for example, we have a semiseparable plus a band matrix $A = S + B$. A wrong approach to investigate the rank structure of the resulting matrix $\hat{A}$, after having performed a sequence of Givens transformations is to decompose both matrices, applying the theorem to both of them, and then adding up the ranks of the different structures. In general, this will lead to constraints on the rank structure, which are correct, but not tight enough. This naive approach gives us the following rank structure for the matrix $\hat{A}$ after having performed a sequence of Givens transformations from bottom to top. Initially we have the following constraints:

$$\mathrm{r}\left(\Sigma_l^{(p)}; S\right) \le p, \qquad \mathrm{r}\left(\Sigma_u^{(-q)}; S\right) \le q,$$
$$\mathrm{r}\left(\Sigma_l^{(-p)}; B\right) = 0, \quad \text{and} \quad \mathrm{r}\left(\Sigma_u^{(q)}; B\right) = 0.$$

After the transformations we get:

$$\mathrm{r}\left(\Sigma_l^{(p-1)}; S\right) \le p, \qquad \mathrm{r}\left(\Sigma_u^{(-q-1)}; S\right) \le q + 1,$$
$$\mathrm{r}\left(\Sigma_l^{(-p-1)}; B\right) = 0, \quad \text{and} \quad \mathrm{r}\left(\Sigma_u^{(q-1)}; B\right) \le 1.$$

Recombining the results now leads us to the observation that the matrix $\hat{A}$ will satisfy the following constraint:

$$\mathrm{r}\left(\Sigma_u^{(q-1)}; \hat{A}\right) \le q + 2.$$

We know however from this chapter that both the matrix B and the matrix S increase in rank, but with the same structure. Hence we know that we should obtain a bound of the form:

$$\mathrm{r}\left(\Sigma_u^{(q-1)}; \hat{A}\right) \le q + 1.$$

The correct approach is to put all constraints together, and then apply the theorem at all constraints at the same time. Our matrix A also satisfies the following constraint:

$$\mathrm{r}\left(\Sigma_u^{(q)}; \hat{A}\right) \le q.$$

The sequence of transformations creates there a structured rank part of the form:

$$\mathrm{r}\left(\Sigma_u^{(q-1)}; \hat{A}\right) \le q + 1,$$

just as we expected.

Moreover we would like to remark that this approach is interesting if one has summations of different structured rank matrices. One can easily decompose these matrices and investigate the structure of the terms separately, and using the approach above, one also gets specific details about the combined structure.

9.1.10   Summary

In fact, one can easily summarize the effect of a sequence of Givens transformations on the structured rank matrix. Let us here summarize in a few bullets what really happens if performing a sequence of transformations.

- The direction of shifting of the rank structures is opposed to the direction of the sequence of Givens transformations. This means, if the Givens ascend, the structures descend. If the Givens go from right to left, the structure goes from left to right, and so on.

- In the case of arbitrary Givens transformations, the part in which one is starting remains of the same rank. The part to which one is going increases in rank. Going from left to right does not change the rank in the left part, but increases by one the rank in the right part.

- Rank-decreasing Givens transformations:

  - One can only perform rank-decreasing Givens transformations in the part where one starts. This means if we go up, with our Givens transformations, we start in the bottom part; hence, we can only perform rank-decreasing operations in the bottom part. If we go from right to left with the Givens transformations, we can only perform rank-decreasing Givens transformations in the right part.

  - The performance of rank-decreasing Givens operations reduces the rank in the part where one starts by one.

  - The performance of rank-decreasing Givens operations has no special effect on the part to which one is going, hence the rank still increases by one.

- Rank-expanding Givens transformation:

  - One can only perform rank-expanding Givens transformations in the part to which one is going. If we go from top to bottom, we can only perform rank-expanding Givens transformations in the bottom part. If we go to the right, we can only perform expanding Givens transformations in the right part.

  - The performance of rank-expanding Givens transformations causes the rank in the part to which one goes to remain the same.

  - Rank-expanding Givens transformations have no special effect on the part where one starts, hence the rank remains the same.

To conclude, we remark that rank-expanding and rank-decreasing Givens transformations can take place at the same time. This is illustrated by the following example and, for example, unitary Hessenberg matrices also satisfy this property.

Example 9.34 Consider the following matrix:

$$
A = \begin{bmatrix}
0 & 1 & 1 & 0 & 0 \\
\frac{1}{\sqrt{2}} & \frac{1}{\sqrt{2}} & \frac{-1}{\sqrt{2}} & \frac{-1}{\sqrt{2}} & 0 \\
\frac{1}{\sqrt{6}} & \frac{1}{\sqrt{6}} & \frac{1}{\sqrt{6}} & \frac{-1}{\sqrt{6}} & -\sqrt{\frac{2}{3}} \\
\frac{1}{\sqrt{6}} & \frac{1}{\sqrt{6}} & \frac{1}{\sqrt{6}} & \frac{2}{\sqrt{6}} & \frac{1}{\sqrt{6}} - \frac{1}{\sqrt{2}} \\
\frac{1}{\sqrt{6}} & \frac{1}{\sqrt{6}} & \frac{1}{\sqrt{6}} & \frac{2}{\sqrt{6}} & \frac{1}{\sqrt{6}} + \frac{1}{\sqrt{2}}
\end{bmatrix}.
$$

Applying the sequence of Givens transformations from bottom to top $G_2^T \dots G_5^T$ on the matrix, where the following Givens transformations are embedded in G_2 up to G_5:

$$
\breve{G}_2 = \begin{bmatrix} 0 & 1 \\ 1 & 0 \end{bmatrix}, \qquad
\breve{G}_3 = \begin{bmatrix} \frac{1}{\sqrt{2}} & \frac{1}{\sqrt{2}} \\ \frac{-1}{\sqrt{2}} & \frac{1}{\sqrt{2}} \end{bmatrix},
$$

$$
\breve{G}_3 = \begin{bmatrix} \frac{1}{\sqrt{3}} & \sqrt{\frac{2}{3}} \\ -\sqrt{\frac{2}{3}} & \frac{1}{\sqrt{3}} \end{bmatrix} \text{ and } \breve{G}_4 = \begin{bmatrix} \frac{1}{\sqrt{2}} & \frac{1}{\sqrt{2}} \\ \frac{-1}{\sqrt{2}} & \frac{1}{\sqrt{2}} \end{bmatrix},
$$

gives us a matrix only having the diagonal and the superdiagonal different from zero. This means that the ascending sequence of Givens transformations decreased the rank in the lower part, and moreover it expanded the part of rank zero in the upper part. Hence the sequence of Givens transformations was both rank-decreasing as well as rank-expanding. ∎

We know now what the effect of a sequence of Givens transformations on a matrix is. In the next section we will combine different sequences to obtain the QR-factorization of a structured rank matrix.

Notes and references

See the notes and references of the next section.

9.2 Making the structured rank matrix upper triangular

In the previous section we discussed the effect of rank-decreasing Givens transformations and of rank-expanding Givens transformations. In the beginning of the book we discussed the QR-factorization of quasiseparable matrices. To compute the QR-factorization of quasiseparable matrices, we first performed a sequence of rank-decreasing transformations from bottom to top, followed by a sequence of rank-expanding transformations, to expand the zero rank structure, such that the matrix became of upper triangular form.

In this section we will discuss the effect of multiple sequences of ascending rank-decreasing Givens transformations, and of multiple sequences of descending rank-expanding Givens transformations. We remark, before continuing, that there are many other patterns of sequences of Givens transformations, to obtain the desired upper triangular structures. These patterns will be discussed in an upcoming section.

Hence, in this section, we will derive the following scheme for annihilating structured rank matrices. First, we will remove the complete rank structure by ascending rank-decreasing Givens transformations; second, we will remove the extra created subdiagonals (if there are any) by descending Givens transformations of rank-expanding form. This means that these Givens transformations will expand the zero structure in the lower triangular part.

9.2.1   Annihilating completely the rank structure

The first step in the computation of the upper triangular matrix R consists of removing the rank structure of the lower part in the matrix. Suppose, in general, that we have a structured rank matrix A satisfying the following structural constraints (with $p \leq r_1$):

$$\mathrm{r}\left(\Sigma_l^{(p)}; A\right) \leq r_1.$$

The performance of a sequence of rank-decreasing Givens transformations from bottom to top leads to a structured rank matrix $A^{(1)}$ of the following form:

$$\mathrm{r}\left(\Sigma_l^{(p-1)}; A^{(1)}\right) \leq r_1 - 1.$$

We remark that after having performed this sequence of transformations, we still have that $p-1 \leq r_1-1$ and hence we can perform a next sequence of transformations.

So, in general, we need r_1 rank-decreasing Givens transformations from bottom to top to obtain a block of rank zero. After having performed r_1 ascending sequences of rank-decreasing Givens transformations, we obtain a matrix $A^{(r_1)}$ of the following form:

$$\mathrm{r}\left(\Sigma_l^{(p-r_1)}; A^{(r_1)}\right) = 0. \tag{9.1}$$

Hence we created the desired zero structure in the bottom part.

9.2.2   Expanding the zero rank structure

The second step in the computation of the upper triangular part is the expansion of the zero part by means of descending sequences of Givens transformations. We only need to expand the zero part until we obtain an upper triangular matrix.

Suppose a structured rank matrix $A = A^{(0)}$ is given, having the following structure (assume hereby $p < 0$, otherwise the matrix is already in upper triangular form)[11]:

$$r\left(\Sigma_l^{(p)}; A\right) = 0.$$

The performance of a descending rank-expanding sequence of Givens transformations leads us to the following matrix.

$$r\left(\Sigma_l^{(p+1)}; A^{(1)}\right) = 0.$$

Hence, to obtain an upper triangular matrix we need to perform $-p$ descending sequences of Givens transformations. This gives us a matrix of the following form:

$$r\left(\Sigma_l^{(0)}; A^{(p)}\right) = 0.$$

This means that the matrix $A^{(p)}$ is in upper triangular form.

9.2.3 Combination of ascending and descending sequences

Let us combine now the results of the previous two subsections, to reduce a structured rank matrix to upper triangular form. In this case we also include the structure of the upper triangular part of the matrix to see how this part behaves under these sequences of ascending and descending transformations.

Unfortunately, we always assumed for a matrix A satisfying the following constraint:

$$r\left(\Sigma_l^{(p)}; A\right) \leq r_1,$$

that p was less or equal to r_1. Does this mean we cannot compute the QR-factorization of matrices for which $p > r_1$ by using this theory?

Let us first take a closer look at these matrices for which $p > r_1$. It turns out that these matrices have to be singular!

Theorem 9.35. *Suppose we have a matrix A and there is an index i, with $1 \leq i \leq n - r_1$, such that*

$$\text{rank}\left[A(i:n, 1:i+r_1)\right] \leq r_1. \tag{9.2}$$

Then the matrix A will be singular.

Proof. The proof consists of straightforward linear algebra arguments. We know that Equation (9.2) is satisfied. Adding the block $A(1:i-1, 1:i+r_1)$ to this

---

[11]If $p < 0$, we can always perform a sequence of rank-expanding Givens transformations to lift up the zero part. More information related to the existence of this rank-expanding sequence of Givens transformations can be found in a forthcoming section (see Section 9.4).

structure can at most increase the rank with $i-1$ because we add $i-1$ rows to the matrix. This gives us

$$\mathrm{rank} \left[\begin{array}{c} A(1:i-1,1:i+r_1) \\ A(i:n,1:i+r_1) \end{array} \right] \leq r_1 + i - 1.$$

Adding the last $n-i-r_1$ columns to this matrix gives us the complete matrix A. The rank can increase at most with $n-i-r_1$, as we add $n-i-r_1$ columns. This gives us for the complete matrix A:

$$\mathrm{rank}(A) \leq r_1 + i - 1 + n - i - r_1 = n - 1.$$

It is clear that our matrix needs to be rank deficient. Hence the matrix is singular.

$\square$

Using this theorem, it is easy to see that structured rank matrices with $p > r_1$ have to be singular.

Corollary 9.36. *Suppose we have a structured rank matrix A satisfying the following constraint:*

$$\mathrm{r} \left(\Sigma_l^{(p)}; A \right) \leq r_1 \quad and \quad \mathrm{r} \left(\Sigma_u^{(q)}; A \right) \leq r_2.$$

If $p > r_1$ or $-q > r_2$, the matrix A will be singular.

Proof. The proof is trivially based on the previous theorem. $\square$

Hence we know now at least that matrices having $p > r_1$ need to be singular. We will make this distinction for computing the QR-factorization. We will first compute the QR-factorization for $p \leq r_1$, then we will solve the problem in the case that $p > r_1$.

The case $p \leq r_1$

Suppose, in general, that we have a structured rank matrix $A = A^{(0)}$, which we want to reduce to upper triangular form. The matrix satisfies the following rank structural constraints (with $p \leq r_1$):

$$\mathrm{r} \left(\Sigma_l^{(p)}; A \right) \leq r_1 \quad and \quad \mathrm{r} \left(\Sigma_u^{(q)}; A \right) \leq r_2.$$

The first part consists of performing r_1 sequences of rank-decreasing Givens transformations from bottom to top. This gives us the following resulting structure for our matrix:

$$\mathrm{r} \left(\Sigma_l^{(p-r_1)}; A^{(r_1)} \right) = 0 \quad and \quad \mathrm{r} \left(\Sigma_u^{(q-r_1)}; A^{(r_1)} \right) \leq r_2 + r_1.$$

Hence we created our zero block in the bottom part. Note that the complete rank of the lower part is now incorporated in the upper triangular structure.

If $(p - r_1) < 0$, we have to eliminate the subdiagonals. We do this by performing $(-p + r_1)$ rank-expanding sequences of Givens transformations from top to bottom, leading to the following structure for the matrix:

$$r\left(\Sigma_l^{(0)}; A^{(2r_1-p)}\right) = 0 \quad \text{and} \quad r\left(\Sigma_u^{(q-p)}; A^{(2r_1-p)}\right) \leq r_2 + r_1.$$

Let us apply these theoretical considerations to some different examples of structured rank matrices. But first we will see how to compute the QR-factorization if $p > r_1$.

The case $p > r_1$

Suppose a structured rank matrix A is given, satisfying the following constraint (with $p > r_1$):

$$r\left(\Sigma_l^{(p)}; A\right) \leq r_1 \quad \text{and} \quad r\left(\Sigma_u^{(q)}; A\right) \leq r_2.$$

It is clear that this matrix naturally also satisfies the following constraints:

$$r\left(\Sigma_l^{(r_1)}; A\right) \leq r_1 \quad \text{and} \quad r\left(\Sigma_u^{(q)}; A\right) \leq r_2.$$

Hence, it is clear that performing r_1 sequences of ascending transformations naturally transforms the matrix to upper triangular form.

So, in general, one needs to perform r_1 ascending sequences of transformations, and if $p \geq r_1$, these sequences are sufficient to bring the matrix to upper triangular form. Let us now consider some examples.

9.2.4 Examples

In this section we will briefly show the rank structure of the R-factor after computing the QR-decomposition in the way described in this section. We list some of the traditional matrices covered in this book. Moreover one can combine the results of the R-factor here with the results on the Q-factor from Theorem 5.3. This leads to a completely characterized QR-factorization.

Semiseparable matrices

Suppose S is a $\{p, q\}$-semiseparable matrix. This means that S is of the following form:

$$r\left(\Sigma_l^{(p)}; S\right) \leq p \quad \text{and} \quad r\left(\Sigma_u^{(-q)}; S\right) \leq q.$$

First p ascending sequences of rank-decreasing Givens transformations are performed. This leads to an upper triangular matrix R:

$$r\left(\Sigma_l^{(0)}; R\right) \leq 0 \quad \text{and} \quad r\left(\Sigma_u^{(-q-p)}; R\right) \leq q + p.$$

This means that our matrix is already in upper triangular form, with the upper triangular part above diagonal $-q-p$ (i.e., the complete upper triangular part) of structured rank at most $q+p$.

Moreover, the resulting matrix is a $\{0, q+p\}$-semiseparable matrix. Hence the R-factor belongs again to the class of semiseparable matrices.

Band matrices

For reducing a band matrix to upper triangular form, we have to remove the subdiagonals by performing as many descending sequences as there are subdiagonals.

Suppose we have the following band matrix:

$$\text{r}\left(\Sigma_l^{(-p)}; B\right) = 0 \quad \text{and} \quad \text{r}\left(\Sigma_u^{(q)}; B\right) = 0.$$

After having performed p sequences of descending rank-expanding Givens transformations for as many as there are subdiagonals, we obtain an upper triangular structured rank matrix R of the following form:

$$\text{r}\left(\Sigma_l^{(0)}; R\right) = 0 \quad \text{and} \quad \text{r}\left(\Sigma_u^{(q+p)}; R\right) = 0.$$

Clearly, as we already knew, our diagonals shift up and we now have a new band matrix, namely, a $\{0, q+p\}$-band matrix. The R-factor is therefore again a band matrix.

Quasiseparable matrices

Suppose we have a quasiseparable matrix A satisfying the following constraints:

$$\text{r}\left(\Sigma_l^{(0)}; A\right) \le p \quad \text{and} \quad \text{r}\left(\Sigma_u^{(0)}; A\right) \le q.$$

After having performed the ascending, and the descending sequences of transformations, we obtain an upper triangular matrix of the form:

$$\text{r}\left(\Sigma_l^{(0)}; R\right) = 0 \quad \text{and} \quad \text{r}\left(\Sigma_u^{(0)}; R\right) = p+q.$$

Clearly, the matrix R is again a quasiseparable matrix, namely, a $\{0, p+q\}$-quasiseparable one.

The reader can try to verify this. Note that we do not have to take into account the diagonal; the structured rank conditions as posed on the quasiseparable matrix regulate this themselves.

Extended/decoupled semiseparable matrices

Suppose S is an extended $\{p, q\}$-semiseparable matrix:

$$\text{r}\left(\Sigma_l^{(1)}; S\right) \le p \quad \text{and} \quad \text{r}\left(\Sigma_u^{(-1)}; S\right) \le q.$$

The resulting QR-factorization gives us for the factor R the following structure:

$$\mathrm{r}\left(\Sigma_l^{(0)}; R\right) = 0 \quad \text{and} \quad \mathrm{r}\left(\Sigma_u^{(-2)}; R\right) \leq p + q.$$

This matrix R also satisfies the following constraints:

$$\mathrm{r}\left(\Sigma_l^{(1)}; R\right) \leq 1 \quad \text{and} \quad \mathrm{r}\left(\Sigma_u^{(-1)}; R\right) \leq p + q,$$

making it an extended $\{1, p + q\}$-semiseparable matrix.

Semiseparable plus band matrices

Even though we formulated our previous results only for rank structures satisfying one constraint for the lower and one constraint for the upper part, we can easily involve more and more difficult constraints for the upper/lower triangular part. A semiseparable plus band matrix is an example of such a matrix with intertwined structures.

Suppose, for example, we have a semiseparable plus a band matrix $A = S + B$, with A a $\{p_1, q_1\}$-semiseparable matrix, and B a $\{p_2, q_2\}$-band matrix. This matrix satisfies the following structural constraints.

$$\mathrm{r}\left(\Sigma_l^{(p_1)}; S\right) \leq p_1, \qquad \mathrm{r}\left(\Sigma_u^{(-q_1)}; S\right) \leq q_1,$$
$$\mathrm{r}\left(\Sigma_l^{(-p_2)}; A\right) \leq p_1 \quad \text{and} \quad \mathrm{r}\left(\Sigma_u^{(q_2)}; A\right) \leq q_1.$$

Performing the transformations for removing the lower rank structure, involving p_1 sequences of ascending Givens transformations, gives us the following result $\tilde{A} = \tilde{S} + \tilde{B}$:

$$\mathrm{r}\left(\Sigma_l^{(0)}; \tilde{S}\right) = 0, \qquad \mathrm{r}\left(\Sigma_u^{(-q_1-p_1)}; \tilde{S}\right) \leq q_1 + p_1,$$
$$\mathrm{r}\left(\Sigma_l^{(-p_2-p_1)}; \tilde{A}\right) = 0 \quad \text{and} \quad \mathrm{r}\left(\Sigma_u^{(q_2-p_1)}; \tilde{A}\right) \leq q_1 + p_1.$$

There are now $p_2 + p_1$ subdiagonals still left; they are removed via descending rank-expanding Givens transformations. The resulting matrix $R = \hat{S} + \hat{B}$ satisfies:

$$\mathrm{r}\left(\Sigma_l^{(0)}; \hat{S}\right) = 0, \qquad \mathrm{r}\left(\Sigma_u^{(-q_1+p_2)}; \hat{S}\right) \leq q_1 + p_1,$$
$$\mathrm{r}\left(\Sigma_l^{(0)}; R\right) = 0 \quad \text{and} \quad \mathrm{r}\left(\Sigma_u^{(q_2+p_2)}; R\right) \leq q_1 + p_1.$$

This matrix still has some of the semiseparable structure expanding underneath the transformed band. This is important information which one can take into consideration when trying to tackle these problems.

We know now for the most popular classes of structured rank matrices how to compute the QR-factorization. Let us briefly discuss now how to solve a system of equations by using this QR-factorization.

9.2.5   Solving systems of equations

Even though the focus of this book is system solving, we have not yet mentioned how to solve systems of equations by using the above-presented techniques.

Solving systems of equations using the techniques presented in this chapter is straightforward. Suppose we have a system of the following form:

$$A\mathbf{x} = \mathbf{b}.$$

When one likes to solve this system via the QR-method, one applies directly all orthogonal transformations to the right-hand side. This gives us the following equation:

$$Q^T A\mathbf{x} = R\mathbf{x} = Q^T \mathbf{b} = \hat{\mathbf{b}}.$$

Then one is left with an upper triangular system of equations which can easily be solved.

Of course the implementation of such a solver might require some time. We will not describe implementations here, as there is a huge variety of representations and structured rank matrices. Hence, focusing attention on only one specific case would not help most people. Using a specific representation of the structured rank matrix, however, the reader should be able to derive all necessary Givens transformations for bringing this matrix to upper triangular form. Moreover, the reader knows in advance what the structure of the upper triangular part will be. Hence using previously discussed techniques in the chapter on the QR-factorization for the easy classes, one should be able to present a representation of this upper triangular part. Finally one should exploit this representation for solving the system of equations. This can be done either via the Levinson-like technique discussed earlier and in a forthcoming chapter (see Chapter 11). Or one can try to apply immediately backward substitution. This is left to the reader.

9.2.6   Other decompositions

We will not go into detail, but clearly the methods discussed above can be combined with Givens sequences going from left to right and the ones going from right to left. In this way, one can predict the structures of the factors R and L in the URV or QL-decompositions, for example.

We briefly present here some possible combinations and the resulting factorizations. The reader will recognize many of these factorizations. The reader can try to perform these sequences of transformations on structured rank matrices, e.g., a quasiseparable matrix, and see that one obtains the desired factorization.

- Combinations of ascending and descending sequences. This is used for computing the QR-factorization.

- Combinations of ascending sequences followed by sequences from right to left. This is used for computing for example the URV-decomposition.

- Combinations of descending sequences, followed by ascending sequences. This can be used for computing the QL-factorization.

- Combinations of sequences from right to left, followed by sequences from left to right, for computing the LQ-factorization.

- Other combinations can lead to other factorizations such as RQ, ULV and so on.

We have now shown in this section how to make a matrix of upper triangular form by using sequences from bottom to top followed by sequences from top to bottom. We performed all sequences one after another. In the next section we will see if it is possible to combine some sequences and to interchange their order. For example, we will prove that it is possible to first perform some sequences from top to bottom, followed by sequences from bottom to top to compute the QR-factorization of the structured rank matrix.

Notes and references

The theory of performing sequences of Givens transformations as discussed in this and the previous section can also be found in a more general framework in the following manuscript.

☞ S. Delvaux and M. Van Barel. A QR-based solver for rank structured matrices. Technical Report TW454, Department of Computer Science, Katholieke Universiteit Leuven, Celestijnenlaan 200A, 3000 Leuven (Heverlee), Belgium, March 2006.

In this manuscript on the QR-factorization, Delvaux and Van Barel develop an effective QR-factorization method for structured rank matrices using the Givens-weight representation as developed in one of their earlier manuscripts [72] (see Section 8.5, Chapter 8, for more information on this representation). The structured rank matrices are characterized by the structure blocks as discussed in the footnotes and references section following Section 8.1 in Chapter 8.

The QR/URV-decomposition of higher order generator representable semiseparable plus diagonal/band matrices was discussed in [56, 209]. More information can be found in Section 5.6 in Chapter 5 in which these methods are explained for the rank 1 case.

The QR-decomposition for quasiseparable matrices was discussed in [96, 83, 84]. More information can be found in Chapter 12.

9.3 Different patterns of annihilation

In Section 9.2, we illustrated how to compute the QR-factorization of structured rank matrices, first, by removing all the rank structures, and second, by annihilating all off-diagonal elements. In this section we will show that there are also different annihilation patterns by interchanging the different Givens sequences and by letting sequences of Givens transformations interact with each other. We remark that the changing of Givens patterns as discussed here will not at all affect the Q or the R factor of the factorization. We will only take a look at the combinations of Givens transformations and how they interact mutually.

To be able to design different patterns of annihilation, and to characterize them, we introduce a new kind of notation. For example, to bring a semiseparable matrix to upper triangular form, we use one sequence of Givens transformations from bottom to top. This means that for a 5×5 matrix the first applied Givens transformation works on the last two rows, followed by a Givens transformation working on rows 3 and 4 and so on.

To depict graphically these Givens transformations with regard to their order and the rows they are acting on, we use the following figure.

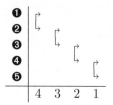

The numbered circles on the vertical axis depict the rows of the matrix. The numbers across the bottom represent in some sense a time line. Let us explain this scheme in more detail. First, a Givens transformation is performed, acting on row 5 and row 4. Second, a Givens transformation is performed acting on row 3 and row 4, and this process continues.

Let us illustrate this graphical representation with a second example. Suppose we have a semiseparable plus diagonal matrix of rank 1. To make this matrix upper triangular, we first perform a sequence of Givens transformations from bottom to top to remove the low rank part of the semiseparable matrix; second, we perform a sequence of Givens transformations from top to bottom to remove the subdiagonal elements of the remaining Hessenberg matrix. Graphically this is depicted as follows:

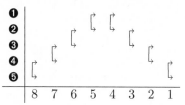

The first four transformations clearly go from bottom to top, whereas the last four transformations go from top to bottom.

Using this notation, we will construct some types of different annihilation patterns. Note that, based on the sequences of Givens transformations as initially designed for bringing the matrix to upper triangular form, we can derive other patterns of Givens transformations leading to the same QR-factorization. Based on this new order of performing Givens transformations we also get new schemes for computing the upper triangular factor R. For some of the newly designed patterns we will illustrate these new annihilation sequences on the matrices.

First, we start with some examples for reducing the rank structure; second, we present some examples for creating the zeros in the generalized Hessenberg forms; and finally, we combine these two chasing patterns to develop some new hybrid variants.

9.3.1 The leaf form for removing the rank structure

In the algorithms we deduced previously, for bringing a matrix to upper triangular form, we worked on each of the different terms in the low rank decomposition of the matrix separately. More precisely, suppose we have a lower triangular semiseparable matrix S of semiseparability rank r, then this matrix can be written as

$$S = S_1 + S_2 + \cdots + S_r,$$

for which all the matrices S_i are lower triangular and of semiseparability rank 1. To remove the rank structure of these matrices, we start, for example, by working on matrix S_1 and reduce its rank. Then we reduce the rank of matrix S_2 and we continue in this fashion.

Let us depict this process of annihilation by our graphical representation. For reasons of simplicity, we work on a 5×5 matrix having the lower triangular part of semiseparability rank 3. Hence we have to perform three sequences of ascending Givens transformations to completely remove the rank structure. We remark that only the Givens transformations necessary for removing the rank structure are depicted; hence, the second and last sequence of transformations consist of less than $n - 1$ transformations.

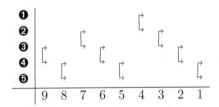

It is clear that the second and third sequence of Givens from bottom to top do not go as far as the first sequence; this is because the low rank structure has shifted down.

We call this sequence of annihilation the leaf form, as we peal off the low rank structure rank by rank, like leaves falling from a tree and taking with every leaf a rank 1 part with them. Leaf one corresponds to transformations 1 to 4, leaf two to transformations 5 to 7, and the last leaf to transformations 8 to 9.

9.3.2 The pyramid form for removing the rank structure

Considering the graphical representation of the previous subsection, we see that in fact the Givens transformation performed in the fifth step can already be performed in step 3. This is because after step 2 no actions are performed anymore on row 4 and row 5, they do not change anymore, and hence the fifth Givens transformation is already uniquely defined.

Let us now compress the graphical representation without, in fact, changing the Givens transformations. This gives us the following graphical representation for performing the same reduction.

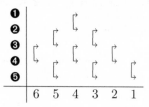

Performing now the transformations in a different order, we get the so-called pyramid form. Let us illustrate in more detail what we mean by this form. We rearrange the compressed form from above in a different order.

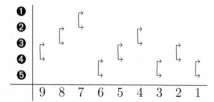

Looking at the first three steps we see that the Givens transformations create some kind of pyramid. Adding the following three steps gives us in the compressed notation a larger pyramid. We start with a small pyramid, which grows every few steps, hence the pyramid form. Let us now show what happens with the matrices if we follow this annihilation pattern. This is a pattern for reducing the lower triangular rank structure of a 5×5 matrix with semiseparability rank in the lower triangular part equal to 3. We consider only the lower triangular part of the matrix S in decomposed form: $S = S_1 + S_2 + S_3$. Assume also that the initial leaf pattern works first on matrix S_1, then on S_2 and finally on S_3 (this order can be shuffled without loss of generality). This means that the pyramid form works on S as follows. The first Givens transformation will make zeros in the last row of S_1, and the second Givens transformation will create zeros in the second to last row of S_1; these Givens transformations have their effect on the matrices S_2 and S_3 and destroy parts of the low rank structure. After these two transformations we have the following situation. We show the structure of the summation:

$$
\begin{bmatrix}
\boxtimes & & & & \\
\boxtimes & \boxtimes & & & \\
\boxtimes & \boxtimes & \boxtimes & & \\
 & & & \times & \\
 & & & & \times
\end{bmatrix}
+
\begin{bmatrix}
\boxtimes & & & & \\
\boxtimes & \boxtimes & & & \\
\boxtimes & \boxtimes & \boxtimes & & \\
\boxtimes & \boxtimes & \boxtimes & \times & \\
\boxtimes & \boxtimes & \boxtimes & \boxtimes & \times
\end{bmatrix}
+
\begin{bmatrix}
\boxtimes & & & & \\
\boxtimes & \boxtimes & & & \\
\boxtimes & \boxtimes & \boxtimes & & \\
\boxtimes & \boxtimes & \boxtimes & \times & \\
\boxtimes & \boxtimes & \boxtimes & \boxtimes & \times
\end{bmatrix}.
$$

Instead of working again on matrix S_1, the pyramid sequence now works on matrix S_2 annihilating its last row. We get the following situation:

$$
\begin{bmatrix}
\boxtimes & & & & \\
\boxtimes & \boxtimes & & & \\
\boxtimes & \boxtimes & \boxtimes & & \\
 & & & \times & \\
 & & & \times & \times
\end{bmatrix}
+
\begin{bmatrix}
\boxtimes & & & & \\
\boxtimes & \boxtimes & & & \\
\boxtimes & \boxtimes & \boxtimes & & \\
 & \boxtimes & \boxtimes & \boxtimes & \times \\
 & & & \times & \times
\end{bmatrix}
+
\begin{bmatrix}
\boxtimes & & & & \\
\boxtimes & \boxtimes & & & \\
\boxtimes & \boxtimes & \boxtimes & & \\
\boxtimes & \boxtimes & \boxtimes & \times & \\
\boxtimes & \boxtimes & \boxtimes & \boxtimes & \times & \times
\end{bmatrix}.
$$

We have now completed all the transformations of the first small pyramid. The fourth transformation works on matrix S_1, the fifth on matrix S_2 and the sixth on matrix S_3. This gives us the following situation:

$$
\begin{bmatrix}
\boxtimes & & & & \\
\boxtimes & \boxtimes & & & \\
 & & \times & & \\
 & & \times & \times & \\
 & & \times & \times & \times
\end{bmatrix}
+
\begin{bmatrix}
\boxtimes & & & & \\
\boxtimes & \boxtimes & & & \\
\boxtimes & \boxtimes & \times & & \\
 & & \times & \times & \\
 & & \times & \times & \times
\end{bmatrix}
+
\begin{bmatrix}
\boxtimes & & & & \\
\boxtimes & \boxtimes & & & \\
\boxtimes & \boxtimes & \times & & \\
\boxtimes & \boxtimes & \times & \times & \\
 & & \times & \times & \times
\end{bmatrix}.
$$

This completes the construction of the second, larger pyramid. To complete the reduction procedure, we need to perform three more Givens transformations, each creating zeros in one of the three matrices. The result will be of the following form:

$$
\begin{bmatrix}
\times & & & & \\
 & \times & & & \\
 & \times & \times & & \\
 & \times & \times & \times & \\
 & \times & \times & \times
\end{bmatrix}
+
\begin{bmatrix}
\times & & & & \\
 & \times & \times & & \\
 & & \times & \times & \\
 & & \times & \times & \times \\
 & & \times & \times & \times
\end{bmatrix}
+
\begin{bmatrix}
\times & & & & \\
 & \times & \times & & \\
 & \times & \times & \times & \\
 & & \times & \times & \times \\
 & & & \times & \times & \times
\end{bmatrix}.
$$

This completes the procedure, and as in the other case, we similarly get a matrix having two subdiagonals different from zero. To bring the matrix completely to upper triangular form, we have to perform some sequences from top to bottom. In the following subsections we check whether we also have flexibility in these descending patterns.

9.3.3 The leaf form for creating zeros

If multiple subdiagonals in the matrix need to be annihilated, different sequences of Givens transformations from bottom to top are needed. If we perform these sequences of transformations, one after each other, we call this again the leaf form, similarly as in Subsection 9.3.1.

For example, if we want to reduce a generalized Hessenberg matrix with two subdiagonals to upper triangular form we need to perform two sequences of Givens transformations, from top to bottom. The following graphical representation illustrates this.

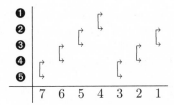

9.3.4 The diamond form for creating zeros

Similarly as in the pyramid case, we can rearrange the transformations, for example, the following rearrangement is called the diamond form, as we create growing diamonds $\diamond$ in the different steps. If more diagonals have to be chased, it is clearer that in every step growing diamonds (thereby neglecting the last transformations in this case from 4 to 6) are created.

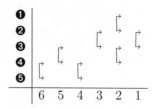

We have now briefly discussed some other patterns for reducing the rank and for creating zeros. One might correctly wonder now that the combination of the Givens transformations from bottom to top and from top to bottom can also be performed in a different order, i.e., first the descending followed by the ascending sequences. In the following sections we illustrate some of these new patterns just by letting the Givens transformations from our standard algorithm interact with each other.

Before being able to deduce some different orders for annihilating the lower triangular part of the structured rank matrix, we need some more information related to Givens transformations, more precisely, on how they interact with each other.

9.3.5   Theorems connected to Givens transformations

In the next subsections, we need to have more flexibility for working with the Givens transformations. In order to do so, we need two lemmas. The first lemma shows us that we can concatenate two Givens transformations acting on the same rows. The second lemma shows us that, under some conditions, we can rearrange the order of some Givens transformations.

Lemma 9.37. *Suppose two Givens transformations G_1 and G_2 are given:*

$$G_1 = \begin{bmatrix} c_1 & -s_1 \\ s_1 & c_1 \end{bmatrix} \text{ and } G_2 = \begin{bmatrix} c_2 & -s_2 \\ s_2 & c_2 \end{bmatrix}.$$

Then we have that $G_1 G_2 = G_3$ is again a Givens transformation. We will call this the fusion of Givens transformations in the remainder of the text.

The proof is trivial. In our graphical schemes, we will depict this as follows:

resulting in

The next lemma is slightly more complicated and changes the direction of three Givens transformations.

Lemma 9.38 (Shift through lemma). *Suppose three 3×3 Givens transformations G_1, G_2 and G_3 are given, such that the Givens transformations G_1 and G_3 act on the first two rows of a matrix, and G_2 acts on the second and third row.*
Then we have that

$$G_1 G_2 G_3 = \hat{G}_1 \hat{G}_2 \hat{G}_3,$$

where $\hat{G}_1$ and $\hat{G}_3$ work on the second and third row and $\hat{G}_2$ works on the first two rows.

Proof. The proof is straightforward, based on the factorization of a 3×3 orthogonal matrix. Suppose we have an orthogonal matrix U. We will now depict a factorization of this matrix U into two sequences of Givens transformations as described in the lemma.

The first factorization of this orthogonal matrix makes the matrix upper triangular in the traditional way. The first Givens transformation $\hat{G}_1^T$ acts on rows 2 and 3 of the matrix U, creating thereby a zero in the lower-left position:

$$\hat{G}_1^T U = \begin{bmatrix} \times & \times & \times \\ \times & \times & \times \\ 0 & \times & \times \end{bmatrix}.$$

The second Givens transformation acts on the first and second rows to create a zero in the second position of the first column:

$$\hat{G}_2^T \hat{G}_1^T U = \begin{bmatrix} \times & \times & \times \\ 0 & \times & \times \\ 0 & \times & \times \end{bmatrix}.$$

Finally the last transformation $\hat{G}_3^T$ creates the last zero to make the matrix of upper triangular form.

$$\hat{G}_3^T \hat{G}_2^T \hat{G}_1^T U = \begin{bmatrix} \times & \times & \times \\ 0 & \times & \times \\ 0 & 0 & \times \end{bmatrix}.$$

Suppose we have chosen all Givens transformations in such a manner that the upper triangular matrix has positive diagonal elements. Because the resulting upper triangular matrix is orthogonal, it has to be the identity matrix; hence, we have the following factorization of the orthogonal matrix U:

$$U = \hat{G}_1 \hat{G}_2 \hat{G}_3. \tag{9.3}$$

Let us consider now a different factorization of the orthogonal matrix U. Perform a first Givens transformation to annihilate the upper-right element of the matrix U where the Givens transformation acts on the first and second row:

$$G_1^T U = \begin{bmatrix} \times & \times & 0 \\ \times & \times & \times \\ \times & \times & \times \end{bmatrix}.$$

Similarly as above, one can continue to reduce the orthogonal matrix to lower triangular form with positive diagonal elements. Hence, one obtains a factorization of the following form:

$$U = G_1 G_2 G_3. \tag{9.4}$$

Combining Equations (9.3) and (9.4) leads to the desired result.   □

Note 9.39. *Two remarks have to be made.*

- *We remark that, in fact, there is more to the proof than we mention here. In fact, the transformations acting on the orthogonal matrix, reducing it to lower triangular form, also have a specific effect on the lower triangular part. Looking in more detail at the lower triangular part, one can see that the first Givens transformation creates a 2×2 low rank block in the lower left corner of the orthogonal matrix. More information on this can be found in Section 9.4, where we will also reconsider this example.*

- *In some sense one can consider the fusion of two Givens transformations as a special case of the shift through lemma. Instead of applying directly the fusion, the reader can put the identity Givens transformation in between these two transformations. Then he can apply the shift through lemma. The final outcome will be identical to applying directly the fusion of these two Givens transformations.*

Graphically we will depict the shift through lemma as follows:

and in the other direction this becomes:

Remark that, if we cannot place the ⌢ or ⌣ arrow at that specific position, we cannot apply the shift through lemma. The reader can verify that, for example, in the following graphical scheme, we cannot use the lemma.

To apply the shift through lemma, in some sense, we need to have some extra place to perform the action. Based on these operations we can interchange the order of the ascending and descending sequences of Givens transformations. Let us mention some of the different patterns.

9.3.6 The ∧-pattern

The ∧-pattern for computing the QR-factorization of a rank structured matrix is, in fact, the standard pattern as described in the previous section. First, we remove the rank structure by performing sequences of Givens transformations from bottom to top. This gives us the following sequences of Givens transformations (e.g., two in this case) ∖∖. Depending on the number of subdiagonals in the resulting matrix, we need to perform some rank-expanding sequences of Givens transformations from top to bottom ∕∕ (two in this case). Combining these Givens transformations from both sequences gives us the following pattern ∕∕∖∖, which we briefly call the ∧-pattern.

Suppose, e.g., that we have a quasiseparable matrix of rank 1. Performing the Givens transformations in the order as described in this subsection, we get the following graphical representation of the reduction.

Graphically this is depicted as follows:

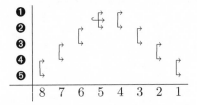

Combining the top two Givens transformations as already depicted in the previous figure, we get the following reduction, with one less Givens transformation.

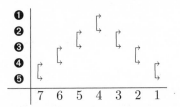

The reader can observe that the top three Givens transformations admit the shift through lemma. In this way we can drag the Givens transformation in position 5 through the Givens transformations in positions 4 and 3. Let us observe what kind of patterns we get in this case.

9.3.7 The ✕-pattern

We will graphically illustrate what happens if we apply the shift through lemma as indicated in the previous subsection. Suppose we have the following graphical reduction scheme for reducing our matrix to upper triangular form. For esthetical reasons in the figures, we assume here that our matrix is of size 6×6. First, we apply the shift through lemma at positions $6, 5$, and 4.

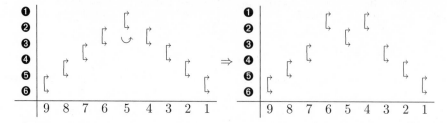

Rearranging slightly the Givens transformations from positions, we can reapply the shift through lemma.

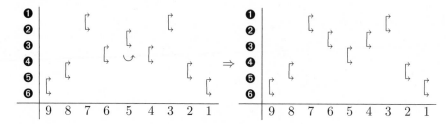

Let us compress the representation above.

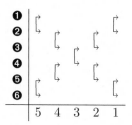

This shows us another pattern of performing the Givens transformations, namely the X-pattern.

Note 9.40. *We remark that the possibility of performing the Givens transformations in this order does not necessarily mean that it will lead to an effective algorithm doing so. In the next section we will investigate these Givens transformations in more detail.*

Continuing to apply the shift through lemma gives us another pattern.

9.3.8 The ∨-pattern

Continuing now this procedure, by applying two more times the shift through lemma, gives us the following graphical representation of a possible reduction of the matrix to upper triangular form.

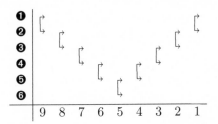

This clearly presents the ∨-pattern for computing the QR-factorization. In case there are more ascending and descending sequences of Givens transformations, one can also shift all of the descending sequences through. In fact this creates an incredible number of possibilities, as shown in the next example.

Example 9.41 Suppose we have a matrix brought to upper triangular form by performing two ascending sequences of Givens transformations and two descending sequences of transformations. The following, not complete list shows some possibilities of combinations of these sequences for making the matrix upper triangular. We start with the ∧-pattern and continuously change the order of the involved transformations to arrive at the ∨-pattern.

- The standard ∧-pattern gives us briefly the following sequences: ⫻⫹.

- In the middle we can create one ✕-pattern: ⋀⋀.

- In the middle we can have one ∨-pattern: ⋀⋀.

- Combinations with ✕-patterns: ✕⋀ or ⋀✕ or ✕✕.

- Combinations following from the previous patterns: ∨⋀ and ⋀∨.

- In the middle one can have one ∧-pattern: ⋁⋁.

- In the middle we can create another ✕-pattern: ⋈.

- The ∨-pattern: ⫻⫻.

Clearly there are already numerous possibilities for 2 ascending and 2 descending sequences. ∎

In the following section we will take a look at the effect of the first sequence of Givens transformations on the matrix if we apply a ∨-pattern for computing the QR-factorization of a structured rank matrix.

9.3.9 Givens transformations in the ∨-pattern

As mentioned in a previous note, there are different ways of performing the Givens transformations to reduce the matrix to upper triangular, but the computation of the Givens transformations is not necessarily effective in terms of computer time.

We will illustrate this more clearly by looking in detail at the ∨-pattern for transforming a matrix to upper triangular form. The ∧-pattern for reducing matrices to upper triangular form was discussed extensively in the sections preceding this one.

We investigate this ∨-pattern via reverse engineering. Suppose we have a ∨-pattern for making a 5×5 lower triangular quasiseparable matrix upper triangular, assume the matrix to be of quasiseparability rank 1. We will now investigate what the effect of the first sequence of descending Givens transformations on this matrix A will be. We have the following equation:

$$\hat{G}_1^T \hat{G}_2^T \hat{G}_3^T \hat{G}_4^T G_3^T G_2^T G_1^T A = R, \tag{9.5}$$

where R is a 5×5 upper triangular matrix. Moreover the first applied sequence of Givens transformations $G_3^T G_2^T G_1^T$ works on the matrix A from top to bottom. More precisely G_1^T acts on row 1 and row 2, G_2^T acts on rows 2 and 3 and so on. The sequence of transformations $\hat{G}_1^T \hat{G}_2^T \hat{G}_3^T \hat{G}_4^T$ works from bottom to top, where $\hat{G}_4^T$ acts on rows 4 and 5, $\hat{G}_3^T$ acts on rows 3 and 4, and so on. Rewriting Equation (9.5), by bringing the ascending sequence of transformations to the right gives us

$$G_3^T G_2^T G_1^T A = \hat{G}_4 \hat{G}_2 \hat{G}_2 \hat{G}_1 R$$
$$= S.$$

Because the sequence of transformations applied on the matrix R goes from top to bottom, we know that these transformations transform the matrix R into a matrix having a lower triangular part of semiseparable form. Hence we have that the transformations from top to bottom, namely $G_3^T G_2^T G_1^T$, lift up in some sense the strictly lower triangular semiseparable structure to a lower triangular semiseparable structure. The following figures denote more precisely what is happening. We start on the left with the matrix A, and we depict what the impact of the transformations $G_3^T G_2^T G_1^T$ needs to be on this matrix to satisfy the equation above. Assume $A_0 = A$. To see more clearly what happens, we already include the upper left and lower right element in the strictly lower triangular semiseparable structure:

$$
\begin{bmatrix}
\boxtimes & & & & \\
\boxtimes & \times & & & \\
\boxtimes & \boxtimes & \times & & \\
\boxtimes & \boxtimes & \boxtimes & \times & \\
\boxtimes & \boxtimes & \boxtimes & \boxtimes & \boxtimes
\end{bmatrix}
\xrightarrow{G_1^T A_0}
\begin{bmatrix}
\boxtimes & & & & \\
\boxtimes & \boxtimes & & & \\
\boxtimes & \boxtimes & \times & & \\
\boxtimes & \boxtimes & \boxtimes & \times & \\
\boxtimes & \boxtimes & \boxtimes & \boxtimes & \boxtimes
\end{bmatrix}
$$

$$\Updownarrow$$

$$A_0 \xrightarrow{G_1^T A_0} A_1.$$

As the complete result needs to be of lower triangular semiseparable form, the transformation G_1^T needs to add one more element into the semiseparable structure. This results in an inclusion of the second diagonal element in the lower triangular rank structure. Givens transformation G_2^T causes the expansion of the low rank structure towards the third diagonal element.

$$
\begin{bmatrix}
\boxtimes & & & & \\
\boxtimes & \boxtimes & & & \\
\boxtimes & \boxtimes & \times & & \\
\boxtimes & \boxtimes & \boxtimes & \times & \\
\boxtimes & \boxtimes & \boxtimes & \boxtimes & \boxtimes
\end{bmatrix}
\xrightarrow{G_2^T A_1}
\begin{bmatrix}
\boxtimes & & & & \\
\boxtimes & \boxtimes & & & \\
\boxtimes & \boxtimes & \boxtimes & & \\
\boxtimes & \boxtimes & \boxtimes & \times & \\
\boxtimes & \boxtimes & \boxtimes & \boxtimes & \boxtimes
\end{bmatrix}
$$

$$\Updownarrow$$

$$A_1 \xrightarrow{G_2^T A_1} A_2.$$

Finally the last Givens transformation G_3^T creates the following structure.

$$
\begin{bmatrix}
\boxtimes & & & & \\
\boxtimes & \boxtimes & & & \\
\boxtimes & \boxtimes & \boxtimes & & \\
\boxtimes & \boxtimes & \boxtimes & \times & \\
\boxtimes & \boxtimes & \boxtimes & \boxtimes & \boxtimes
\end{bmatrix}
\xrightarrow{G_3^T A_2}
\begin{bmatrix}
\boxtimes & & & & \\
\boxtimes & \boxtimes & & & \\
\boxtimes & \boxtimes & \boxtimes & & \\
\boxtimes & \boxtimes & \boxtimes & \boxtimes & \\
\boxtimes & \boxtimes & \boxtimes & \boxtimes & \boxtimes
\end{bmatrix}
$$

$$\Updownarrow$$

$$A_2 \xrightarrow{G_3^T A_2} A_3.$$

Hence the result of applying this sequence of Givens transformations from top to bottom is a matrix which has the lower triangular structure shifted upwards one position. In fact we have performed a rank-expanding sequence of Givens transformations. It is clear that it is not only possible to expand the rank structure in case it equals zero, but it seems to be possible also for higher order structured rank matrices. In the following section we will investigate in more detail the Givens transformations causing the rank-expanding transformations.

Notes and references

The graphical interpretation as presented in this section, which is very useful when developing for example the different annihilation patterns, was discussed for the first time by Delvaux and Van Barel in the following manuscript.

> ☞ S. Delvaux and M. Van Barel. Unitary rank structured matrices. Technical Report TW464, Department of Computer Science, Katholieke Universiteit Leuven, Celestijnenlaan 200A, 3000 Leuven (Heverlee), Belgium, July 2006.

This article discusses the Givens-weight representation for unitary rank structured matrices. Based on some theorems for Givens transformations it is shown that one can create different patterns in the Givens-weight representation for these matrices.

9.4   Rank-expanding sequences of Givens transformations

In the previous section we discovered, due to the changing of the order of sequences of Givens transformations, that more patterns for factorizing structured rank matrices seem to exist than our traditional $\wedge$-pattern. The $\vee$-pattern especially drew our attention. It seemed to be possible to perform a sequence of Givens transformations from top to bottom, such that the lower rank part was expanded. Moreover the reduction of a Hessenberg matrix to upper triangular form seemed to be a special case of this transformation, in which we expanded the rank 0 structure. Let us investigate in more detail the structure of this sequence of rank-expanding Givens transformations. We will investigate here how to compute these Givens transformations and we will investigate the effect of these transformations on general structured rank matrices.

9.4.1   The Givens transformation

In this subsection we will propose an analytic way of computing the Givens transformation for expanding the rank structure. First we will provide a proposition for computing an ascending rank-expanding Givens transformation. In the example following the proposition, we will use the reduction of a Hessenberg matrix to upper triangular form as an example of a rank-expanding sequence of Givens transformations, but from top to bottom.

Proposition 9.42 (Ascending rank-expanding Givens transformation).
Suppose a 2×2 matrix A is given. Then a Givens transformation G exists such that the first row of the matrix $G^T A$, and the row $[e, f]$ are linearly dependent, with e and f two arbitrary elements.

Proof. Suppose we have the matrix A and the Givens transformation G as follows:

$$A = \begin{bmatrix} a & b \\ c & d \end{bmatrix} \text{ and } G = \frac{1}{\sqrt{1+t^2}} \begin{bmatrix} t & -1 \\ 1 & t \end{bmatrix}. \tag{9.6}$$

Assume $[a, b]$ and $[e, f]$ to be linearly independent, otherwise we could have taken the Givens transformation equal to the identity matrix.
Let us compute the product $G^T A$:

$$\frac{1}{\sqrt{1+t^2}} \begin{bmatrix} t & 1 \\ -1 & t \end{bmatrix} \begin{bmatrix} a & b \\ c & d \end{bmatrix} = \frac{1}{\sqrt{1+t^2}} \begin{bmatrix} at+c & bt+d \\ -a+ct & -b+dt \end{bmatrix}.$$

The first row being dependent on $[e, f]$ leads to the following relation:

$$f(at+c) - e(bt+d) = 0.$$

Rewriting this equation towards t gives us the following well-defined equation:

$$t = \frac{ed - cf}{fa - eb}.$$

This equation is well defined, as we assumed $[a, b]$ to be independent of $[e, f]$.

$\Box$

We remark that this type of Givens transformation is also used in the following manuscripts [294, 280]. This theorem defines the rank-expanding Givens transformation for an ascending sequence of transformations. In practice, e.g., for the QR-factorization, we only encountered rank-expanding Givens transformations for descending sequences of Givens transformations. Let us therefore also formulate the following theorem and present the reduction of a Hessenberg matrix to upper triangular form as an example.

Corollary 9.43 (Descending rank-expanding Givens transformation). *Suppose the following 2×2 matrix is given*

$$A = \begin{bmatrix} a & b \\ c & d \end{bmatrix}. \tag{9.7}$$

Then there exists a Givens transformation such that the second row of the matrix $G^T A$ and the row $[e, f]$ are linearly dependent. The value t in the Givens transformation G as in (9.6) is defined as

$$t = \frac{af - be}{cf - de},$$

under the assumption that $cf - de \neq 0$, otherwise one could have taken $G = I_2$.

Let us show that the rank-expanding Givens transformations as we computed them here are a generalization of the transformations used for bringing an upper Hessenberg matrix back to upper triangular form.

Example 9.44 Suppose we have a Hessenberg matrix H and we want to reduce it to upper triangular form. Instead of using the standard Givens transformations, eliminating the subdiagonal elements, we will use here the Givens transformations from Corollary 9.43 to expand the zero rank below the subdiagonal. This is done by a sequence of Givens transformations going from top to bottom.

Suppose we have, for example, the following Hessenberg matrix:

$$H = \begin{bmatrix} 1 & \frac{-1}{\sqrt{6}} & \frac{3}{\sqrt{3}} \\ 1 & \frac{3}{\sqrt{6}} & \frac{-1}{\sqrt{3}} \\ 0 & \frac{2\sqrt{2}}{\sqrt{3}} & \frac{5}{\sqrt{3}} \end{bmatrix}.$$

Computing the first Givens transformation applied on rows 1 and 2 to make part of the transformed second row dependent on

$$[e, f] = \left[0, \frac{2\sqrt{2}}{\sqrt{3}} \right],$$

gives us the following transformation (use the same notation as in Corollary 9.43):

$$t = \frac{af - be}{cf - de}$$
$$= \frac{a}{c}$$
$$= 1.$$

Hence our Givens transformation will be of the following form:

$$\breve{G}_1^T = \frac{1}{\sqrt{2}} \begin{bmatrix} 1 & 1 \\ -1 & 1 \end{bmatrix}.$$

Applying the transformation G_1^T (the 2×2 Givens transformation $\breve{G}_1^T$ is embedded in G_1^T) onto the matrix H annihilates the first subdiagonal element, thereby expanding the zero rank structure below the subdiagonal. One can easily continue this procedure and conclude that the rank-expanding Givens transformations lift up the zero structure and hence create an upper triangular matrix. ∎

Let us investigate in more detail how we can use these Givens transformations, acting on 2×2 matrices, to make them suitable for working on general structured rank matrices. From now on, we focus again on transformations going from bottom to top. The theorems and corollaries as they will be provided here have their analogous formulations in terms of sequences of transformations going from right to left, from left to right and from top to bottom, but they are not included anymore.

9.4.2 Rank-expanding Givens transformations on upper rank 1 structures

Without loss of generality we assume we are working with an upper triangular matrix, for which the strictly upper triangular part is of semiseparable form. We will investigate now the existence of a sequence of rank-expanding Givens transformations, such that the resulting matrix will have the diagonal included in the structure, i.e., that finally the matrix will have the upper triangular part of semiseparable form.

Theorem 9.45. *Suppose S is an upper triangular matrix, with the strictly upper triangular part of semiseparable form. A sequence of rank-expanding $n - 1$ Givens transformations G_i^T as defined in Proposition 9.42, acting on row $i - 1$ and i, is performed on the matrix S:*

$$G_2^T \dots G_n^T S = \hat{S}.$$

The matrix $\hat{S}$ will have the upper triangular part of semiseparable form.

The difference with an arbitrary sequence of Givens transformations is that the rank of the upper part will not increase!

Proof. Our matrix S will be of the following form (we already included the upper left and lower right elements in the semiseparable structure):

$$
S = \begin{bmatrix}
\times & \boxtimes & \boxtimes & \boxtimes & \boxtimes \\
 & \times & \boxtimes & \boxtimes & \boxtimes \\
 & & \times & \boxtimes & \boxtimes \\
 & & & \times & \boxtimes \\
 & & & & \times
\end{bmatrix}
=
\begin{bmatrix}
\boxtimes & \boxtimes & \boxtimes & \boxtimes & \boxtimes \\
 & \times & \boxtimes & \boxtimes & \boxtimes \\
 & & \times & \boxtimes & \boxtimes \\
 & & & \times & \boxtimes \\
 & & & & \boxtimes
\end{bmatrix}.
$$

The Givens transformation for expanding the lower triangular rank structure is based on the lower right 2×2 block, whereas the vector $[e, f]$ is chosen out of the last two columns, above this block. If possible, one chooses the elements $[e, f]$ to be different from zero. Performing this Givens transformation on rows 4 and 5, for expanding the rank structure in row 4, gives us the following result:

$$
G_5^T S = \begin{bmatrix}
\boxtimes & \boxtimes & \boxtimes & \boxtimes & \boxtimes \\
 & \times & \boxtimes & \boxtimes & \boxtimes \\
 & & \times & \boxtimes & \boxtimes \\
 & & & \boxtimes & \boxtimes \\
 & & & \times & \boxtimes
\end{bmatrix}.
$$

In a similar way as one computed transformation G_5^T, one can compute now the transformation G_4^T. Applying transformation G_4^T to include diagonal element $(3, 3)$ into the structure leads to the following situation:

$$
G_4 G_5^T S = \begin{bmatrix}
\boxtimes & \boxtimes & \boxtimes & \boxtimes & \boxtimes \\
 & \times & \boxtimes & \boxtimes & \boxtimes \\
 & & \boxtimes & \boxtimes & \boxtimes \\
 & & \times & \boxtimes & \boxtimes \\
 & & & \times & \boxtimes
\end{bmatrix}.
$$

Performing finally the transformation G_3^T to include element $(2, 2)$ into the structure gives us a matrix having the complete upper triangular part of semiseparable form:

$$
G_3^T G_4^T G_5^T S = \begin{bmatrix}
\boxtimes & \boxtimes & \boxtimes & \boxtimes & \boxtimes \\
 & \boxtimes & \boxtimes & \boxtimes & \boxtimes \\
 & \times & \boxtimes & \boxtimes & \boxtimes \\
 & & \times & \boxtimes & \boxtimes \\
 & & & \times & \boxtimes
\end{bmatrix}.
$$

It turns out that the application of the transformation G_2^T is not essential in this case. □

Note 9.46. *Even though we did not mention it in the preceding proof, for a real example, one always needs to take, if possible, the vector $[e, f]$ different from zero.*

For example, suppose we have the following upper triangular matrix, with the strictly upper triangular part of semiseparable form:

$$S = \begin{bmatrix} \boxtimes & \boxtimes & \boxtimes & \boxtimes & \boxtimes \\ & \times & \boxtimes & \boxtimes & \boxtimes \\ & & \times & 0 & 0 \\ & & & \times & \boxtimes \\ & & & & \boxtimes \end{bmatrix}.$$

When computing the first Givens transformation for expanding the semiseparable structure of the strictly upper triangular part, one should take the two elements above $[0,0]$, with which to make the lower right 2×2 block dependent.

Example 9.47 The reader can try to run the algorithm on the following matrix

$$S = \begin{bmatrix} 0 & \boxtimes & \boxtimes & \boxtimes & \boxtimes \\ & 0 & \boxtimes & \boxtimes & \boxtimes \\ & & 0 & \boxtimes & \boxtimes \\ & & & 0 & \boxtimes \\ & & & & 0 \end{bmatrix},$$

and see what the outcome will be. ∎

Before illustrating how the procedure works on higher order structured rank matrices, we will illustrate here what happens if the matrix is not generator representable. This case is not at all excluded in the previous theorem, but we only want to investigate more precisely what the structure of the Givens transformation and the resulting matrix will be. Suppose we have the following upper triangular matrix S, for which the strictly upper triangular part is semiseparable but not of generator representable form. This means that there is a zero block in the upper right corner of the matrix. Suppose our matrix S is of the following form (the lower triangular part is not essential for our purposes and hence set equal to zero):

$$S = \begin{bmatrix} \times & \boxtimes & \boxtimes & \boxtimes & 0 & 0 & 0 \\ & \times & \boxtimes & \boxtimes & 0 & 0 & 0 \\ & & \times & \boxtimes & 0 & 0 & 0 \\ & & & \times & \boxtimes & \boxtimes & \boxtimes \\ & & & & \times & \boxtimes & \boxtimes \\ & & & & & \times & \boxtimes \\ & & & & & & \times \end{bmatrix}.$$

Applying the first and second rank-expanding transformations to this matrix gives us a matrix of the following form:

$$G_6^T G_7^T S = \begin{bmatrix} \times & \boxtimes & \boxtimes & \boxtimes & 0 & 0 & 0 \\ & \times & \boxtimes & \boxtimes & 0 & 0 & 0 \\ & & \times & \boxtimes & 0 & 0 & 0 \\ & & & \times & \boxtimes & \boxtimes & \boxtimes \\ & & & & \boxtimes & \boxtimes & \boxtimes \\ & & & & \times & \boxtimes & \boxtimes \\ & & & & & \times & \boxtimes \end{bmatrix}.$$

The next Givens transformation works on row 4 and row 5 and makes the element in position $(4, 4)$ includable in the upper triangular semiseparable structure. If all elements remain different from zero, this leads to a contradiction, as the inclusion of this element always results in a rank 2 instead of a rank 1 block, namely the upper right 4×4 block. Hence there is only one possibility left for this Givens transformation, namely the annihilation of the $\boxtimes$ marked elements in the fourth row. This gives us:

$$G_5^T G_6^T G_7^T S = \begin{bmatrix} \times & \boxtimes & \boxtimes & \boxtimes & 0 & 0 & 0 \\ & \times & \boxtimes & \boxtimes & 0 & 0 & 0 \\ & & \times & \boxtimes & 0 & 0 & 0 \\ & & & \boxtimes & 0 & 0 & 0 \\ & & & \times & \boxtimes & \boxtimes & \boxtimes \\ & & & & \times & \boxtimes & \boxtimes \\ & & & & & \times & \boxtimes \end{bmatrix}.$$

Clearly the element is includable in the structure now. What happened is that we expanded the zero block!

Finally the resulting matrix will be of the following form:

$$G_3^T G_4^T G_5^T G_6^T G_7^T S = \begin{bmatrix} \boxtimes & \boxtimes & \boxtimes & \boxtimes & 0 & 0 & 0 \\ \times & \boxtimes & \boxtimes & \boxtimes & 0 & 0 & 0 \\ & \times & \boxtimes & \boxtimes & 0 & 0 & 0 \\ & & \times & \boxtimes & 0 & 0 & 0 \\ & & & \times & \boxtimes & \boxtimes & \boxtimes \\ & & & & \times & \boxtimes & \boxtimes \\ & & & & & \times & \boxtimes \end{bmatrix}.$$

We have proved now that it is possible to expand the rank structure of the strictly upper triangular part if the structured rank of this part equals 1. Let us investigate whether it is also possible if the structured rank is higher than 1.

9.4.3 Existence and the effect on upper rank structures

In this subsection, we will investigate in more detail the existence of rank-expanding sequences of transformations, as well as the effect of such a sequence of transformations on the rank structure of a general structured rank matrix. In the previous subsection we illustrated that such a sequence of rank-expanding transformations existed for upper rank 1 structures. Now we will generalize this to higher order structured rank matrices.

We will illustrate that also for higher order structured rank matrices the rank will not increase if one of its terms in the decomposed form undergoes a rank-expanding sequence of Givens transformations.

To make the process easy to understand we distinguish between two cases: the generator representable semiseparable and the general semiseparable case.

The generator representable case

We start the simple case by working on a matrix which has the strictly upper triangular part of generator representable semiseparable form of semiseparability rank 2. The proof for more general rank structures will proceed in a similar way.

Theorem 9.48. *Suppose S is an upper triangular matrix, for which the strictly upper triangular form is generator representable and of semiseparability rank 2. Suppose we decompose our matrix S as the sum of two matrices S_1 and S_2 such that both these matrices have the strictly upper triangular part of generator representable form.*

Then there exists a sequence of rank-expanding Givens transformations G_i^T, acting on row $i-1$ and i such that the resulting semiseparable matrix $\hat{S}$:

$$G_2^T \ldots G_n^T S = \hat{S},$$

will have the upper triangular part (including the diagonal) of $\{2\}$-semiseparable form.

Proof. Decompose the matrix S as the sum of two matrices S_1 and S_2. The strictly upper triangular part of both matrices S_1 and S_2 is of generator representable semiseparable form of semiseparability rank 1. We can easily modify the subdiagonal elements to incorporate them in the rank structure of the strictly upper triangular part. This is the procedure we will follow here to prove the existence of a rank-expanding sequence of Givens transformations from bottom to top.
Our matrix S is decomposed in the following form: assume all diagonal elements are included in the first matrix (and we already depict that the first matrix has the upper left and lower right element includable in the structure):

$$
S =
\begin{bmatrix}
\boxtimes & \boxtimes & \boxtimes & \boxtimes & \boxtimes \\
 & \times & \boxtimes & \boxtimes & \boxtimes \\
 & & \times & \boxtimes & \boxtimes \\
 & & & \times & \boxtimes \\
 & & & & \boxtimes
\end{bmatrix}
+
\begin{bmatrix}
0 & \boxtimes & \boxtimes & \boxtimes & \boxtimes \\
 & 0 & \boxtimes & \boxtimes & \boxtimes \\
 & & 0 & \boxtimes & \boxtimes \\
 & & & 0 & \boxtimes \\
 & & & & 0
\end{bmatrix}.
$$

Before computing the Givens transformation G_5^T, which will add an element to the structure of the first matrix in the decomposition, we rewrite both terms, such that the second matrix has the fourth diagonal element already includable in the structure (we mark with +, which element changed in the first term of the decomposition). Remark that this is possible because we are working with generator representable semiseparable matrices:

$$
S = \begin{bmatrix}
\boxtimes & \boxtimes & \boxtimes & \boxtimes & \boxtimes \\
 & \times & \boxtimes & \boxtimes & \boxtimes \\
 & & \times & \boxtimes & \boxtimes \\
 & & & + & \boxtimes \\
 & & & & \boxtimes
\end{bmatrix}
+
\begin{bmatrix}
0 & \boxtimes & \boxtimes & \boxtimes & \boxtimes \\
 & 0 & \boxtimes & \boxtimes & \boxtimes \\
 & & 0 & \boxtimes & \boxtimes \\
 & & & \boxtimes & \boxtimes \\
 & & & 0 & 0
\end{bmatrix}.
$$

Computing now the Givens transformation G_5^T of rank-expanding form to add the + in the first matrix into the rank structure and applying it onto the matrix S gives us the following situation:

$$
G_5^T S = \begin{bmatrix}
\boxtimes & \boxtimes & \boxtimes & \boxtimes & \boxtimes \\
 & \times & \boxtimes & \boxtimes & \boxtimes \\
 & & \times & \boxtimes & \boxtimes \\
 & & & \boxtimes & \boxtimes \\
 & & & \times & \boxtimes
\end{bmatrix}
+
\begin{bmatrix}
0 & \boxtimes & \boxtimes & \boxtimes & \boxtimes \\
 & 0 & \boxtimes & \boxtimes & \boxtimes \\
 & & 0 & \boxtimes & \boxtimes \\
 & & & \boxtimes & \boxtimes \\
 & & & \boxtimes & \boxtimes
\end{bmatrix}.
$$

Changing now the third diagonal element and the element below, such that there is an expansion in the semiseparable structure in the second matrix, gives us:

$$
G_5^T S = \begin{bmatrix}
\boxtimes & \boxtimes & \boxtimes & \boxtimes & \boxtimes \\
 & \times & \boxtimes & \boxtimes & \boxtimes \\
 & & + & \boxtimes & \boxtimes \\
 & & + & \boxtimes & \boxtimes \\
 & & & \times & \boxtimes
\end{bmatrix}
+
\begin{bmatrix}
0 & \boxtimes & \boxtimes & \boxtimes & \boxtimes \\
 & 0 & \boxtimes & \boxtimes & \boxtimes \\
 & & \boxtimes & \boxtimes & \boxtimes \\
 & & \boxtimes & \boxtimes & \boxtimes \\
 & & & \boxtimes & \boxtimes
\end{bmatrix}.
$$

Performing now the Givens transformation G_4^T in such a way that diagonal element 3 of the first term is incorporated in the structure gives us:

$$
G_4^T G_5^T S = \begin{bmatrix}
\boxtimes & \boxtimes & \boxtimes & \boxtimes & \boxtimes \\
 & \times & \boxtimes & \boxtimes & \boxtimes \\
 & & \boxtimes & \boxtimes & \boxtimes \\
 & & \times & \boxtimes & \boxtimes \\
 & & & \times & \boxtimes
\end{bmatrix}
+
\begin{bmatrix}
0 & \boxtimes & \boxtimes & \boxtimes & \boxtimes \\
 & 0 & \boxtimes & \boxtimes & \boxtimes \\
 & & \boxtimes & \boxtimes & \boxtimes \\
 & & \boxtimes & \boxtimes & \boxtimes \\
 & & & \boxtimes & \boxtimes
\end{bmatrix}.
$$

Changing now diagonal element 2 and the element below to expand the structure of the second matrix gives us:

$$
G_4^T G_5^T S = \begin{bmatrix}
\boxtimes & \boxtimes & \boxtimes & \boxtimes & \boxtimes \\
 & + & \boxtimes & \boxtimes & \boxtimes \\
 & + & \boxtimes & \boxtimes & \boxtimes \\
 & & \times & \boxtimes & \boxtimes \\
 & & & \times & \boxtimes
\end{bmatrix}
+
\begin{bmatrix}
0 & \boxtimes & \boxtimes & \boxtimes & \boxtimes \\
 & \boxtimes & \boxtimes & \boxtimes & \boxtimes \\
 & \boxtimes & \boxtimes & \boxtimes & \boxtimes \\
 & & \boxtimes & \boxtimes & \boxtimes \\
 & & & \boxtimes & \boxtimes
\end{bmatrix}.
$$

Performing transformation G_3^T leads to:

$$
G_3^T G_4^T G_5^T S =
\begin{bmatrix}
\boxtimes & \boxtimes & \boxtimes & \boxtimes & \boxtimes \\
 & \boxtimes & \boxtimes & \boxtimes & \boxtimes \\
 & & \times & \boxtimes & \boxtimes & \boxtimes \\
 & & & \times & \boxtimes & \boxtimes \\
 & & & & \times & \boxtimes
\end{bmatrix}
+
\begin{bmatrix}
0 & \boxtimes & \boxtimes & \boxtimes & \boxtimes \\
 & \boxtimes & \boxtimes & \boxtimes & \boxtimes \\
 & \boxtimes & \boxtimes & \boxtimes & \boxtimes \\
 & & \boxtimes & \boxtimes & \boxtimes \\
 & & & \boxtimes & \boxtimes
\end{bmatrix}.
$$

Then one should change the first diagonal element and the first subdiagonal element and then perform transformation G_2^T. This gives us finally the following situation.

$$
G_2^T G_3^T G_4^T G_5^T S =
\begin{bmatrix}
\boxtimes & \boxtimes & \boxtimes & \boxtimes & \boxtimes \\
\times & \boxtimes & \boxtimes & \boxtimes & \boxtimes \\
 & \times & \boxtimes & \boxtimes & \boxtimes \\
 & & \times & \boxtimes & \boxtimes \\
 & & & \times & \boxtimes
\end{bmatrix}
+
\begin{bmatrix}
\boxtimes & \boxtimes & \boxtimes & \boxtimes & \boxtimes \\
\boxtimes & \boxtimes & \boxtimes & \boxtimes & \boxtimes \\
 & \boxtimes & \boxtimes & \boxtimes & \boxtimes \\
 & & \boxtimes & \boxtimes & \boxtimes \\
 & & & \boxtimes & \boxtimes
\end{bmatrix}.
$$

It is clear that the summation of both these matrices leads to a matrix having the desired structural constraints, namely the upper triangular part of semiseparability rank 2.   □

Note 9.49. *We proved here in a simple manner how to compute the rank-expanding Givens transformations if the rank equals 2. This does not mean that the way presented here provides a robust algorithm. It is more of a theoretical proof, stating that it is possible to construct such a sequence of rank-expanding Givens transformations.*

The theorem as formulated here is, of course, not valid for general semiseparable matrices, because not every semiseparable matrix is of generator representable form. Let us investigate the more general case.

The general semiseparable case

We will investigate in the next theorem what happens if the semiseparable matrix is not of generator representable semiseparable form. We shall show that the theorem above remains valid. We illustrate the theorem for an upper triangular $\{2\}$-semiseparable matrix, as this illustrates the general case.

Theorem 9.50. *Suppose S is an upper triangular matrix with the strictly upper triangular part of $\{2\}$-semiseparable form. There exists a sequence of $n-1$ rank-expanding Givens transformations G_i^T, acting on row $i-1$ and i of the matrix S, such that*

$$
G_2^T \dots G_n^T S = \hat{S},
$$

and the matrix $\hat{S}$ will have the upper triangular part of $\{2\}$-semiseparable form.

Proof. We know from Chapter 1 that if an upper triangular semiseparable matrix is not of generator representable semiseparable form, it can always be written as a block diagonal matrix, for which all the blocks are of generator representable form. For simplicity we assume in our example that we have a 5×5 matrix written as the sum of two 5×5 matrices:

$$S = S_1 + S_2,$$

where both matrices S_1 and S_2 have the strictly upper triangular part of semiseparable form. We have to consider three specific cases. In the first case one of the two matrices will not be of generator representable form. In the second and third case both matrices are not generator representable, but the second case assumes the block division of both matrices is different, whereas the third case assumes that they have the same block division.

- **Case 1:** Suppose one of the two matrices is not generator representable. Without loss of generality, we can assume matrix S_1 is of semiseparable, but not generator representable form. If we apply our expanding Givens transformations on matrix S_1, this case coincides with the case in Theorem 9.48. As matrix S_2 is generator representable, one can always easily adapt its structure and apply the expanding transformations on matrix S_1. Hence no problems occur in this case.

- **Case 2:** Suppose both of the matrices are not of generator representable form, but they do not have the same block diagonal form. Assume we have the following decomposition of our matrix S:

$$S = \begin{bmatrix} \boxtimes & \boxtimes & \boxtimes & 0 & 0 \\ & \times & \boxtimes & 0 & 0 \\ & & \times & \boxtimes & \boxtimes \\ & & & \times & \boxtimes \\ & & & & \boxtimes \end{bmatrix} + \begin{bmatrix} 0 & \boxtimes & 0 & 0 & 0 \\ & 0 & \boxtimes & \boxtimes & \boxtimes \\ & & 0 & \boxtimes & \boxtimes \\ & & & 0 & \boxtimes \\ & & & & 0 \end{bmatrix}.$$

Similarly as in case 1, we change the elements of the second term in the equation. Leading to the following splitting of the matrix S:

$$S = \begin{bmatrix} \boxtimes & \boxtimes & \boxtimes & 0 & 0 \\ & \times & \boxtimes & 0 & 0 \\ & & \times & \boxtimes & \boxtimes \\ & & + & \boxtimes \\ & & & & \boxtimes \end{bmatrix} + \begin{bmatrix} 0 & \boxtimes & 0 & 0 & 0 \\ & 0 & \boxtimes & \boxtimes & \boxtimes \\ & & 0 & \boxtimes & \boxtimes \\ & & & \boxtimes & \boxtimes \\ & & & 0 & 0 \end{bmatrix}.$$

Applying a rank-expanding Givens transformation, determined by the first term in the sum gives us:

$$G_5^T S = \begin{bmatrix} \boxtimes & \boxtimes & \boxtimes & 0 & 0 \\ & \times & \boxtimes & 0 & 0 \\ & & \times & \boxtimes & \boxtimes \\ & & & \boxtimes & \boxtimes \\ & & & + & \boxtimes \end{bmatrix} + \begin{bmatrix} 0 & \boxtimes & 0 & 0 & 0 \\ & 0 & \boxtimes & \boxtimes & \boxtimes \\ & & 0 & \boxtimes & \boxtimes \\ & & & \boxtimes & \boxtimes \\ & & & \boxtimes & \boxtimes \end{bmatrix}.$$

Similarly as in the first step we expand the structure in the second matrix, giving us the following situation:

$$
G_5^T S = \begin{bmatrix} \boxtimes & \boxtimes & \boxtimes & 0 & 0 \\ & \times & \boxtimes & 0 & 0 \\ & & + & \boxtimes & \boxtimes \\ & & + & \boxtimes & \boxtimes \\ & & & + & \boxtimes \end{bmatrix} + \begin{bmatrix} 0 & \boxtimes & 0 & 0 & 0 \\ & 0 & \boxtimes & \boxtimes & \boxtimes \\ & & \boxtimes & \boxtimes & \boxtimes \\ & & \boxtimes & \boxtimes & \boxtimes \\ & & & \boxtimes & \boxtimes \end{bmatrix}.
$$

The application of a rank-expanding Givens transformation on the first matrix acts now in a special way due to the zero structure (see end of the previous subsection). This gives us the following matrix decomposition:

$$
G_4^T G_5^T S = \begin{bmatrix} \boxtimes & \boxtimes & \boxtimes & 0 & 0 \\ & \times & \boxtimes & 0 & 0 \\ & & \boxtimes & 0 & 0 \\ & & + & \boxtimes & \boxtimes \\ & & & + & \boxtimes \end{bmatrix} + \begin{bmatrix} 0 & \boxtimes & 0 & 0 & 0 \\ & 0 & \boxtimes & \boxtimes & \boxtimes \\ & & \boxtimes & \boxtimes & \boxtimes \\ & & \boxtimes & \boxtimes & \boxtimes \\ & & & \boxtimes & \boxtimes \end{bmatrix}.
$$

To continue, we have to switch our working scheme. Instead of completing again the second matrix in the decomposition now, we complete the first matrix. This gives us the following decomposition:

$$
G_4^T G_5^T S = \begin{bmatrix} \boxtimes & \boxtimes & \boxtimes & 0 & 0 \\ & \boxtimes & \boxtimes & 0 & 0 \\ & \boxtimes & \boxtimes & 0 & 0 \\ & & + & \boxtimes & \boxtimes \\ & & & + & \boxtimes \end{bmatrix} + \begin{bmatrix} 0 & \boxtimes & 0 & 0 & 0 \\ & & + & \boxtimes & \boxtimes & \boxtimes \\ & & + & \boxtimes & \boxtimes & \boxtimes \\ & & & \boxtimes & \boxtimes & \boxtimes \\ & & & & \boxtimes & \boxtimes \end{bmatrix}.
$$

And now we apply a rank-expanding Givens transformation to expand the structure in the second term of the decomposition. This creates extra zeros in the second term.

$$
G_3^T G_4^T G_5^T S = \begin{bmatrix} \boxtimes & \boxtimes & \boxtimes & 0 & 0 \\ & \boxtimes & \boxtimes & 0 & 0 \\ & \boxtimes & \boxtimes & 0 & 0 \\ & & + & \boxtimes & \boxtimes \\ & & & + & \boxtimes \end{bmatrix} + \begin{bmatrix} 0 & \boxtimes & 0 & 0 & 0 \\ & \boxtimes & 0 & 0 & 0 \\ & + & \boxtimes & \boxtimes & \boxtimes \\ & & \boxtimes & \boxtimes & \boxtimes \\ & & & \boxtimes & \boxtimes \end{bmatrix}.
$$

This completes in fact the procedure. The summation of both matrices gives an upper triangular $\{2\}$-semiseparable matrix. We remark that it is interesting to see that the lower rank blocks (zero blocks in the decomposition) shift down and maintain also their zero rank structure.

- **Case 3:** Assume both terms in the summation have the same block decomposition. Instead of starting working directly on the two terms in the sum, we take a closer look at our original matrix. Suppose for this case we have a

7×7 matrix, with a 3×3 zero block in the upper right corner.

$$
S = \begin{bmatrix}
\boxtimes & \boxtimes & \boxtimes & \boxtimes & & & \\
 & \times & \boxtimes & \boxtimes & & & \\
 & & \times & \boxtimes & & & \\
 & & & \times & \boxtimes & \boxtimes & \boxtimes \\
 & & & & \times & \boxtimes & \boxtimes \\
 & & & & & \times & \boxtimes \\
 & & & & & & \boxtimes
\end{bmatrix}.
$$

We can now immediately write this matrix as the sum of two matrices having the strictly upper triangular part of $\{1\}$-semiseparable form, but this will not help us further. Before splitting up the matrix, we observe that we can already also include the element in position $(4,4)$ in the $\{2\}$-semiseparable structure. This gives us in fact the following form for the matrix S:

$$
S = \begin{bmatrix}
\boxtimes & \boxtimes & \boxtimes & \boxtimes & & & \\
 & \times & \boxtimes & \boxtimes & & & \\
 & & \times & \boxtimes & & & \\
 & & & \boxtimes & \boxtimes & \boxtimes & \boxtimes \\
 & & & & \times & \boxtimes & \boxtimes \\
 & & & & & \times & \boxtimes \\
 & & & & & & \boxtimes
\end{bmatrix}.
$$

Decomposing this matrix as the sum of two matrices, both having the same structure (including element $(4,4)$), leads to a decomposition into two matrices having structured rank with regard to this new structure equal to 1:

$$
S = \begin{bmatrix}
\boxtimes & \boxtimes & \boxtimes & \boxtimes & & & \\
 & \times & \boxtimes & \boxtimes & & & \\
 & & \times & \boxtimes & & & \\
 & & & \boxtimes & & & \\
 & & & & \boxtimes & \boxtimes & \boxtimes \\
 & & & & & \times & \boxtimes \\
 & & & & & & \boxtimes
\end{bmatrix}
+
\begin{bmatrix}
\boxtimes & \boxtimes & \boxtimes & & & & \\
 & \times & \boxtimes & & & & \\
 & & \boxtimes & & & & \\
 & & & \boxtimes & \boxtimes & \boxtimes & \boxtimes \\
 & & & & \times & \boxtimes & \boxtimes \\
 & & & & & \times & \boxtimes \\
 & & & & & & \boxtimes
\end{bmatrix}.
$$

This fits perfectly in Case 2.

Moreover, in the final matrix, one can clearly see that all blocks in the upper triangular part of a certain rank have shifted down one position, thereby maintaining their rank.

□

We can formulate this in words as.

Note 9.51.

- *The upper triangular structure has shifted down one position thereby maintaining its rank.*

- *Note that the previous remark is also true for blocks satisfying a certain structure in the upper triangular part.*

- *Again this theorem can be generalized to more hybrid forms of rank structures.*

We can write this in a more general structured rank context as follows.

Corollary 9.52. *Suppose we have a structured rank matrix satisfying the following rank structural constraints (with $q \geq 0$ and $r \geq 1$):*

$$\mathrm{r}\left(\Sigma_u^{(q)}; A\right) \leq r.$$

Performing a sequence of rank-expanding Givens transformations from bottom to top leads to a structured rank matrix of the following form:

$$\mathrm{r}\left(\Sigma_u^{(q-1)}; A\right) \leq r.$$

The restrictions posed in the corollary are somewhat confusing. We assume that the structured rank is higher than or equal to 1. But didn't we use throughout the previous section rank-expanding Givens transformations for expanding a zero rank structure? We can use a sequence of rank-expanding transformations but then q needs to be larger than or equal to 1. This is clear, as a rank-expanding sequence of transformations cannot be performed on the following matrix having $r = 0$ and $q = 0$:

$$\begin{bmatrix} \times & & & & \\ \times & \times & & & \\ \times & \times & \times & & \\ \times & \times & \times & & \\ \times & \times & \times & \times & \times \end{bmatrix}. \tag{9.8}$$

Before formulating a general theorem in terms of structured rank constraints, we will investigate in more detail the existence of rank-expanding sequences of Givens transformations.

Constraints for rank-expanding Givens transformations

It is not always possible to perform rank-expanding Givens transformations. If one has, for example, a lower triangular matrix, with nonzero diagonal elements, one cannot expand anymore the upper triangular zero part of this matrix (see Equation (9.8)). This is logical, the given matrix is nonsingular, and expanding the rank structure of the upper triangular part would make the matrix singular.

It is easy to prove that one can always perform a rank-expanding sequence of Givens transformations, if the following constraint is satisfied, with $q \geq 1 - r$:

$$\mathrm{r}\left(\Sigma_u^{(q)}; A\right) \leq r.$$

For example, if $r = 0$ the matrix can be a lower Hessenberg matrix, if $r = 1$, the matrix can have the strictly upper triangular part of semiseparable form.

But suppose we have, for example, the following matrix:

$$
\begin{bmatrix}
0 & & & & \\
\times & 0 & & & \\
\times & \times & 0 & & \\
\times & \times & 0 & & \\
\times & \times & \times & 0 & 0
\end{bmatrix}. \tag{9.9}
$$

Even though this matrix does not meet the requirements $q \geq r+1$, it is still possible to perform a sequence of rank-expanding Givens transformations to expand the zero rank structure. It is possible to formulate a lemma which, depending on the rank of the global matrix, says that it is possible to expand the rank structure or not. We will not go too deep into detail that is too deep, but in general we can formulate the following theorem.

Theorem 9.53. *Suppose a matrix A of rank r is given. This matrix always needs to satisfy the following constraint:*

$$
\mathrm{r}\left(\Sigma_u^{(r-n-p)}; A\right) \geq p,
$$

for all values of p. A similar equation has to be satisfied for lower triangular rank structures.

Proof. We prove the theorem by contradiction. Suppose we have a matrix A satisfying the following structural constraints:

$$
\mathrm{r}\left(\Sigma_u^{(r-n-p)}; A\right) < p.
$$

To prove that this is in contradiction with our previous assumptions, namely the matrix A being of rank r, we perform $p - 1$ sequences of descending Givens transformations, reducing thereby the rank of the upper triangular part. After having performed these transformations, the reader can verify that we get $n - r + 1$ zero rows on top of the matrix. This means that the rank of the complete matrix can at most be $r - 1$, which is in contradiction with the assumptions. $\square$

The previous theorem clearly shows when it is not possible anymore to perform another sequence of rank-expanding Givens transformations from bottom to top. For example, if a matrix is of rank r and the matrix is of the form:

$$
\mathrm{r}\left(\Sigma_u^{(r-n-p)}; A\right) = p,
$$

then it is not possible anymore to extend the structure of the upper triangular part with one more diagonal, because this would give us the following relation:

$$
\mathrm{r}\left(\Sigma_u^{(r-n-p-1)}; A\right) = p.
$$

Due to the previous theorem, this last equation is in contradiction with the assumption that the matrix is of rank r.

One should be careful, however, not to misinterpret the theorem above. It states when it is surely not possible anymore to perform a sequence of rank-expanding transformations, but it does not state anything about the existence of a sequence of rank-expanding transformations. More precisely, if the condition above is not met, the theorem states that there *might* exist a sequence of rank-expanding Givens transformations.

For example, the following rank-deficient matrix A_1 does not admit a rank-expanding sequence of transformations from bottom to top. The theorem, however, does not provide information concerning this matrix, it states that it *might* be possible to perform a transformation.

The matrix A_2 does however admit another rank-expanding sequence of Givens transformations. Both matrices have the upper triangular part of semiseparable form. The first matrix does not admit expansion of the rank structure by a sequence of Givens transformations, the second one does.

$$A_1 = \begin{bmatrix} 2 & 2 & 2 & 2 \\ 1 & 1 & 1 & 1 \\ 3 & 4 & 1 & 1 \\ 5 & 6 & 7 & 1 \end{bmatrix} \qquad A_2 = \begin{bmatrix} 2 & 2 & 2 & 2 \\ 5 & 1 & 1 & 1 \\ 3 & 4 & 1 & 1 \\ 5 & 6 & 2 & 2 \end{bmatrix}$$

We will not go further into detail, but this theorem can also be translated towards the case of rank-decreasing sequences of Givens transformations. Hence one obtains a theorem stating when it might be possible to create more zeros or stating when it is surely not possible anymore to create more zeros. The theorem can be formulated similarly for a lower triangular rank structure.

Before we reconsider a global theorem for sequences of Givens transformations from bottom to top, we reconsider the shift through lemma.

The shift through lemma revisited

Let us briefly take another look at the proof of the shift through lemma.

The proof was based on the factorization of a 3×3 unitary matrix. First, we considered a QR-decomposition (with R having positive diagonal elements), and based on the fact that R needed to be unitary, we noticed that it had to be equal to the identity matrix.

Second, we investigated the QL-decomposition. Similar reasonings led to the observation that the matrix L also had to be equal to the identity matrix. Let us now reconsider this QL-factorization.

The first Givens transformation was performed to annihilate the upper right element of the matrix U, where the Givens transformation acts on the first and second row:

$$\hat{G}_1^T U = \begin{bmatrix} \times & \times & 0 \\ \times & \times & \times \\ \times & \times & \times \end{bmatrix}.$$

As the final outcome of performing the three Givens transformations needs to be the identity matrix, this Givens transformation also makes the bottom 2×2 block of rank 1. Hence performing this first Givens transformation gives us the following matrix:

$$
\hat{G}_1^T U = \begin{bmatrix} \times & \times & 0 \\ \boxtimes & \boxtimes & \times \\ \boxtimes & \boxtimes & \times \end{bmatrix}.
$$

The second Givens transformation creates a zero in the second row, but also creates, due to the rank 1 structure, two zeros in the bottom row:

$$
\hat{G}_2^T \hat{G}_1^T U = \begin{bmatrix} \times & \times & 0 \\ \times & \times & 0 \\ 0 & 0 & \times \end{bmatrix}.
$$

So this Givens transformation is annihilating in the upper triangular part and rank-expanding in the lower triangular part.

Finally, the last Givens transformation annihilates the remaining off-diagonal elements.

$$
\hat{G}_3^T \hat{G}_2^T \hat{G}_1^T U = \begin{bmatrix} \times & 0 & 0 \\ 0 & \times & 0 \\ 0 & 0 & \times \end{bmatrix}.
$$

Considering only the lower triangular part of the matrix U, on which the transformations are performed, we see that the shift through lemma changes zero-creating Givens transformations to rank-expanding Givens transformations. The $\wedge$-pattern was changed into a $\vee$-pattern, and indeed this created the rank-expanding Givens transformations.

Let us combine all our knowledge gained in this chapter related to sequences of Givens transformations from bottom to top.

9.4.4   Global theorem for sequences from bottom to top

Globally, we have the following theorem for ascending sequences of Givens transformations, covering both the rank-decreasing and rank expanding cases.

Theorem 9.54 (Ascending sequences of Givens transformations). *Suppose we have a structured rank matrix satisfying the following rank structural constraints:*

$$
r\left(\Sigma_l^{(p)}; A \right) \leq r_1 \quad and \quad r\left(\Sigma_u^{(q)}; A \right) \leq r_2.
$$

Performing a sequence of arbitrary Givens transformations from bottom to top leads to a structured rank matrix of the following form:

$$
r\left(\Sigma_l^{(p-1)}; A \right) \leq r_1 \quad and \quad r\left(\Sigma_u^{(q-1)}; A \right) \leq r_2 + 1.
$$

Performing a sequence of rank-decreasing Givens transformations (with $p \leq r_1$) from bottom to top leads to a structured rank matrix of the following form:

$$
r\left(\Sigma_l^{(p-1)}; A \right) \leq r_1 \quad and \quad r\left(\Sigma_u^{(q-1)}; A \right) \leq r_2 + 1.
$$

Performing a sequence of rank-expanding Givens transformations (in case $q \geq 1 - r_2$) from bottom to top leads to a structured rank matrix $\hat{A}$ satisfying the following constraints:

$$\mathrm{r}\left(\Sigma_l^{(p-1)}; A\right) \leq r_1 - 1 \quad and \quad \mathrm{r}\left(\Sigma_u^{(q-1)}; A\right) = r_2.$$

We remark that this theorem can also be formulated for more hybrid structures. More precisely, as already indicated in this chapter, these remarks are in general also valid for blocks satisfying a certain rank structure inside the matrix.

In the next section we will investigate in more detail how sequences of Givens transformations can be performed effectively on Givens-vector represented semiseparable matrices.

9.5 QR-factorization for the Givens-vector representation

In this section we will discuss in more detail the QR-factorization of structured rank matrices for which the structured rank part is represented using the Givens-vector representation. We will make extensive use of the graphical representation of the Givens sequences. We focus in this part towards the Givens-vector representation, as we will see that there is a strong interaction between the sequences of Givens transformations performed on the matrix and the Givens transformations in the representation of the structured rank part. First, we will expand our graphical representation of the Givens transformations towards a graphical representation, also incorporating the structure of the matrix on which the transformations work. We will show how we can easily update the Givens-vector representation in case an arbitrary sequence of transformations is performed on it, and finally we will investigate the rank-expanding sequences of transformations and how to compute them effectively in the case of a Givens-vector representation.

9.5.1   The Givens-vector representation

First of all we will expand our graphical representation. We will graphically depict the Givens-vector representation for the lower triangular part of a 6×6 matrix, as follows.

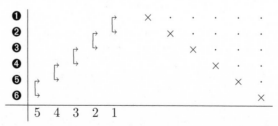

In the previous sections, we were not really interested in the interaction between the Givens transformations and the matrix itself. From now on, we have to take this interaction into consideration. This is logical, as in the computation of the *QR*-factorization, information from the original matrix is necessary. Hence we adapted slightly our graphical scheme to show the interaction between the Givens transformations and a real matrix. The elements marked with · are not essential in the coming analysis, and hence we will not pay attention to them.

To illustrate more precisely what is depicted here, we will construct our semiseparable matrix from this scheme. In fact this is a graphical representation of the reconstruction of a semiseparable matrix from its Givens-vector representation as presented in Section 2.4.

We see that on the left in our graphical representation, the Givens transformations are still shown, whereas on the right we have just the upper triangular part of our semiseparable matrix. Applying now the first Givens transformation gives us the following graphical representation.

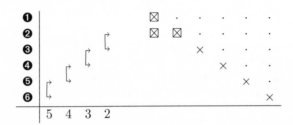

Hence, we removed one Givens transformation, and we filled up the corresponding elements in the matrix. In fact we just performed Givens transformation 1 on the matrix on the right in the graphical representation.

Remark that, in fact, the elements of the upper triangular part also changed, but they are not essential in our analysis, and hence we omitted them in this scheme. The element which is now in position $(2,2)$ of the right matrix corresponds to the second element in our vector **v** of the Givens vector representation. This shows that one should be careful when directly implementing this approach, but the scheme will be very useful for deriving algorithms, as we will show in the forthcoming pages.

To conclude, we perform one more Givens transformation. This gives us the following graphical representation

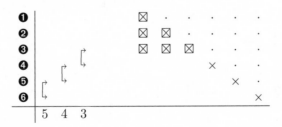

In the following section we will update our Givens-vector representation in an efficient way, when a sequence of arbitrary ascending Givens transformations is applied on our original structured rank matrix.

9.5.2 Applying transformations on the Givens-vector representation

Suppose one is calculating the QR-factorization of a certain structured rank matrix. The structured rank part is decomposed into a sum of matrices all having semiseparability rank 1 and represented with the Givens-vector representation. To compute the QR-factorization we start by annihilating the rank structure of one of the terms in this decomposition. This means that a sequence of arbitrary Givens transformations from bottom to top is performed on the remaining terms in the summation.

In this scheme we will graphically illustrate how one can effectively update the Givens-vector representation when applying sequences of transformations from bottom to top. We would like to derive an easy implementable algorithm for computing the updated Givens-vector representation, if a sequence of ascending Givens transformations is performed on a semiseparable matrix. (Other structured rank matrices can be done similarly, but this illustrates the general case.)

We know that at the end we should obtain a matrix for which the structure in the lower part has shifted down one diagonal position, but its rank is maintained.

Suppose we apply a first Givens transformation on a semiseparable matrix. We use the graphical representation of the Givens-vector representation as presented in the previous subsection.

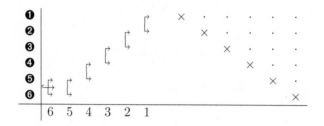

The first Givens transformation can easily fuse with the Givens transformation in position 5. Fusing the transformations in position 5 and position 6, and applying the next ascending transformation on rows 4 and 5, leads to the following scheme.

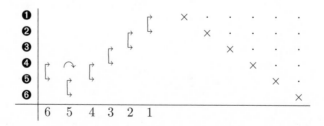

As already depicted in the graphical representation, we apply now the shift through lemma leading to the following scheme.

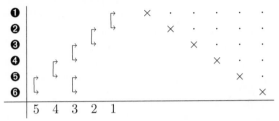

We will apply this newly created Givens transformation in position 3 directly on our matrix. In the next scheme we inserted the next ascending Givens transformation acting on rows 3 and 4, and we applied the transformation in position 3 (below) to the matrix.

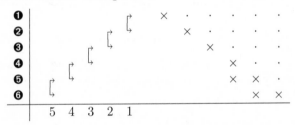

Applying the shift through lemma creates another Givens transformation, which we directly apply to the matrix. This gives us:

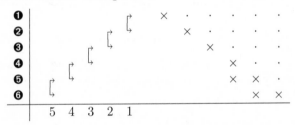

One can clearly see that on the left in the graphical representation, there is still a sequence of Givens transformations from top to bottom, just as in the initial representation, but on the right the matrix has gained an extra subdiagonal.

Continuing this procedure, gives us finally the following resulting matrix.

It is clear that this matrix represents a matrix for which the rank structure has shifted down one position. Moreover, following this approach gives us an easy way of finding the Givens-vector representation of this newly, shifted down, semiseparable part in the matrix.

9.5.3   Computing the rank-expanding Givens transformations

In this section, we will describe how to compute the rank-expanding Givens transformations of a Givens-vector represented matrix in an efficient way. Assume we have a quasiseparable matrix of quasiseparability rank 1 with the Givens-vector representation. We will provide an easy implementable algorithm in this section for computing the rank-expanding Givens transformations.

Graphically, the quasiseparable matrix A, with its Givens-vector representation, is represented as follows:

$$(9.10)$$

This is our initial situation. Finally, after having performed a sequence of descending Givens transformations, we would like to obtain that the complete lower triangular part is of semiseparable form.

We will first prove the existence of a special Givens transformation, such that we can come to the final matrix having the lower triangular part of semiseparable form. Second, we will pour this into an easy implementable algorithm.

Existence of a special Givens transformation

We would like to obtain that our quasiseparable matrix A after having performed one Givens transformation acting on the first and second rows, to be of the following form $G_1^T A$.

$$
G_1^T A = \begin{bmatrix}
\boxtimes & \cdot & \cdot & \cdot & \cdot & \cdot \\
\boxtimes & \boxtimes & \cdot & \cdot & \cdot & \cdot \\
\boxtimes & \boxtimes & \times & \cdot & \cdot & \cdot \\
\boxtimes & \boxtimes & \boxtimes & \times & \cdot & \cdot \\
\boxtimes & \boxtimes & \boxtimes & \boxtimes & \times & \cdot \\
\boxtimes & \boxtimes & \boxtimes & \boxtimes & \boxtimes & \times
\end{bmatrix} .
$$

This means that we have performed the first rank-expanding Givens transformation, such that the element in position $(2,2)$ is includable in the lower structure.

Graphically, one can easily check that this corresponds to the following scheme:

$$(9.11)$$

We will start now, from Scheme 9.10, to apply a transformation on the first and the second row, and investigate which constraints this transformation needs to satisfy to obtain finally Scheme 9.11.

Performing a Givens transformation acting on rows 1 and 2 on the left of the matrix A leads to an adaptation of scheme 9.10 in the following sense.

$$(9.12)$$

As Scheme 9.11 has a zero in the matrix in the first column on position 2, we will also create here this zero before continuing. This slightly changes scheme 9.12, where the new Givens transformation in position 1 was determined to annihilate the element in the first column on position 2:

$$(9.13)$$

Applying now the shift through lemma on the first three Givens transformations, as indicated:

leads to the following Scheme 9.14, in which we marked one element with $\otimes$.

$$(9.14)$$

Assume now that it is possible to choose the initial Givens transformation performed on rows 1 and 2, such that after applying the shift through lemma the Givens transformation in position 1 in Scheme 9.14 annihilates the element $\otimes$. This means that after application of the Givens transformation in position 1, we get the following situation, which is exactly the situation as we wanted to obtain after performing the first rank-expanding Givens transformation.

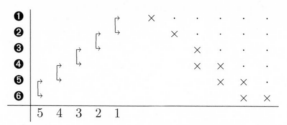

Summarizing, the rank-expanding transformation was chosen in such a way that, after performing the shift through lemma, an element was annihilated in the matrix. Of course the main question is, does such a transformation always exist. If such a transformation exists, one can see that it is fairly easy to compute this rank-expanding Givens transformation, as well as the updated Givens-vector representation.

Hence we want to know whether, based on the two Givens transformations in positions 1 and 2 in Scheme 9.13 and based on the Givens transformation in position 1 in Scheme 9.14, which are all known, one can compute the Givens transformations in positions 2 and 3 in Scheme 9.14 and the initial rank-decreasing Givens transformation. The following lemma solves this problem.

Lemma 9.55. *Suppose three 3×3 Givens transformations G_2, G_3 and $\hat{G}_3$, are given, such that the Givens transformations G_2 and $\hat{G}_3$ act on the second and third rows of a matrix, and G_3 acts on the first and second rows of a matrix.*

Then we have that transformations G_1, $\hat{G}_1$ and $\hat{G}_2$ exist, such that the following equation is satisfied:

$$G_1 G_2 G_3 = \hat{G}_1 \hat{G}_2 \hat{G}_3,$$

where G_1 and $\hat{G}_2$ work on the first and second rows and $\hat{G}_1$, works on the last two rows.

Proof. The proof is a straightforward application of the standard shift through lemma. Rewriting the following equation

$$G_1 G_2 G_3 = \hat{G}_1 \hat{G}_2 \hat{G}_3,$$

in the following sense

$$G_2 G_3 \hat{G}_3^T = G_1^T \hat{G}_1 \hat{G}_2 \hat{G},$$

combines all unknowns to one side and the known Givens transformations to the other side. Hence we can apply the standard shift through lemma. $\square$

It is clear that these transformations always exist. Let us present a working algorithm for performing this reduction.

Algorithm

The real algorithm works in fact in reverse order as the deduction above. First, the Givens transformation is determined to annihilate a specific subdiagonal element. This Givens transformation is dragged through the Givens transformations of the Givens-vector representation. This means that it pops out on the left side of the descending sequence of transformations. Then, one chooses as the rank-expanding Givens transformation the transpose of the Givens transformation just dragged through. In this way both Givens transformations on the left collapse and disappear.

The algorithm can be described as follows.

- Annihilate the first subdiagonal element and expand thereby the Givens-vector representation. Do not forget to apply this transformation also to the upper triangular part.

- For the remaining subdiagonal elements, process them from top to bottom.

  - Annihilate the current subdiagonal element.

  - Apply this Givens transformation also to the upper triangular part.

  - Combine the Givens transformations from the Givens-vector representation, with this newly computed Givens transformation to update the Givens-vector representation by using the shift through lemma.

  - Note that there is one Givens transformation extra now. But this transformation is the transpose of the expanding Givens transformation applied on the corresponding rows and hence collapses and disappears.

One can clearly see that, at fairly low cost, one can compute the rank-expanding Givens transformations.

Let us demonstrate graphically how the algorithm works. Suppose we have already performed one transformation. This means that we are in the same situation as in Scheme 9.11. According to the algorithm we start by choosing our Givens transformation to annihilate element $(4,3)$. This means that after this element $(4,3)$ is annihilated, we get the following situation.

Applying now the shift through lemma updates our representation and gives us the following situation.

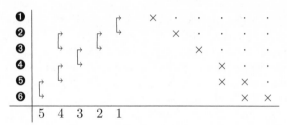

In fact, we did not yet apply a rank-decreasing Givens transformation on rows 2 and 3. Choosing now the rank-expanding Givens transformation as the inverse of the Givens transformation in position 3 (the top transformation) gives us.

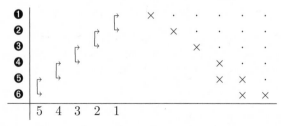

This results in a vanishing of these two first transformations and gives us the resulting scheme.

It is clear that this matrix has an extra element includable in the lower triangular semiseparable structure. Moreover, the reader can see that the performance of the rank-expanding Givens transformation is not really essential anymore. After having applied the shift through lemma, one can simply remove the Givens transformation in position 3.

Finally, after having completed our algorithm, we obtain the following graphical representation of the resulting matrix. Clearly the lower triangular part (including the diagonal) is now of semiseparable form.

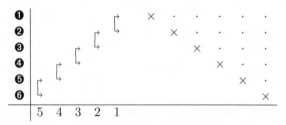

We showed in this section that the Givens-vector representation is quite flexible with regard to the performance of sequences of Givens transformations. In the next section, some results are presented, related to the theoretical considerations discussed in this chapter. They are not necessarily directly linked with the QR-factorization. But they make use of the theory discussed here.

9.6 Extra material

We discussed quite a lot of properties of structured rank matrices with regard to sequences of Givens transformations and the QR-factorization. In this section we provide a blend of material related to the previous theorems and algorithms. We will discuss here briefly how we can interpret the QR-factorization of an unstructured matrix. We show how we can design a QZ-factorization of a matrix, in which the matrix Z is of Hessenberg-like form. A parallel implementation of the QR-algorithm for structured rank matrices is discussed. Finally we conclude this section by proving some results related to the multiplication of structured rank matrices.

9.6.1 A rank-expanding QR-factorization

For computing the QR-factorization of an unstructured matrix, one can use householder transformations or Givens transformations. When using Givens transformations, and choosing the pyramid form for annihilating the subdiagonal entries, one can in fact interpret this pattern as a sequence of rank-expanding Givens transformations. Consider, for example, the following 4×4 matrix A. A first Givens transformation applied on the left reduces the matrix to the following form:

$$A^{(1)} = \begin{bmatrix} \times & \cdot & \cdot & \cdot \\ \times & \times & \cdot & \cdot \\ \times & \times & \times & \cdot \\ 0 & \times & \times & \times \end{bmatrix}.$$

Annihilating now one element in the first column and one in the second column, we obtain the following situation:

$$A^{(3)} = \begin{bmatrix} \times & \cdot & \cdot & \cdot \\ \times & \times & \cdot & \cdot \\ 0 & \times & \times & \cdot \\ 0 & 0 & \times & \times \end{bmatrix}.$$

Hence one can clearly see that we expanded the zero block in the bottom left position of the matrix $A^{(1)}$, and we added one more diagonal to this structure.

To obtain the complete QR-factorization, we need to perform one more sequence of rank-expanding Givens transformations from bottom to top to remove the subdiagonal.

Moreover, a general unstructured matrix A satisfies the following rank constraint:

$$\mathrm{r}\left(\Sigma_l^{(n-1)}; A\right) \leq n - 1.$$

Hence we know that for computing the QR-factorization, in general, we need $n-1$ sequences of ascending rank-decreasing Givens transformations. This is exactly the number of sequences that we needed in the transformations above.

9.6.2   Parallel factorization

Combining the rank-decreasing and the rank-expanding Givens transformations, we can deduce a parallel algorithm for computing the QR-factorization of structured rank matrices. Let us deduce this parallel algorithm for the simple case of a quasiseparable matrix. To compute the QR-factorization of a quasiseparable matrix, we need to use one ascending and one descending sequence of Givens transformations. This standard $\wedge$-pattern is not suitable for parallel implementation. The X-pattern is, however. (For higher order matrices, X X X-patterns are suitable for parallel implementation.) We see that the X-pattern allows us to start working at the top and the bottom at the same time. The only time when there is communication between the upper and the lower part is in the intersection between the descending and the ascending sequence of Givens transformations.

Let us illustrate this technique on a 6×6 matrix. The graphical scheme which will be used is the following one.

Because we will start working on two processors at the same time, we can perform both transformations in column 1 at the same time, followed by both transformations in row 2 at the same time. Then there is communication between the lower and upper part, and we can proceed again in a parallel way. More precisely, the algorithm starts by dividing the matrix into two layers of 3×6 matrices. Both layers are sent to different processors. On the top layer we start the descending sequences of rank-expanding Givens transformations. On the bottom layer we perform ascending sequence of Givens transformations to annihilate the lower triangular semiseparable part. Name the top part A_T and the bottom part A_B.

Let us illustrate what happens on the top part. Our matrix A_T is of the following form, where the $\boxtimes$ denote the elements satisfying the quasiseparable rank structure:

$$A_T = \begin{bmatrix} \times & \times & \times & \times & \times & \times \\ \boxtimes & \times & \times & \times & \times & \times \\ \boxtimes & \boxtimes & \times & \times & \times & \times \end{bmatrix}.$$

First, we perform a rank-expanding Givens transformation on the first two rows of the matrix A_T. This leads to the following structure (Let us denote the Givens transformations acting on the top part with an H):

$$H_1^T A_T = \begin{bmatrix} \boxtimes & \times & \times & \times & \times & \times \\ \boxtimes & \boxtimes & \times & \times & \times & \times \\ \boxtimes & \boxtimes & \times & \times & \times & \times \end{bmatrix}.$$

Second, we add also an element in the third row to the structure by performing a rank-expanding Givens transformation (we remark that, little information is needed from the bottom part for defining this rank-expanding transformation):

$$H_2^T H_1^T A_T = \begin{bmatrix} \boxtimes & \times & \times & \times & \times & \times \\ \boxtimes & \boxtimes & \times & \times & \times & \times \\ \boxtimes & \boxtimes & \boxtimes & \times & \times & \times \end{bmatrix}.$$

At the same time, another processor can be working on the matrix A_B, which has the following structure:

$$\begin{bmatrix} \boxtimes & \boxtimes & \boxtimes & \times & \times & \times \\ \boxtimes & \boxtimes & \boxtimes & \boxtimes & \times & \times \\ \boxtimes & \boxtimes & \boxtimes & \boxtimes & \boxtimes & \times \end{bmatrix}.$$

On the lower part, we start with annihilating the last row:

$$G_6^T A_B = \begin{bmatrix} \boxtimes & \boxtimes & \boxtimes & \times & \times & \times \\ \boxtimes & \boxtimes & \boxtimes & \boxtimes & \times & \times \\ & & & & \times & \times \end{bmatrix}.$$

The Givens transformation G_5^T annihilates elements in the second to last row:

$$G_5^T G_6^T A_B = \begin{bmatrix} \boxtimes & \boxtimes & \boxtimes & \times & \times & \times \\ & & & \times & \times & \times \\ & & & & \times & \times \end{bmatrix}.$$

At this point, we see that we have arrived at the bottom of the matrix A_T and at the top of the matrix A_B; this is the intersection in the X-pattern. Hence we need communication between the two processors. This communication corresponds to the performance of one Givens transformation, based on the bottom row of $H_2^T H_1^T A_T$ and the top row of $G_5^T G_6^T A_B$. Let us illustrate this on the full matrix.

Combining the transformed top and transformed bottom part into one matrix gives us the following matrix $\tilde{A}$:

$$\tilde{A} = \begin{bmatrix} \boxtimes & \times & \times & \times & \times & \times \\ \boxtimes & \boxtimes & \times & \times & \times & \times \\ \boxtimes & \boxtimes & \boxtimes & \times & \times & \times \\ \boxtimes & \boxtimes & \boxtimes & \times & \times & \times \\ & & & & \times & \times & \times \\ & & & & & \times & \times \end{bmatrix}.$$

Performing now the Givens transformation acting on row 3 and row 4, annihilating thereby elements in row 4, gives us:

$$\hat{A} = \tilde{G}\tilde{A} = \begin{bmatrix} \boxtimes & \times & \times & \times & \times & \times \\ \boxtimes & \boxtimes & \times & \times & \times & \times \\ \boxtimes & \boxtimes & \boxtimes & \times & \times & \times \\ & & & \times & \times & \times \\ & & & \times & \times & \times \\ & & & & \times & \times \end{bmatrix}.$$

This ends the communication step. If we divide now again the matrix $\hat{A}$ into a 3×6 matrix $\hat{A}_T$ consisting of the top three rows and a 3×6 matrix $\hat{A}_B$, we can again start working in parallel.

On the upper part we will now perform annihilating Givens transformations, and in the lower part we will now perform rank-expanding Givens transformations. Applying the first annihilating Givens transformation on the matrix $\hat{A}_T$ gives us:

$$\hat{H}_3\hat{A}_T = \begin{bmatrix} \boxtimes & \times & \times & \times & \times & \times \\ \boxtimes & \boxtimes & \times & \times & \times & \times \\ & & \times & \times & \times & \times \end{bmatrix}.$$

The next transformation annihilates the last off-diagonal element in the upper part.

$$\hat{H}_2^T\hat{H}_3^T\hat{A}_T = \begin{bmatrix} \times & \times & \times & \times & \times & \times \\ & \times & \times & \times & \times & \times \\ & & \times & \times & \times & \times \end{bmatrix}.$$

This completes the procedure for reducing the upper part to upper triangular form.

At the same time, the second computer can perform rank-expanding Givens transformations to expand the zero part in the lower part. In fact these Givens transformations are annihilating the off-diagonal elements as in the Hessenberg case. The first Givens transformation acting on $\hat{Q}_B$ creates the following structure:

$$\hat{G}_4^T\hat{A}_B = \begin{bmatrix} 0 & 0 & 0 & \times & \times & \times \\ 0 & 0 & 0 & 0 & \times & \times \\ 0 & 0 & 0 & 0 & \times & \times \end{bmatrix}.$$

The last transformation on the bottom part creates the following structure:

$$\hat{G}_5^T\hat{G}_4^T\hat{A}_B = \begin{bmatrix} 0 & 0 & 0 & \times & \times & \times \\ 0 & 0 & 0 & 0 & \times & \times \\ 0 & 0 & 0 & 0 & 0 & \times \end{bmatrix}.$$

Recombining the matrices again gives us the upper triangular factorization of the matrix computed in a parallel manner.

The interested reader can try to verify that for specific higher order rank structured matrices even more processors can often be used in the parallelization scheme.

9.6.3 QZ-factorization of an unstructured matrix

In this section we will illustrate that one can also easily compute the QZ-factoriza-tion of an arbitrary matrix, where Z is a Hessenberg-like matrix. This means that the lower triangular part of the matrix Z will be of semiseparable form.

Suppose we have a matrix A given, and for some reason we want to compute a factorization of the form QZ, with Q an orthogonal matrix and Z a Hessenberg-like matrix (i.e., the matrix Z has the lower triangular part of semiseparable form). The outcome of our algorithm will give us the factored form of the matrix Q, factored with Givens transformations, and the matrix Z his lower triangular part repre-sented with the Givens-vector representation. We will illustrate the approach for a 4×4 matrix by using our graphical representation. Without any transformations performed on the matrix, the matrix is of the following form.

```
❶ | ×   ·   ·   ·
❷ | ×   ×   ·   ·
❸ | ×   ×   ×   ·
❹ | ×   ×   ×   ×
```

We start the procedure by annihilating the lower left two elements in the first column by two Givens transformations. We do not depict it, but of course the other elements in these rows change too. The following figure represents exactly the same matrix as above, but already partially factored.

```
❶ |              ×   ·   ·   ·
❷ |        ⌐  ⌐  ×   ×   ·   ·
❸ |     ⌐  ⌐  0   ×   ×   ·
❹ |     ⌐       0   ×   ×   ×
   ─────────────
        2   1
```

In fact these two Givens transformations already depict part of the Givens-vector representation of the small lower left block to spread out the structure of the first column downwards again. In the next figure, we annihilated the last element in column 2. Moreover, we will apply the shift through lemma as indicated.

```
❶ |              ×   ·   ·   ·              ❶ |              ×   ·   ·   ·
❷ |     ⌐  ⌐  ⌐  ×   ×   ·   ·              ❷ |  ⌐  ⌐  ⌐  ×   ×   ·   ·
❸ |  ⌐  ⌐  ⌣  ⌐  0   ×   ×   ·  leading to  ❸ |     ⌐     0   ×   ×   ·
❹ |  ⌐     ⌣  ⌐  0   0   ×   ×              ❹ |     ⌐     0   0   ×   ×
   ──────────────                             ──────────────
        3   2   1                                 3 │ 2   1
```

We placed an extra vertical bar in the scheme. This bar indicates the separation between our final matrix Z on the right and our orthogonal matrix Q on the left. It is clear that the right part represents a tiny semiseparable block with the Givens-vector representation, whereas the left part (transformation 3) only represents a Givens transformation, which will be stored later on in the matrix Q. Let us annihilate now the first and second subdiagonal elements and then apply the shift through lemma.

$$
\begin{array}{cccc}
\times & \cdot & \cdot & \cdot \\
0 & \times & \cdot & \cdot \\
0 & 0 & \times & \cdot \\
0 & 0 & \times & \times
\end{array}
\quad\Rightarrow\quad
\begin{array}{cccc}
\times & \cdot & \cdot & \cdot \\
0 & \times & \cdot & \cdot \\
0 & 0 & \times & \cdot \\
0 & 0 & \times & \times
\end{array}
$$

To complete the reduction procedure, we annihilate one more subdiagonal element and apply again the shift through lemma.

$$
\begin{array}{cccc}
\times & \cdot & \cdot & \cdot \\
0 & \times & \cdot & \cdot \\
0 & 0 & \times & \cdot \\
0 & 0 & 0 & \times
\end{array}
$$

This gives us the final result. On the right of the vertical bar we have the Givens-vector representation of the matrix Z. On the left of the vertical bar, we have the matrix Q in factored form.

$$
\begin{array}{cccc}
\times & \cdot & \cdot & \cdot \\
0 & \times & \cdot & \cdot \\
0 & 0 & \times & \cdot \\
0 & 0 & 0 & \times
\end{array}
$$

Of course this is not the only possible way to compute this factorization. One can, for example, first make the complete first column zero, then continue with the second column and so on.

Notes and references

The results on the parallel QR-factorization as presented in this section are discussed more elaborately in the following manuscript. The algorithm is implemented and tests concerning the speed and accuracy of the proposed method are provided. Also indications for parallelizing the QR-factorization of higher order structured rank matrices are presented.

☞ R. Vandebril, M. Van Barel, and N. Mastronardi. A parallel QR-factorization/solver of structured rank matrices. Technical Report TW474, Department of Computer Science, Katholieke Universiteit Leuven, Celestijnenlaan 200A, 3000 Leuven (Heverlee), Belgium, October 2006.

9.7 Multiplication between structured rank matrices

In this section we will investigate the rank structure of the product of two structured rank matrices. We will investigate the products of these matrices, based on their structured rank and based on the results of the QR-decomposition of Chapter 9.

9.7.1   Products of structured rank matrices

We will prove first a general theorem on the multiplication of two structured rank matrices, followed by several examples.

Theorem 9.56. *Suppose A_1 and A_2 are two structured rank matrices satisfying the following constraints:*

$$\mathrm{r}\left(\Sigma_l^{(p_1)}; A_1\right) \leq r_1 \quad \text{and} \quad \mathrm{r}\left(\Sigma_u^{(q_1)}; A_1\right) \leq s_1,$$

$$\mathrm{r}\left(\Sigma_l^{(p_2)}; A_2\right) \leq r_2 \quad \text{and} \quad \mathrm{r}\left(\Sigma_u^{(q_2)}; A_2\right) \leq s_2.$$

The matrix $A = A_1 A_2$ will satisfy the following relations:

$$\mathrm{r}\left(\Sigma_l^{(p_1+p_2)}; A\right) \leq r_1 + r_2 \quad \text{and} \quad \mathrm{r}\left(\Sigma_u^{(q_1+q_2)}; A\right) \leq s_1 + s_2.$$

Proof. The proof is based entirely on performing sequences of Givens transformations (see Chapter 9). We start by proving the statement related to the lower part of the matrix A. Suppose we know the QR-decomposition of the matrix $A_2 = Q_2 R_2$. To compute the QR-decomposition of the matrix A_2, in general, the matrix Q_2^T will consist of r_2 ascending sequences of Givens transformations, followed by $-p_2 + r_2$ sequences of descending Givens transformations, in the case $(p_2 - r_2) < 0$.
This means that the matrix Q_2, the transpose of the matrix Q_2^T, consists of $-p_2 + r_2$ ascending sequences of Givens transformations, followed by r_2 descending sequences of Givens transformations.
Consider the following equation:

$$\begin{aligned} A &= A_1 A_2 \\ &= A_1(Q_2 R_2) = (A_1 Q_2) R_2. \end{aligned}$$

Let us now use the Corollaries from Subsection 9.1.8 for computing the matrix product $A_1 Q_2$. Performing the sequences of Givens transformations of Q_2 on the matrix A_1 corresponds to performing r_2 sequences of transformations from right to left, followed by $-p_2 + r_2$ sequences from left to right.
The result of performing these sequences gives us the following constraint for the lower part on the matrix $A_1 Q_1$:

$$\mathrm{r}\left(\Sigma_l^{(p_1+p_2)}; A_1 Q_1\right) \leq r_1 + r_2.$$

As the matrix R_2 is of upper triangular form, this will not change the rank structure of the lower part; hence, we have that $A = A_1 Q_2 R_2$ satisfies the following relation:

$$\mathrm{r}\left(\Sigma_l^{(p_1+p_2)}; A\right) \leq r_1 + r_2.$$

The relation for the upper triangular part can be proven similar as the one for the lower part. Instead of using the QR-factorization of A_2, we use now the LQ-factorization of the matrix $A_1 = L_1 Q_1$, then perform the transformations of Q_1

onto the matrix A_2 and finally multiply with the matrix L_1.   ☐

Let us now apply these results onto different classes of structured rank matrices.

9.7.2   Examples

In the following examples we investigate the matrix product of two structured rank matrices, where the structured rank matrices are from specific classes.

We do not provide the proof for all classes, as most of them are direct applications of Theorem 9.56.

The product of two semiseparable matrices is again a semiseparable matrix.

Corollary 9.57. *Suppose S_1 and S_2 are two semiseparable matrices of $\{p_1, q_1\}$-semiseparable and $\{p_2, q_2\}$-semiseparable form, respectively. The resulting matrix $S = S_1 S_2$ will in general be a $\{p_1 + p_2, q_1 + q_2\}$-semiseparable matrix.*

A consequence of this corollary is the fact that the multiplication of a $\{p, q\}$-semiseparable matrix with a diagonal matrix ($\{0\}$-semiseparable) is again a $\{p, q\}$-semiseparable matrix.

The product of two quasiseparable matrices is again a quasiseparable matrix.

Corollary 9.58. *Suppose A_1 and A_2 are two quasiseparable matrices, respectively, $\{p_1, q_1\}$-quasiseparable and $\{p_2, q_2\}$-quasiseparable form. The resulting matrix $A = A_1 A_2$ will in general be a $\{p_1 + p_2, q_1 + q_2\}$-quasiseparable matrix.*

The product of two band matrices will again be a band matrix.

Corollary 9.59. *Suppose B_1 and B_2 are a $\{p_1, q_1\}$-band and a $\{p_2, q_2\}$-band matrix, respectively. The resulting matrix $B = B_1 B_2$, will be a $\{p_1 + p_2, q_1 + q_2\}$-band matrix.*

We have proved now that the classes of semiseparable, quasiseparable, and band matrices are closed under multiplication. It is easy to see that in general extended/decoupled semiseparable matrices are not closed under multiplication.

To conclude this section, we would like to pay some attention briefly to the class of higher order generator representable semiseparable matrices. Is the product of two generator representable semiseparable matrices again a generator representable semiseparable matrix, but of higher order?

In general we can formulate the following corollary:

Corollary 9.60. *Suppose S_1 and S_2 are two generator representable semiseparable matrices, respectively, $\{p_1, q_1\}$-generator representable semiseparable and $\{p_2, q_2\}$-generator representable semiseparable. The resulting matrix $S = S_1 S_2$ will in general be a $\{p_1 + p_2, q_1 + q_2\}$-generator representable semiseparable matrix.*

Proof. The proof is almost similar to the proof of Theorem 9.56. Let us focus attention on proving the structure of the lower triangular part. In this case we do need to decompose the matrix A_1 as the sum of a rank p_1 matrix and a strictly upper triangular matrix (the remainder). Then one should perform p_2 sequences of Givens transformations from right to left on these decomposed matrices. Because the matrix A_2 is also generator representable, none of the Givens transformations will be equal to the identity. Therefore the matrix A_1 will gradually have the lower part transformed, such that it will become of $\{p_1 + p_2\}$-generator representable form. The details of the proof are left to the reader. $\square$

By direct calculation, one can also examine that the product of two strict band matrices is again a strict band matrix.

Notes and references

In the manuscript [93] by Eidelman and Gohberg (see also Chapter 12), the structure of the product of two quasiseparable matrices is investigated theoretically. The theoretical result is followed by an algorithmic description on how to compute the quasiseparable representation of the product in the case both quasiseparable matrices were represented with the quasiseparable representation.

9.8   Conclusions

Even though we did not really develop directly applicable algorithms in this chapter, we developed some interesting tools for working with structured rank matrices. Especially the interaction of sequences of Givens transformations with the structure of these matrices was particularly interesting. Moreover the rank-decreasing and rank-expanding sequences of transformations provided tools which led to different patterns of annihilating the lower triangular part in structured rank matrices.

Finally we concluded this chapter by providing a parallel QR-algorithm for structured rank matrices. We concluded by providing a theorem, predicting the rank structure of the product of two structured rank matrices.

Chapter 10

A Gauss solver for higher order structured rank systems

In the previous chapter we studied thoroughly the QR-factorization of structured rank matrices. Special attention was paid to the rank structure of all involved matrices. The matrix Q was factored as a product of Givens transformations and theorems were proven predicting the structure of the upper triangular matrix R. In this chapter we will derive similar theorems for working with Gauss transforms instead of Givens transformations. We already know that the LU-factorization of such a structured rank matrix is not always easy to compute, in particular, pivoting seemed to be troublesome. In this chapter we will go into a little more detail and investigate what the effect of sequences of Gauss transforms (including pivoting) can be on structured rank matrices.

In the first section attention will be given to the basic sequences, namely the sequences from bottom to top. We investigate first the effect of these sequences on the lower triangular structure of structured rank matrices. We will show that the performance of such a sequence of Gauss transforms has no effect on the rank structure of the upper triangular structure. This is very interesting, but unfortunately we cannot guarantee in general the existence of such a sequence of Gauss transforms. Following this ascending sequence of transformations, we discuss a sequence of descending transformations for expanding the rank structure of the lower triangular part. Also these transformations leave the rank structure of the upper triangular part unchanged, but again we cannot guarantee their existence in general.

In Section 10.2, we discuss pivoting. First, we see what the effect of pivoting is on a structured rank matrix. Second, we will combine pivoting with Gauss transforms to guarantee the existence of rank-decreasing or rank-expanding sequences of transformations. We will deduce general theorems predicting the rank structure of a structured rank matrix after having applied a sequence of Gauss transforms involving pivoting or not.

Section 10.3 pays attention to the existence of other Gauss transforms. The Gauss transforms considered in the previous sections were lower Gauss transforms applied on the left side of the matrix. In this section we will briefly investigate upper Gauss transforms and also the application of transformations on the right of a matrix.

Solving systems of equations using sequences of Gauss transforms is the subject of Section 10.4. First, we discuss how to use the above-discussed techniques for efficiently transforming a matrix to upper triangular form. Unfortunately, we will see that using pivoting in the sequences of transformations will generally destroy the lower or upper triangular form of the factors. This means that in general we will not obtain an LU-factorization anymore, but we will see that we can still use this technique for solving systems of equations. To conclude we show also the existence of some other decompositions such as the UL-decomposition and the L_1UL_2-decomposition.

To conclude this chapter we present similarly as we did in the case of the QR-factorizations a graphical scheme to work with Givens transformations. We will also give a generalization of the nullity theorem and some other interesting relations between lower/upper Gauss transforms and pivoting matrices.

✎ *The idea in this chapter is to solve systems of equations by using Gauss transforms. It will be shown that systems can be solved via Gauss transforms, but the computed factorization will not necessarily be linked to the LU-factorization anymore.*

In a similar way as it was proved and shown in the previous chapter, the effect of a sequence of Gauss transforms on a structured rank matrix is stated in this chapter. The types of sequences involved are annihilating sequences of Gauss transforms (Definition 10.2) and their extension to rank-decreasing sequences of Gauss transforms (Subsection 10.1.4). The effect of such a sequence is discussed in Corollary 10.9 followed by a clarifying explanation.

Before being able to explain the effect of a sequence of rank expanding Gauss transformations (Subsection 10.1.7), explanations of the difference between an ascending and a descending sequence of Gauss transforms, discussed in Subsection 10.1.5, and the effect of a zero-creating sequence of Gauss transforms (Subsection 10.1.6) are necessary. Corollary 10.15 summarizes the effect of a rank-expanding sequence of Gauss transforms on the structured rank of a matrix.

The existence of none of both sequences could be guaranteed in general. Hence pivoting is required. The inclusion of pivoting leads to a changed effect on the rank structure. The global effect for either ascending and descending sequences is presented in Corollary 10.19 and Corollary 10.20. A clarifying explanation of the effect is presented in Section 10.3, including also the effect of Gauss transforms applied to the right of the matrix. Based on this knowledge the reader can solve systems of equations by performing Gauss transforms and pivoting: Section 10.4. In this section especially Subsection 10.4.1 and Subsection 10.4.2 are the ones on which to focus.

The final section (Section 10.5) discusses a graphical representation for Gauss transforms, and possible interactions between Gauss transforms and pivoting are for the interested reader.

10.1   A sequence of Gauss transformation matrices without pivoting

Similarly as in the previous chapter, concerned with the QR-decomposition, we will derive here the effect of a sequence of Gauss transforms from bottom to top on structured rank matrices.

There are some essential differences between Gauss transforms and Givens transformations. We would first like to discuss these differences before starting to apply sequences of Gauss transforms on structured rank matrices.

- A first difference between Gauss transforms and Givens transformations is the action on the matrix. An elementary Gaussian transformation matrix changes only one row in the matrix it is acting on. A Givens transformation, however, changes two rows at once.

- A very important difference between the performance of Gauss transforms and the performance of Givens transformations is the numerical stability. Elementary Gaussian transformation matrices are not orthogonal, whereas Givens transformations are. This means that when applying Gauss transformations on a matrix this can have a large effect on the norm of the involved elements. This might lead to numerical instabilities. Therefore pivoting is often used to improve the numerical stability of the method. In our case, however, we will see that pivoting will often destroy the involved rank structure. Therefore one should be very careful when applying the techniques that will be discussed in this chapter. If the matrix is positive definite some theorems are available for predicting stability, but we will not go into these details.

We start as in the previous chapter and investigate in detail the effect of sequences of Gauss transforms on matrices. More precisely we will start by applying ascending sequences of Gauss transforms on matrices.

10.1.1   Annihilating Gauss transforms on lower rank 1 structures

We are already familiar with the concepts of rank structures, as they were frequently used in the previous chapter on the QR-factorization. Hence, we will go faster in this chapter and leave several details to the reader.

Remember that an elementary Gaussian transformation matrix (in brief Gauss transform) is of the following form $M_i = I - \tau_i \mathbf{e}_{i+1} \mathbf{e}_i^T$, with the vectors $\mathbf{e}_i$, the standard basis vectors. In our reasonings we assume that the Gauss transform M_i is working on row $i + 1$ and row i. Remark that the Gauss transforms as defined here, only change row $i + 1$. Row $i + 1$ is replaced by a summation of a multiple of row i and row $i + 1$.

Theorem 10.1. *Suppose S is a lower triangular semiseparable matrix. Suppose a sequence of $n - 1$ Gauss transformations M_i^T, acting on row $i - 1$ and i exists*[12]

---

[12]We will address the existence issue later on.

and is performed on the matrix S for making it upper triangular:

$$M_1 \ldots M_{n-1} S = D.$$

The matrix D will then be a diagonal matrix.

Definition 10.2. *Similarly as for Givens transformations, we can define a sequence of annihilating Gauss transforms. A sequence of Gauss transforms $M_k^T \ldots M_l^T$ (with $2 \leq k \leq l \leq n$) applied on the matrix A is called an annihilating sequence of Gauss transforms, if for all i the application of the Gauss transform M_i^T (acting on rows $i-1$ and i) onto $M_{i+1}^T \ldots M_n^T A$ creates at least one zero in row i in the matrix $M_{i+1}^T \ldots M_n^T A$.[13]*

   A sequence of Gauss transformations is called a rank-annihilating sequence of Gauss transforms, if it annihilates elements in a rank 1 structure.

Proof. We are especially interested in the behavior related to the upper triangular part. In the QR case this part gradually filled up, thereby generally increasing the rank of the upper triangular part. In this case, however, we will see that the strictly upper triangular part remains zero, hence the rank of the strictly upper triangular part is not affected.

Let us illustrate this for a 5×5 lower triangular semiseparable matrix, as this illustrates the general case. Let us denote the lower semiseparable part of the matrix with elements $\boxtimes$. Assume moreover that every row of elements denoted with $\boxtimes$ possesses at least one nonzero element. The demand of nonzeroness makes sure that the Gauss transforms are all well-defined (this issue will be addressed in Note 10.3). The elements that will be annihilated with the Gauss transform are marked with $\otimes$. The invisible elements are assumed to be zero. Let us start with the annihilation of the last row by performing the Gauss transform M_4 (define $S_0 = S$):

$$
\begin{bmatrix}
\boxtimes & & & & \\
\boxtimes & \boxtimes & & & \\
\boxtimes & \boxtimes & \boxtimes & & \\
\boxtimes & \boxtimes & \boxtimes & \boxtimes & \\
\otimes & \otimes & \otimes & \otimes & \boxtimes
\end{bmatrix}
\xrightarrow{M_4 S_0}
\begin{bmatrix}
\boxtimes & & & & \\
\boxtimes & \boxtimes & & & \\
\boxtimes & \boxtimes & \boxtimes & & \\
\boxtimes & \boxtimes & \boxtimes & \boxtimes & \\
& & & & \times
\end{bmatrix}
$$

$$\Updownarrow$$

$$S_0 \xrightarrow{M_4 S_0} S_1.$$

---

[13] We are flexible in this definition, i.e., if the row i is zero, we choose the Gauss transformation equal to the identity matrix.

Continuing this process by annihilating row 4 leads to:

$$\begin{bmatrix} \boxtimes & & & & \\ \boxtimes & \boxtimes & & & \\ \boxtimes & \boxtimes & \boxtimes & & \\ \otimes & \otimes & \otimes & \boxtimes & \\ & & & & \times \end{bmatrix} \xrightarrow{M_3 S_1} \begin{bmatrix} \boxtimes & & & & \\ \boxtimes & \boxtimes & & & \\ \boxtimes & \boxtimes & \boxtimes & & \\ & & & \times & \\ & & & & \times \end{bmatrix}$$

$$\Updownarrow$$

$$S_1 \xrightarrow{M_3 S_1} S_2.$$

It is already clear that this process of annihilating the lower triangular structure will leave the strictly upper triangular zero part untouched. Continuing this process leads to the following diagonal matrix:

$$\begin{bmatrix} \boxtimes & & & & \\ \otimes & \boxtimes & & & \\ & & \times & & \\ & & & \times & \\ & & & & \times \end{bmatrix} \xrightarrow{M_1 S_3} \begin{bmatrix} \times & & & & \\ & \times & & & \\ & & \times & & \\ & & & \times & \\ & & & & \times \end{bmatrix}$$

$$\Updownarrow$$

$$S_3 \xrightarrow{M_1 S_3} S_4 = D.$$

This clearly shows that after performing $n-1$ annihilating Gauss transforms, from bottom to top, we annihilated the complete lower triangular semiseparable structure, and we left the upper triangular rank structure untouched. □

Note 10.3. *Even though the results in the previous proof seem rather obvious, the existence of the Gauss transforms annihilating the lower triangular part is far from obvious. The following, almost trivial example causes the preceding method of successive annihilation of the rows to fail. Consider the matrix A with the following structure:*

$$\begin{bmatrix} \boxtimes & \times & \times & \times & \times \\ \boxtimes & \boxtimes & \times & \times & \times \\ \boxtimes & \boxtimes & \boxtimes & \times & \times \\ 0 & 0 & 0 & 0 & \times \\ \otimes & \otimes & \otimes & \otimes & \boxtimes \end{bmatrix}.$$

The reader can see that it is impossible to construct a Gauss transform acting on rows 4 and 5 to annihilate the bottom row.

Pivoting would solve the problem. Interchanging rows 4 and 5 creates zeros in the last row. As we will see further on, this pivoting does have a strong effect on the rank structure of the matrix. Gauss transforms, in general, do not affect the upper triangular rank structure, but introducing pivoting will create the same effect on the upper triangular rank structure, as in the case of the QR-factorization. Hence, pivoting will generally increase the rank in the upper triangular rank structure. Pivoting will be discussed in Section 10.2.

We would like to remark briefly, that if the lower triangular part of the matrix is semiseparable and the matrix is strongly nonsingular, no pivoting is required. This can be verified easily.

Corollary 10.4. *Suppose we have a structured rank matrix A satisfying the following rank structural constraints (with $p \leq 1$):*

$$\mathrm{r}\left(\Sigma_l^{(p)}; A\right) \leq 1 \quad and \quad \mathrm{r}\left(\Sigma_u^{(p-1)}; A\right) = 0.$$

Performing a sequence of annihilating Gauss transforms from bottom to top leads to a structured rank matrix $\hat{A}$ of the following form:

$$\mathrm{r}\left(\Sigma_l^{(p-1)}; \hat{A}\right) = 0 \quad and \quad \mathrm{r}\left(\Sigma_u^{(p-1)}; \hat{A}\right) = 0.$$

It is clear that the rank structure of the upper triangular part remains the same; it does not increase in rank as was the case for the Givens transformations. As already mentioned, this is because Givens transformations act on two rows at once, whereas a Gauss transform only changes one row at a time.

When decomposing a higher order structured rank matrix as the sum of lower order terms, one might sometimes perform an arbitrary sequence on one of these terms. The effect of an arbitrary sequence of Gauss transforms is discussed in the following subsection.

10.1.2   Arbitrary Gauss transforms on lower rank structures

Let us take a look at the effect of arbitrary Gauss transforms applied on a structured rank matrix.

Theorem 10.5. *Suppose S is a lower triangular $\{p\}$-semiseparable matrix. A sequence of arbitrary $n - 1$ Gauss transforms M_i, acting on row i and $i + 1$, is performed on the matrix S:*

$$M_1 \ldots M_{n-1}S = \hat{S}.$$

The matrix $\hat{S}$ will be lower triangular and have the strictly lower triangular part of $\{p\}$-semiseparable form.

Proof. We will illustrate the proof for a lower triangular 5×5 matrix. We use a similar notation as in Theorem 10.1 consisting of the elements ⊠. Assume that our lower triangular semiseparable matrix S is of $\{p\}$-semiseparable form. This means that all blocks taken out of the part of the matrix marked with the elements ⊠ have rank at most p.

Similarly as in the previous chapter, we can write the matrix as a sum of lower triangular {1}-semiseparable matrices, but we illustrate this here immediately on the complete matrix S.

A first Gauss transform is performed on rows 5 and 4:

$$
\begin{bmatrix}
\boxtimes & & & & \\
\boxtimes & \boxtimes & & & \\
\boxtimes & \boxtimes & \boxtimes & & \\
\boxtimes & \boxtimes & \boxtimes & \boxtimes & \\
\boxtimes & \boxtimes & \boxtimes & \boxtimes & \boxtimes
\end{bmatrix}
\xrightarrow{M_4 S_0}
\begin{bmatrix}
\boxtimes & & & & \\
\boxtimes & \boxtimes & & & \\
\boxtimes & \boxtimes & \boxtimes & & \\
\boxtimes & \boxtimes & \boxtimes & \boxtimes & \\
\boxtimes & \boxtimes & \boxtimes & \boxtimes & \times
\end{bmatrix}
$$

$$
\Updownarrow
$$

$$
S_0 \xrightarrow{M_4 S_0} S_1.
$$

It is clear that the rank of all blocks taken out of the part of the matrix marked with the elements $\boxtimes$ is still maximum of rank p. Continuing this process gives us:

$$
\begin{bmatrix}
\boxtimes & & & & \\
\boxtimes & \boxtimes & & & \\
\boxtimes & \boxtimes & \boxtimes & & \\
\boxtimes & \boxtimes & \boxtimes & \times & \\
\boxtimes & \boxtimes & \boxtimes & \boxtimes & \times
\end{bmatrix}
\xrightarrow{M_3 S_1}
\begin{bmatrix}
\boxtimes & & & & \\
\boxtimes & \boxtimes & & & \\
\boxtimes & \boxtimes & \times & & \\
\boxtimes & \boxtimes & \boxtimes & \times & \\
\boxtimes & \boxtimes & \boxtimes & \boxtimes & \times
\end{bmatrix}
$$

$$
\Updownarrow
$$

$$
S_1 \xrightarrow{M_3 S_1} S_2.
$$

Finally, we obtain a matrix of the following form:

$$
M_1 M_2 M_3 M_4 S =
\begin{bmatrix}
\times & & & & \\
\boxtimes & \times & & & \\
\boxtimes & \boxtimes & \times & & \\
\boxtimes & \boxtimes & \boxtimes & \times & \\
\boxtimes & \boxtimes & \boxtimes & \boxtimes & \times
\end{bmatrix} .
$$

It is clear that the strictly lower triangular part of this matrix is of $\{p\}$-semiseparable form. $\qquad\Box$

This gives us in structured rank context the following corollary.

Corollary 10.6. *Suppose we have a structured rank matrix A satisfying the following rank structural constraints:*

$$
\mathrm{r}\left(\Sigma_l^{(p)}; A\right) \leq r \quad and \quad \mathrm{r}\left(\Sigma_u^{(p-1)}; A\right) = 0.
$$

Performing a sequence of arbitrary Givens transformations from bottom to top leads to a structured rank matrix $\hat{A}$ of the following form:

$$
\mathrm{r}\left(\Sigma_l^{(p-1)}; \hat{A}\right) \leq r \quad and \quad \mathrm{r}\left(\Sigma_u^{(p-1)}; \hat{A}\right) = 0.
$$

We have in fact that the structured part shifts down one position and the strictly upper triangular part stays equal to zero. One might wonder now what will happen with an upper triangular structured rank part, when performing a sequence of Gauss transforms on the matrix.

10.1.3   Gauss transforms on upper rank structures

We noticed in the previous subsections that the rank of the strictly upper triangular part, in the case it was zero, was maintained. Let us take a closer look now at this rank structure, in case it can be higher than zero.

Theorem 10.7. *Suppose S is a matrix having the upper triangular part of $\{p\}$-semiseparable form. A sequence of $n-1$ Gauss transforms M_i, acting on row i and $i+1$, is performed on the matrix S:*

$$M_1 \ldots M_{n-1} S = \hat{S}.$$

The matrix $\hat{S}$ will have the upper triangular part still of $\{p\}$-semiseparable form.

Proof. Let us work directly on the matrix S having its upper triangular part of $\{p\}$-semiseparable form. This means that every block taken out of the part of the matrix marked with the elements $\boxtimes$ has maximum rank p. The elements $\times$ are arbitrary elements. We illustrate this for a 5×5 matrix. The general case is similar. Assume that our matrix S is of the following form:

$$S = \begin{bmatrix} \boxtimes & \boxtimes & \boxtimes & \boxtimes & \boxtimes \\ \times & \boxtimes & \boxtimes & \boxtimes & \boxtimes \\ \times & \times & \boxtimes & \boxtimes & \boxtimes \\ \times & \times & \times & \boxtimes & \boxtimes \\ \times & \times & \times & \times & \boxtimes \end{bmatrix}.$$

Performing the first Gauss transform on the matrix gives us the following situation:

$$M_4 S = \begin{bmatrix} \boxtimes & \boxtimes & \boxtimes & \boxtimes & \boxtimes \\ \times & \boxtimes & \boxtimes & \boxtimes & \boxtimes \\ \times & \times & \boxtimes & \boxtimes & \boxtimes \\ \times & \times & \times & \boxtimes & \boxtimes \\ \times & \times & \times & \times & \boxtimes \end{bmatrix}.$$

Nothing changed. This because the Gauss transform M_4 only changes row 5 and leaves row 4 unchanged. This procedure can easily be continued leading to the following situation:

$$M_1 M_2 M_3 M_4 S = \begin{bmatrix} \boxtimes & \boxtimes & \boxtimes & \boxtimes & \boxtimes \\ \times & \boxtimes & \boxtimes & \boxtimes & \boxtimes \\ \times & \times & \boxtimes & \boxtimes & \boxtimes \\ \times & \times & \times & \boxtimes & \boxtimes \\ \times & \times & \times & \times & \boxtimes \end{bmatrix}.$$

Hence we see that, also for more general structured rank forms, the Gauss transforms leave the upper triangular part untouched.   □

We can write this in a more general rank structured context as follows.

Corollary 10.8. *Suppose we have a structured rank matrix A satisfying the following rank structural constraints:*

$$\mathrm{r}\left(\Sigma_u^{(q)}; A\right) \leq r.$$

Performing a sequence of arbitrary Gauss transforms from bottom to top leads to a structured rank matrix $\hat{A}$ of the following form:

$$\mathrm{r}\left(\Sigma_u^{(q)}; \hat{A}\right) \leq r.$$

In fact we now have all the necessary results to formulate in a general context what the effect of a sequence of Gauss transforms on a structured rank matrix will be.

10.1.4   A sequence of Gauss transforms from bottom to top

The following theorem combines all the information concerning the performance of Gauss transforms from bottom to top on structured rank matrices. Even though we did not mention it, nor did we investigate it, the generalization of an annihilating sequence of Gauss transforms towards a rank-decreasing sequence of Gauss transforms is similar to the QR case and left to the reader. In the following corollary we assume of course that the sequence of rank-decreasing Gauss transforms exists.

Corollary 10.9. *Suppose we have a structured rank matrix A satisfying the following rank structural constraints:*

$$\mathrm{r}\left(\Sigma_l^{(p)}; A\right) \leq r_1 \quad and \quad \mathrm{r}\left(\Sigma_u^{(q)}; A\right) \leq r_2.$$

Performing a sequence of arbitrary Gauss transforms, from bottom to top leads to a structured rank matrix $\hat{A}$ of the following form:

$$\mathrm{r}\left(\Sigma_l^{(p-1)}; \hat{A}\right) \leq r_1 \quad and \quad \mathrm{r}\left(\Sigma_u^{(q)}; \hat{A}\right) \leq r_2.$$

Performing, if it exists, a sequence of rank-decreasing Gauss transforms (with $p \leq r_1$)[14], from top to bottom leads to a structured rank matrix $\hat{A}$ of the following form:

$$\mathrm{r}\left(\Sigma_l^{(p-1)}; \hat{A}\right) \leq r_1 - 1 \quad and \quad \mathrm{r}\left(\Sigma_u^{(q)}; \hat{A}\right) \leq r_2.$$

---

[14]The extra condition $p \leq r_1$ was deduced in the previous chapter in Section 9.1.4. These arguments remain valid in this case.

Let us also formulate this once in words.

Corollary 10.10. *Performing a sequence of Gauss transforms from bottom to top on a structured rank matrix has the following effect on this matrix:*

- *The structure of the lower part shifts down one position.*

  - *The rank of the lower part stays the same in the case of an arbitrary sequence of Givens transformations.*

  - *The rank decreases by one in the case of a rank-decreasing pattern of Givens transformations.*

- *The rank structure of the upper triangular part stays the same.*

Moreover these observations are also valid for different types of structured blocks inside the matrix! Reconsider some of the examples for the QR-decomposition.

In the chapter of the QR-decomposition we did not discuss in detail sequences from top to bottom, as they behaved similarly to sequences from bottom to top. This is because a Givens transformation changes two rows at the same time. The Gauss transforms as discussed in this book, however, do not change two rows at the same time; they use one row and add a multiple of it to another row. Due to this substantial difference, there is an essential difference between descending and ascending sequences of Gauss transforms. Let us look at this difference in more detail.

10.1.5   Ascending and descending Gauss transforms

Let us look in more detail at the difference between a descending and an ascending sequence of elementary Gaussian transformations performed on a matrix.

We will illustrate this by looking at a sequence of Gauss transforms performed from bottom to top on the matrix A. Suppose A is a 4×4 matrix, with rows R_1, R_2, R_3 and R_4. Having Gauss transforms M_3, M_2 and M_1, with parameters τ_3, τ_2 and τ_1 acting, respectively, on rows 3 and 4, rows 2 and 3 and rows 1 and 2, leads to a matrix $\hat{A}$. The matrix $\hat{A}$ is of the following form:

$$
M_1 M_2 M_3 A = \hat{A} = \begin{bmatrix} R_1 \\ R_2 + \tau_1 R_1 \\ R_3 + \tau_2 R_2 \\ R_4 + \tau_3 R_3 \end{bmatrix}.
$$

This means that performing an ascending sequence of Gauss transforms changes a matrix, such that every row is replaced by a new row, containing information from the row itself and the row above. We can say that information transport only has gone from one row to the row below.

Looking now at a sequence of Gauss transforms performed from bottom to top, we use a similar notation as above, but we look now at the matrix $\tilde{A} = M_3 M_2 M_1 A$, which is of the following form:

$$M_3 M_2 M_1 A = \tilde{A} = \begin{bmatrix} R_1 \\ R_2 + \tau_1 R_1 \\ R_3 + \tau_2(R_2 + \tau_1 R_1) \\ R_4 + \tau_3(R_3 + \tau_2(R_2 + \tau_1 R_1)) \end{bmatrix}.$$

We see clearly now that information transport has gone from the top to the bottom.

Note 10.11. *We remark that this difference between an ascending and descending sequence of Gauss transforms does not occur if Givens transformations are performed. Givens transformations from bottom to top are also able to transport information from the bottom row to the top, which is not the case for Gauss transforms.*

We know now why we distinguish between ascending and descending sequences of Gauss transformations. Let us now take a closer look at sequences of transformations from top to bottom.

10.1.6   Zero-creating Gauss transforms

We know, from an earlier chapter, the existence of the zero-creating Gauss transforms for removing the remaining nonzero subdiagonal elements. These transformations are useful, e.g., when reducing an Hessenberg matrix to triangular form. A sequence of zero-creating Gauss transforms has the following effect on a matrix:

Corollary 10.12. *Suppose a structured rank matrix A is given satisfying the following structural constraints (with $p \leq r_1 - 1$)[15]:*

$$\mathrm{r}\left(\Sigma_l^{(p)}; A\right) = 0 \quad and \quad \mathrm{r}\left(\Sigma_u^{(q)}; A\right) \leq r_2.$$

Perform, if they exist[16], a sequence of zero-creating Gauss transforms from top to bottom on the matrix A. This leads to a structured rank matrix $\hat{A}$ of the following form:

$$\mathrm{r}\left(\Sigma_l^{(p+1)}; \hat{A}\right) = 0 \quad and \quad \mathrm{r}\left(\Sigma_u^{(q)}; \hat{A}\right) \leq r_2.$$

Proof. The proof is similar to the proofs discussed previously and hence is left to the reader.   □

The zero-creating Gauss transforms expanded in fact the zero rank structure in the lower triangular part, similarly as in the case of the QR-decomposition. Let us therefore investigate the existence of rank-expanding Gauss transforms.

---

[15]The constraint $p \leq r_1 - 1$ was discussed in the previous chapter in Section 9.4.3.

[16]Zero-creating Gauss transforms do not always exist. Their existence will be addressed in Subsection 10.1.7.

10.1.7   Rank-expanding Gauss transforms

In the previous chapter on the QR-decomposition, we discovered the existence of a rank-expanding sequence of Givens transformations. This sequence of Givens transformations, if performed from bottom to top, expanded the upper triangular rank structure by adding one more diagonal into the structure. A similar transformation exists for Gauss transforms. Unlike the Givens case, a sequence of ascending Gauss transforms cannot expand the rank structure of the upper triangular part[17]. We can however expand the structure in the lower triangular part, when performing a sequence of descending Gauss transforms.

The rank-expanding Gauss transform

Let us analytically prove the existence of a Gauss transform, expanding the rank structure.

Proposition 10.13 (Descending rank-expanding Gauss transformation). *Suppose the following 2×2 matrix is given*

$$A = \begin{bmatrix} a & b \\ c & d \end{bmatrix}. \tag{10.1}$$

Then there exists a Gauss transformation M:

$$M = \begin{bmatrix} 1 & 0 \\ \tau & 1 \end{bmatrix}$$

such that the second row of the matrix MA is linearly dependent on the row $[e, f]$. Assume moreover if $[c, d]$ and $[e, f]$ are linearly independent (otherwise take $\tau = 0$), that also $[a, b]$ is linearly independent of $[e, f]$.

Proof. Assume $[c, d]$ to be linearly independent on the row $[e, f]$, otherwise we can take $\tau = 0$.

Computing MA gives us the following equation:

$$\begin{bmatrix} 1 & 0 \\ \tau & 1 \end{bmatrix} \begin{bmatrix} a & b \\ c & d \end{bmatrix} = \begin{bmatrix} a & b \\ c + \tau a & d + \tau b \end{bmatrix}.$$

The second row being dependent on $[e, f]$ leads to the following relation:

$$f(c + \tau a) - e(d + \tau b) = 0.$$

Rewriting this equation towards τ gives us the following well-defined equation:

$$\tau = \frac{ed - cf}{fa - eb}.$$

---

[17]In Section 10.3, we will introduce another type of Gauss transform which can be used to expand the structure in the upper part.

This equation does not seem to be valid for every value of a, f, e and b. The parameter τ only exists in the case that $fa - eb$ is different from zero. This means in fact that the top two elements need to be linearly independent of the elements $[e, f]$. □

Let us present some examples of matrices for which a rank-expanding Gauss transform does not exist.

Example 10.14 A rank-expanding sequence of Gauss transforms does not exist for the following matrices.

- Suppose we have a 5×5 matrix A, with A of Hessenberg form. We want to extend the strictly lower triangular zero structure by annihilating the subdiagonal. One can clearly see that there is no Gaussian transformation (if the element in position $(3, 2)$ is different from zero), such that the matrix will become upper triangular. The matrix A is of the form:

$$A = \begin{bmatrix} \times & \times & \times & \times & \times \\ 0 & 0 & \times & \times & \times \\ 0 & \times & \times & \times & \times \\ 0 & 0 & \times & \times & \times \\ 0 & 0 & 0 & \times & \times \end{bmatrix}.$$

The matrix above is not strongly nonsingular, but even posing the extra constraint that the matrix should be strongly nonsingular does not solve the problem as the following example illustrates.

- A less trivial example is the following matrix, in which the matrix satisfies the following constraint $r\left(\Sigma_l^{(0)}; A\right) = 1$. The reader can see that it is not possible to expand this rank 1 structure, such that it will include the diagonal:

$$A = \begin{bmatrix} -1 & 1 & 2 \\ 0 & 1 & 3 \\ 1 & -1 & 4 \end{bmatrix}.$$

Moreover this matrix is strongly nonsingular. ∎

Reconsidering these examples, with respect to the rank-expanding Givens transformations, we see that Givens transformations are much more flexible, and no problems are encountered for computing rank-expanding Givens transformations.

The problems encountered in these examples are easily solved by permuting the problem rows. This permutation will always create the desired structure.

Unfortunately, it seems that the existence of expanding Gauss transforms, if no pivoting is allowed, cannot be guaranteed in general. Just as we could not guarantee the existence of a rank-decreasing sequence of Gauss transforms. Let us summarize the action of a sequence of Gauss transformations from top to bottom.

10.1.8 A sequence of Gauss transforms from top to bottom

Let us summarize now the effect of a sequence of Gauss transforms from top to bottom performed on a matrix. We did not examine in detail an arbitrary sequence of Gauss transforms from top to bottom, but the reader can easily deduce the effect himself. We present the results below. Again we assume that the sequence of rank-expanding Gauss transforms exists.

Corollary 10.15. *Suppose a structured rank matrix A is given satisfying the following structural constraints:*

$$\mathrm{r}\left(\Sigma_l^{(p)}; A\right) \le r_1 \quad and \quad \mathrm{r}\left(\Sigma_u^{(q)}; A\right) \le r_2.$$

Performing a sequence of arbitrary Gauss transforms from top to bottom leads to a structured rank matrix $\hat{A}$ of the following form:

$$\mathrm{r}\left(\Sigma_l^{(p+1)}; \hat{A}\right) \le r_1 + 1 \quad and \quad \mathrm{r}\left(\Sigma_u^{(q)}; \hat{A}\right) \le r_2.$$

Perform, if they exist, a sequence of rank-expanding Gauss transforms from top to bottom on the matrix A (assume $p \le r_1 - 1$). This leads to a structured rank matrix $\hat{A}$ of the following form:

$$\mathrm{r}\left(\Sigma_l^{(p+1)}; \hat{A}\right) \le r_1 \quad and \quad \mathrm{r}\left(\Sigma_u^{(q)}; \hat{A}\right) \le r_2.$$

We do not prove these statements, the proofs are left to the reader and are similar to the ones presented in the chapter on the QR-factorization.

In this section we investigated two things: the existence of a sequence of rank-decreasing Gauss transformation matrices and the existence of a sequence of rank-expanding Gauss transformation matrices. Unfortunately we could not provide a positive answer for either sequence of transformations. We noticed, however, that pivoting might solve the problem. In the next section we will investigate in more detail the effect of pivoting on the structure of a matrix, and moreover we will allow pivoting in the sequences of Gauss transforms.

10.2 Effect of pivoting on rank structures

Even though the effect of pivoting on the matrix might seem to be quite destructive, it is not. In the first chapter on the LU-decomposition, we made the reader aware of the destructive effect that pivoting might have on the structured rank matrix. But we also showed that after recombining all permutation matrices the final structure of the LU-decomposition of the nonsingular matrix A was simply the structure of the matrix PA. As the matrix A is of structured rank form, this means that in some sense also the matrix PA will be of structured rank form, shuffled, however, but still of structured rank form.

We will describe now the effect of pivoting on the structure of the matrix A. In fact, the reader can immediately deduce what the effect of a sequence of permutation matrices will be on the structured rank of A, because a permutation, involving only two successive rows, is just a special Givens transformation. One can therefore immediately use the theorems of the previous chapter to predict the structure of the structured rank matrix after having performed a sequence of permutation matrices.

A sequence of permutation matrices will rarely occur in practice, however. Pivoting will occur more or less sporadically in a sequence of Gauss transforms; hence, we will investigate what the effect of a permutation matrix itself is on the structure of the matrix.

10.2.1   The effect of pivoting on the rank structure

The effect of pivoting on a structured rank matrix has already been investigated thoroughly in the preceding chapter. A permutation matrix interchanging two rows is in fact nothing else than a special kind of Givens transformation. We will briefly repeat the effect of pivoting on the rank structure of a matrix, mostly by presenting examples. As our analysis of this book mainly focuses on structured rank matrices obeying a certain lower/upper triangular structure, we will put most effort into analyzing these structures. We will start with two examples and apply pivoting on these matrices.

Example 10.16 Let us consider the following two 4×4 matrices having the elements $\boxtimes$ of the structured rank form. On both matrices we apply a permutation on the left, interchanging thereby the second and third row.

- The first matrix is of the following form:

$$
A = \begin{bmatrix}
\boxtimes & \times & \times & \times \\
\boxtimes & \boxtimes & \boxtimes & \times \\
\boxtimes & \boxtimes & \boxtimes & \times \\
\boxtimes & \boxtimes & \boxtimes & \boxtimes
\end{bmatrix}.
$$

Applying a permutation on the left of this matrix A gives us the matrix PA, which is of the following form:

$$
PA = \begin{bmatrix}
1 & & & \\
& 0 & 1 & \\
& 1 & 0 & \\
& & & 1
\end{bmatrix}
\begin{bmatrix}
\boxtimes & \times & \times & \times \\
\boxtimes & \boxtimes & \boxtimes & \times \\
\boxtimes & \boxtimes & \boxtimes & \times \\
\boxtimes & \boxtimes & \boxtimes & \boxtimes
\end{bmatrix}
=
\begin{bmatrix}
\boxtimes & \times & \times & \times \\
\boxtimes & \boxtimes & \boxtimes & \times \\
\boxtimes & \boxtimes & \boxtimes & \times \\
\boxtimes & \boxtimes & \boxtimes & \boxtimes
\end{bmatrix}.
$$

This matrix PA has essentially the same rank structure as the matrix A. Hence it is clear if one applies pivoting on two rows for which the rank structure has the same extent, that pivoting will not interfere with the structure.

- The following matrix has the lower triangular part of semiseparable form. We represent the matrix as follows:

$$
A = \begin{bmatrix}
\boxtimes & \times & \times & \times \\
\boxtimes & \boxtimes & \times & \times \\
\boxtimes & \boxtimes & \boxtimes & \times \\
\boxtimes & \boxtimes & \boxtimes & \boxtimes
\end{bmatrix}.
$$

The application of pivoting unfortunately will destroy in some sense the structure:

$$PA = \begin{bmatrix} 1 & & & \\ & 0 & 1 & \\ & 1 & 0 & \\ & & & 1 \end{bmatrix} \begin{bmatrix} \boxtimes & \times & \times & \times \\ \boxtimes & \boxtimes & \times & \times \\ \boxtimes & \boxtimes & \boxtimes & \times \\ \boxtimes & \boxtimes & \boxtimes & \boxtimes \end{bmatrix} = \begin{bmatrix} \boxtimes & \times & \times & \times \\ \boxtimes & \boxtimes & \boxtimes & \times \\ \boxtimes & \boxtimes & \times & \times \\ \boxtimes & \boxtimes & \boxtimes & \boxtimes \end{bmatrix}.$$

It is clear that we have now in fact only a strictly lower triangular semiseparable structure left. The structure has shifted down one position. This is not so bad, as we also had this effect for a sequence of Gauss transforms from bottom to top.

∎

Even though we did not provide examples of it, pivoting also has a large impact on the structure of the upper triangular part. Just like in the QR-decomposition, we can state that pivoting shifts the upper structure down and increases its rank by 1.

In the examples considered so far, pivoting was always rather exclusive. Only a few permutation matrices were needed. In other examples, if pivoting was involved, no successive permutation matrices occurred. In the case of the semiseparable matrix, this was due to the strong nonsingularity of the matrix (see Section 4.5 in Chapter 4). The next example shows, however, that in general pivoting can occur even in sequences.

Example 10.17 Suppose we have the following matrix, for which the lower triangular part is of semiseparable form. We would like to reduce this matrix to upper triangular form by a sequence of Gauss transforms in which we now also admit permutation matrices. More precisely we want to construct a sequence from bottom to top consisting only of Gauss transforms or permutation matrices. The matrix A is of the following structured rank form:

$$\begin{bmatrix} 0 & \times & \times & \times & \times \\ 0 & 0 & \times & \times & \times \\ 0 & 0 & 0 & \times & \times \\ 0 & 0 & 0 & 0 & \times \\ \times & \times & \times & \times & \times \end{bmatrix}.$$

The sequence of transformations for bringing this matrix to upper triangular form will consist only of permutation matrices. The matrix discussed here is not strongly nonsingular, but one can easily construct an example in which one wants to remove a strictly lower triangular rank structure in a matrix which is strongly nonsingular.

∎

So it is clear that in general pivoting can occur at any time and not only in isolation, but even in sequences of permutation matrices.

Also if a sequence is performed from top to bottom for expanding the lower triangular rank structure by Gauss transforms pivoting might be essential. The following matrix presents again an almost trivial example in which a sequence of permutation matrices expands the rank structure.

Example 10.18 The following matrix is an upper Hessenberg matrix. Trying to expand the zero rank structure below the subdiagonal by a sequence of Gauss transformation matrices from top to bottom is not possible. Admitting the application of permutation matrices solves the problem. Constructing the sequence of transformation matrices results in a descending sequence, consisting only of permutation matrices:

$$
\begin{bmatrix}
0 & 0 & 0 & 0 & \times \\
\times & \times & \times & \times & \times \\
0 & \times & \times & \times & \times \\
0 & 0 & \times & \times & \times \\
0 & 0 & 0 & \times & \times
\end{bmatrix}.
$$

∎

We can conclude this section on pivoting very easily by stating that it will have the same effect on our structured rank matrix as a Givens transformation would have. No more or no less. In very specific cases, however, it will not destroy the structure, but this depends heavily on the case. Let us now investigate how we can combine pivoting with Gauss transforms.

10.2.2   Combining Gauss transforms with pivoting

In the previous section we clearly stated that in general performing only Gauss transforms was not enough to obtain sequences of rank-decreasing or of rank-expanding Gauss transformations. Using pivoting will solve the problem. Let us take a closer look at this.

Rank-decreasing sequences of transformations

Problems occurred if a situation as depicted below appeared:

$$
\begin{bmatrix}
0 & 0 & 0 & 0 & 0 \\
\boxtimes & \boxtimes & \boxtimes & \boxtimes & \boxtimes
\end{bmatrix}.
$$

In this case it is not possible to perform a Gauss transform to make the bottom row in this example zero. Applying a permutation matrix, however, will interchange both rows and create the desired zero row. With respect to the lower triangular structure, nothing essential changed. The transformation just decreased the rank in the lower triangular structure as we wanted it to. The upper triangular part, however, is different. In general, the rank of the upper triangular part will increase. So we need to take this into consideration.

Rank-expanding sequences of transformations

Suppose we want to apply a rank-expanding Gauss transform on the matrix A:

$$
A = \begin{bmatrix}
a & b \\
c & d
\end{bmatrix},
\tag{10.2}
$$

to make the second row of the transformed matrix dependent on $[e, f]$. We proved that it was not possible to construct such a transformation for the matrix A if $[c, d]$ was linearly independent of $[e, f]$ and $[a, b]$ was linearly dependent on $[e, f]$ (see Theorem 10.13).

But if $[a, b]$ is already dependent on $[e, f]$ a simple permutation would solve this problem. Hence, in the case of rank-expanding sequences of transformations performed from bottom to top, pivoting can also overcome the remaining problems.

Let us examine in the next subsection what the effect on the rank structure will be if we admit pivoting in our sequences.

So we can conclude that allowing pivoting in the sequences of Gauss transforms solves the existence issue. When we allow pivoting, a sequence of rank-decreasing or rank-expanding transformations will always exist. Let us formulate this in a structured rank context.

10.2.3   Sequences of transformations involving pivoting

In the previous section the effect of a sequence of Gauss transforms performed from bottom to top and from top to bottom was discussed. We assumed, however, that no pivoting was involved in these sequences. Now, we will revisit these corollaries and expand them towards sequences of transformations admitting pivoting.

Note that pivoting seems to affect only the rank structure of the upper triangular part. The rank structure of the lower triangular part does not change under the performance of the permutation matrices.

We distinguish between two sequences: the sequence from bottom to top and the sequence from top to bottom. In the following two corollaries, the sequence of Gauss transforms without pivoting does not necessarily exist, but the sequence with pivoting will always exist.

Corollary 10.19 (Ascending sequence). *Suppose we have a structured rank matrix A satisfying the following rank structural constraints:*

$$\mathrm{r}\left(\Sigma_l^{(p)}; A\right) \le r_1 \quad and \quad \mathrm{r}\left(\Sigma_u^{(q)}; A\right) \le r_2.$$

In the next results, the matrix $\hat{A}$ denotes the effect of a sequence without pivoting and the matrix $\hat{A}_P$ denotes the effect (in general) of a sequence with pivoting.

Performing a sequence of arbitrary Gauss transforms from bottom to top leads to a structured rank matrix $\hat{A}$ or $\hat{A}_P$, of the following form:

$$\mathrm{r}\left(\Sigma_l^{(p-1)}; \hat{A}\right) \le r_1 \quad and \quad \mathrm{r}\left(\Sigma_u^{(q)}; \hat{A}\right) \le r_2 \quad or$$

$$\mathrm{r}\left(\Sigma_l^{(p-1)}; \hat{A}_P\right) \le r_1 \quad and \quad \mathrm{r}\left(\Sigma_u^{(q-1)}; \hat{A}_P\right) \le r_2 + 1.$$

Performing a sequence of rank-decreasing Gauss transforms (with $p \le r_1$) from bottom to top leads to a structured rank matrix $\hat{A}$ or $\hat{A}_P$ of the following form:

$$\mathrm{r}\left(\Sigma_l^{(p-1)}; \hat{A}\right) \le r_1 - 1 \quad and \quad \mathrm{r}\left(\Sigma_u^{(q)}; \hat{A}\right) \le r_2 \quad or$$

$$\mathrm{r}\left(\Sigma_l^{(p-1)}; \hat{A}_P\right) \le r_1 - 1 \quad and \quad \mathrm{r}\left(\Sigma_u^{(q-1)}; \hat{A}_P\right) \le r_2 + 1.$$

So if pivoting is involved, the only thing one can guarantee in general is that the structured rank conditions are similar to the ones created by a sequence of Givens transformations.

We remark, however, that if one applies only a few permutation matrices, one can easily keep track of the distortions in the matrix. This means that one does not necessarily need to switch to the general higher order case as defined in the corollary. When keeping track of the distortions, one might come to faster algorithms. This special case is not discussed here, however, as it is heavily case dependent.

Let us formulate the corollary for a descending sequence of transformations.

Corollary 10.20 (Descending sequence). *Suppose a structured rank matrix A is given satisfying the following structural constraints*

$$\mathrm{r}\left(\Sigma_l^{(p)}; A\right) \leq r_1 \quad and \quad \mathrm{r}\left(\Sigma_u^{(q)}; A\right) \leq r_2.$$

In the next results, the matrix $\hat{A}$ denotes the effect of a sequence without pivoting and the matrix $\hat{A}_P$ denotes the effect (in general) of a sequence with pivoting.

Performing a sequence of arbitrary Gauss transforms from top to bottom leads to a structured rank matrix $\hat{A}$ or $\hat{A}_P$ of the following form:

$$\mathrm{r}\left(\Sigma_l^{(p+1)}; \hat{A}\right) \leq r_1 + 1 \quad and \quad \mathrm{r}\left(\Sigma_u^{(q)}; \hat{A}\right) \leq r_2 \quad or$$

$$\mathrm{r}\left(\Sigma_l^{(p+1)}; \hat{A}_P\right) \leq r_1 + 1 \quad and \quad \mathrm{r}\left(\Sigma_u^{(q+1)}; \hat{A}_P\right) \leq r_2.$$

Perform a sequence of rank-expanding Gauss transforms from top to bottom on the matrix A (assume $p \leq r_1 - 1$). This leads to a structured rank matrix $\hat{A}$ of the following form:

$$\mathrm{r}\left(\Sigma_l^{(p+1)}; \hat{A}\right) \leq r_1 \quad and \quad \mathrm{r}\left(\Sigma_u^{(q)}; \hat{A}\right) \leq r_2 \quad or$$

$$\mathrm{r}\left(\Sigma_l^{(p+1)}; \hat{A}_P\right) \leq r_1 \quad and \quad \mathrm{r}\left(\Sigma_u^{(q+1)}; \hat{A}_P\right) \leq r_2.$$

Again we observe that there are no changes with regard to the lower triangular part, only the upper triangular part changes if a sequence involving pivoting is performed. The most important thing to conclude is that, when using pivoting, the structure of the lower triangular part still behaves exactly as we want it to. So if we want to make the matrix upper triangular, we do know exactly how many sequences we need to perform.

Before combining different sequences of Gauss transforms to get an upper triangular matrix, we would like to investigate in more detail other sequences, such as from left to right and so forth.

10.2.4 Numerical stability

To conclude we would like to make a remark on the numerical stability of the proposed LU-decomposition. In our case, pivoting was only used if a zero was encountered. To increase the numerical stability, however, we can chose to do pivoting in every step, such as to use the largest pivot possible for computing the Gauss transforms just as in the regular pivoted LU-decomposition. The only price we pay is that we lose some rank structural properties, but we do know that our structural constraints will be the same as the ones proposed in the QR-algorithm. Hence our factorization loses some rank structure but gains in numerical stability.

10.3   More on sequences of Gauss transforms

In this section we will investigate if other possible sequences of Gauss transforms exist. For example, the section on the QR-factorization showed that sequences also existed from left to right and from right to left. Moreover, it was interesting that these sequences could easily be used for generating other types of decompositions, such as a URV decomposition or also a ULV decomposition. These different sequences open a huge variety for eliminating elements in our matrices.

The most remarkable difference between Gauss transformations and Givens transformations is the fact that Gauss transformations only act on one row at a time. The immediate effect of this is that (in the case of no pivoting), the Gauss transforms only have an effect on the rank structure of the lower triangular part. Even a sequence from bottom to top only seems to be able to have a significant influence on the lower triangular part. This is because the Gaussian transformation matrices we discussed were of lower triangular form.

First, we will define a different kind of Gauss transform, and then we will indicate how we can use this transform in different sequences of transformations.

10.3.1   Upper triangular Gauss transforms

In the previous sections we discussed lower triangular Gauss transformation matrices. These Gauss transforms were embedded in larger matrices to perform an action on two subsequent specified rows. These matrices essentially were of the following form:

$$\begin{bmatrix} 1 & 0 \\ \tau & 1 \end{bmatrix}.$$

These matrices were chosen to be lower triangular to obtain a LU-factorization of a matrix A, with L lower triangular and U upper triangular. If we now omit this constraint, and hence also admit Gaussian transformation matrices of the following kind,

$$\begin{bmatrix} 1 & \tau \\ 0 & 1 \end{bmatrix},$$

we might most likely end up with different factorizations than the LU-factorization.

Let us call this Gauss transform an upper Gauss transform. This will make it clear that this transformation will have the parameter τ in the upper triangular part of the matrix. We will not investigate it in detail, but the reader can easily verify the following statements on upper Gauss transforms[18]:

- An upper Gauss transform can only have a significant influence (rank-decreasing or rank-expanding) on the upper triangular structure.

- A sequence of descending upper Gauss transforms can be chosen in such a way that they perform a rank-decreasing action on the upper triangular structure of a structured rank matrix.

- A sequence of ascending upper Gauss transforms can be chosen in such a way that they perform a sequence of rank-expanding transforms on the upper triangular structure of a structured rank matrix.

We do not repeat the Corollaries 10.19 and 10.20: on the rank structure of structured rank matrices for the upper Gauss transforms, the generalization is easy.

We would like to remark that a sequence of Givens transformations is in fact more powerful than a sequence of Gauss transforms. If we have a sequence of ascending Givens transformations they can be chosen to annihilate the lower triangular rank structure, but they can also be chosen to expand the upper triangular rank structure. So a sequence of Givens transformations can work on both the lower and the upper triangular part of a matrix. A Gauss transform cannot do this. For a sequence from bottom to top, one can only perform rank-decreasing operations in the case of lower Gauss transforms and rank-expanding operations in the case of upper Gauss transforms.

10.3.2 Transformations on the right of the matrix

In this section we will briefly describe the effect of Gauss transforms performed on the right-hand side of the matrix. We distinguish between both upper and lower Gauss transforms.

- An upper (lower) Gauss transform performed on the right-hand side of a matrix can only have a significant influence (rank-decreasing or rank-expanding) on the upper (lower) triangular structure.

- A sequence of upper (lower) Gauss transforms from right to left can be chosen in such a way that they perform a rank-decreasing (rank-expanding) action on the upper (lower) triangular structure of a structured rank matrix.

- A sequence of upper (lower) Gauss transforms from left to right can be chosen in such a way that they perform a sequence of rank-expanding (rank-decreasing) transforms on the upper (lower) triangular structure of a structured rank matrix.

Using the results presented in this section, it is clear that one can combine different schemes of annihilation to obtain different factorizations instead of only the LU-decomposition.

In the next section some different possibilities will be explored.

---

[18]The statements are valid under some mild additional conditions such as the ones in Corollaries 10.19 and 10.20.

10.4   Solving systems with Gauss transforms

One might wonder why we did not choose the title "Computing the LU-factorization" for this section. This is because applying sequences of Gauss transforms, including pivoting, on the matrix will in general not lead to a lower triangular matrix anymore. What we do get is a sequence of transformations bringing the structured rank matrix to upper triangular form.

The involved transformations for bringing the structured rank matrix to upper triangular form are either elementary Gaussian transformation matrices or permutation matrices. These transformations are easily computed and hence we can use this factorization for solving systems of equations.

First, we will derive how sequences of Gauss and permutation matrices can be used to reduce a matrix to upper triangular form. Second, we will discuss in more detail the difference between the LU-factorization and our annihilation scheme presented here. Finally, we will conclude this section by proposing some other decompositions such as the UL-decomposition based on these sequences of Gauss transforms and pivoting matrices.

10.4.1   Making the structured rank matrix upper triangular

In this subsection, we will first discuss how to bring the matrix to upper triangular form. We discuss here only lower Gauss transforms performed on the left side of the matrix. First, we will annihilate the rank structure, and then we will annihilate the remaining nonzero subdiagonal elements.

Annihilating completely the rank structure

To make the structured rank matrix upper triangular one can perform different patterns of annihilation. In this section we discuss the traditional annihilation scheme. First we remove the rank structure and afterwards we remove the remaining nonzero subdiagonal elements. In the next section, we discuss different patterns of annihilation similarly as in the case of the QR-decomposition.

The theoretical results discussed in this section are similar to the ones discussed in the section on the QR-factorization. The sequences of transformations discussed here also involve pivoting. This means that, in general, the structure of the upper triangular part will vary depending on the number of permutation matrices that were applied during the annihilation and rank-expanding sequences of transformations. We could present theorems stating between which bounds the rank structure would lie. If one would be interested in these bounds, one just has to consider the worst-case and the best-case scenario. Worst case meaning that every sequence of transformations involves pivoting and best case means no pivoting was used during the reduction.

Hence, we will only discuss the creation of the lower triangular zero structure in the matrix. In general, the matrix will become upper triangular but we will not specify its rank structure.

Suppose we have structured rank matrix A given, satisfying the following rank structural constraint:

$$r\left(\Sigma_l^{(p)}; A\right) \leq r_1.$$

Performing r_1 rank-decreasing Gauss transforms on this matrix to remove completely the low rank part results in the following structure of the resulting matrix $\hat{A}$:

$$r\left(\Sigma_l^{(p-r_1)}; \hat{A}\right) = 0.$$

Expanding the zero rank structure

Removing nonzero subdiagonal entries can also be done by using Gauss transforms from top to bottom, combined with permutation matrices. Suppose a structured rank matrix A is given, with the following structure (assume hereby $p < 0$, otherwise the matrix is already in upper triangular form):

$$r\left(\Sigma_l^{(p)}; A\right) = 0.$$

Hence, to obtain an upper triangular matrix we need to perform $-p$ descending sequences of Gauss transforms. This gives us a matrix $\hat{A}$ of the following form:

$$r\left(\Sigma_l^{(0)}; \hat{A}\right) = 0.$$

Combining the rank-decreasing sequences with the rank-expanding sequences, we can easily transform our matrix to upper triangular form.

Combination of ascending and descending sequences

Suppose our structured rank matrix satisfies the following constraint:

$$r\left(\Sigma_l^{(p)}; A\right) \leq r_1.$$

To reduce this matrix to upper triangular form we first perform r_1 sequences of rank decreasing Gauss transforms. This leads to a structured rank matrix $\hat{A}$ satisfying the following constraint:

$$r\left(\Sigma_l^{(p-r_1)}; \hat{A}\right) = 0.$$

In case $(p - r_1) < 0$, the matrix is not yet of upper triangular form. To reduce the matrix to upper triangular form, we have to perform $(-p+r_1)$ sequences of Gauss transforms from bottom to top to eliminate the remaining nonzero subdiagonal elements. Performing these sequences on the matrix leads to the following matrix $\tilde{A}$:

$$r\left(\Sigma_l^{(0)}; \tilde{A}\right) = 0.$$

This matrix is clearly of upper triangular form. Hence we transformed a structured rank matrix to upper triangular form only by using Gauss transforms and permutation matrices.

Unfortunately this will not always lead to the LU-decomposition of a matrix. Let us take a closer look at this.

10.4.2 Gauss solver and the LU-factorization

We already mentioned that the annihilation scheme as presented here will not necessarily lead to an LU-factorization of the original matrix. Unfortunately it will not even lead to a pivoted LU-decomposition, in general.

This was illustrated with Example 4.23, in Section 4.6 in Chapter 4. An example was presented in which an ascending sequence of Gauss transforms and a descending sequence of Gauss transforms involving pivoting were used for transforming a quasiseparable matrix to upper triangular form. The combination of the Gauss transforms and the attempt to rewrite the factorization scheme as an LU-decomposition failed. The matrix U was upper triangular, but the matrix L was not lower triangular anymore.

One might correctly wonder what the use of this approach is if one is not able to derive an LU-factorization for the matrix in question. The answer is easy; even though we do not get the desired factorization, the Gauss transforms are useful for solving system of equations. Let us elaborate on this. Suppose a system of equations is presented with a structured rank matrix A:

$$A\mathbf{x} = \mathbf{b}.$$

Just as in the LU-factorization for solving systems of equations, one applies directly all involved transformations on the vector $\mathbf{b}$ as well. In the LU case, the combination of all these transformations on the vector $\mathbf{b}$ leads to a multiplication on the left, with a lower triangular matrix. Here we cannot guarantee that the combined transformation is lower triangular, but since we are interested in solving systems of equations, we don't mind. After applying all transformations we get (Denote with M the combination of all sequences of Gauss transforms performed on the left of the matrix A):

$$M A\mathbf{x} = U\mathbf{x} = M\mathbf{b} = \hat{\mathbf{b}}.$$

The matrix U is upper triangular, by construction, and hence we can easily solve the system of equations.

One might wonder what the advantage of this method might be. In the worst case, every sequence involves pivoting and hence the upper triangular matrix will have the same rank structure as the matrix R coming from the QR-factorization. But in case not every sequence of transformations involves pivoting, the rank structure of the matrix U might be less complex than the rank structure of the matrix R. This is clear when reconsidering the corollaries presented in Subsection 10.2.3. When the structure rank of the upper triangular part is lower, the upper triangular system can of course be solved more quickly.

We did not mention anything about the implementation details of the algorithms involved. This is due to the huge variety of representations and structured rank matrices. It is impossible to cover all classes and varieties. Nevertheless, readers should be able to combine the techniques presented here with the implementation techniques discussed in the previous part of the book. Combining these results, readers should be able to implement a Gauss solver or a factorization exploiting thereby the structured rank of the involved matrices.

10.4.3   Other decompositions

Similarly as in the case of the QR-factorization, sequences of Gauss transforms can be chosen to act on the left and on the right of the matrix to give different types of factorizations. Again, we need to remark that, in general, when pivoting is involved in general the lower or upper triangular form of the transformation matrices will get lost.

Here, however, we will present the ideal case; this means that no pivoting is involved. In this way we can also provide statements concerning the structure of the factors operating on the matrix.

Let us design some other patterns, the techniques applied are similar to the techniques discussed for the QR-factorization.

- The standard factorization discussed before consists of applying first sequences of Gauss transforms from bottom to top, to remove the rank structure in the lower triangular part. Then one removes by sequences from top to bottom the remaining nonzero subdiagonal elements. This gives the standard LU-factorization.

- One can annihilate the upper triangular rank structure with upper Gauss transforms performed on the left side from top to bottom. Afterwards one can remove the remaining nonzero super diagonal elements by performing a sequence of upper Gauss transforms on the left side from bottom to top. This leads to a UL-factorization.

- The LU and the UL factorization can also be obtained by performing transformation on the right-hand side of the matrix.

- Suppose we start similarly as in the LU-decomposition by annihilating from the left the rank structure in the lower triangular part of the matrix. This gives us, in general, a generalized Hessenberg matrix with some nonzero subdiagonals left. Instead of continuing to apply transformations on the left, we can now switch to the other side. Annihilating the subdiagonal elements by sequences of lower Gauss transforms performed from the right to the left leads to a L_1UL_2-factorization of the matrix. The matrices L_1 and L_2 are both lower triangular. The following example will illustrate this decomposition on a quasiseparable matrix.

- In a similar manner as described above one can compute a U_1LU_2-factorization of the matrix in which the matrices U_1 and U_2 are both upper triangular.

- The last two factorizations can also be computed in different ways by starting the annihilation pattern on a different side.

- To conclude, we can say that these factorizations can also be computed by different patterns of annihilation. This will be discussed in a forthcoming section.

Let us illustrate the computation of the $L_1 U L_2$-factorization of a quasiseparable matrix

Example 10.21 In this example we will compute, as described above, the $L_1 U L_2$-factorization of a quasiseparable matrix. Let us consider the following quasiseparable matrix A:

$$A = \begin{bmatrix} 2 & 3 & 3 & 3 \\ 1 & 2 & 3 & 3 \\ 1 & 1 & 2 & 3 \\ 1 & 1 & 1 & 2 \end{bmatrix}.$$

We start by annihilating the strictly lower triangular rank structure by applying two Gauss transforms. First M_3 on the bottom two rows, followed by M_2 acting on rows 2 and 3:

$$M_3 = \begin{bmatrix} 1 & & & \\ 0 & 1 & & \\ 0 & 0 & 1 & \\ 0 & 0 & -1 & 1 \end{bmatrix} \text{ and } M_2 = \begin{bmatrix} 1 & & & \\ 0 & 1 & & \\ 0 & -1 & 1 & \\ 0 & 0 & 0 & 1 \end{bmatrix}.$$

Applying these transformations onto the matrix A results in the following Hessenberg matrix H:

$$M_2 M_3 A = H = \begin{bmatrix} 2 & 3 & 3 & 3 \\ 1 & 2 & 3 & 3 \\ 0 & -1 & -1 & 0 \\ 0 & 0 & -1 & -1 \end{bmatrix}.$$

Let us annihilate now the remaining nonzero subdiagonal elements by lower Gauss transforms performed on the right of the matrix. This sequence is performed from right to left. We apply the transformations $\hat{M}_3$, $\hat{M}_2$, $\hat{M}_1$, defined as follows:

$$\hat{M}_3 = \begin{bmatrix} 1 & & & \\ 0 & 1 & & \\ 0 & 0 & 1 & \\ 0 & 0 & -1 & 1 \end{bmatrix}, \quad \hat{M}_2 = \begin{bmatrix} 1 & & & \\ 0 & 1 & & \\ 0 & -1 & 1 & \\ 0 & 0 & 0 & 1 \end{bmatrix} \text{ and }$$

$$\hat{M}_1 = \begin{bmatrix} 1 & & & \\ -\frac{1}{2} & 1 & & \\ 0 & 0 & 1 & \\ 0 & 0 & 0 & 1 \end{bmatrix}.$$

The resulting upper triangular matrix U will be of the following form:

$$U = \begin{bmatrix} \frac{1}{2} & 3 & 0 & 3 \\ & 2 & 0 & 3 \\ & & -1 & 0 \\ & & & -1 \end{bmatrix}.$$

In fact we calculated now the following factorization for the matrix A:

$$M_2 M_3 A \hat{M}_3 \hat{M}_2 \hat{M}_1 = M A \hat{M} = U.$$

Bringing the matrices M and $\hat{M}$ to the other side leads to the following equalities:

$$A = M^{-1} U \hat{M}^{-1}$$
$$= \begin{bmatrix} 1 & & & \\ 0 & 1 & & \\ 0 & 1 & 1 & \\ 0 & 1 & 1 & 1 \end{bmatrix} \begin{bmatrix} \frac{1}{2} & 3 & 0 & 3 \\ & 2 & 0 & 3 \\ & & -1 & 0 \\ & & & -1 \end{bmatrix} \begin{bmatrix} 1 & & & \\ \frac{1}{2} & 1 & & \\ 0 & 1 & 1 & \\ 0 & 0 & 1 & 1 \end{bmatrix}.$$

This factorization is indeed an $L_1 U L_2$-factorization.   ∎

In this section we showed how to solve systems of equations involving Gauss transforms. Moreover, other types of factorizations were also presented by operating on two sides of the matrix.

In the next section we will present a graphical scheme for Gauss transforms, similarly as we did for the QR-factorization. Moreover, we will investigate whether there is also a shift through lemma for Gauss transforms, and based on this shift through lemma, one can deduce different patterns for computing the desired factorization.

10.5   Different patterns of annihilation

Similarly as in the case of the QR-decomposition, we can design different patterns for annihilating the lower triangular part of structured rank matrices. Note that also in this case we have a kind of shift through lemma. Hence most of the patterns of annihilation as described in the previous chapter can also be used in this case. Therefore we will not discuss everything in a so detailed manner we did for the QR-decomposition.

One should be very careful when applying pivoting, however. Pivoting has a very strong effect on the sequences of transformations. First, we will again introduce the graphical representation we will be using. Then, we will present some patterns of annihilation and rank expansion, followed by the shift through lemma and some other interesting connections for interchanging Gauss transforms and so forth.

10.5.1   The graphical representation

The graphical representation we use for representing the Gauss transforms is almost similar to the one used for the Givens transformations. Givens transformations have the property that they change both rows on which they act. Gauss transforms act on two rows, but, in fact, they change only one row. Hence, we use the following graphical representation for Gauss transforms. On the left we have depicted a lower Gauss transform and on the right we can see an upper Gauss transform.

Suppose we have, for example, a semiseparable matrix S, and we perform a sequence of ascending Gauss transforms acting on this 5×5 matrix for annihilating the lower triangular structure. This is graphically depicted in the left figure. Annihilating its upper triangular structure by a sequence of upper Gauss transforms going from top to bottom is depicted in the right figure.

The upper/lower part (depicted with $\lceil / \lfloor$) denotes which row is also involved in the Gauss transform, and the arrows denote which row is actually changed by the Gaussian transformation.

Similarly as in the previous chapter, the dotted circles depict the rows of the matrix. The bottom numbers represent in some sense a time line.

For example, the annihilation of a quasiseparable matrix via Gaussian transformation matrices (without pivoting) is graphically depicted as follows:

The first four transformations go from bottom to top, whereas the last four transformations go from top to bottom to annihilate the subdiagonal elements.

The last thing for which we do not yet have a symbol is the permutation matrix. In some sense we could just depict it as a standard Givens transformation, but because it is a special kind of Givens transformation we decided to give it a different symbol. With the following symbol we will depict a permutation involving two successive rows.

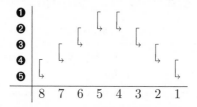

We also depicted two successive permutations, as this is, in fact, a very special operation. Two successive permutations can be combined and they just disappear. Hence the right figure represents, in fact, no real transformation on a matrix. It is just the identity matrix which remains after combining these transformations.

10.5.2 Some standard patterns

Similarly as in the QR-decomposition case, we have different patterns of annihilation. Let us quickly repeat some of the common ones. We assume no pivoting is involved, otherwise we can have too many combinations for the figures. Pivoting is admitted, however, and can easily be combined with the sequences to obtain the different annihilation patterns as they are depicted here.

Suppose we have a 5×5 semiseparable matrix of semiseparability rank 3. For removing the rank structure we have the following patterns.

- **The leaf pattern for removing the rank structure.**
  Graphically this is represented as follows.

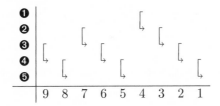

It is called the leaf form as every low rank part is removed one after another. Hence the structured rank is pealed off, one by one.

- **The pyramid form for removing the rank structure.**
  Now we have the following graphical representation of the annihilation process.

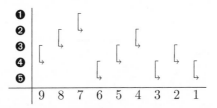

In this case the successive Gauss transforms create a growing pyramid form. The ranks are not pealed of anymore, but removed in the special order as depicted below. The reader can try to depict what is going on in the matrix. This is almost identical (except for the upper triangular part) to the QR-decomposition case.

Similarly as for removing the rank structure, we can create zeros via different patterns. Suppose we have a generalized Hessenberg matrix with two subdiagonals. Then we have the following two annihilation forms.

- **The leaf form for creating zeros.**
  If we perform sequences of transformations annihilating one subdiagonal after another, we call this the leaf form. For our generalized Hessenberg matrix with two subdiagonals this is graphically depicted as follows.

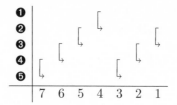

- **The diamond form for creating zeros.**
  Similar to the pyramid case we can reorder the transformations and get the following graphical representation which we call the diamond form:

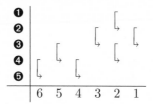

Even though we already mentioned it, one can easily incorporate pivoting in the sequences of transformations presented above. This does not compromise any of the patterns.

10.5.3   Some theorems connected to Gauss transforms

In this section we will show some interesting theorems to deal with Gauss transforms in a more flexible manner. We remark that because there are more different Gauss transforms than Givens transformations (namely the lower and the upper Gauss transform), there are many more combinations possible. Moreover, if we even start involving pivoting, one can get lots of different behaviors. In this section we will present some of the possible combinations (not all of them), so that the reader himself can experiment with these schemes of annihilation.

We remark that we will only formulate some theorems related to lower Gauss transforms. Upper gauss transforms behave in a similar way. Some connections between upper and lower Gauss transforms will be quickly given, however.

Lemma 10.22 (Fusion of Gauss transforms). *Suppose two Gauss transforms* M_1 *and* M_2 *are given:*

$$M_1 = \begin{bmatrix} 1 & \\ \tau_1 & 1 \end{bmatrix} \ and \ M_2 = \begin{bmatrix} 1 & \\ \tau_2 & 1 \end{bmatrix}.$$

Then we have that $M_1 M_2 = M_3$ *is again a Gauss transform with parameter* $\tau_3 = \tau_1 + \tau_2$. *We will call this operation, the fusion of Gauss transforms in the remainder of the text.*

The proof is trivial. In our graphical schemes, we will depict this as follows:

$$
\begin{array}{c|cc}
\mathbf{1} \\
\mathbf{2}
\end{array} \quad \text{resulting in} \quad
\begin{array}{c|c}
\mathbf{1} \\
\mathbf{2} \\
\hline
& 1
\end{array}
$$

The next lemma generalizes the shift through lemma for Givens transformations towards elementary Gaussian transformation matrices. Unfortunately it is not always possible to do so.

Lemma 10.23 (Shift through lemma). *Suppose three 3×3 Gauss transforms M_1, M_2 and M_3 are given, such that the Gauss transforms M_1 and M_3 act on the first two rows of a matrix, and M_2 acts on the second and third row. Denote the Gauss parameters as τ_1, τ_2 and τ_3, with all Gauss parameters different from zero. Then we have that*

$$M_1 M_2 M_3 = \hat{M}_1 \hat{M}_2 \hat{M}_3,$$

where $\hat{M}_1$ and $\hat{M}_3$ work on the second and third row and $\hat{M}_2$, works on the first two rows, with Gauss parameters $\hat{\tau}_1, \hat{\tau}_2$ and $\hat{\tau}_3$

Proof. Constructing both transformation matrices gives us two times a lower triangular matrix. We obtain the following equation:

$$
\begin{bmatrix}
1 & & \\
\tau_1 + \tau_3 & 1 & \\
\tau_2 \tau_3 & \tau_2 & 1
\end{bmatrix}
=
\begin{bmatrix}
1 & & \\
\hat{\tau}_2 & 1 & \\
\hat{\tau}_1 \hat{\tau}_2 & \hat{\tau}_1 + \hat{\tau}_3 & 1
\end{bmatrix}.
$$

This leads to the following equalities:

$$\hat{\tau}_2 = \tau_1 + \tau_3,$$
$$\hat{\tau}_1 \hat{\tau}_2 = \tau_2 \tau_3,$$
$$\hat{\tau}_1 + \hat{\tau}_3 = \tau_2.$$

Suppose we have the ∨-pattern (left-hand side) and we would like to transform this into the ∧-pattern (right-hand side). The reader can verify that this is only possible if $\tau_1 + \tau_3$ is different from zero. When $\tau_1 + \tau_3 \neq 0$, we can write down the following relations:

$$\hat{\tau}_1 = \frac{\tau_2 \tau_3}{\tau_1 + \tau_3},$$
$$\hat{\tau}_2 = \tau_1 + \tau_3,$$
$$\hat{\tau}_3 = \frac{\tau_2 (1 - \tau_3)}{\tau_1 + \tau_3}.$$

This defines the ∧-pattern.

Also working the other way around one should be very carefull. If one wants to transform the ∧-pattern into a ∨-pattern, one has to make sure that $\hat{\tau}_1 + \hat{\tau}_3 \neq 0$.

□

Note 10.24. *In the case it is possible to change the order of the involved Gauss transforms, the same remark with regard to Givens transformations is applicable here. This means that the shift through lemma changes the order between a rank-decreasing and a rank-expanding Gauss transform.*

Graphically we depict it exactly the same as for the Givens transformations as follows (of course, one has the be careful with the applicability of the theorem):

and in the other direction this becomes:

We remark again that if we cannot place the ⌢ or ⌣ arrow at that position, that we cannot apply the shift through lemma.

Suppose we have the ∨-patterns and the shift through lemma is not applicable, in general, we can write down the following graphical representation. This is checked rather easily assuming $\tau_1 + \tau_3 = 0$. In this next graphical representation, a new kind of Gauss transform is introduced. The same interpretation holds for these transformations, the arrow denotes which row is changed, whereas at the other end ⌈ or ⌊ denotes which row is used for adding to the changed row.

Going from the ∧-pattern to the ∨-pattern, in case $\hat{\tau}_1 + \hat{\tau}_3 = 0$, can be written as follows.

In the coming list an alternative factorization to the ones presented above will be presented to interpret also these last schemes as either ∨- or ∧-patterns.

In the following list we present some properties of Gauss transforms which might come in handy when working with these patterns. Some known results will also be repeated. The properties derived here focus on lower Gauss transforms. Several properties can easily be translated towards upper Gauss transforms, but as the transformations are often straightforward we do not cover them.

1. Suppose two Gauss transforms M_1 and M_2 are given, with parameters τ_1 and τ_2. The fusion of these two Gauss transforms leads to another Gauss transform with Gauss parameter $\tau_1 + \tau_2$:

$$\begin{bmatrix} 1 & \\ \tau_1 & 1 \end{bmatrix} \begin{bmatrix} 1 & \\ \tau_2 & 1 \end{bmatrix} = \begin{bmatrix} 1 & \\ \tau_1 + \tau_2 & 1 \end{bmatrix} = \begin{bmatrix} 1 & \\ \tau_3 & 1 \end{bmatrix}.$$

2. Performing two Gauss transforms as depicted below leads, in general, to a bidiagonal matrix:

$$\begin{bmatrix} 1 & & \\ \tau_1 & 1 & \\ & & 1 \end{bmatrix} \begin{bmatrix} 1 & & \\ & 1 & \\ & \tau_2 & 1 \end{bmatrix} = \begin{bmatrix} 1 & & \\ \tau_1 & 1 & \\ & \tau_2 & 1 \end{bmatrix}.$$

In general, this means that an ascending sequence of lower Gauss transforms generates a bidiagonal matrix. The following sequence illustrates the generation of the bidiagonal matrix presented above:

$$\text{generates} \quad \begin{bmatrix} \times & & \\ \times & \times & \\ 0 & \times & \times \end{bmatrix}.$$

3. A sequence of lower Gauss transforms from bottom to top generates a lower semiseparable matrix.

$$\begin{bmatrix} 1 & & \\ & 1 & \\ & \tau_1 & 1 \end{bmatrix} \begin{bmatrix} 1 & & \\ \tau_2 & 1 & \\ & & 1 \end{bmatrix} = \begin{bmatrix} 1 & & \\ \tau_2 & 1 & \\ \tau_2\tau_1 & \tau_1 & 1 \end{bmatrix}.$$

Graphically we have the following:

$$\text{generates} \quad \begin{bmatrix} \boxtimes & & \\ \boxtimes & \boxtimes & \\ \boxtimes & \boxtimes & \boxtimes \end{bmatrix}.$$

4. Two permutation matrices fuse and disappear. This is rather straightforward and has already been discussed. It is clear that:

$$\begin{bmatrix} 0 & 1 \\ 1 & 0 \end{bmatrix} \begin{bmatrix} 0 & 1 \\ 1 & 0 \end{bmatrix} = \begin{bmatrix} 1 & \\ & 1 \end{bmatrix}.$$

Graphically this is depicted as follows:

vanishes.

5. A permutation acting on the left of a lower triangular Gauss transform turns it into an upper Gauss transform with a permutation on the right.

$$\begin{bmatrix} 0 & 1 \\ 1 & 0 \end{bmatrix}\begin{bmatrix} 1 & \\ \tau & 1 \end{bmatrix} = \begin{bmatrix} 1 & \tau \\ & 1 \end{bmatrix}\begin{bmatrix} 0 & 1 \\ 1 & 0 \end{bmatrix}.$$

Graphically this is depicted as follows:

     equals

6. An upper and lower Gauss transform using the same row to add to another row commute. More precisely this leads to:

$$\begin{bmatrix} 1 & \tau_1 & \\ & 1 & \\ & & 1 \end{bmatrix}\begin{bmatrix} 1 & & \\ & 1 & \\ & \tau_2 & 1 \end{bmatrix} = \begin{bmatrix} 1 & & \\ & 1 & \\ & \tau_2 & 1 \end{bmatrix}\begin{bmatrix} 1 & \tau_1 & \\ & 1 & \\ & & 1 \end{bmatrix}.$$

Graphically we obtain the following scheme.

     is identical to

7. An upper and a lower Gaussian transformation acting on the same rows commute up to some extent. We have that:

$$\begin{bmatrix} 1 & \\ \tau_1 & 1 \end{bmatrix}\begin{bmatrix} 1 & \tau_2 \\ 0 & 1 \end{bmatrix} = \begin{bmatrix} 0 & 1 \\ 1 & 0 \end{bmatrix}\begin{bmatrix} 1 & \tau_1 \\ & 1 \end{bmatrix}\begin{bmatrix} 1 & 0 \\ \tau_2 & 1 \end{bmatrix}\begin{bmatrix} 0 & 1 \\ 1 & 0 \end{bmatrix}.$$

Graphically we obtain the following

     equals

8. We discussed the action of the long arrow before; let us see how this action interacts with pivoting. In matrix terms we have the following relation:

$$\begin{bmatrix} 1 & & \\ & 0 & 1 \\ & 1 & 0 \end{bmatrix}\begin{bmatrix} 1 & & \\ & 1 & \\ \tau & & 1 \end{bmatrix} = \begin{bmatrix} 1 & & \\ \tau & 0 & 1 \\ & 1 & 0 \end{bmatrix} = \begin{bmatrix} 1 & & \\ \tau & 1 & \\ & & 1 \end{bmatrix}\begin{bmatrix} 1 & & \\ & 0 & 1 \\ & 1 & 0 \end{bmatrix}.$$

Graphically we can depict this relation as follows.

     equals

The reader can easily try to see that the following graphical scheme also holds.

This last figure represents the problems with pivoting that were obtained in the case pivoting was applied for semiseparable matrices. We tried to shift all the permutation matrices to the left, but this changed the Gauss transforms, ending up eventually with a quasiseparable structure instead of a semiseparable one.

9. There is a kind of commutation between longer and smaller arrows. The following figures are easily verified. In fact we presented them before, when presenting the shift through lemma.

Also the following commutation holds.

10. The last one we will present is just completely graphical and generalizes in some sense the shift through lemma by applying some tricks discussed above. Suppose the shift through lemma is not applicable, we only discuss the transition from the ∨-pattern to the ∧-pattern. This means that we have the following situation.

This can easily be rewritten by putting two permutation matrices in the front. Let us then immediately apply the permutation on the long arrow.

In some sense one can interpret this pattern as a ∧-pattern, involving pivoting, however.

We discussed in the previous list some properties of Gauss transforms and permutation matrices. The list is far from complete, but all the essential tools are provided so that the reader can play with them.

Moreover the reader sees that pivoting also fits nicely into these graphical schemes. One can even try to admit regular Givens transformations into these graphical representations. This leads to hybrid factorizations involving Gauss transforms and Givens transforms.

Note 10.25. *Even though we did not really examine it in more detail, the graphical annihilation schemes based on the Gauss transforms can easily be expanded to illustrate the annihilation of a dense matrix, including pivoting. One has just to make the arrows longer. And with pivoting, one can create longer arrows, pointing to the two rows involved in the interchange action.*

In this section different techniques were presented to handle Gauss transforms in a creative visible way. Learning to deal with these patterns efficiently creates a high flexibility and a fast way of seeing what is possible and what is not.

We do not go into further detail as we did with the QR-factorization. Gaussian transformations are slightly more complicated than the Givens transformations, but roughly speaking one can obtain almost the same results. One can derive parallel factorizations, one can interpret the upper triangular making of a matrix as a sequence of rank-expanding Gauss transforms. If the reader is interested, he can try to derive similar approaches for these transformations. However, we will not go any further into detail.

10.6 Conclusions

In this chapter we tried to compute the LU-factorization of structured rank matrices similarly to the QR-factorization discussed in the previous chapter. We deduced theorems predicting the rank structure of matrices after having performed sequences of rank-decreasing and/or rank-expanding Gauss transforms. Unfortunately we were not able to guarantee in general that these sequences existed. To overcome we had to introduce the use of permutation matrices. Using these permutation matrices we could prove the existence of rank-decreasing and rank-expanding sequences of Gauss transforms. Even though it is often necessary to apply pivoting it was shown that pivoting resulted in two unwanted effects. Pivoting seemed to have a strong impact on the rank structure of the upper triangular structure; instead of leaving the rank unchanged, it increased the rank. Secondly, applying pivoting in sequences in general does not lead anymore to an LU-factorization.

But, even though we did not compute an LU-factorization anymore we showed that it was possible to use this factorization in an efficient way for solving systems of equations. We also discussed other types of sequences of Gauss transforms. We distinguished between upper and lower Gauss transforms, and Gauss transforms acting on the left or on the right side of the matrix. Using this flexibility we could derive other factorizations. We concluded this chapter by providing some techniques for interchanging the order of Gauss transforms and we illustrated these possibilities by a graphical representation.

Chapter 11

A Levinson-like solver for structured rank matrices

The second part of this book was dedicated to solving systems of equations for the easy classes of structured rank matrices, such as semiseparable, quasiseparable, semiseparable plus diagonal and so forth. The chapter on the Levinson-like solver, restricted its examples to these easy classes. To deal with all these problems at once a unified framework was presented, covering the classes of semiseparable, both generator representable as well as Givens-vector representable, quasiseparable of order 1, etc. But also new combinations seemed to fit perfectly in this scheme, e.g., companion matrices, arrowhead matrices, arrowhead matrices plus a tridiagonal matrix.

The general framework as proposed in that chapter is also capable of dealing with more general classes of structured rank matrices, more precisely classes of a higher order structured rank. These examples were naturally omitted in the first part. Here, we revisit the Levinson-like solver and apply it to more general examples. As no more theoretical results are needed to apply the algorithm to these matrices, this chapter is the shortest one in the book.

For a clear understanding of the examples below, one might want to reread Section 6.3 of Chapter 6.

We will cover different examples here. First, a method is proposed to solve higher order generator representable semiseparable systems of equations. Second, we will cover the class of quasiseparable matrices in their general form as defined in Chapter 8. Band matrices also seem to fit perfectly in this scheme. They are covered in Section 11.3. This is followed by some new examples of matrices having an unsymmetric structure. Having an unsymmetric structure means that the upper triangular part might have a different rank structure than the lower triangular part. In Section 11.5 higher order rank structured matrices which are summations of Levinson-conform matrices are presented. The titles of these sections resemble the titles of the sections in Chapter 6. The examples provided here, however, are more focused on the classes of higher order structured rank matrices.

Notes and references are not included for this chapter; they were discussed in Chapter 6.

✎ *The main ideas for solving systems of equations via a Levinson-like tech-nique were already discussed in Chapter 6. This chapter contains examples related to higher order classes of structured rank matrices. The title of the sections men-tions the class of matrices considered.*

11.1   Higher order generator representable semiseparable matrices

In this section we will develop a Levinson-like solver for higher order generator representable semiseparable matrices. This is another, higher order example of the general framework as we discussed it earlier.

Suppose we have the following higher order generator representable semisep-arable matrix $S(U, V, P, Q)$:

$$
S(U,V,P,Q) = \begin{bmatrix} \mathbf{q}_1^T\mathbf{p}_1 & \mathbf{q}_2^T\mathbf{p}_1 & \cdots & \mathbf{q}_n^T\mathbf{p}_1 \\ \mathbf{u}_2^T\mathbf{v}_1 & \mathbf{q}_2^T\mathbf{p}_2 & & \vdots \\ \vdots & & \ddots & \mathbf{q}_n^T\mathbf{p}_{n-1} \\ \mathbf{u}_n^T\mathbf{v}_1 & \cdots & \mathbf{u}_n^T\mathbf{v}_{n-1} & \mathbf{q}_n^T\mathbf{p}_n \end{bmatrix},
$$

for which all the row vectors $\mathbf{u}_i^T$ and $\mathbf{v}_i^T$ are of length p and the row vectors $\mathbf{p}_i^T$ and $\mathbf{q}_i^T$ are of length q. This matrix is called a $\{p, q\}$-generator representable semi-separable matrix. We will prove now that this matrix is simple $\{q, p\}$-Levinson conform.

To make the block decomposition of the matrix $S = S(U, V, P, Q)$ more com-prehensible, we rewrite the matrix S as follows, thereby reordering the inner prod-ucts:

$$
S = \begin{bmatrix} \mathbf{p}_1^T\mathbf{q}_1 & \mathbf{p}_1^T\mathbf{q}_2 & \cdots & \mathbf{p}_1^T\mathbf{q}_n \\ \mathbf{v}_1^T\mathbf{u}_2 & \mathbf{p}_2^T\mathbf{q}_2 & & \vdots \\ \vdots & & \ddots & \mathbf{p}_{n-1}^T\mathbf{q}_n \\ \mathbf{v}_1^T\mathbf{u}_n & \cdots & \mathbf{v}_{n-1}^T\mathbf{u}_n & \mathbf{p}_n\mathbf{q}_n^T \end{bmatrix},
$$

Let us define the matrices R_k and S_k as follows:

$$
R_k = \begin{bmatrix} \mathbf{p}_1^T \\ \vdots \\ \mathbf{p}_k^T \end{bmatrix} \text{ and } S_k = \begin{bmatrix} \mathbf{v}_1^T \\ \vdots \\ \mathbf{v}_k^T \end{bmatrix}.
$$

Defining $P_k = I_q$, $Q_k = I_p$, $\boldsymbol{\eta}_k = \mathbf{v}_k$ and $\boldsymbol{\xi}_k = \mathbf{p}_k$ gives us the following relations (these are Conditions 2 and 3 of Definition 6.4):

$$
R_{k+1} = \begin{bmatrix} R_k \\ \hline \mathbf{p}_{k+1}^T \end{bmatrix} \text{ and } S_{k+1} = \begin{bmatrix} S_k \\ \hline \mathbf{v}_{k+1}^T \end{bmatrix}.
$$

Moreover the upper left $(k+1) \times (k+1)$ block of A is of the following form (Condition 1 of Definition 6.4):

$$A_{k+1} = \left[\begin{array}{c|c} A_k & R_k \mathbf{q}_{k+1} \\ \hline \mathbf{u}_{k+1}^T S_k^T & \mathbf{p}_{k+1} \mathbf{q}_{k+1}^T \end{array} \right].$$

This coefficient matrix satisfies all the properties to come to the desired $\mathcal{O}(pqn)$ Levinson-like solver. Using the operation count as presented in Section 6.3.2, we see that the number of operations is bounded by

$$(6pq + 7q + 4p + 1)(n - 1) - 3pq + 3q - 2p + 1.$$

Based on this algorithm for generator representable semiseparable matrices, it is almost trivial to devise the matrices to come to the Levinson-like solver for quasiseparable matrices.

11.2   General quasiseparable matrices

References concerning the class of quasiseparable matrices can be found in Section 6.4.2. In the current section we will briefly compare the Levinson-like method, with the algorithm presented in [94], for solving quasiseparable systems of equations $A\mathbf{x} = \mathbf{b}$, as this algorithm uses a similar approach. Before comparing both methods, we will briefly indicate how quasiseparable matrices fit into this framework. A general $\{p, q\}$-quasiseparable matrix is of the following form $A = (a_{i,j})_{i,j}$: A general $\{p, q\}$-quasiseparable matrix A with the quasiseparable representation is of the following form

$$a_{i,j} = \left\{ \begin{array}{ll} \mathbf{q}_j^T r_{j-1:i+1} \mathbf{p}_i, & 1 \leq i < j \leq n, \\ d_i, & 1 \leq i = j \leq n, \\ \mathbf{u}_i^T t_{i-1:j+1} \mathbf{v}_j, & 1 \leq j < i \leq n, \end{array} \right.$$

where $r_{j-1:i+1} = r_{j-1} r_{j-2} \ldots r_{i+1}$, $t_{i-1:j+1} = t_{i-1} t_{i-2} \ldots t_{j+1}$, $\mathbf{p}_i, \mathbf{q}_i \in \mathbb{R}^q$, $\mathbf{u}_i, \mathbf{v}_i \in \mathbb{R}^p$, $r_i \in \mathbb{R}^{q \times q}$ and $t_i \in \mathbb{R}^{p \times p}$ for all i.

Combining the techniques presented in Sections 11.1 and 6.4.2, we see that this quasiseparable matrix is $\{q, p\}$-Levinson conform (we do not include the details). The quasiseparable matrix, however, is not a simple Levinson conform anymore, as the matrices P_k and Q_k are $P_k = r_{k+1}$ and $Q_k = t_{k+1}$, and these matrices a_{k+1} and b_{k+1} do not necessarily admit a linear multiplication with a vector. Hence, we are, as in the case of Section 6.3.2, leading to a complexity of

$$n \left[2p^2 q + 2qp^2 + 4q^2 + 4pq + 3p^2 + 5q + 2p + 1 \right] + \mathcal{O}(1),$$

for solving a quasiseparable matrix via the Levinson-like solver as presented in this section.

The method presented in [94] computes, in fact, the generators of the inverse of the quasiseparable matrix and then applies a fast multiplication A^{-1} to the right-hand side $\mathbf{b}$ to compute the solution vector $\mathbf{x}$. Moreover the algorithm produces also the generators for the matrices L and U, in the following factorization of the matrix A:

$$A = L\Lambda U,$$

where L and U, respectively, are lower and upper triangular matrices, with ones on the diagonal, and Λ is a diagonal matrix. A complexity count of the algorithm proposed in [94] gives us the following number of flops:

$$n\left[4p^2q + 4q^2p + 12pq + 8p^2 + 8q^2 + 5p + 5q + 3\right] + \mathcal{O}(1).$$

The method presented here, however, does not explicitly compute the generators of the inverse of the quasiseparable matrix. But implicitly it calculates an upper triangular factorization of the following form (see Section 6.3.3)

$$AU = L\Lambda,$$

where U and L, respectively, are upper and lower triangular with ones on the diagonal, and Λ is a diagonal matrix.

The classes of quasiseparable and generator representable are straightforward classes of structured rank matrices. The Levinson-like method has a much wider impact, however. Also band matrices do fit nicely into this general framework.

11.3   Band matrices

We will now briefly illustrate how we can solve band matrices using the Levinson-like method. There is a huge variety of methods for solving banded systems of equations: from QR-decompositions, LU-decompositions, Gaussian elimination to parallel methods. Some of these methods can be found in [152] and the references therein. Band matrices can also be considered as quasiseparable matrices; for example, a $\{p,q\}$-band matrix is also $\{p,q\}$-quasiseparable. But instead of using the quasiseparable approach, the direct approach gives a much faster algorithm: the quasiseparable approach involves the terms p^2q, pq^2 and pq in the complexity count, whereas the Levinson-like approach only involves the term pq in the complexity count (see Section 11.2, and the complexity count at the end of this section).

Assume we have a $\{p,q\}$-band matrix A of the following form:

$$A = \begin{bmatrix} a_{1,1} & a_{1,2} & \cdots & a_{1,q+1} & 0 & \cdots & 0 \\ a_{2,1} & a_{2,2} & a_{2,3} & \cdots & a_{2,q+2} & & \vdots \\ \vdots & a_{3,2} & a_{3,3} & & & \ddots & 0 \\ a_{p+1,1} & \vdots & & \ddots & & & a_{n-q,n} \\ 0 & a_{p+2,2} & & & & & \vdots \\ \vdots & & \ddots & & & \ddots & a_{n-1,n} \\ 0 & \cdots & 0 & a_{n,n-p} & \cdots & a_{n,n-1} & a_{n,n} \end{bmatrix},$$

with all the $a_{i,j} \in \mathbb{R}$. This is not the most compact representation of the band matrix A, as many of its elements are zero. Let us therefore introduce the matrix $\bar{A}$. (Similarly, we can construct the matrix $\underline{A}$ for the lower triangular part.) Let us denote with $\bar{\mathbf{a}}_i^T$ the ith row out of the matrix $\bar{A}$ (filled up with zeros on the left, to make it of length $q+1$), this means:

$$\begin{cases} \bar{\mathbf{a}}_i^T = [0, \ldots, 0, a_{1,i}, a_{2,i}, \ldots, a_{i-1,i}] & \text{if} \quad i \leq q \\ \bar{\mathbf{a}}_i^T = [a_{i-q,i}, a_{i-q+1,i}, \ldots, a_{i-1,i}] & \text{if} \quad i > q \end{cases}.$$

The column vectors $\bar{\mathbf{a}}_i$ are of length $q+1$. It is shown now that the upper triangular part of the matrix satisfies the desired conditions to be simple $\{q, p\}$-Levinson conform. (The lower triangular part is similar.) Let us define R_k as a $k \times q$ matrix of the following form:

$$R_k = \left[\frac{0}{I_q} \right],$$

where I_q denotes the identity matrix of size q. In the beginning of the algorithm, we have that $k \leq q$. In this case we take only the last k lines of I_q, this means that:

$$R_k = \begin{cases} [0, I_k] & \text{if} \quad k < q \\ I_q & \text{if} \quad k = q \\ \left[\dfrac{0}{I_q} \right] & \text{if} \quad k > q. \end{cases}$$

For every k the matrix P_k is defined as the shift operator P, which is a $q \times q$ matrix of the following form

$$P_k = P = \begin{bmatrix} 0 & & & \cdots & 0 \\ 1 & 0 & & \cdots & 0 \\ 0 & 1 & 0 & & \vdots \\ \vdots & & \ddots & \ddots & \\ 0 & \cdots & 0 & 1 & 0 \end{bmatrix}.$$

Multiplying a matrix on the right with this operator will shift all the columns in this matrix one position to the left and add a trailing zero column. Using this shift matrix and defining for every k the vector $\boldsymbol{\xi}_k^T = [0, 0, \ldots, 0, 1]$ gives us

$$R_{k+1} = \begin{bmatrix} R_k P \\ \boldsymbol{\xi}_k^T \end{bmatrix}.$$

Using the above definitions and defining $\mathbf{c}_k = \bar{\mathbf{a}}_k$ (define S_k and $\mathbf{d}_k$ similarly as for the upper triangular part) leads to the following relations:

$$A_{k+1} = \left[\begin{array}{c|c} A_k & R_k \bar{\mathbf{a}}_{k+1} \\ \hline \mathbf{d}_{k+1}^T S_k^T & a_{k+1,k+1} \end{array} \right].$$

This means that our band matrix is simple $\{q, p\}$-Levinson conform, leading to a solver of complexity $\mathcal{O}(pqn)$. More precisely we know, as the multiplication on the right with the matrix $P_k = P$ does not involve operations (just index shuffling), that the number of operations is bounded by

$$(6pq + 7q + 4p + 1)(n - 1) - 3pq + 3q - 2p + 1,$$

as $\kappa_1 = \kappa_2 = \gamma_1 = \gamma_2 = 0$. This is clearly an upper bound, as the special structure of the matrices R_k is not taken into consideration. Exploiting this structure results in further complexity reduction.

Let us consider now some matrices having different structures in the upper and lower triangular part of the matrix.

11.4 Unsymmetric structures

Let us reconsider some matrices with unsymmetric structures, i.e., where the rank structure of the upper part does not need to be related in any way to the rank structure of the lower part.

- An upper triangular $\{p\}$-semiseparable matrix is simple $\{p, 0\}$-Levinson conform, where p stands for the semiseparability rank of the upper triangular part.

- An upper triangular $\{p\}$-band matrix is simple $\{p, 0\}$-Levinson conform, where p stands for the bandwidth in the upper triangular part.

- A matrix for which the upper triangular part is $\{p\}$-semiseparable, and the lower triangular part is coming from a $\{q\}$-band matrix is simple $\{p, q\}$-Levinson conform. For example, a unitary Hessenberg matrix, this matrix has the upper triangular part of semiseparable form and only one subdiagonal different from zero.

- A matrix which has a band structure in the upper triangular part and, for which the lower triangular part comes from an arrowhead is $\{p, 1\}$-Levinson conform (p denotes the bandwidth).

- Moreover, one can also combine matrices for which the upper or the lower triangular part is quasiseparable. In this case the complexity of the Levinson-like solver changes according to the complexity of the multiplication with the matrices P_k and Q_k. See Section 6.3.2 for a more detailed analysis of the complexity.

Let us reconsider some examples of the summation of Levinson conform matrices.

11.5 Summations of Levinson-conform matrices

The summation of Levinson-conform matrices gives us again a Levinson-conform matrix. Let us present some more examples.

- One can solve now summations of all previously defined Levinson-conform matrices. For example, the sum of a higher order semiseparable matrix plus a band matrix.

- Suppose we have a higher order semiseparable matrix, which is written as the sum of Givens-vector represented semiseparable matrices of semiseparability rank 1. Then it is clear that also the sum of these Givens-vector represented matrices is Levinson conform, and hence can be solved by this method.

- Another interesting example is the one of the split representation. We can have a sum of structured rank matrices, all represented in a different way. So one can sum quasiseparable and semiseparable matrices without any problem.

- Different matrices from statistical applications, which have different semiseparability ranks in the upper and the lower triangular part plus a band matrix can be found in [158]. These matrices fit naturally in this framework.

- An arrowhead matrix with a larger bandwidth is also $\{q+1, p+1\}$-Levinson conform, as it can be written as the sum of an arrowhead matrix and a $\{p, q\}$-band matrix.

11.6 Conclusions

In this chapter the Levinson-like algorithm as presented in Chapter 6 was reconsidered. We applied the method to some different classes of higher order structured rank matrices. We discussed quasiseparable, generator representable semiseparable and band matrices. Moreover also new types of summations and matrices having unsymmetric structures were discussed.

Chapter 12

Block quasiseparable matrices

In this chapter we will give an introduction on the class of block quasiseparable matrices. This is again a generalization of the class of higher order quasiseparable matrices. Quasiseparable matrices are also known under the name sequentially semiseparable matrices and matrices having low Hankel rank. The content of this chapter is based on the papers [93, 94, 96, 53]. In these papers block quasiseparable matrices are defined by their representation. We will do the same here, although as we will show in Section 12.3, this class of matrices could be better defined by its structured rank properties. This puts more focus on the properties of the block quasiseparable matrices and gives more freedom in choosing the most suitable representation depending on the application in which these matrices arise.

In the first section, we will define the class of block quasiseparable matrices. The next section describes how the block upper and block lower triangular part of a block quasiseparable matrix can be written in a factorized form. Section 12.3 gives the connection between block quasiseparable matrices and structured rank matrices. Examples are given in Section 12.4. Section 12.5 describes how the matrix-vector multiplication can be performed in linear complexity as well as solving a block upper or block lower quasiseparable system of linear equations. A linear complexity solver is designed in Section 12.6. The connection between block quasiseparable matrices and descriptor systems is covered in Section 12.7.

✎ *Reading this chapter is not essential for understanding the other chapters of this book. Attention is paid on the definition and use of block quasiseparable matrices. Historically the definition (Definition 12.1) is given based on the representation of this class of matrices. In Section 12.3, we indicate that a definition based on the structured rank properties is to be preferred.*

12.1 Definition

This section is based on the papers [93, 94, 96].

Definition 12.1.   *A block matrix $A = (A_{i,j})$ with $m_i \times n_j$ blocks $A_{i,j}$ with $i, j = 1, 2, \ldots, N$, having $m = sum_i m_i$ rows and $n = sum_j n_j$ columns, is called block quasiseparable if the blocks $A_{i,j}$ can be factorized as*

$$A_{i,j} = \begin{cases} U_i T_{i-1} T_{i-2} \cdots T_{j+1} V_j & 1 \le j < i \le N \\ D_i & 1 \le i = j \le N \\ P_i R_{i+1} R_{i+2} \cdots R_{j-1} Q_j & 1 \le i < j \le N \end{cases}$$

with

$$\begin{cases} U_i & \text{of size } m_i \times r_{i-1}, & i = 2, 3, \ldots, N \\ T_i & \text{of size } r_i \times r_{i-1}, & i = 2, 3, \ldots, N-1 \\ V_i & \text{of size } r_i \times n_i, & i = 1, 2, \ldots, N-1 \\ P_i & \text{of size } m_i \times s_i, & i = 1, 2, \ldots, N-1 \\ R_i & \text{of size } s_{i-1} \times s_i, & i = 2, 3, \ldots, N-1 \\ Q_i & \text{of size } s_{i-1} \times n_i, & i = 2, 3, \ldots, N \\ D_i & \text{of size } m_i \times n_i, & i = 1, 2, \ldots, N. \end{cases} \tag{12.1}$$

The matrices U_i, T_i and V_i are called generators of the lower block quasiseparable part of the matrix while the matrices P_i, R_i and Q_i are called generators of the upper block quasiseparable part of the matrix A. It is easy to see that any matrix can be represented in this way. However, in that case, generically we have to choose r_i and s_i, $i = 1, 2, \ldots, N-1$ rather large. The distinction between block quasiseparable matrices and general dense matrices becomes important when the dimensions r_i and s_i are small with respect to the number of rows $m = \sum_{i=1}^{N} m_i$ and the number of columns $n = \sum_{j=1}^{N} n_j$ of the matrix A.

Note 12.2. *Note that block quasiseparable matrices are defined by their representation.*

12.2   Factorization of the block lower/upper triangular part

This section is based on [96, Section 5]. Consider the following matrices

$$W_1 = \begin{bmatrix} D_1 \\ V_1 \end{bmatrix} \tag{12.2}$$

$$W_k = \begin{bmatrix} U_k & D_k \\ T_k & V_k \end{bmatrix}, \qquad k = 2, 3, \ldots, N-1 \tag{12.3}$$

$$W_N = \begin{bmatrix} U_N & D_N \end{bmatrix}. \tag{12.4}$$

We will embed these matrices into $m \times m$ matrices $\tilde{W}_k$ as

$$\tilde{W}_1 = W_1 \oplus I_{m_2} \oplus \cdots \oplus I_{m_N}$$
$$\tilde{W}_k = I_{m_1} \oplus \cdots \oplus I_{m_{k-1}} \oplus W_k \oplus I_{m_{k+1}} \oplus \cdots \oplus I_{m_N}$$
$$\tilde{W}_N = I_{m_1} \oplus \cdots \oplus I_{m_{N-1}} \oplus W_N.$$

The operation $\oplus$ denotes the direct sum of matrices, i.e.,

$$A \oplus B = \begin{bmatrix} A & \\ & B \end{bmatrix}.$$

It is easy to check that the block lower triangular part A_L of the block quasiseparable matrix A can be factorized as

$$A_L = \tilde{W}_N \tilde{W}_{N-1} \cdots \tilde{W}_1. \tag{12.5}$$

A similar result can be obtained for the block upper triangular part A_U of A.

12.3 Connection to structured rank matrices

This section is based on [93, Section 3]. Consider each of the block submatrices $A(i : N, 1 : i - 1)$ for $i = 2, 3, \ldots, N$:

$$A(i : N, 1 : i - 1) =$$
$$\begin{bmatrix} U_i T_{i-1} \cdots T_2 V_1 & \cdots & U_i T_{i-1} V_{i-2} & U_i V_{i-1} \\ \vdots & & \vdots & \vdots \\ U_N T_{N-1} \cdots T_2 V_1 & \cdots & U_N T_{N-1} \cdots T_{i-1} V_{i-2} & U_N T_{N-1} \cdots T_I V_{i-1} \end{bmatrix}.$$

From the previous section it follows now that this submatrix can be constructed starting from the block matrix

$$\begin{bmatrix} T_{i-1} \cdots T_2 V_1 & \cdots & T_{i-1} V_{i-2} & V_{i-1} \\ 0 & & 0 & 0 \\ \vdots & & \vdots & \vdots \\ 0 & & 0 & 0 \end{bmatrix}.$$

Indeed, multiplying the first and second block row by

$$\begin{bmatrix} U_i & 0 \\ T_i & 0 \end{bmatrix},$$

multiplying the second and third block row of the result by

$$\begin{bmatrix} U_{i+1} & 0 \\ T_{i+1} & 0 \end{bmatrix},$$

and so on, results in the block submatrix $A(i : N, 1 : i - 1)$. Because we started from a matrix having rank less than or equal to the row dimension r_{i-1} of T_{i-1} (or V_{i-1}), the resulting rank can not be larger. Hence,

$$\text{rank}\,(A(i : N, 1 : i - 1)) \leq r_{i-1}, \qquad i = 2, 3, \ldots, N.$$

One can also prove the other way around. Given the block sizes $m_1, m_2, \ldots, m_N$ and $n_1, n_2, \ldots, n_N$, any matrix such that

$$\text{rank}\,(A(i : N, 1 : i - 1)) \leq r_{i-1}, \qquad i = 2, 3, \ldots, N$$

and
$$\text{rank}\,(A(1:i-1,i:N)) \le s_{i-1}, \qquad i = 2,3,\dots,N$$
can be represented as a block quasiseparable matrix defined in 12.1.

Note 12.3. *As we already indicated in the introduction of this chapter, it would have been better to define a block quasiseparable matrix by its rank structure instead of choosing already a specific representation for it, as was done in Definition 12.1.*

12.4   Special cases

- When all sizes r_i are equal to r, the matrix is called lower block quasiseparable of order r.

- When all sizes s_j are equal to s, the matrix is called upper block quasiseparable of order s.

- When a matrix is lower block quasiseparable of order r and upper block quasiseparable of order s, it is called $\{r,s\}$-block quasiseparable.

- When $m_i = n_i$ for $i = 1,2,\dots,N$, all diagonal blocks D_i are square.

- When $m_i = n_i = m$ for $i = 1,2,\dots,N$, all blocks are square and have equal size $m \times m$.

- When $m_i = n_i = m = 1$, the matrix is called a scalar quasiseparable matrix with scalar diagonal blocks $D_i = d_i$.

- When $m_i = n_i = r_k = 1$ for $i = 1,2,\dots,N$ and $k = 1,2,\dots,N-1$, the matrix has semiseparable structure in its strictly lower triangular part.

12.5   Multiplication of a block quasiseparable matrix by a vector

This section is based on [96, Section 3]. Given a block quasiseparable matrix A and a vector $\mathbf{x}$, we can compute the matrix-vector product $A\mathbf{x} = \mathbf{b}$ in a very efficient way by splitting up the block quasiseparable matrix A as

$$A = A_L + A_D + A_U$$

with A_L the block strictly lower triangular part, A_U the block strictly upper triangular part, while A_D represents the block diagonal part of A. Note that the block strictly lower triangular part A_L can be factorized by (12.5) where we take the transformation matrices W_k as in (12.2) and put the D_i equal to zero to exclude the block diagonal part from the block lower triangular matrix A_L. Partitioning the vector $\mathbf{x}$ in the same way as the columns of the matrix A and the vector $\mathbf{b_L}$ as the rows of A, we get the following method to compute

$$\begin{aligned}
\mathbf{b_L} &= A_L \mathbf{x} \\
&= \tilde{W}_N \tilde{W}_{N-1} \cdots \tilde{W}_2 \tilde{W}_1 \mathbf{x}.
\end{aligned}$$

Let us compute $\mathbf{b}_L$ by updating the vector $\mathbf{z}$ which we initialize by the vector $\mathbf{x}$. Hence, $\tilde{W}_1 \mathbf{z}$ has $N + 1$ block components of which the first one is zero, the second one is $V_1 \mathbf{x}_1$ and the others the second and following blocks of $\mathbf{x}$. Multiplying by $\tilde{Y}_k$ for $k = 2, 3, \ldots, N - 1$ boils down to

$$\begin{bmatrix} \mathbf{z}_k \\ \mathbf{z}_{k+1} \end{bmatrix} \leftarrow \begin{bmatrix} U_k & 0 \\ T_k & V_k \end{bmatrix} \begin{bmatrix} \mathbf{z}_k \\ \mathbf{z}_{k+1} \end{bmatrix}.$$

The final step is

$$\begin{bmatrix} \mathbf{z}_N \end{bmatrix} \leftarrow \begin{bmatrix} U_k & 0 \end{bmatrix} \begin{bmatrix} \mathbf{z}_N \\ \mathbf{z}_{N+1} \end{bmatrix}$$

and the $(N + 1)$th component of $\mathbf{z}$ is dropped. A similar method can be followed to compute $\mathbf{b}_U = A_U \mathbf{x}$. Finally the ith block component of $\mathbf{b}_D = A_D \mathbf{x}$ can be easily computed as $D_i X_i$. The final result $\mathbf{b} = A\mathbf{x}$ is the sum of the three components

$$\mathbf{b} = \mathbf{b}_L + \mathbf{b}_D + \mathbf{b}_U.$$

This leads to an algorithm for the matrix-vector multiplication requiring $\mathcal{O}(n)$ flops with n the size of the matrix.

In a very similar way the matrix-vector product $A^T \mathbf{y}$ can be computed. Because the matrix-vector product $A\mathbf{x}$ and $A^T \mathbf{y}$ can be computed in an efficient way, these can be used as building blocks of several iterative methods (e.g., Krylov-subspace based methods), used to solve systems of equations or eigenproblems corresponding to the block quasiseparable matrix A.

Note 12.4. *Suppose all diagonal blocks D_i are square and invertible and consider the block lower triangular matrix A_L which is invertible in this case. Consider the following block lower triangular system*

$$(A_L + A_D)\mathbf{x} = \mathbf{b}.$$

Because we can use the method above to compute the product of the block strictly lower triangular part of A_L with the components of $\mathbf{x}$ which are computed by the forward substitution method, the complete block lower triangular system can be solved in an efficient way. For more details, we refer the interested reader to [96, Section 4].

12.6 Solver for block quasiseparable systems

This section is based on [53]. In the literature several methods have been developed to solve a linear system of block quasiseparable equations, i.e.,

$$A\mathbf{x} = \mathbf{b},$$

with A a block quasiseparable matrix. We will assume that A is nonsingular. Writing down explicitly the block quasiseparable structure of A we get the following system of equations

$$
\begin{bmatrix}
D_1 & P_1 Q_2 & P_1 R_2 Q_3 & \cdots \\
U_2 V_1 & D_2 & P_2 Q_3 & \cdots \\
U_3 T_2 V_1 & U_3 V_2 & D_3 & \cdots \\
\vdots & \vdots & \vdots &
\end{bmatrix}
\begin{bmatrix}
\mathbf{x}_1 \\ \mathbf{x}_2 \\ \mathbf{x}_3 \\ \vdots
\end{bmatrix}
=
\begin{bmatrix}
\mathbf{b}_1 \\ \mathbf{b}_2 \\ \mathbf{b}_3 \\ \vdots
\end{bmatrix}
-
\begin{bmatrix}
0 \\ U_2 \\ U_3 V_2 \\ \vdots
\end{bmatrix} L
\qquad (12.6)
$$

with L initially taken to be equal to a zero matrix. In case the number of block rows N is equal to 1, the system reduces to $D_1 \mathbf{x}_1 = \mathbf{b}_1$ that can be solved using any dense matrix solver. In the remainder of this section, we assume that N is larger than 1. One can consider two cases: in the first case, s_1 is smaller than m_1. In this case, the first $m_1 - s_1$ components of x_1 can be solved and eliminated as follows. Note that the first block row of A except for the first block has as a first factor the matrix P_1. Hence, performing a transformation on the rows of P_1 transforms this first block row accordingly except for the first block. We will use the following orthogonal transformation of P_1:

$$
Z_1^T P_1 = \begin{bmatrix} 0 \\ \hat{P}_1 \end{bmatrix},
$$

with $\hat{P}_1$ of size $(m_1 - s_1) \times (m_1 - s_1)$. Applying this transformation to the right-hand side gives

$$
Z_1^T b_1 = \begin{bmatrix} c_1 \\ d_1 \end{bmatrix}.
$$

Similarly, transforming the columns of V_1 transforms the first block column of A except for the first block. We will use this freedom to transform the first diagonal block into the following form:

$$
Z_1^T D_1 W_1^T = \begin{bmatrix} D_{11} & 0 \\ D_{21} & D_{22} \end{bmatrix}
$$

with D_{11} a $(m_1 - s_1) \times (m_1 - s_1)$ matrix. The orthogonal transformation W_1^T applied to V_1 results in

$$
V_1 W_1^T = \begin{bmatrix} V_{11} & \hat{V}_1 \end{bmatrix}.
$$

When we apply a transformation to the left of the coefficient matrix, we have to transform the unknown vector $\mathbf{x}$ accordingly:

$$
W_1 \mathbf{x}_1 = \begin{bmatrix} \mathbf{z}_1 \\ \hat{\mathbf{x}}_1 \end{bmatrix}.
$$

Hence, applying Z_1^T to the left of the first block row of A and W_1^T to the right of the first block column of A results in the following system of linear equations:

$$
\begin{bmatrix}
D_{11} & 0 & 0 & 0 & \cdots \\
D_{21} & D_{22} & \hat{P}_1 Q_2 & \hat{P}_1 R_2 Q_3 & \cdots \\
U_2 V_{11} & U_2 \hat{V}_1 & D_2 & P_2 Q_3 & \cdots \\
U_3 T_2 V_{11} & U_3 T_2 \hat{V}_1 & U_3 V_2 & D_3 & \cdots \\
\vdots & \vdots & \vdots & \vdots &
\end{bmatrix}
\begin{bmatrix}
\mathbf{z}_1 \\ \hat{\mathbf{x}}_1 \\ \mathbf{x}_2 \\ \mathbf{x}_3 \\ \vdots
\end{bmatrix}
=
\begin{bmatrix}
\mathbf{c}_1 \\ \mathbf{d}_1 \\ \mathbf{b}_2 \\ \mathbf{b}_3 \\ \vdots
\end{bmatrix}
-
\begin{bmatrix}
0 \\ 0 \\ U_2 \\ U_3 V_2 \\ \vdots
\end{bmatrix} L.
$$

Hence, we can solve $\mathbf{z}_1$ using the first block row as

$$D_{11}\mathbf{z}_1 = \mathbf{c}_1.$$

Once we know $\mathbf{z}_1$, we can eliminate it from the set of equations by bringing the first block column multiplied to the right by $\mathbf{z}_1$ to the right-hand side. This can easily be done by computing

$$\hat{\mathbf{b}}_2 = \mathbf{d}_1 - D_{21}\mathbf{z}_1$$

and

$$\hat{L} = L + V_{11}\mathbf{z}_1.$$

The remaining block quasiseparable system is again of the form (12.6) with $n - (m_1 - s_1)$ equations instead of n.

Case 2 handles the case when $s_1 \geq m_1$. In this case, we merge the two first block columns and block rows into one by taking

$$\hat{D}_1 = \begin{bmatrix} D_1 & P_1 Q_2 \\ U_2 V_1 & D_2 \end{bmatrix},$$

$$\hat{P}_1 = \begin{bmatrix} P_1 R_2 \\ P_2 \end{bmatrix},$$

$$\hat{V}_1 = \begin{bmatrix} T_2 V_1 & V_2 \end{bmatrix},$$

$$\hat{b}_1 = \begin{bmatrix} b_1 \\ b_2 - U_2 L \end{bmatrix},$$

$$\hat{L} = V_2 L.$$

By merging the two first block rows and block columns, we decrease the number of block rows and columns by one, i.e., N becomes $N - 1$. Note that the resulting $\hat{P}_1$ parameter for the resulting first block row has now dimension $(m_1 + m_2) \times s_2$.

12.7 Block quasiseparable matrices and descriptor systems

This section is based on [93, Section 2]. Consider the following discrete-time system

$$\mathbf{x}_{k+1} = T_k \mathbf{x}_k + V_k \mathbf{u}_k, \qquad k = 1, 2, \ldots, N-1$$
$$\mathbf{y}_{k-1} = R_k \mathbf{y}_k + Q_k \mathbf{u}_k, \qquad k = N, N-1, \ldots, 2$$
$$\mathbf{v}_k = U_k \mathbf{x}_k + P_k \mathbf{y}_k + D_k \mathbf{u}_k, \qquad k = 1, 2, \ldots, N.$$

Here the matrices U_k, T_k, V_k, P_k, R_k, Q_k, D_k have the dimensions as indicated in (12.1). Using these matrices, the block quasiseparable matrix A can be defined as in Definition 12.1. The vectors $\mathbf{u}_1, \mathbf{u}_2, \ldots, \mathbf{u}_N$ are considered the input of the descriptor system, while the vectors $\mathbf{v}_1, \mathbf{v}_2, \ldots, \mathbf{v}_N$ are the output of the descriptor system. The vectors $\mathbf{x}_k$ and $\mathbf{y}_k$ are considered as state variables having r_k and s_{k-1}

components, respectively. When we take $\mathbf{x}_1$ equal to the zero vector, it is easy to derive the other block components of the vector $\mathbf{x}$:

$$\mathbf{x}_1 = 0$$
$$\mathbf{x}_2 = T_1\mathbf{x}_1 + V_1\mathbf{u}_1 = V_1\mathbf{u}_1$$
$$\mathbf{x}_3 = T_2\mathbf{x}_2 + V_2\mathbf{u}_2 = T_2V_1\mathbf{u}_1 + V_2\mathbf{u}_2$$
$$\mathbf{x}_4 = T_3\mathbf{x}_3 + V_3\mathbf{u}_3 = T_3T_2V_1\mathbf{u}_1 + T_3V_2\mathbf{u}_2 + V_3\mathbf{u}_3$$
$$\cdots$$
$$\mathbf{x}_N = T_{N-1}T_{N-2}\cdots T_2V_1\mathbf{u}_1 + T_{N-1}\cdots T_3V_2\mathbf{u}_2 + \cdots + V_{N-1}\mathbf{u}_{N-1}.$$

Hence, it is clear that the kth component of $A_L\mathbf{u}$ is equal to $U_k\mathbf{x}_k$ with A_L the block strictly lower triangular part of A. When we assume that $\mathbf{y}_N$ is the zero vector, the same reasoning can be made for the block upper triangular part using the state vectors $\mathbf{y}_k$. Therefore, the input-output operator, i.e., the operator connecting the input vector $\mathbf{u}$ to the output vector $\mathbf{v}$ is given by the block quasiseparable matrix A.

Notes and reference

☞ P. Dewilde and A.-J. van der Veen. *Time-varying systems and computations*. Kluwer Academic Publishers, Boston, June 1998.

☞ P. Dewilde and A.-J. van der Veen. Inner-outer factorization and the inversion of locally finite systems of equations. *Linear Algebra and its Applications*, 313:53–100, February 2000.

In the book and the related paper, Dewilde and van der Veen make a detailed study of time-varying systems and the corresponding algorithms. Besides a lot of other interesting results, the book contains a definition of block quasiseparable matrices which are called matrices having low Hankel rank. Several algorithms are designed related to this class of matrices. Because the matrices can be infinite, the book exhibits a nice interplay between complex function theory and linear algebra.

☞ Y. Eidelman. Fast recursive algorithm for a class of structured matrices. *Applied Mathematics Letters*, 13:57–62, 2000.

In this paper, Eidelman develops an order $\mathcal{O}(n^2)$ algorithm to solve a linear system of equations $A\mathbf{x} = \mathbf{b}$ with A a strongly regular $n \times n$ matrix that is scalar lower (or upper) quasiseparable of order 1. Numerical experiments indicate a loss in accuracy compared to Gaussian elimination with partial pivoting. It can be expected that for matrices being almost not strongly regular, the numerical behavior can be much worse.

☞ Y. Eidelman and I. C. Gohberg. On a new class of structured matrices. *Integral Equations and Operator Theory*, 34:293–324, 1999.

☞ Y. Eidelman and I. C. Gohberg. Fast inversion algorithms for a class of block structured matrices. *Contemporary Mathematics*, 281:17–38, 2001.

In these papers, Eidelman and Gohberg define scalar quasiseparable matrices where all matrices T_i are square and have the same size, and similarly for the matrices R_i. The connection between block quasiseparable matrices and descriptor systems is explained under more general boundary conditions for the state vectors as taken in Section 12.7. In this more general case, the state vectors $\mathbf{x}_1$, $\mathbf{x}_N$, $\mathbf{y}_1$ and $\mathbf{y}_N$ satisfy the boundary conditions

$$M_1 \begin{bmatrix} \mathbf{x}_1 \\ \mathbf{y}_1 \end{bmatrix} + M_2 \begin{bmatrix} \mathbf{x}_N \\ \mathbf{y}_N \end{bmatrix} = 0.$$

In this paper the connection between block quasiseparable matrices and structured rank matrices is studied. The product of two block quasiseparable matrices is again block quasiseparable. In this paper the generators of the block quasiseparable representation of the product are given in terms of the generators of the block quasiseparable representation of the two factors. It is also shown how the matrix-vector product can be computed efficiently. The inverse of a block quasiseparable matrix is again block quasiseparable and its rank structure is given in terms of the rank structure of the initial block quasiseparable matrix. An algorithm is designed to compute the generators of the inverse A^{-1} based on the generators of a strongly regular block quasiseparable matrix A. This algorithm is specified in case all matrices T_i and R_i are equal to the identity matrix of the appropriate size as well as for the special case of a band matrix. The numerical experiments indicate a small loss of accuracy compared to Gaussian elimination with partial pivoting for the examples taken in this paper. It is clear that the algorithm will suffer from severe numerical problems when one takes a block quasiseparable matrix that is almost not strongly regular. On the other hand, the computational complexity is linear in the size n of the matrix while Gaussian elimination with partial pivoting is of order $\mathcal{O}(n^3)$.

☞ Y. Eidelman and I. C. Gohberg. A modification of the Dewilde-van der Veen method for inversion of finite structured matrices. *Linear Algebra and its Applications*, 343-344:419–450, April 2002.

Eidelman and Gohberg present several algorithms concerning block quasiseparable matrices inspired by the book of Dewilde and van der Veen [83]. It is shown how any matrix can be represented as a block quasiseparable matrix. The matrix-vector multiplications can be done in $\mathcal{O}(n)$ flops as well as solving a block lower triangular (or, similarly, a block upper triangular) quasiseparable system of linear equations. The factorization of the block lower triangular part of a block quasiseparable matrix as explained in Section 12.2, is inspired by this paper. The authors also design a linear complexity algorithm to compute the block QR-factorization of a block quasiseparable matrix A in the form

$$A = VUR$$

with U and V orthogonal matrices and R a block upper triangular quasiseparable matrix. Moreover, V is also block lower triangular and quasiseparable, while U is block upper triangular and quasiseparable. The generators of these three block quasiseparable matrices V, U and R can be computed using $\mathcal{O}(n)$ flops based on the generators of the block quasiseparable matrix A. This factorization can then be used to solve the corresponding linear system of equations. These results are specialized in case the block quasiseparable matrix A is scalar, i.e., $m_i = n_i = 1$ for $i = 1, 2, \ldots, N$. The numerical results indicate a good numerical accuracy. However, the execution time is a fixed factor larger than one multiplied by the execution time of previously existing methods.

☞ S. Chandrasekaran, P. Dewilde, M. Gu, T. Pals, and A.-J. van der Veen. Fast stable solver for sequentially semi-separable linear systems of equations. *Lecture Notes in Computer Science*, 2552:545–554, 2002.

Chandrasekaran, Dewilde, Gu, Pals and van der Veen present in this paper a backward stable method for solving sequentially semiseparable systems of equations. This method has been explained in Section 12.6. For a more recent paper on sequentially semiseparable representations and some corresponding algorithms, we refer the reader to [52].

Tyrtyshnikov defines in [270] (See also Section 8.2 in Chapter 8) the class of weakly semiseparable matrices and studies an upper bound for their mosaic rank. For more information on the concept of mosaic rank, we refer to [269, 268] of the same author.

12.8 Conclusions

In this chapter we have given a concise overview of block quasiseparable matrices. We have defined this class based on its representation. The block upper and block lower triangular part can be nicely factorized. We have shown the structured rank properties of block quasiseparable matrices. A linear complexity matrix-vector multiplication was designed as well as a linear complexity solver for the corresponding system of linear equations. The link with descriptor systems was concisely described.

Chapter 13

$\mathcal{H}$, $\mathcal{H}^2$ and hierarchically semiseparable matrices

In this chapter we will give a brief overview of some other classes of structured rank matrices. In Section 13.1 the class of $\mathcal{H}$-matrices or hierarchical matrices is defined. It is shown how these matrices can be used to solve integral equations. A subclass of the $\mathcal{H}$-matrices satisfying additional relations between the different subblocks is the class of $\mathcal{H}^2$-matrices as introduced in Section 13.2. In Section 13.3 a similar type of structured rank matrices is defined, more precisely, the class of hierarchically semiseparable matrices.

✎ *Reading this chapter is not essential for understanding the other chapters of this book. The reader can get an idea of the hierarchical rank structures by looking at the Figures 13.2 and 13.3.*

13.1 $\mathcal{H}$-matrices or hierarchical matrices

This section is based on [39]. We will introduce the concept by considering an example.

Example 13.1 (Fredholm integral equation of the first kind) Given the function $f : [0,1] \to \mathbb{R}$. Let us look for the function $u : [0,1] \to \mathbb{R}$ such that it satisfies the following integral equation

$$\int_0^1 \log|x - y|u(y)dy = f(x), \qquad x \in [0,1]. \tag{13.1}$$

The kernel function $g(x,y) = \log|x - y|$ has a singularity when $x = y$. The graph of $-\log|z|$ with $z \in [-2,2]$ is given in Figure 13.1.

In Galerkin's method, Equation (13.1) is projected onto the (n-dimensional) space $V_n = \{\phi_0, \phi_1, \ldots, \phi_{n-1}\}$, i.e,

$$\int_0^1 \int_0^1 \phi_i(x) \log|x - y|u(y)dydx = \int_0^1 \phi_i(x)f(x)dx, \qquad i = 0, 1, \ldots, n-1. \tag{13.2}$$

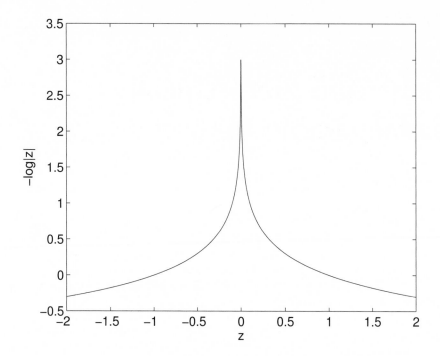

Figure 13.1. *Graph of* $-\log|z|$ *for* $-2 \le z \le 2$.

We look now for an approximation $u_n \approx u$ where u_n is also an element of V_n, i.e.,

$$u_n = \sum_{j=0}^{n-1} \mu_j \phi_j.$$

Plugging this into (13.2) results in

$$\sum_{j=0}^{1} \left[\int_0^1 \int_0^1 \phi_i(x) \log|x-y|\phi_j(y)dydx \right] \mu_j = \int_0^1 \phi_i(x)f(x)dx, \quad i = 0,1,\ldots,n-1.$$

It is clear that the problem reduces to a system of n linear equations in the n unknown coefficients μ_j for $j = 0,1,\ldots,n-1$:

$$G\mathbf{u} = \mathbf{f},$$

with

$$G_{ij} = \int_0^1 \int_0^1 \phi_i(x) \log|x-y|\phi_j(y)dydx \text{and} f_i = \int_0^1 \phi_i(x)f(x)dx.$$

∎

We want to approximate the matrix G by a matrix H such that operations on H can be performed more efficiently while maintaining a good approximation of G, i.e., $\|G - H\|$ should be of the order of the discretization error.

A possible choice for the class of matrices $\mathcal{H}$ from which we take H is the class of hierarchical matrices.

Definition 13.2. *We say that a matrix H belongs to the class of hierarchical matrices $\mathcal{H}$ when the matrix is hierarchically partitioned and its subblocks away from the main diagonal have a restricted rank.*

We clarify this based on the example above.

Example 13.3 (continued) Suppose I_x and I_y are nonempty subintervals of the interval $[0,1]$ whose midpoint is m_x and m_y, respectively. Let us measure the distance between two intervals as

$$\text{distance}(I_x, I_y) = |m_x - m_y|$$

and the diameter of an interval as the length of the interval. It can be proven (see, e.g., [39, Lemma 1.3]) that the kernel function $g(x,y) = \log|x - y|$ can be approximated by a degenerate kernel $\tilde{g}(x,y) = \sum_{\nu=0}^{k-1} g_\nu(x) h_\nu(y)$ for small values of k and $x \in I_x$ and $y \in I_y$ when the so-called admissibility condition is fulfilled, i.e., when $\text{diameter}(I_x) \leq \text{distance}(I_x, I_y)$. When this admissibility condition is fulfilled, it can be shown that there exists an approximation such that

$$|g(x,y) - \tilde{g}(x,y)| \leq 3^{-k}.$$

In words, the admissibility condition says that the kernel function can be approximated well using a degenerate kernel function of rank k for a larger interval of x-values when the x-values are taken further away from the y-values. Choosing now localized basis functions ϕ_i, a similar property will be valid for the matrix G built up by integral values based on $g(x,y)$. Indeed, if we choose, e.g., the functions $\phi_i \in V_n$ as piecewise constant functions in $[0,1]$ as follows

$$\phi_i(x) = 1 \qquad \text{for } \frac{i}{n} \leq x \leq \frac{i+1}{n}$$
$$= 0 \qquad \text{otherwise,}$$

the elements G_{ij} can be computed as

$$G_{ij} = \int_0^1 \int_0^1 \phi_i(x) \log|x - y| \phi_j(y) dy dx$$
$$= \int_{i/n}^{(i+1)/n} \int_{j/n}^{(j+1)/n} \log|x - y| dy dx.$$

Hence, the further the distance between i and j the same approximation $\tilde{g}(x,y)$ for $g(x,y)$ can be used when computing the integral for more values of i and j. This means that the submatrix $G(i : i+k, j : j+l)$ can be approximated by a low rank matrix of the form

$$G(i : i+r, j : j+s) = AB$$

with

$$A(l,t) = \int_{(i+l)/n}^{(i+l+1)/n} g_m(x) dx$$

$$B(t,m) = \int_{(j+m)/n}^{(j+m+1)/n} h_m(y) dy.$$

Let us consider Figure 13.2. In this figure, the patterned "filled" subblocks indicate

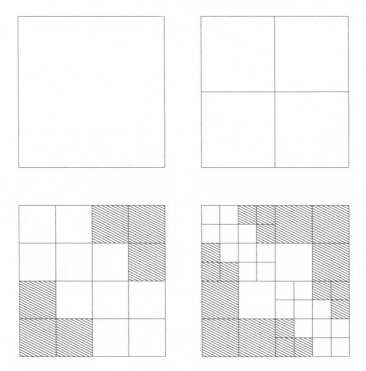

Figure 13.2. *Hierarchical matrices at different levels of hierarchical partitioning.*

submatrices that satisfy the admissibility criterion, i.e., they can be represented very well by a low rank matrix of rank k. The "unfilled" matrices do not satisfy the admissibility criterion, i.e., they can be further subdivided or they can be approximated by a dense generically full rank matrix. Note that in general there is no direct connection between the low rank representation of the different subblocks that satisfy the admissibility criterion. We will see in the next sections that such a connection is required for the class of $\mathcal{H}^2$-matrices and hierarchically semiseparable matrices.

13.2 $\mathcal{H}^2$-matrices

An $\mathcal{H}^2$-matrix is an $\mathcal{H}$-matrix where the column (and row) spaces of the low rank subblocks are related by a column tree (row tree) as indicated in Figure 13.3. For

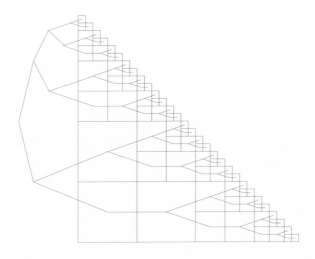

Figure 13.3. *Column tree of the lower triangular part of an $\mathcal{H}^2$-matrix.*

more details we refer the interested reader to [171].

13.3 Hierarchically semiseparable matrices

This section is based on [58]. As for hierarchical matrices, the matrix A is partitioned in 4 subblocks. Here, the off-diagonal subblocks are approximated by low rank matrices:

$$A = \begin{bmatrix} D_{1;1} & U_{1;1}B_{1;1,2}V_{1;2}^T \\ U_{1;2}B_{1;2,1}V_{1;1}^T & D_{1;2} \end{bmatrix}.$$

Each of the diagonal blocks is then partitioned in the same way, resulting in

$$A = \begin{bmatrix} \begin{bmatrix} D_{2;1} & U_{2;1}B_{2;1,2}V_{2;2}^T \\ U_{2;2}B_{2;2,1}V_{2;1}^T & D_{2;2} \end{bmatrix} & U_{1;1}B_{1;1,2}V_{1;2}^T \\ U_{1;2}B_{1;2,1}V_{1;1}^T & \begin{bmatrix} D_{2;3} & U_{2;3}B_{2;3,4}V_{2;4}^T \\ U_{2;4}B_{2;4,3}V_{2;3}^T & D_{2;4} \end{bmatrix} \end{bmatrix}.$$

Moreover, there are connections between the column spaces and row spaces of all off-diagonal blocks. More precisely,

$$U_{1;1} = \begin{bmatrix} U_{2;1}R_{2;1} \\ U_{2;2}R_{2;2} \end{bmatrix}$$

$$U_{1;2} = \begin{bmatrix} U_{2;3}R_{2;3} \\ U_{2;4}R_{2;4} \end{bmatrix}$$

$$V_{1;1} = \begin{bmatrix} V_{2;1}W_{2;1} \\ V_{2;2}W_{2;2} \end{bmatrix}$$

$$V_{1;2} = \begin{bmatrix} V_{2;3}W_{2;3} \\ V_{2;4}W_{2;4} \end{bmatrix}.$$

In [58], an algorithm is designed to solve linear systems having a hierarchically semiseparable matrix similar to the algorithm described in 12.6. The complexity of this algorithm is also linear in the number of equations.

13.4 Other classes of structured rank matrices

Recently there has been great interest in extensions, generalizations and modifications of the classes of structured rank matrices that have already been covered. For the definition of piecewise separable matrices and their linear complexity matrix-vector multiplication, we refer the interested reader to [271]. The class of weakly semiseparable matrices was introduced by Tyrtyshnikov in [270]

Notes and references

☞ S. Börm, L. Grasedyck, and W. Hackbusch. Hierarchical matrices. Technical Report 21, Max-Planck-Institute for Mathematics in the Sciences, Inselstrasse 22, 04103 Leipzig, Germany, 2003.

In these lecture notes Börm, Grasedyck and Hackbusch give a nice overview of the definition and use of hierarchical matrices. Their definition is based on the concept of a cluster tree, i.e., a tree corresponding to the hierarchical partitioning of a matrix as shown in Figure 13.2. They explain how the class of $\mathcal{H}$-matrices can be used when solving integral or elliptic partial differential equations. Attention is given to the fast computation and multiplication of two hierarchical matrices. Also the class of $\mathcal{H}^2$-matrices is defined and studied. On the web, the reader can find the latest version of these lecture notes.

☞ S. Börm, L. Grasedyck, and W. Hackbusch. Introduction to hierarchical matrices with applications. Technical Report 18, Max-Planck-Institute for Mathematics in the Sciences, Inselstrasse 22, 04103 Leipzig, Germany, 2002.

☞ S. Börm, L. Grasedyck, and W. Hackbusch. Introduction to hierarchical matrices with applications. *Engineering Analysis with Boundary Elements*, 27:405–422, 2003.

☞ W. Hackbusch and B. N. Khoromskij. A sparse $\mathcal{H}$-matrix arithmetic, part I: general complexity estimates. *Journal of Computational and Applied Mathematics*, 125(1-2):479–501, 2000.

☞ W. Hackbusch and B. N. Khoromskij. A sparse $\mathcal{H}$-matrix arithmetic, part II: Application to multi-dimensional problems. *Computing*, 64(1):21–47, 2000.

More detailed information about the properties of, computations on and applications with $\mathcal{H}$-matrices can be found in these papers.

☞ S. Chandrasekaran and M. Gu. A fast and stable solver for recursively semi-separable systems of linear equations. *Contemporary Mathematics*, 281:39–53, 2001.

In this paper, Chandrasekaran and Gu define the class of recursively semiseparable matrices and describe a solver for the corresponding system of equations. This class is in fact a special case of the class of $\mathcal{H}$-matrices.

☞ W. Hackbusch, B. N. Khoromskij, and S. A. Sauter. On H^2-matrices. In H. Bungartz and L. Horsten, editors, *Lectures on Applied Mathematics*, pages 9–29. Springer-Verlag, Berlin, 2000.

In this paper, Hackbusch et al. define the class of $\mathcal{H}^2$-matrices as a subset of the class of $\mathcal{H}$-matrices (see Section 13.2). It is shown that such a matrix can be represented using $\mathcal{O}(n)$ floating point numbers and that the matrix-vector multiplication requires only $\mathcal{O}(n)$ flops. The authors also show for an example how the integral operator can be approximated accurately by an $\mathcal{H}^2$-matrix.

☞ S. Chandrasekaran, M. Gu, and W. Lyons. A fast adaptive solver for hierarchically semiseparable representations. *Calcolo*, 42(3-4):171–185, 2005.

In this paper Chandrasekaran, Gu and Lyons define the class of hierarchically semiseparable matrices. Based on an implicit ULV decomposition, an algorithm to solve the corresponding system is designed. A summary of this algorithm is given in Section 12.1.

☞ E. E. Tyrtyshnikov. Piecewise separable matrices. *Calcolo*, 42(3-4):243–248, 2005.

In this paper Tyrtyshnikov defines piecewise separable and piecewise quasiseparable matrices. He shows that the matrix-vector multiplication can be performed in both cases in $\mathcal{O}(pn)$ flops where n is the order of the matrix and p the number of "boundaries" used in the definition.

13.5 Conclusions

In this chapter we have given a brief overview of some other classes of structured rank matrices. The class of $\mathcal{H}$-matrices or hierarchical matrices was defined and situated in the context of solving integral equations. Also the classes of $\mathcal{H}^2$-matrices and hierarchically semiseparable matrices were introduced.

Chapter 14

Inversion of structured rank matrices

In this chapter we will discuss some references related to the inversion of structured rank matrices. This chapter focuses attention on higher order structured rank matrices such as semiseparable, quasiseparable, generalized Hessenberg, Hessenberg-like and generator representable semiseparable matrices. There exist further generalizations of these classes, for example, to block band, block semiseparable, block quasiseparable and so on. Also for these matrices an overview of references related to inversion is presented.

In this chapter several references are discussed. As the literature related to inversion is quite extensive, we decided to present an overview. We start by discussing the inversion of band Toeplitz matrices. Standard techniques for computing the inverse of tridiagonal Toeplitz matrices and also the explicit formulas for this inverse are presented. In Section 14.2 the inversion of Hessenberg-like and generalized Hessenberg matrices is discussed. In the following section the inversion of semiseparable matrices is discussed. A subdivision is made between the classes of generator representable semiseparable and semiseparable.

Section 14.4 discusses the inversion of block band and block semiseparable matrices. Finally we discussed also the inversion methods given for quasiseparable matrices, and briefly we discussed two manuscripts on the generalized inverse of structured rank matrices.

✎ *The interesting parts in this chapter depend on the interest of the reader. The content discusses inversion of matrices related to structured rank matrices. Let us highlight, in our opinion, the most significant results of this chapter.*

The first section presents results related to the inversion of (band/tridiagonal) Toeplitz matrices. The inversion formula presented based on the roots of Equation (14.2) is a general one for tridiagonal Toeplitz matrices. Following this explicit formula for computing the inverse a derivation is presented, proving for some types of Toeplitz matrices the fluctuating structure as seen in Example 14.4.

An explicit formula for inverting an irreducible Hessenberg matrix is presented in Algorithm 14.5. A generalization towards blocks is presented after this algorithm.

When inverting semiseparable and/or generator representable semiseparable matrices of higher order, there are always different possibilities for doing so. Most of the methods pose some constraints. Hence it is not straightforward to define one method as the method of choice. The idea of inverting a broad band matrix by block division is shown in Figure 14.3. A parallel scheme for inverting broad band matrices is explored in Theorem 14.11. An important and interesting idea for inverting block band matrices is the trick of shifting them down. This method is presented starting with Equation (14.7).

Articles discussing inversion methods for quasiseparable matrices are mentioned in Section 14.5. No methods are explicitly explored.

To conclude an interesting theorem providing a nullity theorem for generalized inverses is presented in Theorem 14.12.

14.1   Banded Toeplitz matrices

Even though Toeplitz matrices are not the central subject of the book, we will briefly discuss some results related to the inverse of banded Toeplitz matrices. We know that the inverse of a band Toeplitz matrix will be of semiseparable form. Moreover a Toeplitz matrix has much fewer parameters than a general band matrix, hence this has its effect on the inverse. In a lot of manuscripts inversion techniques designed for inverting band and tridiagonal matrices are tested on band Toeplitz matrices.

When looking at the surface plot of the inverse of a Toeplitz matrix (see, e.g., Figure 14.1) on can see a clear oscillation. This means that the elements oscillate in some sense too. Results presented in this section will provide answers to this oscillating behavior.

Similarly as in the other sections, we present here some information on references. The references discussed here are ordered according to their date of publication.

The manuscripts [263, 264] by Torii provide explicit inversion formulas for tridiagonal matrices, via the solution of a difference equation. These results are applied on a tridiagonal Toeplitz matrix. The results are the same as the one which will be discussed later in this text, in the manuscript [85], by Dow. More information on this article can be found in Section 2.7 in Chapter 2.

☞ E. G. Kounias. An inversion technique for certain patterned matrices. *Journal of Mathematical Analysis and Applications*, 21:695–698, 1968.

Kounias provides explicit formulas for inverting some patterned matrices. For example, a tridiagonal Toeplitz matrix and a matrix completely filled with one element plus a diagonal are inverted. The technique that is applied uses finite difference equations. Let us briefly illustrate the approach. Suppose one wants to invert the tridiagonal Toeplitz matrix T (only the nonzero elements are shown):

$$T = \begin{bmatrix} a & c & & & & & \\ b & a & c & & & & \\ & b & a & c & & & \\ & & b & a & & \ddots & \\ & & & & \ddots & \ddots & c \\ & & & & & b & a \end{bmatrix}. \qquad (14.1)$$

The author inverts this matrix by considering the following matrix equation:

$$T\mathbf{z} = \mathbf{w},$$

where $\mathbf{z} = [z_1, \ldots, z_n]^T$ and $\mathbf{w} = [w^1, \ldots, w^n]^T$ (with w a complex variable).

If one is able to find $z_i = f(w, i)$, without inverting A, and then takes the coefficient of w^j in the expansion of this function, then one obtains the (i, j)th element of T^{-1}. This is solved by solving the following difference equation:

$$bz_{i-1} + az_i + cz_{i+1} = w^i, \text{ for } i = 1, 2, \ldots, n.$$

The boundary conditions are $z_0 = z_{n+1} = 0$. This equation can be solved via standard techniques and computing the inverse of $T^{-1} = S$ in this fashion gives us the following solution:

$$s_{i,j} = \begin{cases} \gamma \left[\left(m_1^{n+1} - m_2^{n+1} \right) \left(m_1^{i-j} - m_2^{i-j} \right) - \left(m_1^i - m_2^i \right) \left(m_1^{n-j+1} - m_2^{n-j+1} \right) \right] \\ \qquad \text{for } j < i \\ \gamma \left[- \left(m_1^i - m_2^i \right) \left(m_1^{n-j+1} - m_2^{n-j+1} \right) \right] \\ \qquad \text{for } j \geq i. \end{cases}$$

The values m_1 and m_2 are the roots of the following equation:

$$cx^2 + ax + b = 0,$$

namely:

$$m_1 = \frac{-a + \sqrt{a^2 - 4bc}}{2c} \text{ and } m_2 = \frac{-a - \sqrt{a^2 - 4bc}}{2c},$$

and γ is the following constant

$$\frac{1}{\left(m_1^{n+1} - m_2^{n+1} \right) (m_1 - m_2) c}.$$

These are explicit formulas, but it is not immediately clear that they provide the factored form in terms of the generators of the semiseparable matrix. Only the upper triangular part clearly shows how one can easily derive the generators. The approach presented by Kounias remains valid if the two roots m_1 and m_2 are equal

to each other. The formulas do change, however. A more general formulation is
presented later in the text.

In [273], Uppuluri and Carpenter present an exact formula to compute the
inverse of a specific covariance matrix of the following form

$$
T = \begin{bmatrix}
1+\alpha^2 & \alpha & & & \\
\alpha & 1+\alpha^2 & \alpha & & \\
& \alpha & 1+\alpha^2 & \alpha & \\
& & \ddots & \ddots & \ddots
\end{bmatrix},
$$

which is a symmetric tridiagonal Toeplitz matrix. This was mentioned previously
in Section 3.2 in Chapter 3.

> ☞ E. L. Allgower. Exact inverse of certain band matrices. *Numerische Math-
>   ematik*, 21:279–284, 1973.

Allgower provides a method for calculating the inverse of banded Toeplitz matrices.
He proves that in the inverse matrix T^{-1} it is sufficient to calculate the elements
in the band surrounding the diagonal, where the size of this band equals the band
size of the original band matrix. He is able to compute these elements, and finally
deduce the remaining elements in this matrix based on these few known elements
in this band.

The manuscript [49], by Capovani, discusses the inversion of nonsymmetric
tridiagonal matrices. The author applies it to Toeplitz matrices. This manuscript
was discussed in Section 2.7 in Chapter 2. The paper [274] was also discussed there.
In this manuscript, the authors Uppuluri and Carpenter apply the same technique
as proposed above by Kounias. But instead of using the power series expansion to
find the components of w^i, the authors use slightly changed formulas and obtain a
polynomial of degree r in w^i. This makes it more easy to obtain the coefficients of
w^i.

> ☞ W. D. Hoskins and P. J. Ponzo. Some properties of a class of band matrices.
>   *Mathematics of Computation*, 26:393–400, 1972.

Hoskins and Ponzo provide formulas for calculating the inverse of a specific $\{p\}$-band
matrix of dimension n, with the binomial coefficients in the expansion of $(x-1)^{2p}$
in each row and column. Such a matrix arises for example in the solution of partial
differential equations. For example, a 5×5 $\{2\}$-band matrix of this special form
looks like:

$$
T = \begin{bmatrix}
6 & -4 & 1 & & \\
-4 & 6 & -4 & 1 & \\
1 & -4 & 6 & -4 & 1 \\
& 1 & -4 & 6 & -4 \\
& & 1 & -4 & 6
\end{bmatrix}.
$$

The authors investigate for this kind of Toeplitz band matrices, the structure of the
inverse, the infinity norm of the inverse and an upper triangular matrix U such that
TU is of lower triangular form.

☞ L. Rehnqvist. Inversion of certain symmetric band matrices. *BIT*, 12:90–98, 1972.

Rehnqvist presents here a method for inverting a very specific Toeplitz band matrix, whose elements smoothly decay in size towards the edges (with edges we mean the upper right and lower left corner) of the matrix. More precisely the elements of the matrix are of the following form (for a certain value of $k < n$, with n the size of the matrix):

$$T = \begin{cases} k - |i - j| & \text{if } k > |i - j| \\ 0 & \text{if } k \leq |i - j|. \end{cases}$$

For example the following matrices are of this specific form:

$$T_1 = \begin{bmatrix} 2 & 1 & 0 & 0 \\ 1 & 2 & 1 & 0 \\ 0 & 1 & 2 & 1 \\ 0 & 0 & 1 & 2 \end{bmatrix} \text{ and } T_1 = \begin{bmatrix} 3 & 2 & 1 & 0 \\ 2 & 3 & 2 & 1 \\ 1 & 2 & 3 & 2 \\ 0 & 1 & 2 & 3 \end{bmatrix}.$$

The author computes the inverse via a multiplication with another invertible Toeplitz matrix which leads to a rather sparse matrix. Then a reordering of the elements and a block division is made. We do not go into more detail as the method is rather specific and several cases have to be considered to compute the desired solution.

☞ W. F. Trench. Inversion of Toeplitz band matrices. *Mathematics of Computation*, 28(128):1089–1095, 1974.

Trench presents a method for inverting $\{p, q\}$-banded Toeplitz matrices by exploiting the banded structure. The method computes the first and the last column of the inverse and uses a technique for Toeplitz matrices for constructing the other elements in the matrix. The solution is computed by solving two difference equations of order $p + q$.

☞ W. D. Hoskins and G. E. McMaster. On the inverses of a class of Toeplitz matrices of band width five. *Linear and Multilinear Algebra*, 4(2):103–106, 1976.

☞ W. D. Hoskins and G. E. McMaster. Properties of the inverses of a set of band matrices. *Linear and Multilinear Algebra*, 5(3):183–196, 1977.

☞ W. D. Hoskins and G. E. McMaster. On the infinity norm of the inverse of a class of Toeplitz matrices of band width five. *Linear and Multilinear Algebra*, 6(2):153–156, 1978.

The summary of the paper of 1976, as provided by the authors Hoskins and McMaster is: for a specific symmetric positive definite Toeplitz matrix of band width five and order n, those regions where the elements of the inverse alternate in sign are determined. More precisely, the following pentadiagonal Toeplitz matrix is discussed (this is a 5×5 example):

$$T_1 = \begin{bmatrix} a & b & 1 & 0 & 0 \\ b & a & b & 1 & 0 \\ 1 & b & a & b & 1 \\ 0 & 1 & b & a & b \\ 0 & 0 & 1 & b & a \end{bmatrix}.$$

The authors prove that if $b^2 \geq 4(a - 2)$ and $2b \leq a + 2$ then the elements of the matrix alternate in sign. More precisely, the sign of the elements $(s_{i,j})$ of $S = T_1^{-1}$ satisfy $s_{i,j} = (-1)^{i+j}$. Remark that this condition only is proved in one direction. In the second manuscript the authors investigate which properties must be satisfied for the matrix

$$T_2 = \begin{bmatrix} a & b & 1 & 0 & 0 \\ 1 & a & b & 1 & 0 \\ 0 & 1 & a & b & 1 \\ 0 & 0 & 1 & a & b \\ 0 & 0 & 0 & 1 & a \end{bmatrix},$$

such that the elements of the inverse alternate in sign. Then, they also reconsider the matrix T_1 and compute the norm of its inverse without explicitly forming the inverse matrix. In the manuscript of 1978, the authors provide bounds on the infinity norm of the inverse of the Toeplitz band matrix T_1.

The manuscript [14], discussed in Section 3.2 in Chapter 3, inverts a tridiagonal Toeplitz matrix as an example.

☞ T. N. E. Greville. On a problem concerning band matrices with Toeplitz inverses. In *Proceedings of the eigth Manitoba Conference on Numerical Mathematics and Computing*, pages 275–283, 1978.

☞ T. N. E. Greville and W. F. Trench. Band matrices with Toeplitz inverses. *Linear Algebra and its Applications*, 27:199–209, 1979.

☞ T. N. E. Greville and W. F. Trench. Band matrices with Toeplitz inverses. *Transactions of the Twenty-Fourth Conference of Army Mathematicians (Univ. Virginia, Charlottesville, Va., 1978)*, pages 365–375, 1979.

We first consider the manuscripts of 1979 (both articles contain similar results). The results in these manuscripts were already known in 1978, but got published in 1979. The manuscript published in 1978 is based on the results in the two manuscripts, published afterwards in 1979. As can happen, the order of publication does not correspond in this case, with the order in which the results were developed. The manuscripts of 1979 poses constraints on a banded matrix, such that its inverse will be a Toeplitz matrix. Strictly banded matrices having a Toeplitz inverse are found in the articles [162, 266] by Trench and Greville. Suppose we have a $\{p, q\}$-band matrix $B \in \mathbb{R}^{(n+1) \times (n+1)}$, let us define $f_i(x)$ as follows[19]:

$$f_i(x) = \sum_{j=0}^{n} b_{ij} x^j,$$

which is the generating function of the ith row of B. The matrix B has as inverse

---

[19]In this example we consider matrices with indices $0, 1, \ldots, n$. This is to have a consistent notation with regard to Toeplitz matrices.

B^{-1} a Toeplitz matrix if and only if:

$$f_i(x) = \sum_{j=0}^{n} b_{i,j} x^j = \begin{cases} x^i g(x) \displaystyle\sum_{\mu=0}^{i} b_\mu x^{-\mu}, & 0 \le i \le p-1, \\[2mm] x^i g(x) h(1/x), & p \le i \le n-q, \\[2mm] x^i h(1/x) \displaystyle\sum_{\nu=0}^{n-i} a_\nu x^\nu, & n-q+1 \le i \le n, \end{cases}$$

where $a_0 b_0 \ne 0$ and

$$g(x) = \sum_{\nu=0}^{q} a_\nu x^\nu \text{ and } h(x) = \sum_{\mu=0}^{p} b_\mu x^\mu,$$

such that $g(x)$ and $x^p h(1/x)$ are relatively prime. These formulas are proved and finally also a method is given for inverting this band matrix. Let us take a closer look at the resulting formulas. We know, due to our background that the inverse of this matrix will be of $\{p,q\}$-semiseparable form, this structure is reflected in the result.

To invert this matrix, the authors assume to be working with a strict band matrix. In this way we will get a generator representable semiseparable Toeplitz matrix. Suppose the Toeplitz matrix $T = (t_{j-i})_{i,j}$ is the inverse of a $\{p,q\}$-band matrix satisfying the property above. The elements $[t_{p-1}, t_{p-2}, \ldots, t_{-q}]$ can be computed by solving the system:

$$\sum_{\nu=0}^{q} a_\nu t_{j-\nu} = b_0^{-1} \delta_{j0}, \qquad 0 \le j \le p-1,$$
$$\sum_{\mu=0}^{p} b_\mu t_{-j+\mu} = 0, \qquad 1 \le j \le r,$$

where δ_{ij} denotes the Kronecker delta. Having computed these elements the other elements are of the following form:

$$t_j = -a_0^{-1} \sum_{\nu=1}^{q} a_\nu t_{j-\nu}, \qquad p \le j \le n,$$
$$t_{-j} = -b_0^{-1} \sum_{\mu=1}^{p} b_\mu t_{-j+\mu}, \qquad q \le j \le n.$$

It is clear that these last elements are of generator representable form.

In the manuscript of 1978, the authors expand some of these results. The matrix B has also some special property called quasi-Toeplitz and the authors try to reconstruct the matrix B, in case not all elements in this quasi-Toeplitz matrix are provided. A quasi-Toeplitz matrix is of Toeplitz form, except some submatrices in the edges are not necessarily of Toeplitz form anymore.

In the manuscript [186], the author Jain presents another method for inverting band Toeplitz matrices. The method uses circular decomposition but does not really exploit the semiseparable structure of the inverse.

The manuscript [172] is discussed in Section 14.3 in this chapter. As an example for the method presented for inverting band matrices, some Toeplitz matrices are considered.

The article [15] by Barrett and Feinsilver was discussed in Section 1.5 of Chapter 1. It deals with the structure of the inverse of a band matrix, thereby investigating the rank of this inverse matrix. The authors also investigate which Toeplitz matrices have a banded inverse. For example the following band matrix B and the Toeplitz matrix T are each others inverses.

$$
B =
\begin{bmatrix}
24 & 18 & 3 & & & \\
8 & 30 & 19 & 3 & & \\
 & 8 & 30 & 19 & 3 & \\
 & & 8 & 30 & 19 & 3 \\
 & & & 8 & 30 & 18 \\
 & & & & 8 & 24
\end{bmatrix},
$$

$$
T =
\begin{bmatrix}
\frac{3}{55} & \frac{-17}{440} & \frac{39}{1760} & \frac{-83}{7040} & \frac{171}{28160} & \frac{-347}{112640} \\
\frac{-1}{55} & \frac{3}{55} & \frac{-17}{440} & \frac{39}{1760} & \frac{-83}{7040} & \frac{171}{28160} \\
\frac{1}{165} & \frac{-1}{55} & \frac{3}{55} & \frac{-17}{440} & \frac{39}{1760} & \frac{-83}{7040} \\
\frac{-1}{495} & \frac{1}{165} & \frac{-1}{55} & \frac{3}{55} & \frac{-17}{440} & \frac{39}{1760} \\
\frac{1}{1485} & \frac{-1}{495} & \frac{1}{165} & \frac{-1}{55} & \frac{3}{55} & \frac{-17}{440} \\
\frac{-1}{4455} & \frac{1}{1485} & \frac{-1}{495} & \frac{1}{165} & \frac{-1}{55} & \frac{3}{55}
\end{bmatrix}.
$$

In the article [242] a method based on solving the associated difference equation, is presented for solving band matrices. The results are applied to compute the inverse of Toeplitz band matrices.

The article [206] by Lewis was discussed extensively earlier. It discusses the inversion of tridiagonal matrices and was discussed in Section 2.7 in Chapter 2 and also in more detail in Chapter 7 in Subsection 7.2.3.

The manuscript [100] by Eijkhout and Polman provides decay rates for the inverse of band matrices satisfying an additional constraint. For example, diagonally dominant Toeplitz matrices satisfy this constraint. It was discussed before in Section 7.5 in Chapter 7.

The manuscript [215] of Meurant discusses the inversion of symmetric irreducible tridiagonal and symmetric block tridiagonal matrices, with invertible off-diagonal blocks. Moreover explicit formulas are presented for inverting symmetric (block) tridiagonal Toeplitz matrices, with minus (identity matrices) ones on the off-diagonals. Also decay rates for the inverses of these symmetric (block) Toeplitz matrices are presented (see Section 7.5 in Chapter 7).

   ☞ C. M. Da Fonseca and J. Petronilho. Explicit inverses of some tridiagonal matrices. *Linear Algebra and its Applications*, 325:7–21, 2001.

☞ C. M. Da Fonseca and J. Petronilho. Explicit inverse of a tridiagonal k-Toeplitz matrix. *Numerische Mathematik*, 100:457–482, 2005.

In these manuscripts the authors invert so-called tridiagonal k-Toeplitz matrices (see [155]), based on results for inverting tridiagonal matrices. The matrices are in some sense a slight generalization of Toeplitz matrices. The manuscript of 2001 discusses inversion formulas for tridiagonal 2-Toeplitz and tridiagonal 3-Toeplitz matrices. The paper of 2005 discusses the general tridiagonal k-Toeplitz case. A tridiagonal k-Toeplitz matrix is of the following form (we illustrate this for $k = 3$):

$$
T = \begin{bmatrix}
a_1 & c_1 \\
b_1 & a_2 & c_2 \\
 & b_2 & a_3 & c_3 \\
 & & b_3 & a_1 & c_1 \\
 & & & b_1 & a_2 & c_2 \\
 & & & & b_2 & a_3 & c_3 \\
 & & & & & b_3 & a_1 & c_1 \\
 & & & & & & \ddots & \ddots & \ddots
\end{bmatrix}.
$$

☞ M. Dow. Explicit inverses of Toeplitz and associated matrices. *Australian & New Zealand Industrial and Applied Mathematics Journal*, 44(E):185–215, 2003.

Dow discusses the inversion of Toeplitz and associated matrices. The inverses are obtained by solving the associated difference equation. The author inverts tridiagonal and band Toeplitz matrices and provides their representation in terms of the generators. The inverse of a tridiagonal Toeplitz matrix (see Equation (14.1)), was already provided by Kounias. Let us distinguish here now briefly between the three types of roots m_1 and m_2 of the equation:

$$
cx^2 + bx + a = 0. \tag{14.2}
$$

Let us denote, similarly as before $S = T^{-1}$. We distinguish between the three cases:

- **Unequal roots.** This coincides with the case discussed by Kounias.

$$
s_{i,j} = \begin{cases}
-\dfrac{\left(m_1^i - m_2^i\right)\left(m_1^{n+1-j} - m_2^{n+1-j}\right)}{c\left(m_1 - m_2\right)\left(m_1^{n+1} - m_2^{n+1}\right)} & \text{for } i < j, \\[4ex]
-\dfrac{\left(m_1^j - m_2^j\right)\left(m_1^{n+1-i} - m_2^{n+1-i}\right)}{c\left(m_1 - m_2\right)\left(m_1^{n+1} - m_2^{n+1}\right)}\left(\dfrac{b}{c}\right)^{i-j} & \text{for } i \geq j.
\end{cases}
$$

- **Equal roots.** Denote the single root as $m = -a/2c$. Then we have:

$$
s_{i,j} = \begin{cases}
-\dfrac{2m^{i-j}i(j - n - 1)}{a(n+1)} & \text{for } i < j, \\[4ex]
-\dfrac{2m^{i-j}j(i - n - 1)}{a(n+1)} & \text{for } i \geq j.
\end{cases}
$$

- **Complex roots.** In the case of complex roots we have for $m = \sqrt{b/c}$ and $\cos(\theta) = -a/2mc$:

$$
s_{i,j} = \begin{cases}
-\dfrac{m^{i-j}\sin(i\theta)\sin((n+1-j)\theta)}{\sqrt{bc}\sin\theta\sin((n+1)\theta)} & \text{for } i < j, \\[3mm]
-\dfrac{m^{i-j}\sin(j\theta)\sin((n+1-i)\theta)}{\sqrt{bc}\sin\theta\sin((n+1)\theta)} & \text{for } i \geq j.
\end{cases}
$$

If $c < 0$ the sign of $s_{i,j}$ changes.

☞ P. G. Martinsson, V. Rokhlin, and M. Tygert. A fast algorithm for the inversion of general Toeplitz matrices. *Computers & Mathematics with Applications*, 50:741–752, 2005.

Martinsson, Rokhlin and Tygert present in this manuscript a method for inverting general (not necessarily banded) Toeplitz matrices. They use the Fourier representation of the involved Toeplitz matrix. They prove that this Fourier representation is of structured rank form, which is exploited in their method.

To conclude let us reconsider some figures of the inverses of Toeplitz matrices. In fact, two properties are interacting with each other. First, we have the decay of the magnitude of the elements of the matrix towards its edges, and second, we have the oscillating property of the elements. Let us take a closer look at these properties. Several of these results are also valid for general tridiagonal matrices and so forth. But we limit ourselves to the Toeplitz case. We will first briefly discuss some properties, such that the inverse of a matrix is sign regular. A matrix A is called sign regular if the matrix $\left[(-1)^{i+j}a_{ij}\right]_{ij}$ has all its elements positive or negative. The results presented here are based on [131]. The definition of (totally) nonnegative was given in Chapter 3.

Theorem 14.1. *If a nonsingular matrix A is totally nonnegative, its inverse A^{-1} is sign regular.*

Proof. The matrix A is totally nonnegative, meaning that all the minors of square submatrices are positive. Based on the determinantal formulas that the inverse $C = A^{-1} = (c_{ij})_{ij}$ can be written as $(-1)^{i+j}M_{ij}$, where M_{ij} denotes the minor of the submatrix of A having removed row i and column j, the result is straightforward.
□

It remains to find an easy condition, stating whether a tridiagonal Toeplitz matrix is totally nonnegative or not. Based on the formulas for the minors of a tridiagonal matrix as given in Section 3.1.3, which are also valid for nonsymmetric tridiagonal matrices, one can easily prove the following.

Theorem 14.2. *A tridiagonal matrix T is totally nonnegative if all its elements and all the leading principal minors of the matrix are nonnegative.*

The decay of the elements in the inverse of a tridiagonal matrix was discussed earlier. These results can be found in Chapter 7, Section 7.5. These decay bounds for diagonally dominant Toeplitz matrices are explicitly computed in the manuscript [222].

We will now prove that a diagonal dominant tridiagonal Toeplitz matrix inherits this special behavior. This means that the inverse of this matrix will be sign regular and have a decay of its elements towards the edges.

Theorem 14.3. *Suppose we have a diagonal dominant Toeplitz matrix T, with nonnegative elements, then this matrix will be totally nonnegative.*

Proof. Let us denote the upper left $i \times i$ submatrix of T as T_i. Denote the diagonal elements with a, the superdiagonal elements with c and the subdiagonal elements with b.

Due to the diagonal dominance we have that

$$|a| \geq |b| + |c|.$$

As all the elements of the matrix T are nonnegative, we only need to prove that the leading principal minors are nonnegative. We know that the determinants of a tridiagonal matrix satisfy a recurrence relation. More precisely we have the following relation

$$\det\left(T_{i+1}\right) = a \det\left(T_i\right) - bc \det\left(T_{i-1}\right),$$

for $i = 1, \ldots, n - 1$ (define $\det\left(T_0\right) = 1$). Based on this relation we will prove the statement by induction.

We will prove for every $i = 0, \ldots, n - 1$, that $\det\left(T_{i+1}\right) \geq 0$ and that

$$\frac{\det\left(T_{i+1}\right)}{a} \geq \det\left(T_i\right).$$

- **Case $i = 0$.** It is clear that $\det\left(T_1\right) = a \geq 0$. Moreover the relation

  $$\frac{\det\left(T_1\right)}{a} = 1 \geq \det\left(T_0\right) = 1,$$

  is also clearly satisfied.

- **Case $i = l$.** We assume now that the relations hold for $i = 0, \ldots, l - 1$ and we will prove the relation for $i = l$. Hence we know by induction that the following relations are satisfied for all $i = 0, \ldots, l - 1$:

  $$\det\left(T_{i+1}\right) \geq 0 \tag{14.3}$$

  $$\frac{\det\left(T_{i+1}\right)}{a} \geq \det\left(T_i\right). \tag{14.4}$$

  We will first prove that $\det\left(T_{l+1}\right) \geq 0$, based on the relation for computing the determinant:

  $$\det\left(T_{l+1}\right) = a \det\left(T_l\right) - bc \det\left(T_{l-1}\right). \tag{14.5}$$

Substituting $\det(T_l)$ in the equation above by using the inequality (14.4) for $i = l - 1$ and finally inequality (14.4) for $i = l - 2$ leads us to the following inequalities (due to the diagonal dominance we have that $a^2 - bc \geq 0$):

$$\det(T_{l+1}) = a \det(T_l) - bc \det(T_{l-1})$$
$$\geq \left(a^2 - bc\right) \det(T_{l-1})$$
$$\geq 0.$$

Using Equation (14.5), it is straightforward that

$$\det(T_{l+1}) \geq a \det(T_l),$$

as $bc \det(T_{l-1})$ is nonnegative. This proves the second relation. Hence it is clear that both relations are satisfied for $i = 0, \ldots, l$, which ends the inductive proof.

$\square$

We know now that the inverse of a Toeplitz matrix, with nonnegative elements, and which is diagonally dominant, has a sign regular inverse. This means that all elements on the same super(sub)diagonal have the same sign.

We would like to remark that this proof can easily be adapted to tridiagonal matrices; the reader can try to do this himself.

Example 14.4 We will present now some examples of figures of inverses of Toeplitz matrices. The inverse of a diagonally dominant Toeplitz matrix is shown on the left of Figure 14.1, for a matrix having $a = 2, b = 1$ and $c = 1$.

The right matrix of Figure 14.1 shows the inverse of a Toeplitz matrix having diagonal elements equal to 1 and sub- and superdiagonal elements equal to 2. This matrix is not diagonally dominant and neither the oscillation nor the decay property are satisfied.

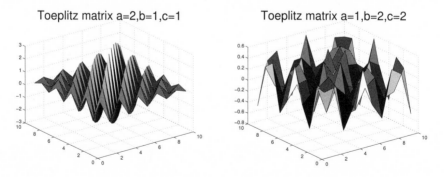

Figure 14.1. *A surfplot of the inverse of different Toeplitz matrices.*

The left matrix of Figure 14.2 has two different values for b and c. This results in a different decay rate for both sides of the matrix. The right matrix shows a

Toeplitz matrix, with negative off-diagonal elements. This means that the matrix will not be sign regular. But one can easily prove that, due to the nonpositive off-diagonal elements, the inverse matrix must have all elements nonnegative. These elements satisfy the decay property, and one gets a smooth image. We slightly turned the image to have a better view. ∎

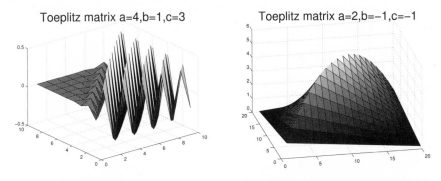

Figure 14.2. *A surfplot of the inverse of different Toeplitz matrices.*

Much more information and references related to Toeplitz matrices can be found in the following book by Heinig and Rost.

> ☞ G. Heinig and K. Rost. *Algebraic Methods for Toeplitz-like Matrices and Operators*. Mathematical Research. Akademie Verlag-Berlin, 1984.

In this section we discussed different methods for inverting Toeplitz matrices and also some special properties such as sign regularity. In the next section we will focus on the class of generalized Hessenberg and Hessenberg-like matrices.

14.2 Inversion of (generalized) Hessenberg matrices

Let us eleborate on some manuscripts related to the inversion of Hessenberg and generalized Hessenberg matrices.

> ☞ E. Asplund. Inverses of matrices a_{ij} which satisfy $a_{ij} = 0$ for $j > i + p$. *Mathematica Scandinavica*, 7:57–60, 1959.

E. Asplund formulates for the first time a theorem stating that the inverse of an invertible lower $\{p\}$-Hessenberg matrix (with no restrictions on the elements and called a $\{p\}$-band matrix in the paper) is an invertible lower $\{p\}$-Hessenberg-like matrix (with the definition of the ranks of subblocks) and vice versa. A $\{p\}$-Hessenberg-like matrix is referred to as a Green's matrix of order p in the paper. No explicit formulas for the inverse are presented, only theoretical results about the structure. More precisely, he also proves that if the elements on the pth superdiagonal of the lower $\{p\}$-Hessenberg matrix are different from zero, then the inverse is a generator representable lower $\{p\}$-Hessenberg-like matrix.

☞ Y. Ikebe. On inverses of Hessenberg matrices. *Linear Algebra and its Applications*, 24:93–97, 1979.

Ikebe provides in this paper an algorithm for inverting an upper Hessenberg matrix under the assumption that the subdiagonal elements are different from zero. It is proved that the inverse matrix has the lower triangular part representable with two generators. An extension towards block Hessenberg matrices is included. The proof is interesting and simple and hence we provide it here.

Suppose we have an upper Hessenberg matrix given, with nonzero subdiagonal elements $H = (h_{i,j})_{i,j}$. We will prove now that the lower triangular part of its inverse H^{-1} will be of generator representable form. Denote with $\mathbf{h}_1^T$ the first row of the matrix H. The matrices R_i are defined as follows:

$$R_i = H(2 : i + 1, 1 : i).$$

This means that these matrices R_i ($i = 1, \ldots, n - 1$), are all upper triangular and invertible, as the subdiagonal elements are different from zero. For example

$$R_1 = H(2 : 2, 1 : 1) = [h_{2,1}]$$

$$R_1 = H(2 : 3, 1 : 2) = \begin{bmatrix} h_{2,1} & h_{2,2} \\ 0 & h_{3,2} \end{bmatrix}.$$

Denote the first column of the inverse of H as $\mathbf{u}$. We will prove now that the lower triangular part of the matrix is of generator representable semiseparable form. Consider the first $i - 1$ components of row i of the matrix product $H^{-1}H$. Denote the elements of H^{-1} as $\beta_{i,j}$. This gives us the following equations:

$$0 = \mathbf{e}_i^T \left(H^{-1} \begin{bmatrix} \mathbf{h}_1(1 : i)^T \\ R_{i-1} \\ 0 \end{bmatrix} \right)$$

$$= \mathbf{e}_i^T \left(H^{-1}(1 : n, 1 : i - 1) \begin{bmatrix} \mathbf{h}_1(1 : i - 1)^T \\ R_{i-1} \end{bmatrix} \right)$$

$$= u_i \mathbf{h}_1(1 : i - 1)^T + [\beta_{i,2}, \ldots, \beta_{i,i}] R_{i-1}.$$

This can be rewritten as:

$$[\beta_{i,2}, \ldots, \beta_{i,i}] = -u_i \mathbf{h}_1(1 : i - 1)^T R_{i-1}^{-1}.$$

Because R_{i-1} is the $(i-1)$th leading principal submatrix of R_{n-1} and the matrices are upper triangular, we have also that R_{i-1}^{-1} is a leading principal submatrix of R_{n-1}^{-1}. This means that $\mathbf{h}_1(1 : i - 1)^T R_{i-1}^{-1}$ contains the first $i - 1$ components of the vector $\mathbf{h}_1(1 : n - 1)^T R_{n-1}^{-1}$. Defining simply $\mathbf{v}^T = [1, -\mathbf{h}_1(1 : n - 1)^T R_{n-1}^{-1}]$ proves that our lower triangular part of H^{-1} is of the form $\beta_{i,j} = u_i v_j$.

This procedure can be implemented in a simple way as follows:

Algorithm 14.5. *Compute the generators of the lower triangular part of H^{-1}, with H an irreducible Hessenberg matrix.*

1. *Choose a nonzero number as value of v_1 (e.g., $v_1 = 1$)*

2. $v_2 = -h_{2,1}^{-1} h_{1,1} v_1$

3. $v_i = -h_{i,i-1}^{-1} \left(\sum_{k=1}^{i-1} h_{k,i-1} v_k \right)$ *for $i = 3, \ldots, n$*

4. $u_n = \left(\sum_{k=1}^{n} h_{k,n} v_k \right)^{-1}$

5. $u_{n-1} = -h_{n,n-1}^{-1} h_{n,n} u_n$

6. $u_i = h_{i+1,i}^{-1} \left(\sum_{k=i+1}^{n} u_k h_{i+1,k} \right)$ *for $i = n-2, \ldots, 2, 1$.*

In the article the author also suggests that applying this method twice, for the lower and the upper triangular part of a tridiagonal matrix, leads to an inversion method for tridiagonal matrices.

☞ W. L. Cao and W. J. Stewart. A note on inverses of Hessenberg-like matrices. *Linear Algebra and its Applications*, 76:233–240, 1986.

A generalization of the papers [184, 310] is presented by Cao and Stewart for Hessenberg matrices with a larger bandwidth and block Hessenberg matrices, for which the blocks do not necessarily need to have the same dimension. The elements/blocks on the extreme subdiagonal need to be invertible. The implementation described is suitable for parallel computations. The results are in some sense a generalization of the results presented by Ikebe. Let us show the idea for $\{p\}$-generalized Hessenberg matrices. The results are extendable in a natural manner to block Hessenberg matrices with invertible blocks on the $\{p\}$th-block diagonal.

Suppose H is an irreducible $\{p\}$-generalized Hessenberg matrix. With the matrix H block partitioned as follows:

$$H = \left[\begin{array}{cc} H_{11} & H_{12} \\ H_{21} & H_{22} \end{array} \right],$$

with H_{21} of dimension $(n-p) \times (n-p)$. This means that the matrix H_{21} is an upper triangular matrix, with invertible diagonal elements. Using a block elimination step, we get the following equality

$$\left[\begin{array}{cc} I & -H_{11}H_{21}^{-1} \\ 0 & I \end{array} \right] \left[\begin{array}{cc} H_{11} & H_{12} \\ H_{21} & H_{22} \end{array} \right] = \left[\begin{array}{cc} 0 & H_{12} - H_{11}H_{21}^{-1}H_{22} \\ H_{21} & H_{22} \end{array} \right]. \qquad (14.6)$$

Using the Laplace expansion theorem (to compute the determinant via expansion via rows or columns) applied to the first p rows, for computing the determinant, we obtain that:

$$\det(H) = -\det(H_{21}) \det\left(H_{12} - H_{11}H_{21}^{-1}H_{22} \right).$$

Because H is invertible we also now that

$$\det \left(H_{12} - H_{11} H_{21}^{-1} H_{22} \right) \neq 0.$$

This leads to the observation that in the right-hand side of Equation (14.6), both the upper right and lower left blocks are invertible. Therefore, we can apply the following lemma to the right-hand side.

Lemma 14.6. *Suppose the matrices A and B are invertible, then we have the following equality:*

$$\left[\begin{array}{cc} 0 & B \\ A & C \end{array} \right]^{-1} = \left[\begin{array}{cc} -A^{-1}CB^{-1} & A^{-1} \\ B^{-1} & 0 \end{array} \right].$$

Proof. The proof is by straightforward calculation.   □

Applying the lemma to Equation (14.6) gives us the following equalities (denote with $\hat{H} = H_{12} - H_{11} H_{21}^{-1} H_{22}$):

$$
\begin{aligned}
H^{-1} &= \left[\begin{array}{cc} -H_{21}^{-1} H_{22} \hat{H}^{-1} & H_{21}^{-1} \\ \hat{H}^{-1} & 0 \end{array} \right] \left[\begin{array}{cc} I & -H_{11} H_{21}^{-1} \\ 0 & I \end{array} \right] \\
&= \left[\begin{array}{cc} -H_{21}^{-1} H_{22} \hat{H}^{-1} & H_{21}^{-1} H_{22} \hat{H}^{-1} H_{11} H_{21}^{-1} + H_{21}^{-1} \\ \hat{H}^{-1} & -\hat{H}^{-1} H_{11} H_{21}^{-1} \end{array} \right] \\
&= \left[\begin{array}{c} -H_{21}^{-1} H_{22} \hat{H}^{-1} \\ \hat{H}^{-1} \end{array} \right] \left[I, -H_{11} H_{21}^{-1} \right] + \left[\begin{array}{cc} 0 & H_{21}^{-1} \\ 0 & 0 \end{array} \right] \\
&= U V^T + \left[\begin{array}{cc} 0 & H_{21}^{-1} \\ 0 & 0 \end{array} \right].
\end{aligned}
$$

This clearly shows that the lower triangular part is of generator representable semiseparable form. Moreover, because the matrix H_{21} is upper triangular, we can clearly see that the structure of the generators expands also to the $(p-1)$th superdiagonal.

> ☞ L. Elsner. Some observations on inverses of band matrices and low rank perturbations of triangular matrices. *Acta Technica Academiae Scientiarum Hungaricae*, 108(1-2):41–48, 1997.

Elsner investigates in more detail the inverses of band and Hessenberg matrices. The different cases, concerning generator representable semiseparable matrices and semiseparable matrices for which the special subblocks have low rank, are included. More precisely the following two problems are addressed in the manuscript. (The manuscript considers lower Hessenberg matrices; we translated this to upper Hessenberg matrices.)

- **Problem I.** Characterize all nonsingular $n \times n$ matrices A whose inverse is of the following form:

$$A^{-1} = R + \mathbf{u}\mathbf{v}^T,$$

for which R is an upper triangular matrix.

- **Problem II.** Characterize all nonsingular $n \times n$ matrices A whose inverse is of the following form:

$$A^{-1} = R + UV^T,$$

for which R is an upper triangular matrix and U and V are both two matrices of size $n \times p$.

One can already obtain a lot of information using the Sherman-Morrison-Woodbury formula in the case R is invertible. This gives us for the matrix A the following equalities:

$$A = \left(R + \mathbf{u}\mathbf{v}^T\right)^{-1}$$
$$= R^{-1} - (1 + \mathbf{v}^T R^{-1} \mathbf{u})^{-1} R^{-1} \mathbf{u}\mathbf{v}^T R^{-1}.$$

This shows that if R is invertible the matrix A is of the same form as A^{-1}.[20] But if R is strictly upper triangular, the inverse will be an irreducible Hessenberg matrix. These problems are addressed in this manuscript.

In the next section we will consider the classes of semiseparable and band matrices. We will distinguish between generator representable semiseparable and general semiseparable matrices.

14.3 Inversion of higher order semiseparable and band matrices

The inversion techniques for higher order semiseparable and band matrices can easily be divided into two different classes, the band and semiseparable versus the strict band and generator representable semiseparable case. Both cases do belong in this section; hence, we divided this section into two subsections. In the first subsection we discuss the generator representable ones, in the second subsection we discuss the general band and semiseparable matrices without restrictions.

Manuscripts which were restricted to the class of inverting tridiagonal and band Toeplitz matrices are not included anymore in this section.

14.3.1 Strict band and generator representable semiseparable matrices

The manuscript [160] by Greenberg and Sarhan poses a constraint on matrices such that their inverse is of band form. Their condition corresponds with the demand that the matrix is of generator representable semiseparable form. Some specific matrices are inverted. For more information, see Section 3.2 in Chapter 3.

The manuscript [7] by Asplund discusses the inversion of tridiagonal matrices, via solving the associated difference equation (see Section 2.2 in Chapter 2). A remark is made on the solving of band matrices.

---

[20]We already know that the rank below the diagonal is maintained due to the nullity theorem.

In [24] Bevilacqua and Capovani present results concerning the inversion of band and block band matrices. More details can be found in Section 2.7 in Chapter 2.

☞ L. Berg. Auflösung von gleichungssystemen mit einer bandmatrix. *Zeitschrift für Angewandte Mathematik und Mechanik*, 57:373–380, 1977. (In German).

Berg provides in this paper for the first time an explicit technique for inverting a band matrix that is not symmetric and has different bandwidths for the upper and lower parts. We could already assume these results based on the paper by Asplund, but here explicit formulas are given. A disadvantage is again the strong assumption that the elements on the extreme super- and subdiagonals have to be different from zero.

☞ T. Oohashi. Some representation for inverses of band matrices. *TRU Mathematics*, 14(2):39–47, 1978.

Oohashi proves that the elements of the inverse of a band matrix can be expressed in terms of the solution of a homogeneous difference equation, related to the original band matrix. In this way explicit formulas for calculating the inverse are obtained. The band matrix does not need to have the same lower and upper band size, but the elements on the extreme diagonals need to be different from zero. The results are an extension of the results proved in [264] for tridiagonal matrices.

In the manuscript [310] by Yamammoto and Ikebe, discussed previously in Section 2.7 in Chapter 2, a method for inverting band matrices was proposed.

☞ S. Cei, M. Galli, M. Soccio, and P. Zellini. On some parallel algorithms for inverting tridiagonal and pentadiagonal matrices. *Calcolo*, 17:303–319, 1981.

The authors propose in this manuscript some parallel algorithms for computing the inverse of tridiagonal and pentadiagonal matrices.

☞ F. Romani. On the additive structure of the inverses of banded matrices. *Linear Algebra and its Applications*, 80:131–140, 1986.

Romani investigates the conditions on a symmetric band matrix, such that its inverse can be written as the sum of inverses of irreducible tridiagonal matrices. The author names this property the *additive* structure of the inverse. The author shows that the inverse of a $\{p\}$-band matrix satisfies the low rank conditions below and above the diagonal. (Nothing is mentioned about the fact that the low rank blocks cross the diagonal.) The author remarks however, that this condition is not enough to write the inverse as the sum of the inverses of irreducible tridiagonal matrices. In fact the author is searching for a condition such that the inverse of the band matrix is of generator representable form.

Let us investigate this in more detail. First, the author considers only symmetric $\{p\}$-band matrices. Let us briefly repeat the definition of proper matrices as proposed in this manuscript.

Definition 14.7. *A nonsingular $\{p\}$-band matrix B is said to be proper if any submatrix obtained by deleting p consecutive rows and p consecutive columns is nonsingular.*

Further on in the text the author proves the following theorem, based on this definition of being proper.

Theorem 14.8. *Suppose B is a nonsingular $\{p\}$-band matrix of proper form. The inverse B^{-1} satisfies the following additive structure:*

$$B^{-1} = \sum_{k=1}^{p} S_k,$$

for which every S_k is the inverse of a symmetric irreducible tridiagonal.

We know, however, that the assumption that the band matrix is strict is enough to have the inverse matrix of generator representable form. One can prove that irreducible band matrices are always of proper form. Let us briefly show that a tridiagonal matrix which is not irreducible is also not proper.

Example 14.9 Consider the following tridiagonal matrix T:

$$T = \begin{bmatrix} a_1 & c_1 & & \\ b_1 & a_2 & c_2 & \\ & 0 & a_3 & c_3 \\ & & b_3 & a_4 \end{bmatrix}.$$

Removing from this matrix the first row and the last column. This clearly leads to a matrix which is singular, and hence T is not proper. ∎

The author extends these results to the nonsymmetric case. Finally the author also wants to provide conditions for the other direction. More precisely he wants to address the following problem: "When is the sum of inverses of irreducible tridiagonal matrices the inverse of a pentadiagonal matrix?"

He proves the following theorem, based on the Sherman-Morrison-Woodbury formula solving this problem for the addition of two generator representable semiseparable matrices.

Theorem 14.10. *Suppose S_1 and S_2 are two nonsingular generator representable semiseparable matrices, such that S_1^{-1} and $-S_2^{-1}$ have the same off-diagonal elements and different diagonal elements. This means that $S_1^{-1} + S_2^{-1}$ is an invertible diagonal matrix. Then we have that $(S_1 + S_2)^{-1}$ is a pentadiagonal matrix.*

☞ P. Rózsa. Band matrices and semi-separable matrices. *Coloquia Mathematica Societatis János Bolyai*, 50:229–237, 1986.

☞ P. Rózsa. On the inverse of band matrices. *Integral Equations and Operator Theory*, 10:82–95, 1987.

Even though the manuscript of 1987 is published later than the other manuscript, the results of the article of 1986 use the ones of 1987. In the article of 1987 the author defines for the first time, to our knowledge, the class of $\{p, q\}$-semiseparable matrices (in our book these are the $\{p, q\}$-generator representable semiseparable matrices). The author names a symmetric $\{1\}$-semiseparable matrix a *separable* matrix in his paper. He generalizes the name of one-pair matrix for a symmetric $\{1\}$-semiseparable matrix towards the nonsymmetric case by naming it a *semi-pair* matrix. Furthermore their generalization to higher order generator representable matrices is called $\{p, q\}$-semiseparable, as is often done. The author proves similarly as Asplund did, by a block division of the original matrix, that the inverse of a strict $\{p, q\}$-band matrix is a $\{p, q\}$-generator representable semiseparable matrix. A method is presented for computing the inverse of the band matrix, based on solving the associated difference equation. The difference equation is reconsidered in case of inverting a band Toeplitz matrix.

In the article of 1986, Rózsa investigates sufficient conditions to be placed on a semiseparable matrix to have a nonsingular $\{p, q\}$-semiseparable matrix. The conditions state that the low rank structure cannot expand above the $(p - 1)$th superdiagonal and below the $(-q + 1)$th subdiagonal. These conditions are the same as the ones provided in Chapter 9, Section 9.2.3.

    ☞ P. Rózsa, R. Bevilacqua, F. Romani, and P. Favati. On band matrices and
        their inverses. *Linear Algebra and its Applications*, 150:287–295, 1991.

A new proof is included by Rózsa, Bevilacqua, Romani and Favati stating that the inverse of a generator representable $\{p, q\}$-semiseparable matrix is a strict $\{p, q\}$-band matrix. This new proof leads to a recursive scheme for calculating the inverse. More precisely the authors make use of the results presented in [243] for inverting block tridiagonal matrices (see upcoming section on inverting block matrices). Consider for example a $\{p, q\}$-band matrix B. Using this almost block tridiagonal matrix (see Figure 14.3), the authors rewrite the matrix slightly and directly compute the generators of its inverse. More details on this method and the ideas behind the proof are presented in the upcoming section on the inverses of block tridiagonal and block band matrices. More precisely, the manuscript [243] contains all this information.

14.3.2   Band and semiseparable matrices

Let us consider now the class of semiseparable and band matrices, without constraints on the extreme super(sub)diagonal elements.

    ☞ D. Szynal and J. Szynal. À propos de l'inversion des matrices généralisées
        de Jacobi. *Aplikace Matematiky*, 17:28–32, 1972. (In French).

D. Szynal and J. Szynal present two theorems for the existence of the inverse of a Jacobi matrix and two methods for calculating the inverse. The results can be rewritten in block form such that they can be applied to specific band matrices. No specific demands of symmetry or nonzeroness of the elements have to be fulfilled. The inverse matrices are not represented in a specific way.

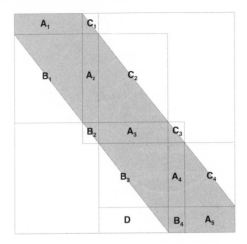

Figure 14.3. *A tridiagonal block division of a band matrix.*

The manuscript [180] by Hoskins and Thurgur was discussed in Section 2.13 in Chapter 2, and discusses the inverse, the computation of the determinant and the infinity norm of a very specific band matrix (see also [179]).

☞ S. B. Haley. Solution of band matrix equations by projection-recurrence. *Linear Algebra and its Applications*, 32:33–48, 1980.

Haley splits the band system $B\mathbf{x} = \mathbf{u}$ into two blocks, which can be divided once more via recurrence. This leads to a method for calculating explicit inverses of banded matrices. As example tridiagonal and Toeplitz tridiagonal matrices are considered.

☞ R. Bevilacqua, G. Lotti, and F. Romani. Storage compression of inverses of band matrices. *Computers & Mathematics with Applications*, 20(8):1–11, 1990.

☞ R. Bevilacqua, F. Romani, and G. Lotti. Parallel inversion of band matrices. *Computers and Artificial Intelligence*, 9(5):493–501, 1990.

Bevilacqua, Lotti and Romani, present in the first manuscript, two types of algorithms for reducing the total amount of storage locations needed by the inverse of a band matrix. More precisely, given the inverse of an irreducible generalized Hessenberg, band or block tridiagonal matrix, their method extracts from these full dense matrices the generator or block generator representation. The method gradually builds up the generators one by one.

The same authors provide in the second manuscript a method for parallel inversion of band matrices. No other assumption is required than the nonsingularity of some of the principal submatrices. The algorithm for computing the inverse in this way was also presented in the first article. Let us briefly show the key theorem for the parallel method.

Theorem 14.11. *Suppose we have a $\{p\}$-band matrix B given, partitioned as follows:*

$$B = \left[\begin{array}{c|c} B_{11} & B_{12} \\ \hline B_{21} & B_{22} \end{array} \right],$$

where B_{11} and B_{22} are two square matrices of dimension $n_1 \times n_1$ and $n_2 \times n_2$, where $n_2 = n - n_1$. Assume also that $n_1 \geq p$ and $n_2 \geq p$. With

$$B_{12} = \left[\begin{array}{cc} 0 & 0 \\ \hat{B}_{12} & 0 \end{array} \right] \text{ and } B_{21} = \left[\begin{array}{cc} 0 & \hat{B}_{21} \\ 0 & 0 \end{array} \right],$$

where both matrices $\hat{B}_{12}$ and $\hat{B}_{21}$ are of dimension $p \times p$.

If B, B_{11} and B_{22} are nonsingular, then there exist four $p \times p$ matrices A_{ij} $(i,j = 1,2)$, $M_{11} \in \mathbb{R}^{n_1 \times p}$, $M_{22} \in \mathbb{R}^{n_2 \times p}$, $N_{11} \in \mathbb{R}^{p \times n_1}$ and $N_{22} \in \mathbb{R}^{p \times n_2}$ such that:

$$B^{-1} = \left[\begin{array}{cc} B_{11}^{-1} & \\ & B_{22}^{-1} \end{array} \right] + \left[\begin{array}{cc} M_{11} & \\ & M_{22} \end{array} \right] \left[\begin{array}{cc} A_{11} & A_{12} \\ A_{21} & A_{22} \end{array} \right] \left[\begin{array}{cc} N_{11} & \\ & N_{22} \end{array} \right].$$

Proof. The matrices M_{11}, M_{22}, N_{11} and N_{22} are submatrices of the matrices B_{11}^{-1} and B_{22}^{-1}. Divide the matrices as follows:

$$B_{11}^{-1} = B^{(1)} = \left[\begin{array}{c|c} B_{11}^{(1)} & B_{12}^{(1)} \\ \hline B_{21}^{(1)} & B_{22}^{(1)} \end{array} \right] \text{ and } B_{22}^{-1} = B^{(2)} = \left[\begin{array}{c|c} B_{11}^{(2)} & B_{12}^{(2)} \\ \hline B_{21}^{(2)} & B_{22}^{(2)} \end{array} \right],$$

with $B_{22}^{(1)}$ and $B_{11}^{(2)}$ both of dimension $p \times p$. The matrices M_{11} and M_{22} consist of the last p columns of $B^{(1)}$ and the first p columns of $B^{(2)}$, respectively. This gives us:

$$M_{11} = \left[\begin{array}{c} B_{12}^{(1)} \\ B_{22}^{(1)} \end{array} \right] \text{ and } M_{22} = \left[\begin{array}{c} B_{11}^{(2)} \\ B_{21}^{(2)} \end{array} \right].$$

The matrices N_{11} and N_{22}, respectively, consist of the last p rows of $B^{(1)}$ and the first p rows of $B^{(2)}$. This gives us:

$$N_{11} = \left[B_{21}^{(1)}, B_{22}^{(1)} \right] \text{ and } M_{22} = \left[B_{11}^{(2)}, B_{12}^{(2)} \right].$$

Defining the remaining matrices A_{ij} as follows:

$$A_{11} = \hat{B}_{12} B_{11}^{(2)} \left(I_p - \hat{B}_{21} B_{22}^{(1)} \hat{B}_{12} B_{11}^{(2)} \right)^{-1} \hat{B}_{21},$$

$$A_{12} = - \left(I_p - \hat{B}_{12} B_{11}^{(2)} \hat{B}_{21} B_{22}^{(1)} \right)^{-1} \hat{B}_{12},$$

$$A_{21} = - \left(I_p - \hat{B}_{21} B_{22}^{(1)} \hat{B}_{12} B_{11}^{(2)} \right)^{-1} \hat{B}_{21},$$

$$A_{22} = \hat{B}_{21} B_{22}^{(1)} \left(I_p - \hat{B}_{12} B_{11}^{(2)} \hat{B}_{21} B_{22}^{(1)} \right)^{-1} \hat{B}_{12}.$$

By direct calculation, we obtain now the result. □

It is clear that this factorization of the inverse opens the possibilities of parallelization. The inverse of the band matrices B_{11} and B_{22} can be computed independently, and both computations can be sent to a different processor. This leads to a divide-and-conquer approach for computing the inverse of the band matrix. The authors implemented this method and provide numerical tests for this algorithm. Also adaptations are presented for minimizing the storage of information needed in the algorithm. The method was also adapted to solve systems of equations via this scheme in a parallel way.

The manuscript [101] was discussed earlier in this chapter in the section on inverting Hessenberg matrices. Elsner investigates in more detail the conditions to obtain generator representable inverses of band and Hessenberg matrices.

14.4   Block matrices

Several of the algorithms for inverting band and semiseparable matrices can be generalized to a block version. For example, the generalization towards a block tridiagonal matrix with square invertible blocks is straightforward. Let us present here some references on the inversion of these matrices.

☞ S. Schechter. Quasi tridiagonal matrices and type insensitive difference equations. *Quarterly of Applied Mathematics*, 18:285–295, 1960.

Schechter provides a method for solving systems of equations with nonsymmetric block tridiagonal matrices (called quasi-tridiagonal matrices in their manuscript). Initially, the only constraint is that the blocks on the diagonal are square. Remark that this does not necessarily mean that all matrices are square. The solver is based on a block LU-decomposition of the matrix. The presented formulas admit an easy inversion of the factors L and U and can hence be used to compute the inverse of the block tridiagonal matrix (see the manuscript [234] by Rizvi).

Schlegel and Kershaw present explicit methods for inverting very specific tridiagonal matrices in their manuscripts [249, 198] (see Section 7.2 in Chapter 7). They also briefly discuss the applicability of their methods with regard to block tridiagonal matrices.

☞ A. Kounadis. Inversion de matrices tridiagonales symétriques. *Revue Française d' Automatique, Informatique et de Recherce Opérationelle*, 7:91–100, 1973. (In French).

Kounadis deduces in this paper a recursive formula for inverting symmetric block tridiagonal matrices for which all the blocks are of the same dimension. The recursive method consists of first computing the block first row and the block last column of the inverse. Based on this information an algorithm is provided for computing the remaining elements of the matrix. In block form, the inverse of the matrix will be of the following form:

$$
\begin{bmatrix}
U_1 & & & & & \\
U_2 & U_2\Sigma^{-1}V_1^T & & & & \\
U_3 & U_3\Sigma^{-1}V_1^T & U_3\Sigma^{-1}V_2^T & & & \\
\vdots & \vdots & \vdots & \ddots & & \\
U_{n-1} & U_{n-1}\Sigma^{-1}V_1^T & U_{n-1}\Sigma^{-1}V_2^T & \dots & U_{n-1}\Sigma^{-1}V_{n-1}^T & \\
\Sigma & V_1^T & V_2^T & \dots & V_{n-1} & V_n
\end{bmatrix}.
$$

It is clear that this form can easily be rewritten in terms of the generator representation, by, e.g., slightly redefining the first column of the matrix. As an example, this technique is also applied to the class of tridiagonal matrices.

The authors Bevilacqua and Capovani present in [24] some results on the inversion of band and block band matrices (see Section 2.7 in Chapter 2).

☞ V. N. Singh. The inverse of a certain block matrix. *Bulletin of the Australian Mathematical Society*, 20:161–163, 1979.

Singh gives an explicit formula for inverting lower block bidiagonal matrices with invertible blocks on the subdiagonal.

☞ S. A. H. Rizvi. Inverses of quasi-tridiagonal matrices. *Linear Algebra and its Applications*, 56:177–184, 1984.

Rizvi generalizes in this manuscript several results of Barrett and Feinsilver towards the block tridiagonal case (quasi-tridiagonal, as in the manuscript by Schechter). First he inverts a lower quasi-bidiagonal matrix. Based on the LU-decomposition of [248] (the manuscript discussed above by Schechter) a method for inverting a quasi-tridiagonal matrix is presented. Furthermore the author generalizes the *triangle* property of Barrett and Feinsilver. This property characterizes the low rank part in the inverse of a tridiagonal matrix. The generalization is named the *quasi-triangle* property. Based on this property necessary and sufficient conditions are provided such that the inverse of a matrix is a quasi-tridiagonal matrix. Also a method for computing the determinant of a matrix satisfying this quasi-triangle property is presented.

☞ R. Mattheij and M. Smooke. Estimates for the inverse of tridiagonal matrices arising in boundary-value problems. *Linear Algebra and its Applications*, 73:33–57, 1986.

Mattheij and Smooke provide in this paper a method for deriving the inverse of block tridiagonal matrices, as coming from nonlinear boundary value problems, in terms of their Green's functions. The block tridiagonal matrices need to have invertible blocks on the extreme diagonals.

☞ R. Bevilacqua, B. Codenotti, and F. Romani. Parallel solution of block tridiagonal linear systems. *Linear Algebra and its Applications*, 104:39–57, 1988.

Bevilacqua, Codenotti and Romani present a method to solve a block tridiagonal system in a parallel way, by exploiting the structure of the inverse of the block tridiagonal matrix.

☞ P. Rózsa, R. Bevilacqua, P. Favati, and F. Romani. On the inverse of block tridiagonal matrices with applications to the inverses of band matrices and block band matrices. *Operator Theory: Advances and Applications*, 40:447–469, 1989.

Rózsa, Bevilacqua, Favati and Romani present generalizations of their previous papers towards methods for computing the inverse of block tridiagonal and block band matrices. The results are very general and the blocks do not all need to be square or of the same dimension. The paper contains a lot of interesting references connected to the theory of semiseparable and tridiagonal matrices. Let us summarize some of the most interesting results of the manuscript.

An important theorem, used throughout several manuscripts of the authors proves the following statement:

The inverse of a submatrix of a given nonsingular matrix, if its exists, is equal to the Schur complement for the corresponding block of the inverse of the given matrix.

Mathematically we can formulate this as follows. Suppose we have an invertible matrix A and its inverse $B = A^{-1}$, both partitioned as follows:

$$A = \left[\begin{array}{c|c} A_{11} & A_{21} \\ \hline A_{12} & A_{22} \end{array}\right] \text{ and } B = \left[\begin{array}{c|c} B_{11} & B_{21} \\ \hline B_{12} & B_{22} \end{array}\right]. \tag{14.7}$$

Then it is proved in the manuscript that the following relations hold, in case A_{12} and A_{21} are invertible:

$$A_{12}^{-1} = B_{21} - B_{22}B_{12}^{-1}B_{11} \tag{14.8}$$

$$A_{21}^{-1} = B_{12} - B_{11}B_{21}^{-1}B_{22}. \tag{14.9}$$

These formulas are used extensively in most of the manuscripts of these authors. Let us illustrate how. Suppose we have a block tridiagonal matrix T of the following form (we illustrate this here for a 4×4 block matrix):

$$T = \left[\begin{array}{cccc} A_1 & C_1 & & \\ B_1 & A_2 & C_2 & \\ & B_2 & A_3 & C_3 \\ & & B_3 & A_4 \end{array}\right],$$

with all matrices A_{2i-1} of size $p \times q$ and the matrices A_{2i} of size $q \times p$. In this case the block matrix has a $\{p, q\}$ partitioning. This means that the first block row has p rows, the second block row has q rows and so forth. Furthermore we assume that the blocks on the extreme diagonals are invertible.

To invert this block tridiagonal matrix, the following *bordering* trick is applied. Define the matrix $\hat{T}$ as follows:

$$
\hat{T} = \begin{bmatrix}
I_q \\
A_1 & C_1 \\
B_1 & A_2 & C_2 \\
& B_2 & A_3 & C_3 \\
& & B_3 & A_4 & I_q
\end{bmatrix}.
$$

This matrix is lower triangular, and has two block subdiagonals. It is easy to generalize its inverse, which will be of block generator representable form[21]:

$$
\hat{T} = \begin{bmatrix}
U_1 \\
U_2 & L_{21} \\
U_3 & L_{31} & L_{32} \\
U_4 & L_{41} & L_{42} & L_{43} \\
-U_5 & U_5 V_1^T & U_5 V_2^T & U_5 V_3^T & U_5 V_4^T
\end{bmatrix}.
$$

Using now Equation (14.9) leads to a way to compute the solution and the generators of the upper triangular part of the inverse of the block tridiagonal matrix (Consider the block matrices T_{ij}^{-1}):

$$
T_{ij}^{-1} = L_{ij} - U_i V_j^T,
$$

because $L_{ij} = 0$ if $i \leq j$.

In a similar way we can transform T to an upper triangular matrix having two block superdiagonals different from zero, then invert this matrix and apply Equation (14.9) to obtain the generators of the lower triangular part.

Combining these results leads to the block generator representation of the inverse of this block tridiagonal matrix. This resulting matrix is called a block $\{q, p\}$ semiseparable matrix in their manuscript. The $\{q, p\}$ refers to the partitioning of the matrix and does not refer to the semiseparability rank. The block semiseparability rank is 1 in this case.[22]

The authors expand their analysis towards the case in which the sub(super)-diagonal blocks are not necessarily invertible. In the simple semiseparable case of semiseparability rank 1, we know that the diagonal belongs to the lower as well as the upper triangular structure. This property is generalized towards the block case. Based on these observations the authors prove that, in general, the inverse of a $\{p, q\}$ block tridiagonal matrix (without constraints on the sub(super)diagonal blocks) is a $\{q, p\}$ block semiseparable matrix. Moreover they also clearly distinguish between the generator representable and the non-generator representable case.

To conclude they discuss the class of band matrices which admit such a $\{p, q\}$ block partitioning in tridiagonal form. More precisely, suppose we have a $\{p, q\}$-band matrix of dimension $n = k(p+q)$, with k an integer. Under these assumptions the band matrix can be considered as a $\{p, q\}$ block tridiagonal matrix. We illustrate

---

[21]We use the same notational conventions as the authors. They chose the notation for convenience with the implementation.

[22]For consistency with the assumptions in the book we adapted their definition of a $\{p, q\}$ block partitioning by first on the rows instead of the columns.

this on a 12×12 $\{1,3\}$-band matrix. This matrix can be partitioned as a $\{1,3\}$ block tridiagonal matrix as follows:

$$
\left[
\begin{array}{cccc|cc|cc|cc|cc}
\times & \times & \times & \times & & & & & & & & \\
\times & \times & \times & \times & \times & & & & & & & \\
 & \times & \times & \times & \times & \times & & & & & & \\
 & & \times & \times & \times & \times & \times & & & & & \\
\hline
 & & & \times & \times & \times & \times & \times & & & & \\
 & & & & \times & \times & \times & \times & \times & & & \\
 & & & & & \times & \times & \times & \times & \times & & \\
 & & & & & & \times & \times & \times & \times & \times & \\
\hline
 & & & & & & & \times & \times & \times & \times & \times \\
 & & & & & & & & \times & \times & \times & \times \\
 & & & & & & & & & \times & \times & \times \\
 & & & & & & & & & & \times & \times \\
\end{array}
\right]
$$

Hence one can compute the block generators of the inverse of this $\{1,3\}$ block tridiagonal matrix. These generators will coincide with the generators of the $\{1,4\}$-semiseparable matrix, which is the inverse of this $\{1,4\}$-band matrix.

The last section of the manuscript discusses some results on the inverses of block tridiagonal matrices, for which the elements are blocks as well.

The manuscript [27], by Bevilacqua, Lotti and Romani, discusses a method for extracting the block generators out of a dense matrix, which is the inverse of a block tridiagonal. More details can be found in Section 14.3.

The article [215] of Meurant discusses the inversion of symmetric block tridiagonal matrices with invertible off-diagonal blocks. Decay rates for the elements of the inverse are given. As a special example also explicit formulas for inverting a block tridiagonal Toeplitz matrix are given (see Section 7.5 in Chapter 7).

14.5 Quasiseparable matrices

In this section we will discuss the inversion of higher order semiseparable plus band matrices and quasiseparable matrices.

> ☞ I. C. Gohberg and M. A. Kaashoek. Time varying linear systems with boundary conditions and integral operators, I. the transfer operator and its properties. *Integral Equations and Operator Theory*, 7:325–391, 1984.

In this article by Gohberg and Kaashoek higher order generator representable quasiseparable matrices arose as input-output maps from discrete linear systems. A method for inverting these matrices was presented, under one restriction: The external coefficients should be nonvanishing. In fact this corresponds with the fact that the diagonal cannot be incorporated into the low rank structure in the upper or lower part of the matrix.

> ☞ I. C. Gohberg, T. Kailath, and I. Koltracht. Linear complexity algorithms for semiseparable matrices. *Integral Equations and Operator Theory*, 8(6):780–804, 1985.

☞ I. C. Gohberg, T. Kailath, and I. Koltracht. A note on diagonal innovation
matrices. *Acoustics Speech and Signal Processing*, 7:1068–1069, 1987.

In these articles the authors present an order $\mathcal{O}(n)$ algorithm for calculating the
inverse of a generator representable quasiseparable matrix, which is strongly nonsingular. They name their matrices semiseparable, but their structure does not
include the diagonal. Hence these matrices are essentially quasiseparable. An extra
limitation, namely that the low rank parts are of generator representable form gives
us a generator representable quasiseparable matrix of quasiseparability rank 1. In
the manuscript of 1987 an order $\mathcal{O}(p^2 n)$ method is proposed for computing an *LDU*-
factorization of a quasiseparable matrix, where p stands for the quasiseparability
rank of the matrix. This *LDU*-factorization can then be used for computing the
inverse of the matrices or to compute the solution of a linear system. Furthermore
it is also proved that the L and U factors inherit some of the rank properties of the
original matrix.

☞ P. Favati, G. Lotti, F. Romani, and P. Rózsa. Generalized band matrices
and their inverses. *Calcolo*, 28:45–92, 1991.

Favati, Lotti, Romani and Rózsa consider here a more general class of structured
rank matrices, namely the class of generalized band matrices. Generalized band matrices have a $\{p\}$-lower triangular part and a $\{q\}$-upper triangular part of generator
representable form, whereas the remainder is arbitrary. This class can be seen as
a generalization of the generator representable quasiseparable matrices, towards a
broader band, instead of only a diagonal. The paper contains an overview of results
related to these matrices. Most of the results are theoretical in nature, providing information on the structure of the inverses, etc. The results are applicable to several
classes of matrices such as band, strict band, semiseparable, block band, matrices
with low rank corrections and so forth. The article applies known techniques from
previous manuscripts [27, 28] to this new class of matrices.

The manuscript [245] by Rózsa and others was discussed in Section 8.3 and
discusses further, the inversion of generalized band matrices.

☞ Y. Eidelman and I. C. Gohberg. Fast inversion algorithms for diagonal
plus semiseparable matrices. *Integral Equations and Operator Theory*,
27(2):165–183, 1997.

☞ Y. Eidelman and I. C. Gohberg. Inversion formulas and linear complexity
algorithm for diagonal plus semiseparable matrices. *Computers & Mathematics with Applications*, 33(4):69–79, August 1997.

Eidelman and Gohberg present an order $\mathcal{O}(n)$ algorithm for calculating the inverse
of a generator representable quasiseparable matrix. In the manuscript of 1996 only
the case of quasiseparability rank 1 is considered; the manuscript of 1997 considers
the more general case. The method consists of a slightly adapted version of the
method presented in [145]. Here the assumption that the diagonal cannot be incorporated into the structure is omitted. Also details are given on how this algorithm
can be used to solve systems of equations.

☞ I. Koltracht. Linear complexity algorithm for semiseparable matrices. *Integral Equations and Operator Theory*, 29(3):313–319, 1997.

In this manuscript, Koltracht presents a method for inverting higher order generator representable quasiseparable matrices. The quasiseparable matrices are transformed into narrow banded matrices, for which standard algorithms can be applied.

☞ Y. Eidelman and I. C. Gohberg. A look-ahead block Schur algorithm for diagonal plus semiseparable matrices. *Computers & Mathematics with Applications*, 35(10):25–34, 1997.

Eidelman and Gohberg present a look-ahead recursive algorithm to compute the triangular factorization of generator representable quasiseparable matrices. This factorization is used for solving systems of equations. The method can also be used to compute the inverse of the involved generator representable quasiseparable matrix.

In the manuscript [93], the authors Eidelman and Gohberg omit the restriction of working with generator representable quasiseparable matrices. They switch to the general quasiseparable case with their specific representation. A factorization and an inversion algorithm are presented. More information can be found in Chapter 12.

In the manuscript [89], Eidelman develops an order $\mathcal{O}(n^2)$ algorithm to solve a linear system of equations based on a decomposition of the original quasiseparable matrix, which can be used for inverting matrices. This algorithm is revisited in the manuscript [94] to reduce its complexity to linear form (see also Chapter 12).

Eidelman and Gohberg present in [96] several algorithms concerning quasiseparable matrices inspired by the book of Dewilde and van der Veen [83]. The article also contains an inversion formula. More information on this manuscript can be found in Chapter 12.

In the manuscript [118] Fiedler discusses the class of *generalized* Hessenberg matrices. This class corresponds to the class of matrices having weakly lower triangular rank equal to 1. In fact the lower triangular part of the matrix is of quasiseparable form. Formulas are presented for obtaining the elements of the inverse, which is also of this form.

14.6 Generalized inverses

In the manuscript by Robinson, the nullity theorem as presented by Fiedler and Markham is generalized towards the Moore-Penrose inverse.

☞ D. W. Robinson. Nullities of submatrices of the Moore-Penrose inverse. *Linear Algebra and its Applications*, 94:127–132, 1987.

More precisely the following theorem is proven.

Theorem 14.12 (Generalized nullity theorem). *Suppose we have the following matrix $A \in \mathbb{R}^{n \times m}$ of rank r partitioned as*

$$A = \left[\begin{array}{cc} A_{11} & A_{12} \\ A_{21} & A_{22} \end{array} \right]$$

with A_{11} of size $p \times q$. The Moore-Penrose inverse $B = A^\dagger = (A^T A)^{-1} A^T \in \mathbb{R}^{m \times n}$ of A is partitioned as

$$B = \begin{bmatrix} B_{11} & B_{12} \\ B_{21} & B_{22} \end{bmatrix}$$

with B_{11} of size $q \times p$.

Then we have the following bounds for the nullities $\mathrm{n}(A_{11})$ and $\mathrm{n}(B_{22})$:

$$r - n \leq \mathrm{n}(B_{22}) - \mathrm{n}(A_{11}) \leq m - r.$$

In the following two manuscripts, the authors discuss the rank structure of generalized inverses of rectangular band matrices.

> ☞ R. Bevilacqua and E. Bozzo. Generalized inverses of band matrices. Technical Report TR-05-09, Department of Computer Science, University of Pisa, Largo Bruno Pontecorvo 3, 56127 Pisa, Italy, March 2005.

> ☞ R. Bevilacqua, E. Bozzo, G. M. Del Corso, and D. Fasino. Rank structure of generalized inverses of rectangular banded matrices. *Calcolo*, 42(3-4):157–169, 2005.

Consider the Moore-Penrose equations in the matrix variable X:

1. $AXA = A$,

2. $XAX = X$,

3. $(AX)^T = AX$,

4. $(XA)^T = XA$.

A matrix satisfying, for example, conditions 1 and 2 is called a $\{1,2\}$-inverse by the authors. The Moore-Penrose inverse satisfies all the conditions above and is defined as:

$$A^\dagger = (A^T A)^{-1} A^T.$$

In the above-mentioned manuscripts, the authors investigate the rank structure of the Moore-Penrose inverse of a rectangular band matrix, as well as of inverses satisfying some (not necessarily) all of the conditions above.

The author Bapat generalizes in the following manuscript the nullity theorem of Fiedler towards different forms of generalized inverses.

> ☞ R. B. Bapat. Outer inverses: Jacobi type identities and nullities of submatrices. *Linear Algebra and its Applications*, 361:107–120, 2003.

Many more references on generalized inverses and their relations with vanishing minors can be found in the following book.

> ☞ A. Ben-Israel and T. N. E. Greville. *Generalized inverses: theory and applications*. CMS Books in Mathematics. Springer, second edition, 2003.

14.7 Conclusions

In this chapter several methods related to the inversion of structured rank matrices were discussed. Different types of matrices were discussed such as Toeplitz, Hessenberg, band and semiseparable, block band and block semiseparable and quasiseparable.

Chapter 15

Concluding remarks & software

15.1 Software

As already mentioned, several of the proposed methods were implemented by the authors in MATLAB and are freely available for download at the following site:
`http://www.cs.kuleuven.be/~mase/books/`

The package containing several routines related to semiseparable matrices is called SSPACK, the Semiseparable Software Package. As we will continue the development of the package, it is unwise to list here all the available routines. It is better that people interested in the package check the site on a regular base.

We will provide an example of a routine to illustrate the global package. For example, the routine:
MULSS: MULtiply a vector with a SemiSeparable matrix
Providing MATLAB the following command:
`>> help MULSS`
will provide the following output:

```
%  MULSS   performs an O(n) matrix vector multiplication
%
%    B=MULSS(G,d,X) is the vector B=SS*X where SS is the
%    symmetric semiseparable matrix constructed from
%    the Givens-vector representation, denoted with G,d
%
%    B=MULSS(G,d,G1,d1,X) performs the multiplication
%    SS*X where SS is the nonsymmetric semiseparable matrix
%    where the lower triangular part is constructed with the
%    Givens-vector representation G,d, and the upper triangular
%    part is constructed from left to right with the
%    Givens-vector representation G1,d1
%
%
%    Software of the MaSe - Group
```

531

```
%     mase@cs.kuleuven.ac.be
%     Revision Date: 16/12/2003
```

Hence, it is clear exactly what this function does. The package contains functions for building representations out of a dense matrix and vica versa, it contains an implementation of the QR-solver for semiseparable plus diagonal matrices. It contains some Levinson-like algorithms and many more.

15.2 Conclusions

In the first part of the book the simpler classes of structured rank matrices were defined, i.e., the matrices having structured rank 1. Examples of these classes were studied in detail: semiseparable, quasiseparable, Hessenberg, tridiagonal, Hessenberg-like, semiseparable plus diagonal, ... A clear distinction was made between the definition and different possible representations for each of these classes.

In the second part several methods were designed to solve the corresponding system of equations in linear complexity.

The third part was dedicated to the generalization of the definitions, representations and solution methods to the higher order cases, i.e., where the structured ranks are larger than one.

The companion volume of this book will be devoted to the application of structured rank matrices in solving eigenvalue/eigenvector, singular value, singular vector problems. For example, for general symmetric dense matrices the classical way to approximate all eigenvalues is to first reduce the matrix to a similar tridiagonal form and then to apply the implicit QR-algorithm to the tridiagonal matrix. This reduction to tridiagonal form is done because each iteration step of the implicit QR-algorithm then requires only $O(n)$ flops for an $n \times n$ matrix. Because the inverse of a nonsingular tridiagonal matrix is semiseparable, it is natural to ask if a similar reduction procedure can be developed into semiseparable form. The answer to this question turns out to be positive. In the second volume we will develop in detail different reduction algorithms as well as the implicit QR-algorithms connected to each of the structured rank forms. Also inverse eigenvalue problems will be tackled as well as divide-and-conquer algorithms, the connection to orthogonal rational functions, and so on.

We hope that these two volumes give the reader a thorough introduction to the field of structured rank matrices and can lead to a standardization of the notation in this strongly evolving field of research and applications.

Bibliography

[1] A. C. Aitken, *Determinants and matrices*, fourth ed., Oliver and Boyd, Edinburgh, 1939. {**276, 533**}

[2] E. L. Allgower, *Criteria for positive definiteness of some band matrices*, Numerische Mathematik **16** (1970), 157–162. {**89, 533**}

[3] _____, *Exact inverse of certain band matrices*, Numerische Mathematik **21** (1973), 279–284. {**89, 502, 533**}

[4] G. S. Ammar, D. Calvetti, W. B. Gragg, and L. Reichel, *Polynomial zerofinders based on Szegő polynomials*, Journal of Computational and Applied Mathematics **127** (2001), 1–16. {**244, 245, 533**}

[5] G. S. Ammar, D. Calvetti, and L. Reichel, *Continuation methods for the computation of zeros of Szegő polynomials*, Linear Algebra and its Applications **249** (1996), 125–155. {**244, 245, 533**}

[6] E. Asplund, *Inverses of matrices a_{ij} which satisfy $a_{ij} = 0$ for $j > i + p$*, Mathematica Scandinavica **7** (1959), 57–60. {**11, 308, 310, 511, 533**}

[7] S. O. Asplund, *Finite boundary value problems solved by Green's matrix*, Mathematica Scandinavica **7** (1959), 49–56. {**11, 57, 63, 272, 515, 533**}

[8] R. Baltzer, *Theorie und anwendung der determinanten*, fifth ed., S. Hirzel, 1881, (In German). {**105, 116, 533**}

[9] T. Banachiewicz, *Sur l'inverse d'un cracovien et une solution générale d'un système d'équations linéaires*, Comptes Rendus Mensuels des Séances de la Classe des Sciences Mathématiques et Naturelles de l'Académie Polonaise des Sciences et des Lettres **4** (1937), 3–4, (In French). {**276, 533**}

[10] R. B. Bapat, *Outer inverses: Jacobi type identities and nullities of submatrices*, Linear Algebra and its Applications **361** (2003), 107–120. {**50, 528, 533**}

[11] J. Baranger and M. Duc-Jacquet, *Matrices tridiagonales symétriques et matrices factorisables*, Revue Française d' Automatique, Informatique et de Recherce Opérationelle **5** (1971), no. R-3, 61–66, (In French). {**64, 272, 533**}

[12] S. Barnett, *Polynomials and linear control systems*, Marcel Dekker Inc, 1983.
      {**243, 244, 534**}

[13] W. W. Barrett, *A theorem on inverses of tridiagonal matrices*, Linear Algebra
      and its Applications **27** (1979), 211–217. {**37, 48, 130, 310, 317, 534**}

[14] W. W. Barrett and P. J. Feinsilver, *Gaussian families and a theorem on
      patterned matrices*, Journal of Applied Probability **15** (1978), 514–522. {**48,
      123, 130, 273, 504, 534**}

[15] _____ , *Inverses of banded matrices*, Linear Algebra and its Applications **41**
      (1981), 111–130. {**6, 37, 48, 130, 272, 310, 317, 506, 534**}

[16] F. L. Bauer, *Optimally scaled matrices*, Numerische Mathematik **5** (1963),
      73–87. {**277, 278, 534**}

[17] A. Ben-Israel and T. N. E. Greville, *Generalized inverses: theory and appli-
      cations*, second ed., CMS Books in Mathematics, Springer, 2003. {**528, 534**}

[18] L. Berg, *Auflösung von gleichungssystemen mit einer bandmatrix*, Zeitschrift
      für Angewandte Mathematik und Mechanik **57** (1977), 373–380, (In German).
      {**516, 534**}

[19] W. J. Berger and E. Saibel, *On the inversion of continuant matrices*, Journal
      of the Franklin Institute **256** (1953), 249–253. {**11, 260, 272, 534**}

[20] A. Berman and R. J. Plemmons, *Nonnegative matrices in the mathematical
      sciences*, SIAM, Philadelphia, 1994. {**279, 534**}

[21] R. Bevilacqua and E. Bozzo, *Generalized inverses of band matrices*, Tech.
      Report TR-05-09, Department of Computer Science, University of Pisa, Largo
      Bruno Pontecorvo 3, 56127 Pisa, Italy, March 2005. {**528, 534**}

[22] R. Bevilacqua, E. Bozzo, and G. M. Del Corso, *Transformations to rank
      structures by unitary similarity*, Tech. Report TR-04-19, Università di Pisa,
      F. Buonarroti 2, 56127 Pisa, Italy, November 2004. {**139, 534**}

[23] R. Bevilacqua, E. Bozzo, G. M. Del Corso, and D. Fasino, *Rank structure of
      generalized inverses of rectangular banded matrices*, Calcolo **42** (2005), no. 3-
      4, 157–169. {**528, 534**}

[24] R. Bevilacqua and M. Capovani, *Proprietà delle matrici a banda ad elementi
      ed a blocchi*, Bolletino Unione Matematica Italiana **5** (1976), no. 13-B, 844–
      861, (In Italian). {**86, 89, 272, 516, 522, 534**}

[25] R. Bevilacqua, B. Codenotti, and F. Romani, *Parallel solution of block tridi-
      agonal linear systems*, Linear Algebra and its Applications **104** (1988), 39–57.
      {**522, 534**}

[26] R. Bevilacqua and G. M. Del Corso, *Structural properties of matrix unitary reduction to semiseparable form*, Calcolo **41** (2004), no. 4, 177–202. {**65, 136, 535**}

[27] R. Bevilacqua, G. Lotti, and F. Romani, *Storage compression of inverses of band matrices*, Computers & Mathematics with Applications **20** (1990), no. 8, 1–11. {**519, 525, 526, 535**}

[28] R. Bevilacqua, F. Romani, and G. Lotti, *Parallel inversion of band matrices*, Computers and Artificial Intelligence **9** (1990), no. 5, 493–501. {**519, 526, 535**}

[29] D. Bindel, J. W. Demmel, W. Kahan, and O. Marques, *On computing Givens rotations reliably and efficiently*, ACM Transactions on Mathematical Software **28** (2002), no. 2, 206–238. {**187, 535**}

[30] D. A. Bini, F. Daddi, and L. Gemignani, *On the shifted QR iteration applied to companion matrices*, Electronic Transactions on Numerical Analysis **18** (2004), 137–152. {**138, 535**}

[31] D. A. Bini, Y. Eidelman, L. Gemignani, and I. C. Gohberg, *Fast QR eigenvalue algorithms for Hessenberg matrices which are rank-one perturbations of unitary matrices*, SIAM Journal on Matrix Analysis and its Applications **29** (2007), no. 2, 566–585. {**139, 535**}

[32] D. A. Bini, L. Gemignani, and V. Y. Pan, *QR-like algorithms for generalized semiseparable matrices*, Tech. Report 1470, Department of Mathematics, University of Pisa, Largo Bruno Pontecorvo 5, 56127 Pisa, Italy, 2004. {**21, 79, 81, 82, 137, 535**}

[33] _____, *Fast and stable QR eigenvalue algorithms for generalized companion matrices and secular equations*, Numerische Mathematik **100** (2005), no. 3, 373–408. {**139, 535**}

[34] D. A. Bini and V. Y. Pan, *Polynomial and matrix computations, vol. 1: Fundamental algorithms*, Birkhäuser, Boston, 1994. {**242, 535**}

[35] E. Bodewig, *Matrix calculus*, second ed., North-Holland, Amsterdam, 1959. {**276, 535**}

[36] _____, *Comparison of some direct methods for computing determinants and inverse matrices*, Koninklijke Nederlandse Akademie van Wetenschappen, Proceedings of the Section of Sciences, vol. 50, 2003, pp. 49–57. {**276, 535**}

[37] C. F. Borges and W. B. Gragg, *A parallel divide and conquer algorithm for the generalized real symmetric definite tridiagonal eigenproblem*, Numerical Linear Algebra and Scientific Computing (Berlin) (L. Reichel, A. Ruttan, and R. S. Varga, eds.), de Gruyter, 1993, pp. 11–29. {**237, 244, 535**}

[38] S. Börm, L. Grasedyck, and W. Hackbusch, *Introduction to hierarchical matrices with applications*, Tech. Report 18, Max-Planck-Institute for Mathematics in the Sciences, Inselstrasse 22, 04103 Leipzig, Germany, 2002. {**140, 496, 536**}

[39] ———, *Hierarchical matrices*, Tech. Report 21, Max-Planck-Institute for Mathematics in the Sciences, Inselstrasse 22, 04103 Leipzig, Germany, 2003. {**491, 493, 496, 536**}

[40] ———, *Introduction to hierarchical matrices with applications*, Engineering Analysis with Boundary Elements **27** (2003), 405–422. {**496, 536**}

[41] R. A. Brualdi and J. J. Q. Massey, *More on structure-ranks of matrices*, Linear Algebra and its Applications **183** (1993), 193–199. {**49, 536**}

[42] B. Bukhberger and G. A. Emel'yanenko, *Methods of inverting tridiagonal matrices*, Computational Mathematics and Mathematical Physics (translated from Zhurnal Vychislitel'noĭ Matematiki i Matematicheskoĭ Fiziki) **13** (1973), 546–554. {**65, 272, 536**}

[43] A. Bultheel, A. M. Cuyt, W. Van Assche, M. Van Barel, and B. Verdonk, *Generalizations of orthogonal polynomials*, Journal of Computational and Applied Mathematics **179** (2005), no. 1-2, 57–95. {**133, 536**}

[44] A. Bultheel, M. Van Barel, and P. Van gucht, *Orthogonal basis functions in discrete least-squares rational approximation*, Journal of Computational and Applied Mathematics **164-165** (2004), 175–194. {**133, 134, 140, 536**}

[45] R. L. Burden and J. D. Faires, *Numerical analysis*, fourth ed., PWS-Kent publishing company, 1988. {**131, 536**}

[46] W. S. Burnside and A. W. Panton, *An introduction to determinants*, ch. from The theory of equations, Hodges, Figgis, & Co. and Longmans, Green & Co., 1899. {**105, 116, 536**}

[47] D. Calvetti, S. Kim, and L. Reichel, *The restarted QR-algorithm for eigenvalue computation of structured matrices*, Journal of Computational and Applied Mathematics **149** (2002), 415–422. {**244, 245, 536**}

[48] W. L. Cao and W. J. Stewart, *A note on inverses of Hessenberg-like matrices*, Linear Algebra and its Applications **76** (1986), 233–240. {**513, 536**}

[49] M. Capovani, *Sulla determinazione della inversa delle matrici tridiagonali e tridiagonali a blocchi*, Calcolo **7** (1970), 295–303, (In Italian). {**88, 272, 502, 536**}

[50] ———, *Su alcune proprietà delle matrici tridiagonali e pentadiagonali*, Calcolo **8** (1971), 149–159, (In Italian). {**88, 272, 536**}

[51] S. Cei, M. Galli, M. Soccio, and P. Zellini, *On some parallel algorithms for inverting tridiagonal and pentadiagonal matrices*, Calcolo **17** (1981), 303–319. {**516, 537**}

[52] S. Chandrasekaran, P. Dewilde, M. Gu, T. Pals, and X. Sun, *Some fast algorithms for sequentially semiseparable representations*, SIAM Journal on Matrix Analysis and its Applications **27** (2005), no. 2, 341–364. {**490, 537**}

[53] S. Chandrasekaran, P. Dewilde, M. Gu, T. Pals, and A.-J. van der Veen, *Fast stable solver for sequentially semi-separable linear systems of equations*, Lecture Notes in Computer Science **2552** (2002), 545–554. {**80, 82, 132, 140, 481, 485, 489, 537**}

[54] S. Chandrasekaran and M. Gu, *Fast and stable eigendecomposition of symmetric banded plus semi-separable matrices*, Linear Algebra and its Applications **313** (2000), 107–114. {**8, 63, 65, 138, 537**}

[55] _____, *A fast and stable solver for recursively semi-separable systems of linear equations*, Contemporary Mathematics **281** (2001), 39–53. {**132, 140, 497, 537**}

[56] _____, *Fast and stable algorithms for banded plus semiseparable systems of linear equations*, SIAM Journal on Matrix Analysis and its Applications **25** (2003), no. 2, 373–384. {**90, 201, 204, 344, 387, 537**}

[57] _____, *A divide and conquer algorithm for the eigendecomposition of symmetric block-diagonal plus semi-separable matrices*, Numerische Mathematik **96** (2004), no. 4, 723–731. {**63, 65, 135, 537**}

[58] S. Chandrasekaran, M. Gu, and W. Lyons, *A fast adaptive solver for hierarchically semiseparable representations*, Calcolo **42** (2005), no. 3-4, 171–185. {**495–497, 537**}

[59] M. T. Chu and G. H. Golub, *Inverse eigenvalue problems: Theory, algorithms and applications*, Numerical Mathematics & Scientific Computations, Oxford University Press, 2006. {**122, 537**}

[60] D. Colton and R. Kress, *Inverse acoustic and electromagnetic scattering theory*, second ed., Springer-Verlag, 1998. {**140, 537**}

[61] P. Concus, G. H. Golub, and G. Meurant, *Block preconditioning for the conjugate gradient method*, SIAM Journal on Scientific and Statistical Computation **6** (1985), 220–252. {**263, 270, 272, 280, 284, 537**}

[62] C. Corduneanu, *Integral equations and applications*, Cambridge University Press, 1991. {**131, 537**}

[63] C. E. Cullis, *Matrices and determinoids*, Readership lectures, vol. 1, Cambridge, University Press, 1913. {**106, 116, 537**}

[64] J. J. M. Cuppen, *A divide and conquer method for the symmetric tridiagonal eigenproblem*, Numerische Mathematik **36** (1981), 177–195. {**237, 244, 538**}

[65] C. M. Da Fonseca and J. Petronilho, *Explicit inverses of some tridiagonal matrices*, Linear Algebra and its Applications **325** (2001), 7–21. {**506, 538**}

[66] _____, *Explicit inverse of a tridiagonal k-Toeplitz matrix*, Numerische Mathematik **100** (2005), 457–482. {**507, 538**}

[67] C. De Boor, *A bound on the L_∞-norm of the L_2-approximation by splines in term of a global mesh ratio*, Mathematics of Computation **30** (1976), 687–694. {**284, 538**}

[68] _____, *Odd degree spline interpolation at a biinfinite knot sequence*, Approximation Theory (R. Schaback and K. Scherer, eds.), Lecture Notes in Mathematics, vol. 556, Springer, Heidelberg, 1976, pp. 30–53. {**283, 538**}

[69] _____, *Dichotomies for band matrices*, SIAM Journal on Numerical Analysis **17** (1980), 894–907. {**283, 538**}

[70] S. Delvaux and M. Van Barel, *Orthonormal rational function vectors*, Numerische Mathematik **100** (2005), no. 3, 409–440. {**134, 140, 538**}

[71] _____, *The explicit QR-algorithm for rank structured matrices*, Tech. Report TW459, Department of Computer Science, Katholieke Universiteit Leuven, Celestijnenlaan 200A, 3000 Leuven (Heverlee), Belgium, May 2006. {**140, 344, 538**}

[72] _____, *A Givens-weight representation for rank structured matrices*, Tech. Report TW453, Department of Computer Science, Katholieke Universiteit Leuven, Celestijnenlaan 200A, 3000 Leuven (Heverlee), Belgium, March 2006, (To appear in SIMAX). {**73, 79, 139, 179, 192, 344, 387, 538**}

[73] _____, *A Hessenberg reduction algorithm for rank structured matrices*, Tech. Report TW460, Department of Computer Science, Katholieke Universiteit Leuven, Celestijnenlaan 200A, 3000 Leuven (Heverlee), Belgium, May 2006. {**140, 344, 538**}

[74] _____, *A QR-based solver for rank structured matrices*, Tech. Report TW454, Department of Computer Science, Katholieke Universiteit Leuven, Celestijnenlaan 200A, 3000 Leuven (Heverlee), Belgium, March 2006. {**344, 387, 538**}

[75] _____, *Rank structures preserved by the QR-algorithm: the singular case*, Journal of Computational and Applied Mathematics **189** (2006), 157–178. {**21, 137, 297, 538**}

[76] _____, *Structures preserved by matrix inversion*, SIAM Journal on Matrix Analysis and its Applications **28** (2006), no. 1, 213–228. {**44, 49, 297, 317, 538**}

[77] _____ , *Structures preserved by Schur complementation*, SIAM Journal on Matrix Analysis and its Applications **28** (2006), no. 1, 229–252. {**249, 250, 253, 255, 539**}

[78] _____ , *Structures preserved by the QR-algorithm*, Journal of Computational and Applied Mathematics **187** (2006), no. 1, 29–40. {**21, 137, 297, 539**}

[79] _____ , *Unitary rank structured matrices*, Tech. Report TW464, Department of Computer Science, Katholieke Universiteit Leuven, Celestijnenlaan 200A, 3000 Leuven (Heverlee), Belgium, July 2006. {**399, 539**}

[80] S. Demko, *Inverses of band matrices and local convergence of spline projections*, SIAM Journal on Numerical Analysis **14** (1977), no. 4, 616–619. {**282, 539**}

[81] S. Demko, W. F. Moss, and P. W. Smith, *Decay rates for inverses of band matrices*, Mathematics of Computation **43** (1984), 491–499. {**283, 539**}

[82] J. W. Demmel, *Applied numerical linear algebra*, SIAM, 1997. {**27, 183, 539**}

[83] P. Dewilde and A.-J. van der Veen, *Time-varying systems and computations*, Kluwer Academic Publishers, Boston, June 1998. {**80, 82, 132, 140, 194, 387, 488, 489, 527, 539**}

[84] _____ , *Inner-outer factorization and the inversion of locally finite systems of equations*, Linear Algebra and its Applications **313** (2000), 53–100. {**57, 80, 82, 140, 194, 387, 488, 539**}

[85] M. Dow, *Explicit inverses of Toeplitz and associated matrices*, Australian & New Zealand Industrial and Applied Mathematics Journal **44** (2003), no. E, 185–215. {**500, 507, 539**}

[86] W. J. Duncan, *Some devices for the solution of large sets of simultaneous linear equations (with an appendix on the reciprocation of partitioned matrices)*, The London, Edinburgh and Dublin Philosophical Magazine and Journal of Science **7** (1944), no. 35, 660–670. {**276, 539**}

[87] J. Durbin, *The fitting of time series in models*, Review of the International Statistical Institute **28** (1960), 233–243. {**207, 210, 539**}

[88] H. Dym and I. C. Gohberg, *Extensions of band matrices with band inverses*, Linear Algebra and its Applications **36** (1981), 1–24. {**342, 539**}

[89] Y. Eidelman, *Fast recursive algorithm for a class of structured matrices*, Applied Mathematics Letters **13** (2000), 57–62. {**82, 140, 233, 344, 488, 527, 539**}

[90] Y. Eidelman and I. C. Gohberg, *Fast inversion algorithms for diagonal plus semiseparable matrices*, Integral Equations and Operator Theory **27** (1997), no. 2, 165–183. {**90, 344, 526, 539**}

[91] _____, *Inversion formulas and linear complexity algorithm for diagonal plus semiseparable matrices*, Computers & Mathematics with Applications **33** (1997), no. 4, 69–79. {**90, 344, 526, 540**}

[92] _____, *A look-ahead block Schur algorithm for diagonal plus semiseparable matrices*, Computers & Mathematics with Applications **35** (1997), no. 10, 25–34. {**255, 344, 527, 540**}

[93] _____, *On a new class of structured matrices*, Integral Equations and Operator Theory **34** (1999), 293–324. {**9, 11, 49, 80, 82, 93, 132, 140, 235, 344, 433, 481, 483, 487, 488, 527, 540**}

[94] _____, *Fast inversion algorithms for a class of block structured matrices*, Contemporary Mathematics **281** (2001), 17–38. {**49, 233, 236, 475, 476, 481, 488, 527, 540**}

[95] _____, *Algorithms for inversion of diagonal plus semiseparable operator matrices*, Integral Equations and Operator Theory **44** (2002), no. 2, 172–211. {**140, 540**}

[96] _____, *A modification of the Dewilde-van der Veen method for inversion of finite structured matrices*, Linear Algebra and its Applications **343-344** (2002), 419–450. {**80, 140, 194, 344, 387, 481, 482, 484, 485, 489, 527, 540**}

[97] _____, *Fast inversion algorithms for a class of structured operator matrices*, Linear Algebra and its Applications **371** (2003), 153–190. {**140, 540**}

[98] Y. Eidelman, I. C. Gohberg, and V. Olshevsky, *Eigenstructure of order-one-quasiseparable matrices. three-term and two-term recurrence relations*, Linear Algebra and its Applications **405** (2005), 1–40. {**139, 540**}

[99] _____, *The QR iteration method for Hermitian quasiseparable matrices of an arbitrary order*, Linear Algebra and its Applications **404** (2005), 305–324. {**139, 540**}

[100] V. Eijkhout and B. Polman, *Decay rates of inverses of banded M-matrices that are near to Toeplitz matrices*, Linear Algebra and its Applications **109** (1988), 247–277. {**283, 506, 540**}

[101] L. Elsner, *Some observations on inverses of band matrices and low rank perturbations of triangular matrices*, Acta Technica Academiae Scientiarum Hungaricae **108** (1997), no. 1-2, 41–48. {**310, 514, 521, 540**}

[102] L. Elsner and C. Giersch, *Metabolic control analysis: Separable matrices and interdependence of control coefficients*, Journal of theoretical Biology **193** (1998), 649–661. {**140, 540**}

[103] S. M. Fallat, M. Fiedler, and T. L. Markham, *Generalized oscillatory matrices*, Linear Algebra and its Applications **359** (2003), 79–90. {**112, 314, 540**}

[104] D. Fasino, *Rational Krylov matrices and QR-steps on Hermitian diagonal-plus-semiseparable matrices*, Numerical Linear Algebra with Applications **12** (2005), no. 8, 743–754. {**21, 63, 65, 136, 541**}

[105] D. Fasino and L. Gemignani, *Structural and computational properties of possibly singular semiseparable matrices*, Linear Algebra and its Applications **340** (2001), 183–198. {**8, 84, 87, 89, 541**}

[106] ———, *A Lanczos type algorithm for the QR-factorization of regular Cauchy matrices*, Numerical Linear Algebra with Applications **9** (2002), 305–319. {**133, 541**}

[107] ———, *Direct and inverse eigenvalue problems, for diagonal-plus-semiseparable matrices*, Numerical Algorithms **34** (2003), 313–324. {**63, 65, 133, 134, 541**}

[108] D. Fasino, N. Mastronardi, and M. Van Barel, *Fast and stable algorithms for reducing diagonal plus semiseparable matrices to tridiagonal and bidiagonal form*, Contemporary Mathematics **323** (2003), 105–118. {**90, 138, 344, 541**}

[109] P. Favati, G. Lotti, F. Romani, and P. Rózsa, *Generalized band matrices and their inverses*, Calcolo **28** (1991), 45–92. {**526, 541**}

[110] M. Fiedler, *Special matrices and their applications in numerical mathematics*, Martinus Nijhoff, Dordrecht, 1986. {**279, 541**}

[111] ———, *Structure ranks of matrices*, Linear Algebra and its Applications **179** (1993), 119–127. {**48, 71, 162, 296, 310–313, 317, 541**}

[112] ———, *Basic matrices*, Linear Algebra and its Applications **373** (2003), 143–151. {**66, 68–71, 156, 162, 178, 296, 310, 314, 317, 541**}

[113] ———, *Complementary basic matrices*, Linear Algebra and its Applications **384** (2004), 199–206. {**71, 541**}

[114] M. Fiedler and T. L. Markham, *Completing a matrix when certain entries of its inverse are specified*, Linear Algebra and its Applications **74** (1986), 225–237. {**6, 37, 38, 40, 48, 71, 159, 162, 272, 310, 317, 541**}

[115] ———, *Rank-preserving diagonal completions of a matrix*, Linear Algebra and its Applications **85** (1987), 49–56. {**48, 273, 310, 311, 317, 339, 541**}

[116] ———, *Generalized totally nonnegative matrices*, Linear Algebra and its Applications **345** (2002), 9–28. {**10, 112, 314, 541**}

[117] M. Fiedler and V. Pták, *On matrices with non-positive off-diagonal elements and positive principal minors*, Czechoslovak Mathematical Journal **12** (1962), no. 87, 382–400. {**112, 541**}

[118] M. Fiedler and Z. Vavřín, *Generalized Hessenberg matrices*, Linear Algebra and its Applications **380** (2004), 95–105. {**66, 68–71, 156, 163, 178, 296, 297, 310, 314, 317, 527, 542**}

[119] G. E. Forsythe and E. G. Straus, *On best conditioned matrices*, Proceedings of the American Mathematical Society **6** (1955), 340–345. {**277, 278, 542**}

[120] R. A. Frazer, W. J. Duncan, and A. R. Collar, *Elementary matrices and some applications to dynamics and differential equations*, Cambridge University Press, Cambridge, 1938. {**276, 542**}

[121] _____ , *Elementary matrices*, Cambridge, university press, 1965. {**123, 542**}

[122] R. W. Freund and H. Zha, *Formally biorthogonal polynomials and a look-ahead Levinson algorithm for general Toeplitz systems*, Linear Algebra and its Applications **188/189** (1993), 255–303. {**245, 255, 542**}

[123] F. R. Gantmacher, *On non symmetric Kellogg kernels*, Doklady Acadademii Nauk (USSR) **1** (1936), no. 10, 3–5. {**113, 542**}

[124] F. R. Gantmacher and M. G. Kreĭn, *On a special class of determinants related to Kellog integral kernels*, Matematicheskiĭ Sbornik **40** (1933), 501–508. {**112, 542**}

[125] _____ , *Sur les matrices oscillatoires*, Comptes Rendus Mathématique Académie des Sciences Paris **201** (1935), 577–579, (In French). {**110, 112, 542**}

[126] _____ , *Sur les matrices oscillatoires et complètement non négatives*, Compositio Mathematica **4** (1937), 445–476, (In French). {**10, 110, 112, 272, 542**}

[127] _____ , *Oscillyacionye matricy i yadra i malye kolebaniya mehaničeskih sistem. [Oscillation matrices and kernels and small oscillations of mechanical systems.]*, Moscow-Leningrad, 1941, (In Russian). {**10, 110, 542**}

[128] _____ , *Oscillyacionye matricy i yadra i malye kolebaniya mehaničeskih sistem. [Oscillation matrices and kernels and small oscillations of mechanical systems.]*, second ed., Gosudarstv. Isdat. Tehn.-Teor. Lit., Moscow-Leningrad, 1950, (In Russian). {**10, 110, 542**}

[129] _____ , *Oszillationsmatrizen, oszillationskerne und kleine schwingungen mechanischer systeme*, Wissenschaftliche Bearbeitung der deutschen Ausgabe: Alfred Stöhr. Mathematische Lehrbücher und Monographien, I. Abteilung, Bd. V. Akademie-Verlag, Berlin, 1960, (In German). {**11, 110, 542**}

[130] _____ , *Oscillation matrices and kernels and small vibrations of mechanical systems*, Tech. Report AEC-tr-448, Off. Tech. Doc., Dept. Commerce, Washington, DC, 1961. {**11, 110, 542**}

[131] _____, *Oscillation matrices and kernels and small vibrations of mechanical systems*, revised ed., AMS Chelsea Publishing, Providence, Rhode Island, 2002. {**11, 39, 105, 110–112, 114–116, 120–122, 272, 508, 543**}

[132] W. Gautschi, *Numerical analysis, an introduction*, Birkhäuser, 1997. {**131, 543**}

[133] L. Gemignani, *A unitary Hessenberg QR-based algorithm via semiseparable matrices*, Journal of Computational and Applied Mathematics **184** (2005), 505–517. {**30, 36, 68, 69, 139, 543**}

[134] L. Gemignani and D. Fasino, *Fast and stable solution of banded-plus-semiseparable linear systems*, Calcolo **39** (2002), no. 4, 201–217. {**84, 87, 90, 170, 543**}

[135] W. L. Gifford, *A short course in the theory of determinants*, Macmillan and Co., 1893. {**106, 116, 543**}

[136] _____, *Determinants*, fourth ed., Mathematical monographs, vol. 3, John Wiley and Sons, 1906. {**106, 116, 543**}

[137] P. E. Gill, G. H. Golub, W. Murray, and D. Saunders, *Methods for modifying matrix factorizations*, Mathematics of Computation **28** (1974), 505–535. {**189, 190, 543**}

[138] G. M. L. Gladwell, *The inverse problem for the vibrating beam*, Proceedings of the Royal Society of London A **393** (1984), 277–295. {**122, 543**}

[139] _____, *The inverse mode problem for lumped mass systems*, The Quarterly Journal of Mechanics and Applied Mathematics **39** (1986), no. 2, 297–307. {**122, 140, 543**}

[140] _____, *Inverse problems in vibration*, Applied Mechanics Reviews **39** (1986), no. 7, 1013–1018. {**140, 543**}

[141] _____, *Inverse vibration problems for finite element models*, Journal of Computational Analysis and Applications **13** (1997), 311–322. {**140, 543**}

[142] _____, *Inverse finite element vibration problems*, Journal of Sound and Vibration **211** (1999), no. 2, 309–324. {**140, 543**}

[143] _____, *On the reconstruction of a damped vibrating system from two complex spectra, part I: theory*, Journal of Sound and Vibration **240** (2001), no. 2, 203–217. {**140, 543**}

[144] G. M. L. Gladwell and J. A. Gbadeyan, *On the inverse problem of the vibrating string or rod*, The Quarterly Journal of Mechanics and Applied Mathematics **38** (1985), no. 1, 169–174. {**123, 140, 543**}

[145] I. C. Gohberg and M. A. Kaashoek, *Time varying linear systems with bound-ary conditions and integral operators, I. the transfer operator and its proper-ties*, Integral Equations and Operator Theory **7** (1984), 325–391. {**140, 344, 525, 526, 544**}

[146] ———, *Minimal representations of semiseparable kernels and systems with separable boundary conditions*, Journal of Mathematical Analysis and Appli-cations **124** (1987), no. 2, 436–458. {**342, 544**}

[147] I. C. Gohberg, M. A. Kaashoek, and H. J. Woerdeman, *A note on extensions of band matrices with invertible maximal and submaximal blocks*, Linear Algebra and its Applications **150** (1991), 157–166. {**342, 544**}

[148] I. C. Gohberg, T. Kailath, and I. Koltracht, *Linear complexity algorithms for semiseparable matrices*, Integral Equations and Operator Theory **8** (1985), no. 6, 780–804. {**90, 255, 344, 525, 544**}

[149] ———, *Efficient solution of linear systems of equations with recursive struc-ture*, Linear Algebra and its Applications **80** (1986), 81–113. {**221, 544**}

[150] ———, *A note on diagonal innovation matrices*, Acoustics Speech and Signal Processing **7** (1987), 1068–1069. {**90, 255, 344, 526, 544**}

[151] G. H. Golub, *Comparison of the variance of minimum variance and weighted least squares regression coefficients*, Annals of Mathematical Statistics **34** (1963), no. 3, 984–991. {**278, 544**}

[152] G. H. Golub and C. F. Van Loan, *Matrix computations*, third ed., The Johns Hopkins University Press, Baltimore, Maryland, 1996. {**26, 27, 45, 72, 102, 150, 152–154, 162, 163, 171, 183, 187, 205, 207, 210, 231, 240, 244, 476, 544**}

[153] G. H. Golub and J. M. Varah, *On the characterization of the best L_2-scaling of a matrix.*, SIAM Journal on Numerical Analysis **11** (1974), 472–479. {**277, 278, 544**}

[154] R. A. Gonzales, J. Eisert, I. Koltracht, M. Neumann, and G. Rawitscher, *Inte-gral equation method for the continuous spectrum radial Schrödinger equation*, Journal of Computational Physics **134** (1997), 134–149. {**132, 140, 544**}

[155] M. J. C. Gover and S. Barnett, *Inversion of Toeplitz matrices which are not strongly non-singular*, IMA Journal of Numerical Analysis **5** (1985), 101–110. {**507, 544**}

[156] W. B. Gragg, *The QR algorithm for unitary Hessenberg matrices*, Journal of Computational and Applied Mathematics **16** (1986), 1–8. {**30, 544**}

[157] W. B. Gragg and L. Reichel, *A divide and conquer algorithm for the unitary eigenproblem*, Hypercube multiprocessors 1987 (Philadelphia) (M. T. Heath, ed.), SIAM, 1987, pp. 639–647. {**30, 544**}

[158] F. A. Graybill, *Matrices with applications in statistics*, Wadsworth international group, Belmont, California, 1983. {**105, 123, 127–129, 140, 273, 274, 479, 545**}

[159] F. A. Graybill, C. D. Meyer, and R. J. Painter, *Note on the computation of the generalized inverse of a matrix*, SIAM Review **8** (1966), no. 4, 522–524. {**123, 545**}

[160] B. G. Greenberg and A. E. Sarhan, *Matrix inversion, its interest and application in analysis of data*, Journal of the American Statistical Association **54** (1959), 755–766. {**89, 123, 128, 130, 273, 317, 515, 545**}

[161] L. Greengard and V. Rokhlin, *On the numerical solution of two-point boundary value problems*, Communications on Pure and Applied Mathematics **44** (1991), 419–452. {**140, 545**}

[162] T. N. E. Greville, *Moving-weighted-average smoothing extended to the extremities of the data*, Tech. Report MRC 1786, Mathematics, Research Center, University of Wisconsin, Madison, 1977. {**504, 545**}

[163] _____, *On a problem concerning band matrices with Toeplitz inverses*, Proceedings of the eigth Manitoba Conference on Numerical Mathematics and Computing, 1978, pp. 275–283. {**504, 545**}

[164] T. N. E. Greville and W. F. Trench, *Band matrices with Toeplitz inverses*, Linear Algebra and its Applications **27** (1979), 199–209. {**504, 545**}

[165] _____, *Band matrices with Toeplitz inverses*, Transactions of the Twenty-Fourth Conference of Army Mathematicians (Univ.Virginia, Charlottesville, Va., 1978) (1979), 365–375. {**504, 545**}

[166] G. J. Groenewald, M. A. Petersen, and A. C. M. Ran, *Characterization of integral operators with semi–separable kernel with symmetries*, Journal of Functional Analysis **219** (2005), 255–284. {**140, 545**}

[167] W. H. Gustafson, *A note on matrix inversion*, Linear Algebra and its Applications **57** (1984), 71–73. {**37, 47, 272, 317, 545**}

[168] L. Guttman, *A generalized simplex for factor analysis*, Psychometrika **20** (1955), 173–195. {**129, 545**}

[169] W. Hackbusch and B. N. Khoromskij, *A sparse $\mathcal{H}$-matrix arithmetic, part I: general complexity estimates*, Journal of Computational and Applied Mathematics **125** (2000), no. 1-2, 479–501. {**140, 496, 545**}

[170] _____, *A sparse $\mathcal{H}$-matrix arithmetic, part II: Application to multidimensional problems*, Computing **64** (2000), no. 1, 21–47. {**140, 497, 545**}

[171] W. Hackbusch, B. N. Khoromskij, and S. A. Sauter, *On H^2-matrices*, Lectures on Applied Mathematics (H. Bungartz and L. Horsten, eds.), Springer-Verlag, Berlin, 2000, pp. 9–29. {**495, 497, 545**}

[172] S. B. Haley, *Solution of band matrix equations by projection-recurrence*, Linear Algebra and its Applications **32** (1980), 33–48. {**506, 519, 546**}

[173] G. Heinig and K. Rost, *Algebraic methods for toeplitz-like matrices and operators*, Mathematical Research, Akademie Verlag-Berlin, 1984. {**511, 546**}

[174] H. V. Henderson and S. R. Searle, *On deriving the inverse of a sum of matrices*, SIAM Review **23** (1981), no. 1, 53–59. {**274, 275, 546**}

[175] R. A. Horn and C. R. Johnson, *Topics in matrix analysis*, Cambridge University Press, Cambridge, 1991. {**279, 546**}

[176] W. D. Hoskins and G. E. McMaster, *On the inverses of a class of Toeplitz matrices of band width five*, Linear and Multilinear Algebra **4** (1976), no. 2, 103–106. {**503, 546**}

[177] _____, *Properties of the inverses of a set of band matrices*, Linear and Multilinear Algebra **5** (1977), no. 3, 183–196. {**503, 546**}

[178] _____, *On the infinity norm of the inverse of a class of Toeplitz matrices of band width five*, Linear and Multilinear Algebra **6** (1978), no. 2, 153–156. {**503, 546**}

[179] W. D. Hoskins and P. J. Ponzo, *Some properties of a class of band matrices*, Mathematics of Computation **26** (1972), 393–400. {**105, 502, 519, 546**}

[180] W. D. Hoskins and M. C. Thurgur, *Determinants and norms for the inverses of a set of band matrices*, Utilitas Mathematica **3** (1973), 33–47. {**105, 519, 546**}

[181] H. Hotelling, *Further points on matrix calculation and simultaneous equations*, Annals of Mathematical Statistics **14** (1943), 440–441. {**276, 546**}

[182] _____, *Some new methods in matrix calculation*, Annals of Mathematical Statistics **14** (1943), 1–34. {**276, 546**}

[183] A. S. Householder, *A survey of closed methods for inverting matrices*, Journal of the Society for Industrial and Applied Mathematics **5** (1957), 155–169. {**276, 546**}

[184] Y. Ikebe, *On inverses of Hessenberg matrices*, Linear Algebra and its Applications **24** (1979), 93–97. {**512, 513, 546**}

[185] Kh. D. Ikramov and L. Elsner, *On matrices that admit unitary reduction to band form*, Mathematical Notes **64** (1998), no. 6, 753–760. {**135, 546**}

[186] A. K. Jain, *Fast inversion of banded Toeplitz matrices by circular decompositions*, IEEE Transactions on Acoustics, Speech and Signal Processing **26** (1978), no. 2, 121–126. {**506, 546**}

[187] F. Jossa, *Risoluzione progressiva di un sistema di equazioni lineari, analogia con un problema meccanico*, Reale Accadademia di Scienze Fisiche e Matematiche, Società Reale di Napoli **4** (1940), no. 10, 346–352, (In Italian). {**276, 547**}

[188] M. A. Kaashoek and H. J. Woerdeman, *Unique minimal rank extensions of triangular operators*, Journal of Mathematical Analysis and Applications **131** (1988), no. 2, 501–516. {**342, 547**}

[189] ———, *Minimal lower separable representations: characterization and construction*, Operator Theory: Advances and Applications **41** (1989), 329–344. {**342, 547**}

[190] T. Kailath, S.-Y. Kung, and M. Morf, *Displacement ranks of matrices and linear equations*, Journal of Mathematical Analysis and Applications **68** (1979), no. 2, 395–407. {**246, 547**}

[191] T. Kailath and A. H. Sayed (eds.), *Fast reliable algorithms for matrices with structure*, SIAM, Philadelphia, PA, USA, May 1999. {**205, 210, 231, 246, 247, 547**}

[192] W. J. Kammerer and G. W. Reddien, Jr., *Local convergence of smooth cubic spline interpolants*, SIAM Journal on Numerical Analysis **9** (1972), 687–694. {**284, 547**}

[193] S.-Y. Kang, I. Koltracht, and G. Rawitscher, *High accuracy method for integral equations with discontinuous kernels*, 1999, http://arxiv.org/abs/math.NA/9909006. {**130, 132, 140, 272, 547**}

[194] S. Karlin, *Total positivity*, vol. 1, Stanford University Press, Stanford, California, 1968. {**135, 547**}

[195] A. Kavčić and M. F. Moura, *Matrices with banded inverses: inversion algorithms and factorization of Gauss-Markov processes*, IEEE Transactions on Information Theory **46** (2000), no. 4, 1495–1509. {**140, 547**}

[196] O. D. Kellogg, *The oscillation of functions of an orthogonal set*, American Journal of Mathematics **38** (1916), 1–5. {**9, 113, 547**}

[197] ———, *Orthogonal functions sets arising from integral equations*, American Journal of Mathematics **40** (1918), 145–154. {**9, 113, 547**}

[198] D. Kershaw, *The explicit inverses of two commonly occuring matrices*, Mathematics of Computation **23** (1969), no. 105, 189–191. {**272, 521, 547**}

[199] ———, *Inequalities on the elements of the inverse of a certain tridiagonal matrix*, Mathematics of computation **24** (1970), 155–158. {**282, 547**}

[200] I. Koltracht, *Linear complexity algorithm for semiseparable matrices*, Integral Equations and Operator Theory **29** (1997), no. 3, 313–319. {**527, 547**}

[201] A. Kounadis, *Inversion de matrices tridiagonales symétriques*, Revue Française d' Automatique, Informatique et de Recherce Opérationelle **7** (1973), 91–100, (In French). **{273, 521, 548}**

[202] E. G. Kounias, *An inversion technique for certain patterned matrices*, Journal of Mathematical Analysis and Applications **21** (1968), 695–698. **{123, 130, 500, 548}**

[203] M. G. Kreĭn, *On the nodes of harmonic oscillations of mechanical systems of a special form*, Matematicheskiĭ Sbornik **41** (1934), 339–348, (Russian). **{112, 548}**

[204] J.-Y. Lee and L. Greengard, *A fast adaptive numerical method for stiff two-point boundary value problems*, SIAM Journal on Scientific Computing **18** (1997), no. 2, 403–429. **{140, 548}**

[205] N. Levinson, *The Wiener RMS error criterion in filter desing and prediction*, Journal of Mathematical Physics **25** (1947), 261–278. **{205, 208, 210, 548}**

[206] J. W. Lewis, *Inversion of tridiagonal matrices*, Numerische Mathematik **38** (1982), 333–345. **{84, 86, 89, 267, 269, 271, 273, 506, 548}**

[207] R. K. Mallik, *The inverse of a tridiagonal matrix*, Linear Algebra and its Applications **325** (2001), 109–139. **{90, 272, 548}**

[208] P. G. Martinsson, V. Rokhlin, and M. Tygert, *A fast algorithm for the inversion of general Toeplitz matrices*, Computers & Mathematics with Applications **50** (2005), 741–752. **{508, 548}**

[209] N. Mastronardi, S. Chandrasekaran, and S. Van Huffel, *Fast and stable two-way algorithm for diagonal plus semi-separable systems of linear equations*, Numerical Linear Algebra with Applications **8** (2001), no. 1, 7–12. **{90, 204, 344, 387, 548}**

[210] _____, *Fast and stable algorithms for reducing diagonal plus semiseparable matrices to tridiagonal and bidiagonal form*, BIT **41** (2003), no. 1, 149–157. **{8, 90, 138, 344, 548}**

[211] N. Mastronardi, M. Van Barel, and E. Van Camp, *Divide and conquer algorithms for computing the eigendecomposition of symmetric diagonal-plus-semiseparable matrices*, Numerical Algorithms **39** (2005), no. 4, 379–398. **{63, 65, 135, 244, 548}**

[212] N. Mastronardi, M. Van Barel, and R. Vandebril, *A Levinson-like algorithm for symmetric positive definite semiseparable plus diagonal matrices*, Tech. Report TW423, Department of Computer Science, Katholieke Universiteit Leuven, Celestijnenlaan 200A, 3000 Leuven (Heverlee), Belgium, March 2005. **{219, 244, 548}**

[213] R. Mattheij and M. Smooke, *Estimates for the inverse of tridiagonal matrices arising in boundary-value problems*, Linear Algebra and its Applications **73** (1986), 33–57. {**522, 549**}

[214] J. J. McDonald, R. Nabben, M. Neumann, H. Schneider, and M. J. Tsatsomeros, *Inverse tridiagonal Z-matrices*, Linear and Multilinear Algebra **45** (1998), no. 1, 75–97. {**284, 549**}

[215] G. Meurant, *A review of the inverse of symmetric tridiagonal and block tridiagonal matrices*, SIAM Journal on Matrix Analysis and its Applications **13** (1992), 707–728. {**49, 273, 283, 506, 525, 549**}

[216] B. Mityagin, *Quadratic pencils and least-squares piecewise-polynomial approximation*, Mathematics of Computation **40** (1983), 283–300. {**284, 549**}

[217] D. Moskovitz, *The numerical solution of Laplace's and Poisson's equation*, Quarterly of Applied Mathematics **2** (1944), 148–163. {**140, 549**}

[218] A. P. Mullhaupt and K. S. Riedel, *Banded matrix representation of triangular input normal pairs*, IEEE Transactions on Automatic Control **46** (2001), no. 12, 2018–2022. {**140, 549**}

[219] _____, *Low grade matrices and matrix fraction representations*, Numerical Linear Algebra with Applications **342** (2002), 187–201. {**140, 302, 308, 339, 549**}

[220] K. C. Mustafi, *The inverse of a certain matrix with an application*, Annals of Mathematical Statistics **38** (1967), 1289–1292. {**64, 123, 130, 273, 549**}

[221] R. Nabben, *Decay rates of the inverse of nonsymmetric tridiagonal and band matrices*, SIAM Journal on Matrix Analysis and its Applications **20** (1999), no. 3, 820–837. {**84, 89, 270, 278, 280, 284, 549**}

[222] _____, *Two sided bounds on the inverse of diagonally dominant tridiagonal matrices*, Linear Algebra and its Applications **287** (1999), 289–305. {**84, 89, 278, 281, 282, 284, 509, 549**}

[223] E. Neuman, *The inversion of certain band matrices*, Roczniki Polskiego Towarzystwa Matematycznego **3** (1977), no. 9, 15–24, (In Russian). {**170, 261, 549**}

[224] T. Oohashi, *Some representation for inverses of band matrices*, TRU Mathematics **14** (1978), no. 2, 39–47. {**88, 516, 549**}

[225] D. V. Oulette, *Schur complements and statistics*, Master's thesis, McGill University, Montreal, 1978. {**276, 549**}

[226] V. Y. Pan, *Structured matrices and polynomials. unified superfast algorithms*, Birkhäuser Springer, 2001. {**245, 549**}

[227] E. Pascal, *Die determinanten*, Teubner, B.G., 1900, (In German). {**106, 116, 550**}

[228] J. Petersen and A. C. M. Ran, *LU- versus UL-factorization of integral operators with semiseparable kernel*, Integral Equations and Operator Theory **50** (2004), 549–558. {**130, 132, 140, 550**}

[229] B. Plestenjak, E. Van Camp, and M. Van Barel, *A Cholesky LR algorithm for the positive definite symmetric diagonal-plus-semiseparable eigenproblem*, Tech. Report 971, Institute for Mathematics, Physics and Mechanics, Jadranska 19, 1000 Ljubljana, Slovenia, March 2005. {**170, 550**}

[230] R. M. Pringle and A. A. Rayner, *Generalized inverse matrices with applications to statistics*, Griffins's statistical monographs and courses, vol. 28, Griffin, 1971. {**123, 550**}

[231] Y. M. Ram and G. M. L. Gladwell, *Constructing a finite element model of a vibratory rod from eigendata*, Journal of Sound and Vibration **169** (1994), no. 2, 229–237. {**123, 140, 550**}

[232] L. Rehnqvist, *Inversion of certain symmetric band matrices*, BIT **12** (1972), 90–98. {**503, 550**}

[233] M. Reiss, *Beiträge zur theorie der determinanten*, B.G. Teubner, Leipzig, 1867, (In German). {**106, 116, 550**}

[234] S. A. H. Rizvi, *Inverses of quasi-tridiagonal matrices*, Linear Algebra and its Applications **56** (1984), 177–184. {**521, 522, 550**}

[235] D. W. Robinson, *Nullities of submatrices of the Moore-Penrose inverse*, Linear Algebra and its Applications **94** (1987), 127–132. {**527, 550**}

[236] L. Rodman and H. J. Woerdeman, *Perturbations, singular values, and ranks of partial triangular matrices*, SIAM Journal on Matrix Analysis and its Applications **16** (1995), no. 1, 278–288. {**342, 550**}

[237] C. A. Rohde, *Generalized inverses of partitioned matrices*, Journal of the Society for Industrial & Applied Mathematics **13** (1965), no. 4, 1033–1035. {**123, 550**}

[238] F. Romani, *On the additive structure of the inverses of banded matrices*, Linear Algebra and its Applications **80** (1986), 131–140. {**516, 550**}

[239] S. N. Roy, B. G. Greenberg, and A. E. Sarhan, *Evaluation of determinants, characteristic equations and their roots for a class of patterned matrices*, Journal of the Royal Statistical Society. Series B. Statistical Methodology **22** (1960), 348–359. {**64, 105, 123, 130, 550**}

[240] S. N. Roy and A. E. Sarhan, *On inverting a class of patterned matrices*, Biometrika **43** (1956), 227–231. {**11, 64, 130, 158, 273, 274, 550**}

[241] P. Rózsa, *Band matrices and semi-separable matrices*, Coloquia Mathematica Societatis János Bolyai **50** (1986), 229–237. {**517, 551**}

[242] _____ , *On the inverse of band matrices*, Integral Equations and Operator Theory **10** (1987), 82–95. {**506, 517, 551**}

[243] P. Rózsa, R. Bevilacqua, P. Favati, and F. Romani, *On the inverse of block tridiagonal matrices with applications to the inverses of band matrices and block band matrices*, Operator Theory: Advances and Applications **40** (1989), 447–469. {**518, 523, 551**}

[244] P. Rózsa, R. Bevilacqua, F. Romani, and P. Favati, *On band matrices and their inverses*, Linear Algebra and its Applications **150** (1991), 287–295. {**518, 551**}

[245] P. Rózsa, F. Romani, and R. Bevilacqua, *On generalized band matrices and their inverses*, Proceedings of the Cornelius Lanczos International Centenary Conference (Philadelphia PA) (D. J. Brown, M. T. Chu, D. C. Ellison, and R. J. Plemmons, eds.), Proceedings in Applied Mathematics, vol. 73, SIAM Press, 1994, pp. 109–121. {**49, 317, 526, 551**}

[246] Y. Saad, *Iterative methods for sparse linear systems*, second ed., SIAM, January 2000. {**128, 245, 551**}

[247] A. E. Sarhan, E. Roberts, and B. G. Greenberg, *Modified square root method of matrix inversion*, Tech. Report 3, U.S. Army office of ordnance research project, February 195. {**123, 551**}

[248] S. Schechter, *Quasi tridiagonal matrices and type insensitive difference equations*, Quarterly of Applied Mathematics **18** (1960), 285–295. {**521, 522, 551**}

[249] P. Schlegel, *The explicit inverse of a tridiagonal matrix*, Mathematics of Computation **24** (1970), no. 111, 665–665. {**272, 521, 551**}

[250] I. Schur, *Über potenzreihen, die im innern des einheitskreises beschränkt sind.I*, Journal für die Reine und Angewandte Mathematik **147** (1917), 205–232, (In German). {**276, 551**}

[251] R. F. Scott, *Theory of determinants and their applications*, second ed., Cambridge University Press, 1904. {**106, 116, 551**}

[252] J. Sherman and K. E. Morrison, *Adjustment of an inverse matrix corresponding to changes in the elements of a given column or a given row of the original matrix*, Annals of Mathematical Statistics **20** (1949), 621–621. {**275, 551**}

[253] J. Sherman and W. J. Morrison, *Adjustment of an inverse matrix corresponding to a change in one element of a given matrix*, Annals of Mathematical Statistics **21** (1950), 124–127. {**275, 551**}

[254] P. N. Shivakumar, D. R. Williams, Q. Ye, and C. A. Marinov, *On two-sided bounds related to weakly diagonally dominant M-matrices with application to digital circuit dynamics*, SIAM Journal on Matrix Analysis and its Applications **17** (1996), 298–312. {**281, 552**}

[255] V. N. Singh, *The inverse of a certain block matrix*, Bulletin of the Australian Mathematical Society **20** (1979), 161–163. {**159, 522, 552**}

[256] H. P. Jr. Starr, *On the numerical solution of one-dimensional integral and differential equations*, Ph.D. thesis, Yale University, 1992, Research Report YALEU/DCS/RR-888. {**132, 140, 552**}

[257] G. W. Stewart, *An updating algorithm for subspace tracking*, IEEE Transactions on Signal Processing **40** (1992), 1535–1541. {**203, 552**}

[258] ———, *Matrix algorithms, vol I basic decompositions*, SIAM, 1998. {**152, 153, 183, 552**}

[259] ———, *Stability properties of several variants of the unitary Hessenberg QR-algorithm in structured matrices in mathematics*, Computer Science and Engineering, II (Boulder, CO, 1999), Contemp. Math., vol. 281, Amer. Math. Soc., Providence, RI, 2001, pp. 57–72. {**30, 552**}

[260] G. Strang and T. Nguyen, *The interplay of ranks of submatrices*, SIAM Review **46** (2004), no. 4, 637–646. {**40, 49, 310, 317, 552**}

[261] D. Szynal and J. Szynal, *À propos de l'inversion des matrices généralisées de Jacobi*, Aplikace Matematiky **17** (1972), 28–32, (In French). {**518, 552**}

[262] T. Ting, *A method of solving a system of linear equations whose coefficients form a tridiagonal matrix*, Quarterly of Applied Mathematics **22** (1964), no. 2, 105–106. {**158, 552**}

[263] T. Torii, *Inversion of tridiagonal matrices and the stability of tridiagonal systems of linear equations*, Information processing in Japan (Joho Shori) **6** (1966), 41–46, (In Japanese). {**88, 272, 500, 552**}

[264] ———, *Inversion of tridiagonal matrices and the stability of tridiagonal systems of linear equations*, Technology Reports of the Osaka University **16** (1966), 403–414. {**88, 272, 500, 516, 552**}

[265] L. N. Trefethen and D. Bau, *Numerical linear algebra*, SIAM, 1997. {**27, 152, 153, 179, 183, 552**}

[266] W. F. Trench, *Weighting coefficients for the prediction of stationary time series for the finite past*, SIAM Journal on Applied Mathematics **15** (1967), 1502–1510. {**504, 552**}

[267] ———, *Inversion of Toeplitz band matrices*, Mathematics of Computation **28** (1974), no. 128, 1089–1095. {**503, 552**}

[268] E. E. Tyrtyshnikov, *Mosaic-skeleton approximations*, Calcolo **33** (1996), 47–58. {**490, 553**}

[269] _____, *Mosaic ranks and skeletons*, Numerical Analysis and Its Applications (L. Vulkov, J. Wasniewski, and P. Y. Yalamov, eds.), Lecture Notes in Computer Science, vol. 1196, Springer-Verlag, 1997, pp. 505–516. {**490, 553**}

[270] _____, *Mosaic ranks for weakly semiseparable matrices*, Large-Scale Scientific Computations of Engineering and Environmental Problems II (M. Griebel, S. Margenov, and P. Y. Yalamov, eds.), Notes on numerical fluid mechanics, vol. 73, Vieweg, 2000, pp. 36–41. {**11, 308, 490, 496, 553**}

[271] _____, *Piecewise separable matrices*, Calcolo **42** (2005), no. 3-4, 243–248. {**496, 497, 553**}

[272] Y. Ukita, *Characterization of 2-type diagonal matrices with an application to order statistics*, Journal of the Hokkaido College of Art and Literature **6** (1955), 66–75. {**89, 129, 130, 553**}

[273] V. R. R. Uppuluri and J. A. Carpenter, *The inverse of a matrix occurring in first-order moving-average models*, Sankhyā The Indian Journal of Statistics Series A **31** (1969), 79–82. {**130, 273, 502, 553**}

[274] _____, *An inversion method for band matrices*, Journal of Mathematical Analysis and Applications **31** (1970), 554–558. {**64, 502, 553**}

[275] R. A. Usmani, *Explicit inverse of a band matrix with application to computing eigenvalues of a Sturm-Liouville system*, Linear Algebra and its Applications **86** (1987), 189–198. {**140, 273, 553**}

[276] _____, *Inversion of a tridiagonal Jacobi matrix*, Linear Algebra and its Applications **212/213** (1994), 413–414. {**273, 553**}

[277] _____, *Inversion of Jacobi's tridiagonal matrix*, Computers & Mathematics with Applications **27** (1994), no. 8, 59–66. {**273, 553**}

[278] F. Valvi, *Explicit presentation of the inverses of some types of matrices*, Journal of the Institute of Mathematics and its Applications **19** (1977), no. 1, 107–117. {**64, 130, 273, 553**}

[279] M. Van Barel, D. Fasino, L. Gemignani, and N. Mastronardi, *Orthogonal rational functions and diagonal plus semiseparable matrices*, Advanced Signal Processing Algorithms, Architectures, and Implementations XII (F. T. Luk, ed.), Proceedings of SPIE, vol. 4791, 2002, pp. 167–170. {**30, 63, 65, 133, 134, 140, 553**}

[280] _____, *Orthogonal rational functions and structured matrices*, SIAM Journal on Matrix Analysis and its Applications **26** (2005), no. 3, 810–829. {**30, 133, 134, 140, 401, 553**}

[281] M. Van Barel, E. Van Camp, and N. Mastronardi, *Orthogonal similarity transformation into block-semiseparable matrices of semiseparability rank k*, Numerical Linear Algebra with Applications **12** (2005), 981–1000. {**136, 554**}

[282] M. Van Barel, R. Vandebril, and N. Mastronardi, *An orthogonal similarity reduction of a matrix into semiseparable form*, SIAM Journal on Matrix Analysis and its Applications **27** (2005), no. 1, 176–197. {**14, 65, 77, 79, 136, 554**}

[283] E. Van Camp, *Diagonal-plus-semiseparable matrices and their use in numerical linear algebra*, Ph.D. thesis, Department of Computer Science, Katholieke Universiteit Leuven, Celestijnenlaan 200A, 3000 Leuven (Heverlee), Belgium, May 2005. {**192, 199, 554**}

[284] E. Van Camp, N. Mastronardi, and M. Van Barel, *Two fast algorithms for solving diagonal-plus-semiseparable linear systems*, Journal of Computational and Applied Mathematics **164-165** (2004), 731–747. {**8, 79, 88, 92, 95, 153, 190, 192, 194, 199, 204, 554**}

[285] E. Van Camp, M. Van Barel, R. Vandebril, and N. Mastronardi, *An implicit QR-algorithm for symmetric diagonal-plus-semiseparable matrices*, Tech. Report TW419, Department of Computer Science, Katholieke Universiteit Leuven, Celestijnenlaan 200A, 3000 Leuven (Heverlee), Belgium, March 2005. {**137, 554**}

[286] R. Vandebril, *Semiseparable matrices and the symmetric eigenvalue problem*, Ph.D. thesis, Dept. of Computer Science, K.U.Leuven, Celestijnenlaan 200A, 3000 Leuven, May 2004. {**192, 199, 554**}

[287] R. Vandebril, G. H. Golub, and M. Van Barel, *A small note on the scaling of positive definite semiseparable matrices*, Numerical Algorithms **41** (2006), 319–326. {**278, 554**}

[288] R. Vandebril, N. Mastronardi, and M. Van Barel, *A Levinson-like algorithm for symmetric strongly nonsingular higher order semiseparable plus band matrices*, Journal of Computational and Applied Mathematics **198** (2007), 75–97. {**219, 244, 554**}

[289] ———, *Solving linear systems with a levinson-like solver*, Electronic Transactions on Numerical Analysis **26** (2007), 243–269. {**233, 244, 554**}

[290] R. Vandebril and M. Van Barel, *Necessary and sufficient conditions for orthogonal similarity transformations to obtain the Arnoldi(Lanczos)-Ritz values*, Linear Algebra and its Applications **414** (2006), 435–444. {**136, 554**}

[291] ———, *A note on the nulllity theorem*, Journal of Computational and Applied Mathematics **189** (2006), 179–190. {**48, 162, 186, 272, 317, 554**}

[292] R. Vandebril, M. Van Barel, and N. Mastronardi, *A QR-method for computing the singular values via semiseparable matrices*, Numerische Mathematik **99** (2004), 163–195. {**79, 137, 153, 555**}

[293] _____, *An implicit Q theorem for Hessenberg-like matrices*, Mediterranean Journal of Mathematics **2** (2005), 59–275. {**137, 555**}

[294] _____, *An implicit QR-algorithm for symmetric semiseparable matrices*, Numerical Linear Algebra with Applications **12** (2005), no. 7, 625–658. {**77, 79, 137, 401, 555**}

[295] _____, *A note on the representation and definition of semiseparable matrices*, Numerical Linear Algebra with Applications **12** (2005), no. 8, 839–858. {**27, 71, 79, 555**}

[296] _____, *A parallel QR-factorization/solver of structured rank matrices*, Tech. Report TW474, Department of Computer Science, Katholieke Universiteit Leuven, Celestijnenlaan 200A, 3000 Leuven (Heverlee), Belgium, October 2006. {**430, 555**}

[297] _____, *Matrix Computations and Semiseparable Matrices Volume II: Spectral Methods*, Johns Hopkins University Press, 2008. {**133, 555**}

[298] R. Vandebril, E. Van Camp, M. Van Barel, and N. Mastronardi, *On the convergence properties of the orthogonal similarity transformations to tridiagonal and semiseparable (plus diagonal) form*, Numerische Mathematik **104** (2006), 205–239. {**136, 555**}

[299] _____, *Orthogonal similarity transformation of a symmetric matrix into a diagonal-plus-semiseparable one with free choice of the diagonal*, Numerische Mathematik **102** (2006), 709–726. {**136, 555**}

[300] N. Veraverbeke, *Kanstheorie en statistiek*, Course Notes, Limburgs Universitair Centrum, 1997, (In Dutch). {**125, 555**}

[301] T. L. Wang and W. B. Gragg, *Convergence of the shifted QR algorithm, for unitary Hessenberg matrices*, Mathematics of Computation **71** (2002), no. 240, 1473–1496. {**30, 555**}

[302] _____, *Convergence of the unitary QR algorithm with unitary Wilkinson shift*, Mathematics of Computation **72** (2003), no. 241, 375–385. {**30, 555**}

[303] F. Waugh, *A note concerning Hotelling's method of inverting a partitioned matrix*, Annals of Mathematics and Statistics **16** (1945), 216–217. {**276, 555**}

[304] H. J. Woerdeman, *The lower order of lower triangular operators and minimal rank extensions*, Integral Equations and Operator Theory **10** (1987), 859–879. {**325, 339, 555**}

[305] _____, *Matrix and operator extensions*, CWI Tract, vol. 68, Centre for Mathematics and Computer Science, Amsterdam, 1989. {**318, 340, 555**}

[306] _____, *Minimal rank completions for block matrices*, Linear Algebra and its Applications **121** (1989), 105–122. {**339, 556**}

[307] _____, *Toeplitz minimal rank completions*, Linear Algebra and its Applications **202** (1994), 267–278. {**342, 556**}

[308] _____, *A matrix and its inverse: revisiting minimal rank completions*, 2006, http://arxiv.org/abs/math.NA/0608130. {**318, 341, 556**}

[309] M. A. Woodbury, *Inverting modified matrices*, Memorandum Report 42, Statistical Research Group, Princeton, N.J., 1950. {**276, 556**}

[310] T. Yamamoto and Y. Ikebe, *Inversion of band matrices*, Linear Algebra and its Applications **24** (1979), 105–111. {**89, 513, 516, 556**}

Author/Editor Index

AITKEN, A. C., **276**, **533**

ALLGOWER, E. L., **89**, **502**, **533**

AMMAR, G. S., **244**, **245**, **533**

ASPLUND, E., **11**, **308**, **310**, **511**, **533**

ASPLUND, S. O., **11**, **57**, **63**, **272**, **515**, **533**

BALTZER, R., **105**, **116**, **533**

BANACHIEWICZ, T., **276**, **533**

BAPAT, R. B., **50**, **528**, **533**

BARANGER, J., **64**, **272**, **533**

BARNETT, S., **243**, **244**, **534**

Barnett, S., *see* GOVER, M. J. C., 507, 544

BARRETT, W. W., **6**, **37**, **48**, **123**, **130**, **272**, **273**, **310**, **317**, **504**, **506**, **534**

Bau, D., *see* TREFETHEN, L. N., 27, 152, 153, 179, 183, 552

BAUER, F. L., **277**, **278**, **534**

BEN-ISRAEL, A., **528**, **534**

BERG, L., **516**, **534**

BERGER, W. J., **11**, **260**, **272**, **534**

BERMAN, A., **279**, **534**

BEVILACQUA, R., **65**, **86**, **89**, **136**, **139**, **272**, **516**, **519**, **522**, **525**, **526**, **528**, **534**, **535**

Bevilacqua, R., *see* RÓZSA, P., 49, 317, 518, 523, 526, 551

BINDEL, D., **187**, **535**

BINI, D. A., **21**, **79**, **81**, **82**, **137**–**139**, **242**, **535**

BODEWIG, E., **276**, **535**

BORGES, C. F., **237**, **244**, **535**

BÖRM, S., **140**, **491**, **493**, **496**, **536**

BOZZO, E., *see* BEVILACQUA, R., 139, 528, 534

BROWN, D. J., **49**, **317**, **526**, **551**

BRUALDI, R. A., **49**, **536**

BUKHBERGER, B., **65**, **272**, **536**

BULTHEEL, A., **133**, **134**, **140**, **536**

BUNGARTZ, H., **495**, **497**, **545**

BURDEN, R. L., **131**, **536**

BURNSIDE, W. S., **105**, **116**, **536**

CALVETTI, D., **244**, **245**, **536**

Calvetti, D., *see* AMMAR, G. S., 244, 245, 533

CAO, W. L., **513**, **536**

CAPOVANI, M., **88**, **272**, **502**, **536**

Capovani, M., *see* BEVILACQUA, R., 86, 89, 272, 516, 522, 534

Carpenter, J. A., *see* UPPULURI, V. R. R., 64, 130, 273, 502, 553

CEI, S., **516**, **537**

CHANDRASEKARAN, S., **8**, **63**, **65**, **80**, **82**, **90**, **132**, **135**, **138**, **140**, **201**, **204**, **344**, **387**, **481**, **485**, **489**, **490**, **495**–**497**, **537**

Chandrasekaran, S., *see* MASTRONARDI, N., 8, 90, 138, 204, 344, 387, 548

CHU, M. T., **122**, **537**

Chu, M. T., *see* BROWN, D. J., 49, 317, 526, 551

Codenotti, B., *see* BEVILACQUA, R., 522, 534

Collar, A. R., *see* FRAZER, R. A., 123, 276, 542

COLTON, D., **140**, **537**

CONCUS, P., **263, 270, 272, 280, 284, 537**
CORDUNEANU, C., **131, 537**
CULLIS, C. E., **106, 116, 537**
CUPPEN, J. J. M., **237, 244, 538**
Cuyt, A. M., *see* BULTHEEL, A., 133, 536

DA FONSECA, C. M., **506, 507, 538**
Daddi, F., *see* BINI, D. A., 138, 535
DE BOOR, C., **283, 284, 538**
Del Corso, G. M., *see* BEVILACQUA, R., 65, 136, 139, 528, 534, 535
DELVAUX, S., **21, 44, 49, 73, 79, 134, 137, 139, 140, 179, 192, 249, 250, 253, 255, 297, 317, 344, 387, 399, 538, 539**
DEMKO, S., **282, 283, 539**
DEMMEL, J. W., **27, 183, 539**
Demmel, J. W., *see* BINDEL, D., 187, 535
DEWILDE, P., **57, 80, 82, 132, 140, 194, 387, 488, 489, 527, 539**
Dewilde, P., *see* CHANDRASEKARAN, S., 80, 82, 132, 140, 481, 485, 489, 490, 537
DOW, M., **500, 507, 539**
Duc-Jacquet, M., *see* BARANGER, J., 64, 272, 533
DUNCAN, W. J., **276, 539**
Duncan, W. J., *see* FRAZER, R. A., 123, 276, 542
DURBIN, J., **207, 210, 539**
DYM, H., **342, 539**

EIDELMAN, Y., **9, 11, 49, 80, 82, 90, 93, 132, 139, 140, 194, 233, 235, 236, 255, 344, 387, 433, 475, 476, 481–485, 487–489, 526, 527, 539, 540**
Eidelman, Y., *see* BINI, D. A., 139, 535

EIJKHOUT, V., **283, 506, 540**
Eisert, J., *see* GONZALES, R. A., 132, 140, 544
Ellison, D. C., *see* BROWN, D. J., 49, 317, 526, 551
ELSNER, L., **140, 310, 514, 521, 540**
Elsner, L., *see* IKRAMOV, KH. D., 135, 546
Emel'yanenko, G. A., *see* BUKHBERGER, B., 65, 272, 536

Faires, J. D., *see* BURDEN, R. L., 131, 536
FALLAT, S. M., **112, 314, 540**
FASINO, D., **8, 21, 63, 65, 84, 87, 89, 90, 133, 134, 136, 138, 344, 541**
Fasino, D., *see* BEVILACQUA, R., *see* GEMIGNANI, L., *see* VAN BAREL, M., 30, 63, 65, 84, 87, 90, 133, 134, 140, 170, 401, 528, 534, 543, 553
FAVATI, P., **526, 541**
Favati, P., *see* RÓZSA, P., 518, 523, 551
Feinsilver, P. J., *see* BARRETT, W. W., 6, 37, 48, 123, 130, 272, 273, 310, 317, 504, 506, 534
FIEDLER, M., **6, 10, 37, 38, 40, 48, 66, 68–71, 112, 156, 159, 162, 163, 178, 272, 273, 279, 296, 297, 310–314, 317, 339, 527, 541, 542**
Fiedler, M., *see* FALLAT, S. M., 112, 314, 540
FORSYTHE, G. E., **277, 278, 542**
FRAZER, R. A., **123, 276, 542**
FREUND, R. W., **245, 255, 542**

Galli, M., *see* CEI, S., 516, 537
GANTMACHER, F. R., **10, 11, 39, 105, 110–116, 120–122, 272, 508, 542, 543**
GAUTSCHI, W., **131, 543**

Gbadeyan, J. A., *see* GLADWELL, G.
    M. L., 123, 140, 543
GEMIGNANI, L., 30, 36, 68, 69, 84,
    87, 90, 139, 170, 543
Gemignani, L., *see* BINI, D. A., *see*
    FASINO, D., *see* VAN BA-
    REL, M., 8, 21, 30, 63, 65,
    79, 81, 82, 84, 87, 89, 133,
    134, 137–140, 401, 535, 541,
    553
Giersch, C., *see* ELSNER, L., 140, 540
GIFFORD, W. L., 106, 116, 543
GILL, P. E., 189, 190, 543
GLADWELL, G. M. L., 122, 123,
    140, 543
Gladwell, G. M. L., *see* RAM, Y. M.,
    123, 140, 550
GOHBERG, I. C., 90, 140, 221, 255,
    342, 344, 525, 526, 544
Gohberg, I. C., *see* BINI, D. A., *see*
    DYM, H., *see* EIDELMAN,
    Y., 9, 11, 49, 80, 82, 90, 93,
    132, 139, 140, 194, 233, 235,
    236, 255, 342, 344, 387, 433,
    475, 476, 481–485, 487–489,
    526, 527, 535, 539, 540
GOLUB, G. H., 26, 27, 45, 72, 102,
    150, 152–154, 162, 163,
    171, 183, 187, 205, 207,
    210, 231, 240, 244, 277,
    278, 476, 544
Golub, G. H., *see* CHU, M. T., *see*
    CONCUS, P., *see* GILL, P.
    E., *see* VANDEBRIL, R., 122,
    189, 190, 263, 270, 272, 278,
    280, 284, 537, 543, 554
GONZALES, R. A., 132, 140, 544
GOVER, M. J. C., 507, 544
GRAGG, W. B., 30, 544
Gragg, W. B., *see* AMMAR, G. S., *see*
    BORGES, C. F., *see* WANG,
    T. L., 30, 237, 244, 245, 533,
    535, 555
Grasedyck, L., *see* BÖRM, S., 140,
    491, 493, 496, 536
GRAYBILL, F. A., 105, 123, 127–

129, 140, 273, 274, 479,
    545
GREENBERG, B. G., 89, 123, 128,
    130, 273, 317, 515, 545
Greenberg, B. G., *see* ROY, S. N.,
    see SARHAN, A. E., 64, 105,
    123, 130, 550, 551
GREENGARD, L., 140, 545
Greengard, L., *see* LEE, J.-Y., 140,
    548
GREVILLE, T. N. E., 504, 545
Greville, T. N. E., *see* BEN-ISRAEL,
    A., 528, 534
GRIEBEL, M., 11, 308, 490, 496,
    553
GROENEWALD, G. J., 140, 545
Gu, M., *see* CHANDRASEKARAN, S.,
    8, 63, 65, 80, 82, 90, 132,
    135, 138, 140, 201, 204, 344,
    387, 481, 485, 489, 490, 495–
    497, 537
GUSTAFSON, W. H., 37, 47, 272,
    317, 545
GUTTMAN, L., 129, 545

HACKBUSCH, W., 140, 495–497,
    545
Hackbusch, W., *see* BÖRM, S., 140,
    491, 493, 496, 536
HALEY, S. B., 506, 519, 546
HEATH, M. T., 30, 544
HEINIG, G., 511, 546
HENDERSON, H. V., 274, 275, 546
HORN, R. A., 279, 546
Horsten, L., *see* BUNGARTZ, H., 495,
    497, 545
HOSKINS, W. D., 105, 502, 503,
    519, 546
HOTELLING, H., 276, 546
HOUSEHOLDER, A. S., 276, 546

IKEBE, Y., 512, 513, 546
Ikebe, Y., *see* YAMAMOTO, T., 89,
    513, 516, 556
IKRAMOV, KH. D., 135, 546

JAIN, A. K., 506, 546

Johnson, C. R., *see* HORN, R. A., 279, 546
JOSSA, F., **276**, **547**

KAASHOEK, M. A., **342**, **547**
Kaashoek, M. A., *see* GOHBERG, I. C., 140, 342, 344, 525, 526, 544
Kahan, W., *see* BINDEL, D., 187, 535
KAILATH, T., **205**, **210**, **231**, **246**, **247**, **547**
Kailath, T., *see* GOHBERG, I. C., 90, 221, 255, 344, 525, 526, 544
KAMMERER, W. J., **284**, **547**
KANG, S.-Y., **130**, **132**, **140**, **272**, **547**
KARLIN, S., **135**, **547**
KAVČIĆ, A., **140**, **547**
KELLOGG, O. D., **9**, **113**, **547**
KERSHAW, D., **272**, **282**, **521**, **547**
Khoromskij, B. N., *see* HACKBUSCH, W., 140, 495–497, 545
Kim, S., *see* CALVETTI, D., 244, 245, 536
KOLTRACHT, I., **527**, **547**
Koltracht, I., *see* GOHBERG, I. C., *see* GONZALES, R. A., *see* KANG, S.-Y., 90, 130, 132, 140, 221, 255, 272, 344, 525, 526, 544, 547
KOUNADIS, A., **273**, **521**, **548**
KOUNIAS, E. G., **123**, **130**, **500**, **548**
KREĬN, M. G., **112**, **548**
Kreĭn, M. G., *see* GANTMACHER, F. R., 10, 11, 39, 105, 110–112, 114–116, 120–122, 272, 508, 542, 543
Kress, R., *see* COLTON, D., 140, 537
Kung, S.-Y., *see* KAILATH, T., 246, 547

LEE, J.-Y., **140**, **548**
LEVINSON, N., **205**, **208**, **210**, **548**
LEWIS, J. W., **84**, **86**, **89**, **267**, **269**, **271**, **273**, **506**, **548**
Lotti, G., *see* BEVILACQUA, R., *see*

FAVATI, P., 519, 525, 526, 535, 541
LUK, F. T., **30**, **63**, **65**, **133**, **134**, **140**, **553**
Lyons, W., *see* CHANDRASEKARAN, S., 495–497, 537

MALLIK, R. K., **90**, **272**, **548**
Margenov, S., *see* GRIEBEL, M., 11, 308, 490, 496, 553
Marinov, C. A., *see* SHIVAKUMAR, P. N., 281, 552
Markham, T. L., *see* FALLAT, S. M., *see* FIEDLER, M., 6, 10, 37, 38, 40, 48, 71, 112, 159, 162, 272, 273, 310, 311, 314, 317, 339, 540, 541
Marques, O., *see* BINDEL, D., 187, 535
MARTINSSON, P. G., **508**, **548**
Massey, J. J. Q., *see* BRUALDI, R. A., 49, 536
MASTRONARDI, N., **8**, **63**, **65**, **90**, **135**, **138**, **204**, **219**, **244**, **344**, **387**, **548**
Mastronardi, N., *see* FASINO, D., *see* VAN BAREL, M., *see* VAN CAMP, E., *see* VANDEBRIL, R., 8, 14, 27, 30, 63, 65, 71, 77, 79, 88, 90, 92, 95, 133, 134, 136–138, 140, 153, 190, 192, 194, 199, 204, 219, 233, 244, 344, 401, 430, 541, 553–555
MATTHEIJ, R., **522**, **549**
McDONALD, J. J., **284**, **549**
McMaster, G. E., *see* HOSKINS, W. D., 503, 546
MEURANT, G., **49**, **273**, **283**, **506**, **525**, **549**
Meurant, G., *see* CONCUS, P., 263, 270, 272, 280, 284, 537
Meyer, C. D., *see* GRAYBILL, F. A., 123, 545
MITYAGIN, B., **284**, **549**
Morf, M., *see* KAILATH, T., 246, 547

Morrison, K. E., *see* SHERMAN, J., 275, 551

Morrison, W. J., *see* SHERMAN, J., 275, 551

MOSKOVITZ, D., **140, 549**

Moss, W. F., *see* DEMKO, S., 283, 539

Moura, M. F., *see* KAVČIĆ, A., 140, 547

MULLHAUPT, A. P., **140, 302, 308, 339, 549**

Murray, W., *see* GILL, P. E., 189, 190, 543

MUSTAFI, K. C., **64, 123, 130, 273, 549**

NABBEN, R., **84, 89, 270, 278, 280–282, 284, 509, 549**

Nabben, R., *see* MCDONALD, J. J., 284, 549

NEUMAN, E., **170, 261, 549**

Neumann, M., *see* GONZALES, R. A., *see* MCDONALD, J. J., 132, 140, 284, 544, 549

Nguyen, T., *see* STRANG, G., 40, 49, 310, 317, 552

Olshevsky, V., *see* EIDELMAN, Y., 139, 540

OOHASHI, T., **88, 516, 549**

OULETTE, D. V., **276, 549**

Painter, R. J., *see* GRAYBILL, F. A., 123, 545

Pals, T., *see* CHANDRASEKARAN, S., 80, 82, 132, 140, 481, 485, 489, 490, 537

PAN, V. Y., **245, 549**

Pan, V. Y., *see* BINI, D. A., 21, 79, 81, 82, 137, 139, 242, 535

Panton, A. W., *see* BURNSIDE, W. S., 105, 116, 536

PASCAL, E., **106, 116, 550**

PETERSEN, J., **130, 132, 140, 550**

Petersen, M. A., *see* GROENEWALD, G. J., 140, 545

Petronilho, J., *see* DA FONSECA, C. M., 506, 507, 538

Plemmons, R. J., *see* BERMAN, A., *see* BROWN, D. J., 49, 279, 317, 526, 534, 551

PLESTENJAK, B., **170, 550**

Polman, B., *see* EIJKHOUT, V., 283, 506, 540

Ponzo, P. J., *see* HOSKINS, W. D., 105, 502, 519, 546

PRINGLE, R. M., **123, 550**

Pták, V., *see* FIEDLER, M., 112, 541

RAM, Y. M., **123, 140, 550**

Ran, A. C. M., *see* GROENEWALD, G. J., *see* PETERSEN, J., 130, 132, 140, 545, 550

Rawitscher, G., *see* GONZALES, R. A., *see* KANG, S.-Y., 130, 132, 140, 272, 544, 547

Rayner, A. A., *see* PRINGLE, R. M., 123, 550

Reddien, Jr. , G. W., *see* KAMMERER, W. J., 284, 547

REHNQVIST, L., **503, 550**

REICHEL, L., **237, 244, 535**

Reichel, L., *see* AMMAR, G. S., *see* CALVETTI, D., *see* GRAGG, W. B., 30, 244, 245, 533, 536, 544

REISS, M., **106, 116, 550**

Riedel, K. S., *see* MULLHAUPT, A. P., 140, 302, 308, 339, 549

RIZVI, S. A. H., **521, 522, 550**

Roberts, E., *see* SARHAN, A. E., 123, 551

ROBINSON, D. W., **527, 550**

RODMAN, L., **342, 550**

ROHDE, C. A., **123, 550**

Rokhlin, V., *see* GREENGARD, L., *see* MARTINSSON, P. G., 140, 508, 545, 548

ROMANI, F., **516, 550**

Romani, F., *see* BEVILACQUA, R., *see* FAVATI, P., *see* RÓZSA, P., 49, 317, 518, 519, 522, 523, 525, 526, 534, 535, 541, 551

Rost, K., *see* HEINIG, G., 511, 546

ROY, S. N., **11**, **64**, **105**, **123**, **130**, **158**, **273**, **274**, **550**

RÓZSA, P., **49**, **317**, **506**, **517**, **518**, **523**, **526**, **551**

Rózsa, P., *see* FAVATI, P., 526, 541

Ruttan, A., *see* REICHEL, L., 237, 244, 535

SAAD, Y., **128**, **245**, **551**

Saibel, E., *see* BERGER, W. J., 11, 260, 272, 534

SARHAN, A. E., **123**, **551**

Sarhan, A. E., *see* GREENBERG, B. G., *see* ROY, S. N., 11, 64, 89, 105, 123, 128, 130, 158, 273, 274, 317, 515, 545, 550

Saunders, D., *see* GILL, P. E., 189, 190, 543

Sauter, S. A., *see* HACKBUSCH, W., 495, 497, 545

Sayed, A. H., *see* KAILATH, T., 205, 210, 231, 246, 247, 547

SCHABACK, R., **283**, **538**

SCHECHTER, S., **521**, **522**, **551**

Scherer, K., *see* SCHABACK, R., 283, 538

SCHLEGEL, P., **272**, **521**, **551**

Schneider, H., *see* MCDONALD, J. J., 284, 549

SCHUR, I., **276**, **551**

SCOTT, R. F., **106**, **116**, **551**

Searle, S. R., *see* HENDERSON, H. V., 274, 275, 546

SHERMAN, J., **275**, **551**

SHIVAKUMAR, P. N., **281**, **552**

SINGH, V. N., **159**, **522**, **552**

Smith, P. W., *see* DEMKO, S., 283, 539

Smooke, M., *see* MATTHEIJ, R., 522, 549

Soccio, M., *see* CEI, S., 516, 537

STARR, H. P. JR., **132**, **140**, **552**

STEWART, G. W., **30**, **152**, **153**, **183**, **203**, **552**

Stewart, W. J., *see* CAO, W. L., 513, 536

STRANG, G., **40**, **49**, **310**, **317**, **552**

Straus, E. G., *see* FORSYTHE, G. E., 277, 278, 542

Sun, X., *see* CHANDRASEKARAN, S., 490, 537

SZYNAL, D., **518**, **552**

Szynal, J., *see* SZYNAL, D., 518, 552

Thurgur, M. C., *see* HOSKINS, W. D., 105, 519, 546

TING, T., **158**, **552**

TORII, T., **88**, **272**, **500**, **516**, **552**

TREFETHEN, L. N., **27**, **152**, **153**, **179**, **183**, **552**

TRENCH, W. F., **503**, **504**, **552**

Trench, W. F., *see* GREVILLE, T. N. E., 504, 545

Tsatsomeros, M. J., *see* MCDONALD, J. J., 284, 549

Tygert, M., *see* MARTINSSON, P. G., 508, 548

TYRTYSHNIKOV, E. E., **11**, **308**, **490**, **496**, **497**, **553**

UKITA, Y., **89**, **129**, **130**, **553**

UPPULURI, V. R. R., **64**, **130**, **273**, **502**, **553**

USMANI, R. A., **140**, **273**, **553**

VALVI, F., **64**, **130**, **273**, **553**

Van Assche, W., *see* BULTHEEL, A., 133, 536

VAN BAREL, M., **14**, **30**, **63**, **65**, **77**, **79**, **133**, **134**, **136**, **140**, **401**, **553**, **554**

Van Barel, M., *see* BULTHEEL, A., *see* DELVAUX, S., *see* FASINO, D., *see* MAS-TRONARDI, N., *see* VAN CAMP, E., *see* VANDEBRIL, R., *see* PLESTENJAK, B., 8, 21, 27, 44, 48, 49, 63, 65, 71, 73, 77, 79, 88, 90, 92, 95, 133–140, 153, 162, 170, 179, 186, 190, 192, 194, 199, 204, 219, 233, 244, 249, 250, 253,

255, 272, 278, 297, 317, 344,
387, 399, 401, 430, 536, 538,
539, 541, 548, 550, 554, 555

VAN CAMP, E., **8**, **79**, **88**, **92**, **95**,
137, **153**, **190**, **192**, **194**,
199, **204**, **554**

Van Camp, E., *see* MASTRONARDI,
N., *see* VAN BAREL, M.,
see VANDEBRIL, R., *see*
PLESTENJAK, B., 63, 65,
135, 136, 170, 244, 548, 550,
554, 555

van der Veen, A.-J., *see* CHAN-
DRASEKARAN, S., *see*
DEWILDE, P., 57, 80, 82,
132, 140, 194, 387, 481, 485,
488, 489, 527, 537, 539

Van gucht, P., *see* BULTHEEL, A.,
133, 134, 140, 536

Van Huffel, S., *see* MASTRONARDI,
N., 8, 90, 138, 204, 344, 387,
548

Van Loan, C. F., *see* GOLUB, G. H.,
26, 27, 45, 72, 102, 150, 152–
154, 162, 163, 171, 183, 187,
205, 207, 210, 231, 240, 244,
476, 544

VANDEBRIL, R., **27**, **48**, **71**, **77**, **79**,
133, **136**, **137**, **153**, **162**,
186, **192**, **199**, **219**, **233**,
244, **272**, **278**, **317**, **401**,
430, **554**, **555**

Vandebril, R., *see* MASTRONARDI,
N., *see* VAN BAREL, M., *see*
VAN CAMP, E., 14, 65, 77,
79, 136, 137, 219, 244, 548,
554

Varah, J. M., *see* GOLUB, G. H., 277,
278, 544

Varga, R. S., *see* REICHEL, L., 237,
244, 535

Vavřín, Z., *see* FIEDLER, M., 66, 68–
71, 156, 163, 178, 296, 297,
310, 314, 317, 527, 542

VERAVERBEKE, N., **125**, **555**

Verdonk, B., *see* BULTHEEL, A., 133,

536

VULKOV, L., **490**, **553**

WANG, T. L., **30**, **555**

Wasniewski, J., *see* VULKOV, L., 490,
553

WAUGH, F., **276**, **555**

Williams, D. R., *see* SHIVAKUMAR, P.
N., 281, 552

WOERDEMAN, H. J., **318**, **325**, **339–
342**, **555**, **556**

Woerdeman, H. J., *see* GOHBERG, I.
C., *see* KAASHOEK, M. A.,
see RODMAN, L., 342, 544,
547, 550

WOODBURY, M. A., **276**, **556**

Yalamov, P. Y., *see* GRIEBEL, M., *see*
VULKOV, L., 11, 308, 490,
496, 553

YAMAMOTO, T., **89**, **513**, **516**, **556**

Ye, Q., *see* SHIVAKUMAR, P. N., 281,
552

Zellini, P., *see* CEI, S., 516, 537

Zha, H., *see* FREUND, R. W., 245,
255, 542

Subject Index

$\|\cdot\|_1$, 27
$\|\cdot\|_2$, 26
$\|\cdot\|_F$, 27
$\|\cdot\|_\infty$, 26, 27
$\|\cdot\|_{max}$, 27
$\supsetneq$, 13
X-pattern, 395
V-pattern, 396
∧-pattern, 395
{0}-semiseparable matrix, 301

$A(\alpha; \beta)$, xvii, 39, 294
$A(i, j)$, 17
$A(i : j, k : l)$, xvii, 17
$A = (a_{i,j})_{i,j}$, xvii
$\alpha \times \beta$, xvii, 294
$\mathbf{a}(i : j)$, 17
$\mathcal{A}_{inv}$, 13
$\mathcal{A}_{sym}$, 13
$\overline{\mathcal{A}}$, 18
Adding a diagonal to the structure, 298
Additive structure, 516
Annihilating Givens, 351
Annihilating rank structures, 380
Arrowhead matrix, 237

B, xvii
Backward substitution, 152–153
Band matrices
    as quasiseparable matrices, 307
    inversion, 515–521
    {p, q}-band, 306
Basic matrices, 69, 71, 178, 296
Bidiagonal
    decomposition with Gauss transforms, 154
    inverse, 155
    unit, 154
Binomial distribution, 125
Block off-diagonal rank, 311
Block off-diagonal structure, 311
Block quasiseparable, 481–490
    definition, 482
    descriptor systems, 487
    factorization, 482
    multiplication, 484
    solver, 485
    structured rank matrices, 483
Block semiseparable matrix, 521–525
Boundary value problems, 140

Cardinality of a set, 39
Cauchy matrix, 114
Changing representations, 101–104
Characteristic polynomial, 139
Cholesky decomposition, 170
Closure
    $\mathcal{Q}$, 32
    $\mathcal{S}$, 32
    $\mathcal{S} + \mathcal{D}$, 32
    $\mathcal{S}^{(g)}$, 32
    $\mathcal{T}$, 32
    $\mathcal{T}^i$, 32
    generator representable semiseparable, 32
    irreducible tridiagonal, 32
    quasiseparable, 24
    quasiseparables having symmetric rank structure, 35
    $\mathcal{S}_{sym} + \mathcal{D}$, 22
    semiseparable, 32
    semiseparable plus diagonal, 24, 32, 35

tridiagonal, 32
Closure $\overline{\mathcal{A}}$, 18
Companion matrices, 139, 243
Complementary basic matrices, 71, 178
Complete basic matrices, 69, 71, 178
Completion problem, 340
Condition number, 277
Continuant matrix, 260
Continuum, 111
Convergence, 26
Counteridentity matrix, 208
Covariance, 124
    examples, 128–129
    matrices, 123–130

D, xvii
det (A), xvii
diag $(\cdot)$, xvii
$\mathcal{D}$, xvii, 12, 31
Decay rates of semiseparable, 278–284
Decomposition (sum) of
    extended semiseparable matrices, 339
    Hessenberg-like matrices, 338
    quasiseparable matrices, 338
    semiseparable and related matrices, 338–339
    semiseparable matrices, 339
    structured rank, 336
Decoupled representation, 93–95
Descriptor systems, 487
Determinant of a semiseparable matrix, 104–105
Diagonal matrix of type r, 130
Diagonal-subdiagonal representation, 65–71, 156
Diagonally dominant by columns, 280
Diagonally dominant by rows, 280
Diamond pattern, 391
Differentiation, 140
Discretization of integral equations, 130–133
Displacement rank, 246
Distribution
    binomial, 125

continuous, 124
covariance, 124
discrete, 124
expected value, 124
exponential, 128
mean, 124
multinomial, 125
Divide and conquer, 135, 519
Durbin algorithm, 207

Exchange matrices, 208
Expanding zero rank structure, 380
Expected value, 124
Exponential distribution, 128
Extended $\{p,q\}$-generator representable semiseparable matrix, 304
Extended $\{p,q\}$-semiseparable matrix, 304
Extended basic matrices, 71, 178
Extended diagonal position, 314
Extended semiseparable matrices, 304
Extension of symmetric relations, 31–34

Fellow matrix, 244
Flam, 54
Flop, 54
Fredholm integral equation, 491
Full QR-factorization, 183

G, xvii
Galerkin's method, 491
Gauss and Givens transformations, 470
Gauss elementary transformation, 151
Gauss solver (higher order) , 435–470
Gauss transform, 151
Gauss transformations
    annihilating, 437
    annihilating rank structure, 456
    arbitrary, 440
    ascending vs. descending, 444
    expanding rank structure, 457
    from bottom to top, 437, 443

LU-factorization, 458
L_1UL_2-decomposition, 459
on the right, 455
other decompositions, 459
other sequences, 454
pivoting, 448–454
rank-expanding, 446, 448
system solving, 456
UL-decomposition, 459
U_1LU_2-decomposition, 459
upper Gauss transform, 454
zero-creating, 445
Gauss-Markov processes, 140
Gauss-vector representation, 179
Gaussian elimination, 149–180
Generalized Hessenberg, 306
Generalized inverses, 527–528
Generalized nullity theorem, 527
Generator definition, 8
Generator representable, 28, 318
block semiseparable, 17
closure, 19
misunderstandings, 13–14
numerical problem, 14–16
pointwise closure, 19
relation to semiseparable, 16–21
Generator representable semiseparable matrix, 515–518
Generator representation, 59–65, 84–90
Givens and Gauss transformations, 470
Givens transformations, 72, 73
annihilating, 351
annihilating sequence, 351
arbitrary sequence, 354
ascending sequence, 370
fusion, 392
on lower rank structures, 356
on rank structures, 369, 415
on upper rank 1 structures, 363
on upper rank structures, 367
other decompositions, 386
rank annihilating sequence, 351
rank-annihilating, 351
rank-decreasing, 359

rank-expanding, 400
sequence from bottom to top, 350–379
sequence from left to right, 374
sequence from right to left, 373
sequence from top to bottom, 372
sequence on band matrices, 376
sequence on extended semiseparable matrices, 376
sequence on extended semiseparable matrices, 376
sequence on quasiseparable matrices, 376
sequence on semiseparable matrices, 375
sequence on semiseparable plus band matrices, 377
shift through lemma, 392
summary of the effect, 378
updating Givens-vector representation, 418
Givens-vector representation, 71–79, 90–92
examples, 74
quasiseparable matrices, 78
retrieving, 74–77
swapping the representation, 77–78
Gnomonic matrices, 260
Gnomonic symmetry, 260
Graphical Givens-vector representation, 416
Green's kernel, 109, 130, 135
Green's matrix, 63

H, xvii
$\mathcal{H}$-matrices, 140, 491–497
$\mathcal{H}^2$-matrices, 491–497
Harmonic oscillation, 111
Hessenberg matrix, 511–515
Hessenberg-like, 306
Hierarchically partitioned, 493
Hierarchically semiseparable, 491–497
Higher order quasiseparable, 301
Higher order semiseparable, 300

Higher order structured rank, 293–
        345

I_k, xvii
inv, xvii, 13
Implicit Q-theorem, 137
Influence
        coefficients, 112
        function of a string, 120
        matrix, 112
Inner product, 133
Integral equations, 130, 140
Integration, 140
Inverse eigenvalue problem, 133–134
Inverse of, 257–285
        band, 515–521
        bidiagonal, 154–156
        block band, 521–525
        block off-diagonal rank, 311
        block semiseparable, 521–525
        decay rates, 278–284
        direct methods, 261–273
        generalized Hessenberg, 511–515
        generator representable semisep-
            arable, 42, 515–518
        Hessenberg, 511–515
        Hessenberg-like, 511–515
        irreducible tridiagonal, 42
        LU-factorization, 259
        lower bidiagonal matrices, 315
        lower semiseparable, 156–157
        lower triangular rank, 313
        lower triangular semiseparable
            matrices, 315
        off-diagonal rank, 311
        one-pair matrices, 116–120
        $\{p, q\}$-semiseparable
            matrices, 312
        $\{p, q\}$-semiseparable   matrices,
            312
        QR-factorization, 258
        quasiseparable, 43–44, 70, 525–
            527
        rank k plus block diagonal matri-
            ces, 312
        rank k plus diagonal matrices,
            311
        relations, 37–50
        scaling of semiseparable, 277
        semiseparable, 40–43, 515–521
        semiseparable plus block diago-
            nal matrices, 312
        semiseparable plus diagonal, 44–
            47
        semiseparable plus diagonal ma-
            trices, 312
        semiseparable   plus   diagonal,
            with invertible diagonal, 44
        strict band, 515–518
        structured rank matrices,   40,
            310–318
        summary of the relations, 47
        symmetric tridiagonal, 261–264
        Toeplitz, 500–511
        triangular semiseparable,   153–
            159
        tridiagonal, 40–43, 267
        unitary Hessenberg, 44
        via factorization, 258–261
        via Levinson, 259
Irreducible tridiagonal, 11

Jacobi matrix, 109, 111, 115

k-Toeplitz, 507
Kernel
        Green's, 109, 130, 135
        oscillation, 113
        semiseparable, 135
        separable, 135
Kronecker delta, 267

L, xvii
LQ-decomposition, 203, 386
LU-decomposition, 151–152
        nullity theorem, 159–163
        numerical stability, 178–179
        of structured rank matrices, 159,
            160
        quasiseparable, 171–178
        representation, 179

semiseparable, 163–170
LU-decomposition for quasiseparable, 70
Leaf pattern, 389, 391
Levinson, 205–256
    arrowhead matrices, 237–238
    band, 476–478
    companion matrices, 242–243
    dense matrices, 240
    errors in structures, 241–242
    fellow matrices, 244
    framework, 219–234
    generator representable, 474
    generator representable semiseparable plus diagonal, 210–219
    Givens-vector represented, 234–235
    Levinson-like solver, 223–224
    quasiseparable, 475
    quasiseparable represented, 235–236
    summations of Levinson-conform matrices, 240–241, 478
    Toeplitz, 206–210
    Trench, 210
    tridiagonal, 236–237
    unsymmetric structures, 239, 478
    upper triangular factorization, 209, 218
    upper triangular matrices, 239–240
    Yule-Walker like problem, 211
    Yule-Walker-like system, 222–223
Levinson conform, 220
Levinson conform solver, 229–231
Levinson-like solver, 223–224
    higher order, 473–479
Lgrade, 308
Look ahead, 232–234
Low grade matrices, 308
Lower $\{p\}$-Hessenberg-like matrices, 305
Lower order, 340
Lower semiseparable

decomposition with Gauss transforms, 156
inverse, 157
matrices, 306
unit, 156
Lower triangular part, 7
Lower triangular structure, 295

M, xvii
M-matrices, 278
min rank(A), 340
Matrice factorisable, 64
Matrices
    basic, 69, 71, 178
    block semiseparable, 521–525
    Cauchy, 114
    companion, 139
    complementary basic, 71, 178
    complete basic, 69, 71, 178
    counteridentity, 208
    covariance, 123–130
    diagonal of type r, 130
    exchange, 208
    extended basic, 71, 178
    generator representable semiseparable, 515–518
    Green's, 63
    $\mathcal{H}$-matrices, 140, 491
    $\mathcal{H}^2$-matrices, 491
    Hessenberg, 511–515
    hierarchical matrices, 491
    influence, 112
    Jacobi, 109, 111, 115
    low grade, 308
    minor, 112
    one-pair, 110, 115, 116
    oscillation, 110–123
    patterned, 64, 123
    persymmetric, 208
    Quasi-Toeplitz, 505
    Quasiseparable, 525–527
    semi-pair, 518
    semiseparable, 515–521
    single-pair, 110
    Toeplitz, 64, 207, 500–511
    totally nonnegative, 113

totally positive, 113
Vandermonde, 114
Matrix vector multiplication, 98–100
Maximal invertible rank structure, 301
Mean of distribution, 124
Minimal rank, 340
Minor of a matrix, 112
Minus operator for sets, 39
Moore-Penrose equations, 528
Moore-Penrose inverse, 528
Mosaic rank, 308
Multinomial distribution, 125
Multiplication of
    band, 432
    generator representable semisep-
        arable, 432
    matrix and a vector, 98–100
    of structured rank matrices, 430
    quasiseparable, 432
    semiseparable, 432
    strict band, 433

Nested multishift, 135
Nonnegative matrix, 279
Norm
    definition, 26
    inequalities, 26
Norm inequalities, 27
Notation, xvii–xviii
Nullity of a matrix, 38
Nullity theorem, 37–40, 341
    LU-decomposition, 159, 160
    pivoted LU-decomposition, 162
    QR-decomposition, 184

$\mathcal{O}(n)$, 16
Off-diagonal rank, 311
Off-diagonal structure, 311
One-pair matrix, 110, 115, 116
Operator theory, 140
Optimal scaling, 277
Orthogonal decompositions, 199–204
Orthogonal rational functions, 133–134
Oscillation kernel, 113

Oscillation matrices, 110–123
    definition, 114
    eigenvectors and eigenvalues, 121
    examples, 114
    properties, 114
Oscillation properties, 111
Other decompositions, 386

p-lower triangular structure, 295
P, xvii
$\{p, q\}$-Quasiseparable matrix, 301, 316
$\{p, q\}$-Semiseparable matrix, 300, 316, 317
$\{p, q\}$-band matrix, 306
$\{p, q\}$-generator representable semi-
    separable matrix, 302, 316
$\{p\}$-Hessenberg-like matrices, 305, 317
$\{p\}$-generalized Hessenberg, 306
$\{p\}$-semiseparable matrix, 300
$\{p_1, p_2\}$-Levinson conform, 220
$\{p_1\}$-Levinson conform, 220
p-upper triangular structure, 295
Partition of a set, 311
Patterned matrices, 64, 123
Patterns
    blend of patterns, 397
    diamond form, 391
    leaf form, 389, 391
    of annihilation, 387
    of Givens transformations, 387
    pyramid form, 389
    V-pattern, 396
    ∧-pattern, 395
    X-pattern, 395
Persymmetric, 208
Pivot, 151
Pivoted LU-decomposition
    of structured rank matrices, 162
Pivoting, 152, 448–454
Pointwise closure, see Closure
Proper band matrix, 516
Property A, 277
Pure structure, 297
Pyramid pattern, 389

Q, xviii

QL-decomposition, 203, 386

QR-algorithm, 137

QR-factorization, 181–204
    band, 384
    decoupled semiseparable matrices, 384
    extended semiseparable matrices, 384
    Givens-vector representation, 416
    higher order, 381
    nullity theorem, 183–186
    of structured rank matrices, 184
    parallel factorization, 426
    QZ-factorization, 429
    quasiseparable, 384
    semiseparable, 186–192, 383
    semiseparable plus band, 385
    solving systems, 386
    structured rank matrices, 347–433
    unstructured matrix, 425

QZ-factorization, 429

$Q^{(s)}$, xviii

$\mathcal{Q}$, xviii, 31

$\mathcal{Q}_{sym}$, 12

$\mathcal{Q}^{(g)}$, xviii, 87

$\mathcal{Q}^{(g)}_{sym}$, 63

$\mathcal{Q}^{(s)}$, 34

Quasi-Toeplitz, 505

Quasi-triangle property, 522

Quasi-tridiagonal, 522

Quasiseparability rank, 295

Quasiseparable, 9, 29
    inner matrix relations, 302
    relation to semiseparable plus diagonal, 21–25
    higher order, 301
    LU-decomposition, 171–178
    matrices, 80, 140, 525–527
    not representable as semiseparable plus diagonal, 23
    $\{p, q\}$-quasiseparable, 301, 316
    representation, 79–82, 92–93

symmetric rank structure, 34

R, xviii

RQ-decomposition, 202, 386

rank (A), xvii

$r(\Sigma; A)$, xvii

$r_{\mathcal{A}}$, xvii, 56

$r_{\mathcal{Q}^{(g)}_{sym}}$, 63

$r_{\mathcal{Q}^{(g,d)}}$, 87

$r_{\mathcal{S}_{sym}}$, 73

$r_{\mathcal{S}^{(g)}_{sym}}$, 60

$r_{\mathcal{S}^{(g)}}$, 84

$r_{\mathcal{S}^{(g,d)}_{sym}}$, 62

$r_{\mathcal{S}^{(g,d)}}$, 87

$r_{\mathcal{S}^{(q)}_{sym}}$, 80, 81

$r_{\mathcal{S}^{(s)}_{sym}}$, 67

$r_{\mathcal{S}}$, 90

$r_{\mathcal{T}}$, 56

Rank structure, 297

Rank-annihilating Givens, 351

Rank-decreasing Givens, 359

Rank-expanding
    existence, 412
    Givens, 400, 401, 420
    on upper rank 1 structures, 402

Ranks of matrices
    block off-diagonal rank, 311
    off-diagonal rank, 311
    weakly lower triangular block, 312
    weakly upper triangular block, 312

Rational interpolation, 140

Rectangular distribution, 128

Recurrence relation, 133

Recursively semiseparable, 132

Reduced QR-factorization, 183

Reduction algorithms, 135

Relations between
    definitions, 31–37
    symmetric definitions, 12–27
    unsymmetric definitions, 31–37

Removing a diagonal from the structure, 298

Removing subdiagonals, 380

Representation, 56, 342–344
    changing between, 101–104
    decoupled, 93–95
    definition, 56–57
    diagonal-subdiagonal, 58, 65–71, 156
    examples, 82–83
    Gauss-vector, 179
    generator, 58–65, 84–90
    Givens-vector, 58, 71–79, 90–92
    higher order Givens-vector, 343
    'just' a representation, 57–58
    map, 56
    of semiseparables, 53–107
    quasiseparable, 59, 79–82, 92–93
    quasiseparable higher order, 343
    split representations, 344
    summary, 95–97
    swapping of the Givens-vector representation, 77–78
    tridiagonal matrices, 56
Representations, 55–59
Retrieving Givens-vector representation, 74–77
Row pivoting, 152, 166

S, xvii
$S(\mathbf{u}, \mathbf{v})$, xviii, 9
$S(\mathbf{u}, \mathbf{v}, \mathbf{p}, \mathbf{q})$, xviii, 28
$S(i : j, k : l)$, 8
Σ, xvii
$\mathcal{S}$, xviii, 31
$\mathcal{S} + D$, 21
$\mathcal{S}_{sym}$, 12
$\mathcal{S}^{(d)}$, xviii, 31
$\mathcal{S}^{(d)}_{sym}$, 12
$\mathcal{S}^{(g)}$, xviii, 31
$\mathcal{S}^{(g)}_{sym}$, 12
$\mathcal{S}^{(g,d)}$, xviii, 62
$\mathcal{S}^{(s)}_{sym}$, xviii, 68
$\overline{\mathcal{S}}^{(g)}_{sym}$, 18
$\overline{\mathcal{S}}^{(s)}$, 35
sym, xviii, 13
$s_{\mathcal{A}}$, xvii, 56
$s_{\mathcal{S}_{sym}}$, 75

Scaling semiseparable, 277
Schur
    algorithm, 245
    complement, 245
    general framework, 253
    reduction, 247
Semi-pair matrix, 518
Semiseparability rank, 295
Semiseparable, 28
Semiseparable matrices
    Hessenberg-like, 306
Semiseparable form, 10
Semiseparable kernel, 135
Semiseparable matrices, 5–51, 301, 515–521
    $\{0\}$-semiseparable matrix, 301
    applications, 140
    as discretization matrices, 130
    backward substitution, 152–153
    block generator representable, 17
    closure, 19
    comments, 134–140
    covariance matrices, 123
    eigenvalue problems, 135–140
    extended, 304
    extended $\{p, q\}$-generator representable semiseparable, 304
    extended $\{p, q\}$-semiseparable, 304
    generator definition, 8
    generator representable, 28
    Givens-vector representation, 71
    higher order, 300
    inner relations, 301
    LU-decomposition, 151–152, 163–170
    lower $\{p\}$-Hessenberg-like, 305
    lower semiseparable, 306
    not generator representable, 17
    $\{p, q\}$-generator representable semiseparable, 302
    $\{p, q\}$-semiseparable, 300, 316, 317
    $\{p\}$-Hessenberg-like, 305, 317
    $\{p\}$-semiseparable, 300
    pointwise closure, 19

problem matrices , 83
recursively, 132
relation to generator representable, 16–21
representation with diagonal and subdiagonal, 67
representations, 53–107
$S(\mathbf{u}, \mathbf{v})$, 9
semiseparability rank p, 300
semiseparable, 28, 301
sequentially, 80, 132, 140
symmetric definitions, 7–12
symmetric generator representable, 8
symmetric semiseparable, 7
unsymmetric definitions, 28–30
unsymmetric generator representable, 28
unsymmetric quasiseparable, 29
unsymmetric semiseparable, 28
upper $\{p\}$-Hessenberg-like, 305
upper semiseparable, 306
Semiseparable plus diagonal
relation to quasiseparable, 21–25
Separable kernel, 135
Separable matrix, 518
Sequence of Gauss transformations, *see* Gauss transformations
Sequence of Givens transformations, *see* Givens transformations
Sequentially semiseparable, 80, 132, 140
Sets
a partition of, 311
cardinality, 39
difference, 39
minus, 39
Sherman-Morrison-Woodbury formula, 274
Shift element, 297
Shift matrix, 477
Shift through lemma, 392
Sign regular, 508
Signal processing, 140
Simple Levinson conform, 220

Simple Levinson conform solver, 222–231
Single-pair matrix, 110
Singular structured rank matrices, 381, 382
Skeletons, 308
Statistics, 140
Strict band matrix, 316
Strictly lower triangular part, 7
Strictly lower triangular structure, *see* weakly lower triangular structure
Strictly upper triangular part, 7
Strictly upper triangular structure, *see* weakly upper triangular structure
Structure, 294
block off-diagonal structure, 311
lower triangular, 295
off-diagonal structure, 311
p-lower triangular, 295
p-upper triangular, 295
strictly lower triangular, *see* weakly lower triangular
strictly upper triangular, *see* weakly upper triangular
subdiagonal, *see* weakly lower triangular
superdiagonal, *see* weakly upper triangular
upper triangular, 295
weakly lower triangular, 295
weakly lower triangular block structure, 312
weakly upper triangular, 295
weakly upper triangular block structure, 312
Structure block, 297
Structured rank, 294, 295
Structured rank matrices, 7, 294–297
Subdiagonal structure, *see* weakly lower triangular structure
Subspace iteration, 135
Summation of
extended semiseparable matrices, 339

Hessenberg-like matrices, 338
quasiseparable matrices, 338
semiseparable matrices, 339
Superdiagonal structure, *see* weakly
upper triangular structure
Swapping the Givens-vector representation, 77–78
Symmetric
Generator representable, 8
quasiseparable, 9
semiseparable, 7
semiseparable plus diagonal, 9
Symmetric rank structure, 34–36
Symmetric relations, 25–26
Symmetric semiseparable
$S(\mathbf{u}, \mathbf{v})$, 9
Szegö polynomials, 244

T, xviii
$\mathcal{T}$, xviii, 31
$\mathcal{T}_{sym}$, 12
$\mathcal{T}^{(i)}$, xviii, 31
$\mathcal{T}^{(i)}_{sym}$, 12
tril(), 8
tril(A, p), xviii
triu(), 8
triu(A, p), xviii
k-Toeplitz, 507
$t_{i:j}$, xviii, 93
Thin QR-factorization, 183
Toeplitz matrices, 246
Toeplitz matrix, 64, 207, 500–511
Totally nonnegative matrices, 113
Totally positive matrices, 113
Transformation
Gauss, 151
Givens, 72
Trapezoidal rule, 131
Tree of an $\mathcal{H}^2$-matrix, 495
Trench algorithm, 210
Triangle property, 522
Triangular factorization, 231–232
Tridiagonal
irreducible, 11
matrix, 236
representation, 56

U, xviii
ULV-decomposition, 386
URV-decomposition, 386
$\mathbf{u}(i : j)$, xviii
$\mathbf{u}^T = [u_1, u_2, \ldots, u_n]$, xviii
Ugrade, 308
Unit bidiagonal, 154
Unit lower semiseparable, 156
Unitary Hessenberg, 44
Unsymmetric
generator representable, 28
quasiseparable, 29
relations summary, 36–37
semiseparable, 28
structures, 239
Upper $\{p\}$-Hessenberg-like matrices,
305
Upper semiseparable
matrices, 306
Upper triangular factorization, 209,
231–232
Upper triangular making, 379–387
Upper triangular matrix, 239
Upper triangular part, 7
Upper triangular structure, 295

V-pattern, 396
Vandermonde matrix, 114

∧-pattern, 395
Weakly lower triangular block rank,
312
Weakly lower triangular block structure, 312
Weakly lower triangular structure,
295
Weakly upper triangular block rank,
312
Weakly upper triangular block structure, 312
Weakly upper triangular structure,
295

X-pattern, 395

Yule-Walker, 207

Yule-Walker-like problem, 211
Yule-Walker-like system, 222–223

Z, xviii
Z-matrices, 278